A Modern Approach to Oceanography

A Modern Approach to Oceanography

Edited by Suzy Bullock

SYRAWOOD
PUBLISHING HOUSE
New York

Published by Syrawood Publishing House,
750 Third Avenue, 9th Floor,
New York, NY 10017, USA
www.syrawoodpublishinghouse.com

A Modern Approach to Oceanography
Edited by Suzy Bullock

International Standard Book Number: 978-1-68286-530-9 (Hardback)

Cataloging-in-Publication Data

A modern approach to oceanography / edited by Suzy Bullock.
p. cm.
Includes bibliographical references and index.
ISBN 978-1-68286-530-9
1. Oceanography. 2. Marine sciences. 3. Earth sciences. 4. Aquatic ecology.
I. Bullock, Suzy.
GC11.2 .M63 2018
551.46--dc23

Printed in China.

TABLE OF CONTENTS

Preface IX

Chapter 1 **Seasonality in submesoscale turbulence** 1
Jörn Callies, Raffaele Ferrari, Jody M. Klymak and Jonathan Gula

Chapter 2 **Detecting anthropogenic footprints in sea level rise** 9
Sönke Dangendorf, Marta Marcos, Alfred Müller, Eduardo Zorita, Riccardo Riva, Kevin Berk and Jürgen Jensen

Chapter 3 **Anthropogenic effects are associated with a lower persistence of marine food webs** 18
Luis J. Gilarranz, Camilo Mora and Jordi Bascompte

Chapter 4 **Decadal predictions of the North Atlantic CO_2 uptake** 23
Hongmei Li, Tatiana Ilyina, Wolfgang A. Müller and Frank Sienz

Chapter 5 **No inter-gyre pathway for sea-surface temperature anomalies in the North Atlantic** 30
Nicholas P. Foukal and M. Susan Lozier

Chapter 6 **Drying of Indian subcontinent by rapid Indian Ocean warming and a weakening land-sea thermal gradient** 36
Mathew Koll Roxy, Kapoor Ritika, Pascal Terray, Raghu Murtugudde, Karumuri Ashok and B. N. Goswami

Chapter 7 **Constraints on soluble aerosol iron flux to the Southern Ocean at the Last Glacial Maximum** 46
T. M. Conway, E. W. Wolff, R. Röthlisberger, R. Mulvaney and H. E. Elderfield

Chapter 8 **Evidence for ice-free summers in the late Miocene central Arctic Ocean** 55
Ruediger Stein, Kirsten Fahl, Michael Schreck, Gregor Knorr, Frank Niessen, Matthias Forwick, Catalina Gebhardt, Laura Jensen, Michael Kaminski, Achim Kopf, Jens Matthiessen, Wilfried Jokat and Gerrit Lohmann

Chapter 9 **Destructive tsunami-like wave generated by surf beat over a coral reef during Typhoon Haiyan** 68
Volker Roeber and Jeremy D. Bricker

Chapter 10 **Morphodynamics of submarine channel inception revealed by new experimental approach** 77
Jan de Leeuw, Joris T. Eggenhuisen and Matthieu J. B. Cartigny

Chapter 11 **Increased reservoir ages and poorly ventilated deep waters inferred in the glacial Eastern Equatorial Pacific** 84
Maria de la Fuente, Luke Skinner, Eva Calvo, Carles Pelejero and Isabel Cacho

Chapter 12 **Modelled drift patterns of fish larvae link coastal morphology to seabird colony distribution**....95
Hanno Sandvik, Robert T. Barrett, Kjell Einar Erikstad, Mari S. Myksvoll, Frode Vikebø, Nigel G. Yoccoz, Tycho Anker-Nilssen, Svein-Håkon Lorentsen, Tone K. Reiertsen, Jofrid Skarðhamar, Mette Skern-Mauritzen and Geir Helge Systad

Chapter 13 **Stratified coastal ocean interactions with tropical cyclones**....103
S. M. Glenn, T. N. Miles, G. N. Seroka, Y. Xu, R. K. Forney, F. Yu, H. Roarty, O. Schofield and J. Kohut

Chapter 14 **Pliocene cooling enhanced by flow of low-salinity Bering Sea water to the Arctic Ocean**....112
Keiji Horikawa, Ellen E. Martin, Chandranath Basak, Jonaotaro Onodera, Osamu Seki, Tatsuhiko Sakamoto, Minoru Ikehara, Saburo Sakai and Kimitaka Kawamura

Chapter 15 **Radiocarbon constraints on the extent and evolution of the South Pacific glacial carbon pool**....121
T. A. Ronge, R. Tiedemann, F. Lamy, P. Köhler, B. V. Alloway, R. De Pol-Holz, K. Pahnke, J. Southon and L. Wacker

Chapter 16 **Observing mesoscale eddy effects on mode-water subduction and transport in the North Pacific**....132
Lixiao Xu, Peiliang Li, Shang-Ping Xie, Qinyu Liu, Cong Liu and Wendian Gao

Chapter 17 **Tracking ocean heat uptake during the surface warming hiatus**....141
Wei Liu, Shang-Ping Xie and Jian Lu

Chapter 18 **Mid Pleistocene foraminiferal mass extinction coupled with phytoplankton evolution**....150
Sev Kender, Erin L. McClymont, Aurora C. Elmore, Dario Emanuele, Melanie J. Leng and Henry Elderfield

Chapter 19 **Cyanobacterial symbionts diverged in the late Cretaceous towards lineage-specific nitrogen fixation factories in single-celled phytoplankton**....158
Francisco M. Cornejo-Castillo, Ana M. Cabello, Guillem Salazar, Patricia Sánchez-Baracaldo, Gipsi Lima-Mendez, Pascal Hingamp, Adriana Alberti, Shinichi Sunagawa, Peer Bork, Colomban de Vargas, Jeroen Raes, Chris Bowler, Patrick Wincker, Jonathan P. Zehr, Josep M. Gasol, Ramon Massana and Silvia G. Acinas

Chapter 20 **North Atlantic deep water production during the last glacial maximum**....167
Jacob N. W. Howe, Alexander M. Piotrowski, Taryn L. Noble, Stefan Mulitza, Cristiano M. Chiessi and Germain Bayon

Chapter 21 **A global reanalysis of storm surges and extreme sea levels**....175
Sanne Muis, Martin Verlaan, Hessel C. Winsemius, Jeroen C. J. H. Aerts and Philip J. Ward

Chapter 22 **Gains and losses of coral skeletal porosity changes with ocean acidification acclimation** 186
Paola Fantazzini, Stefano Mengoli, Luca Pasquini, Villiam Bortolotti, Leonardo Brizi, Manuel Mariani, Matteo Di Giosia, Simona Fermani, Bruno Capaccioni, Erik Caroselli, Fiorella Prada, Francesco Zaccanti, Oren Levy, Zvy Dubinsky, Jaap A. Kaandorp, Pirom Konglerd, Jörg U. Hammel, Yannicke Dauphin, Jean-Pierre Cuif, James C. Weaver, Katharina E. Fabricius, Wolfgang Wagermaier, Peter Fratzl, Giuseppe Falini and Stefano Goffredo

Chapter 23 **Radiocarbon evidence for enhanced respired carbon storage in the Atlantic at the Last Glacial Maximum** 193
E. Freeman, L. C. Skinner, C. Waelbroeck and D. Hodell

Chapter 24 **Mesoscale atmosphere ocean coupling enhances the transfer of wind energy into the ocean** 201
D. Byrne, M. Münnich, I. Frenger and N. Gruber

Chapter 25 **Mechanisms of increased *Trichodesmium* fitness under iron and phosphorus co-limitation in the present and future ocean** 209
Nathan G. Walworth, Fei-Xue Fu, Eric A. Webb, Mak A. Saito, Dawn Moran, Matthew R. McIlvin, Michael D. Lee and David A. Hutchins

Chapter 26 **Spatial and temporal changes in cumulative human impacts on the world's ocean** 220
Benjamin S. Halpern, Melanie Frazier, John Potapenko, Kenneth S. Casey, Kellee Koenig, Catherine Longo, Julia Stewart Lowndes, R. Cotton Rockwood, Elizabeth R. Selig, Kimberly A. Selkoe and Shaun Walbridge

Permissions

List of Contributors

Index

PREFACE

The study of the physical, chemical and biological aspects of the ocean is known as oceanography. Some of the topics covered under the subject of oceanography are ocean currents, waves, ecosystem dynamics, geology of the sea floor, etc. The subject of oceanography is mainly divided into geological oceanography, chemical oceanography, biological oceanography and physical oceanography. It is very important in today's time, as it helps us in developing a better understanding of the ocean. Different approaches, evaluations, methodologies and advanced studies on oceanography have been included in this book. From theories to research to practical applications, case studies related to all contemporary topics of relevance to this field have been included herein. This book is a vital tool for all researching and studying this field.

The information contained in this book is the result of intensive hard work done by researchers in this field. All due efforts have been made to make this book serve as a complete guiding source for students and researchers. The topics in this book have been comprehensively explained to help readers understand the growing trends in the field.

I would like to thank the entire group of writers who made sincere efforts in this book and my family who supported me in my efforts of working on this book. I take this opportunity to thank all those who have been a guiding force throughout my life.

Editor

1

Seasonality in submesoscale turbulence

Jörn Callies[1], Raffaele Ferrari[2], Jody M. Klymak[3] & Jonathan Gula[4]

Although the strongest ocean surface currents occur at horizontal scales of order 100 km, recent numerical simulations suggest that flows smaller than these mesoscale eddies can achieve important vertical transports in the upper ocean. These submesoscale flows, 1–100 km in horizontal extent, take heat and atmospheric gases down into the interior ocean, accelerating air-sea fluxes, and bring deep nutrients up into the sunlit surface layer, fueling primary production. Here we present observational evidence that submesoscale flows undergo a seasonal cycle in the surface mixed layer: they are much stronger in winter than in summer. Submesoscale flows are energized by baroclinic instabilities that develop around geostrophic eddies in the deep winter mixed layer at a horizontal scale of order 1–10 km. Flows larger than this instability scale are energized by turbulent scale interactions. Enhanced submesoscale activity in the winter mixed layer is expected to achieve efficient exchanges with the permanent thermocline below.

[1] MIT/WHOI Joint Program in Oceanography, Massachusetts Institute of Technology, Building 54-1615, 77 Massachusetts Avenue, Cambridge/Woods Hole, Massachusetts 02139, USA. [2] Massachusetts Institute of Technology, Cambridge, Massachusetts 02139, USA. [3] University of Victoria, Victoria, British Columbia V8W 3P6, Canada. [4] University of California, Los Angeles, California 90095, USA. Correspondence and requests for materials should be addressed to J.C. (email: joernc@mit.edu).

The ocean supports motions on a wide range of scales. The most energetic flows are eddies with horizontal scales of a few hundred kilometres, which are the oceanic analogues of the atmospheric cyclones and anticyclones. These mesoscale eddies are dynamically well understood and are routinely observed with satellite altimeters[1]. Classic theories predict that the kinetic energy in the mesoscale eddy field should decay very rapidly at smaller scales[2] and be associated with weak vertical velocities. Recent numerical modelling studies suggest, however, that there is a very dynamic near-surface eddy field on scales between 1 and 100 km, which drives vertical velocities one or two orders of magnitude larger than those of order 1 m per day associated with the divergence of mesoscale flows[3,4]. These submesoscale flows consist of narrow horizontal currents associated with strong horizontal gradients in buoyancy.

High-resolution numerical simulations suggest that submesoscale flows are much stronger in winter than in summer[5–7]. This is illustrated here with snapshots from a simulation of the Gulf Stream region with a horizontal resolution of 750 m. More detail on the simulation is given in the Methods section. The snapshots of surface buoyancy gradients show that fronts are strong in winter and much less pronounced in summer, except for the Gulf Stream front that persists throughout the year (Fig. 1). This seasonality provides important clues to the dynamics of submesoscale turbulence.

One mechanism that has been proposed to energize submesoscale flows is mesoscale-driven surface frontogenesis[3]: the presence of the ocean surface allows strain fields at the edges of mesoscale eddies to sharpen surface buoyancy gradients[8,9]. Unlike in the ocean interior, upwelling of dense water and downwelling of light water cannot counteract the increase of lateral buoyancy gradients at the surface, where the vertical transport of waters must vanish. The resulting surface fronts have been shown to break up in secondary roll-up instabilities, which drive a nonlinear cascade of surface buoyancy variance and energize the entire range of scales below the mesoscale[10]. Surface quasi-geostrophic theory has been used to qualitatively illustrate these dynamics and predicts surface kinetic and potential energy spectra to scale like $k^{-5/3}$ (ref. 11), where k is the horizontal wavenumber. Surface quasi-geostrophic flows are predicted to be confined to a shallow surface layer; they decay exponentially away from the surface, with smaller-scale motions decaying more rapidly than larger-scale ones. Energy spectra are therefore expected to fall off with wavenumber more rapidly away from the surface. Surface quasi-geostrophic dynamics are modified by ageostrophic effects that accelerate frontogenesis[12], lead to true frontal collapse and thereby produce k^{-2} energy spectra[13].

Another mechanism that has been proposed to energize submesoscale flows in the mixed layer is baroclinic instability[14,15]. This instability allows perturbations to extract energy from the lateral buoyancy gradients generated by mesoscale stirring or by spatial variations in atmospheric forcing. Eady edge waves can propagate both along the surface and along the sharp increase in stratification at the base of the mixed layer[10,16–19]. Edge waves with scales close to the mixed-layer deformation radius, Nh/f, propagate at the same speed along the surface and the base of the mixed layer, and can therefore constructively interfere (N is the density stratification of the mixed layer, h is its depth, f is the Coriolis parameter). This interference results in an amplification of these disturbances on a time scale $N/f\Lambda$ (ref. 16) (Λ is the geostrophic shear in the mixed layer), which causes water from the denser side of the lateral buoyancy gradient to slide under the more buoyant waters. The most unstable mode in the Eady model, which assumes that N and Λ are constant through the mixed layer and that the base of the mixed layer is rigid, has a wavelength of $4Nh/f$ and a growth rate of $0.3\ f\Lambda/N$[16,20]. In deep mixed layers of a few hundred metres and for typical values of $f \sim 10^{-4}\,s^{-1}$, $\Lambda \sim 10^{-4}\,s^{-1}$ and $N \sim 10^{-3}\,s^{-1}$, disturbances therefore grow on horizontal scales of order 1–10 km and time scales of order 1 day (ref. 15). Once these disturbances have grown to finite amplitude, turbulent scale interactions exchange kinetic energy between scales. If the flow is sufficiently constrained by the Earth's rotation, kinetic energy is preferentially transferred to larger scales, filling in the spectrum between the mixed-layer deformation radius and the mesoscale[7,15,21,22]. A major prediction of this scenario, which distinguishes it from mesoscale-driven surface frontogenesis, is that the full depth of the mixed layer is energized and that the energy decays rapidly below the base of the mixed layer.

The energization of the submesoscale by such mixed-layer instabilities depends crucially on the mixed-layer depth. Mixed-layer baroclinic instabilities are powered by the potential energy stored in lateral buoyancy gradients. The deeper the mixed layer, for a given lateral buoyancy gradient, the more potential energy is

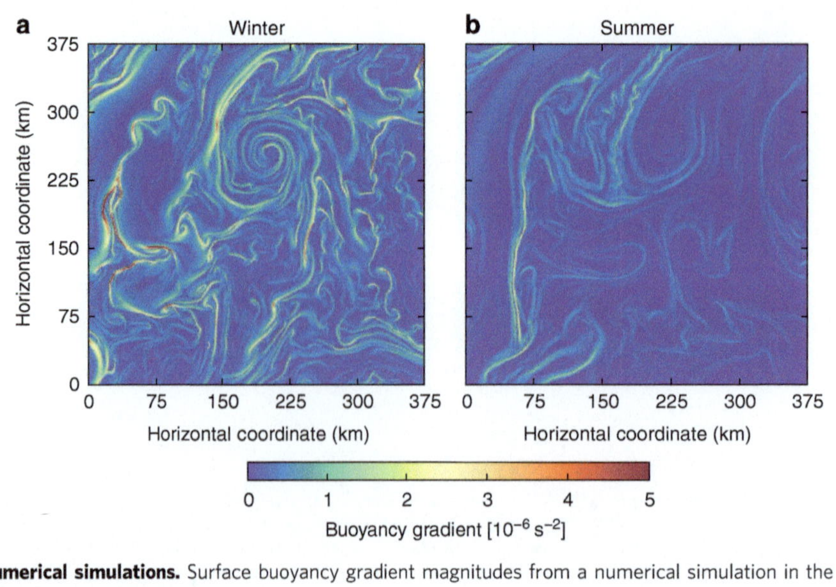

Figure 1 | Seasonality in numerical simulations. Surface buoyancy gradient magnitudes from a numerical simulation in the Gulf Stream region in (**a**) winter (February 15) and (**b**) summer (August 15).

released by the slumping isopycnals and the more energetic the resulting submesoscale flows. Given that lateral buoyancy gradients are fairly constant throughout the year, the kinetic energy generated by the instabilities is thus expected to peak in winter, when mixed layers are deepest. In summer, when mixed layers are shallow, the instabilities can easily be damped out. Although the growth rate of mixed-layer instabilities is independent of the mixed-layer depth h, the damping of mixed layer motions by atmospheric forcing is faster in shallower mixed layers[23]. In addition, the mixed-layer deformation radius scales with h, so the instability scale is smaller when the mixed layer is shallow. As smaller-scale flows can more easily escape the influence of rotation, kinetic energy created in shallow mixed layers may cascade to smaller rather than larger scales, so that the instability fails to energize the range between the mixed-layer deformation radius and the mesoscale[24,25].

The instability mechanism therefore implies a seasonal cycle in submesoscale turbulence. In winter, when the mixed layer is deep, instabilities grow on a time scale of about 1 day and at a horizontal scale between 1 and 10 km, subsequently energizing the entire submesoscale range through turbulent scale interactions. In summer, the instability is either damped out or fails to energize the submesoscale range because of the lack of a strong inverse cascade, when the mixed layer deformation radius is very small. The mesoscale-driven surface frontogenesis mechanism, on the other hand, only depends on the presence of a mesoscale eddy field, which does not undergo a strong seasonal cycle. The strong seasonality at the submesoscale visible in numerical simulations (Fig. 1) supports the instability mechanism.

Observational confirmation of a seasonal cycle in submesoscale turbulence, however, is so far lacking. Numerical models used for studying submesoscale dynamics typically do not explicitly resolve small-scale processes in the mixed layer. In summer, for example, when the mixed layer is shallow, the models do not resolve the instability scale of the baroclinic instability in the mixed layer, which can be as small as 10–100 m. High-resolution satellite observations show a seasonal cycle in sea surface temperature gradients in the Gulf Stream region[26], but it remains unclear whether the temperature variations are reflected in density or compensated by salinity variations[27], whether the temperature signal is accompanied by along-front flows, whether these flows are in geostrophic balance and what the subsurface structure of these flows is.

In-situ observations of velocity and buoyancy in the upper ocean are used in this study to conclusively show that submesoscale flows undergo a strong seasonal cycle on scales of 1–100 km. An enhancement of the observed submesoscale flows throughout the winter mixed layer and the flows' leading-order horizontal non-divergence are argued to be further evidence for the instability mechanism.

Results

Data. We present data from the western subtropical North Atlantic, where the mixed layer is known to undergo a strong seasonal cycle[28]. The data were collected as part of two separate observational programmes: the Oleander project along a transect between Elisabeth, New Jersey, and Hamilton, Bermuda, occupied weekly in 2005–2013, and the Lateral Mixing Experiment (LatMix) along several straight transects off Cape Hatteras in June 2011 (summer) and just south of the Gulf Stream extension in March 2012 (winter)[29] (Fig. 2). We consider only the Oleander data southwest of the Gulf Stream extension and separate the transects into winter ones (January through March, when the mixed layers

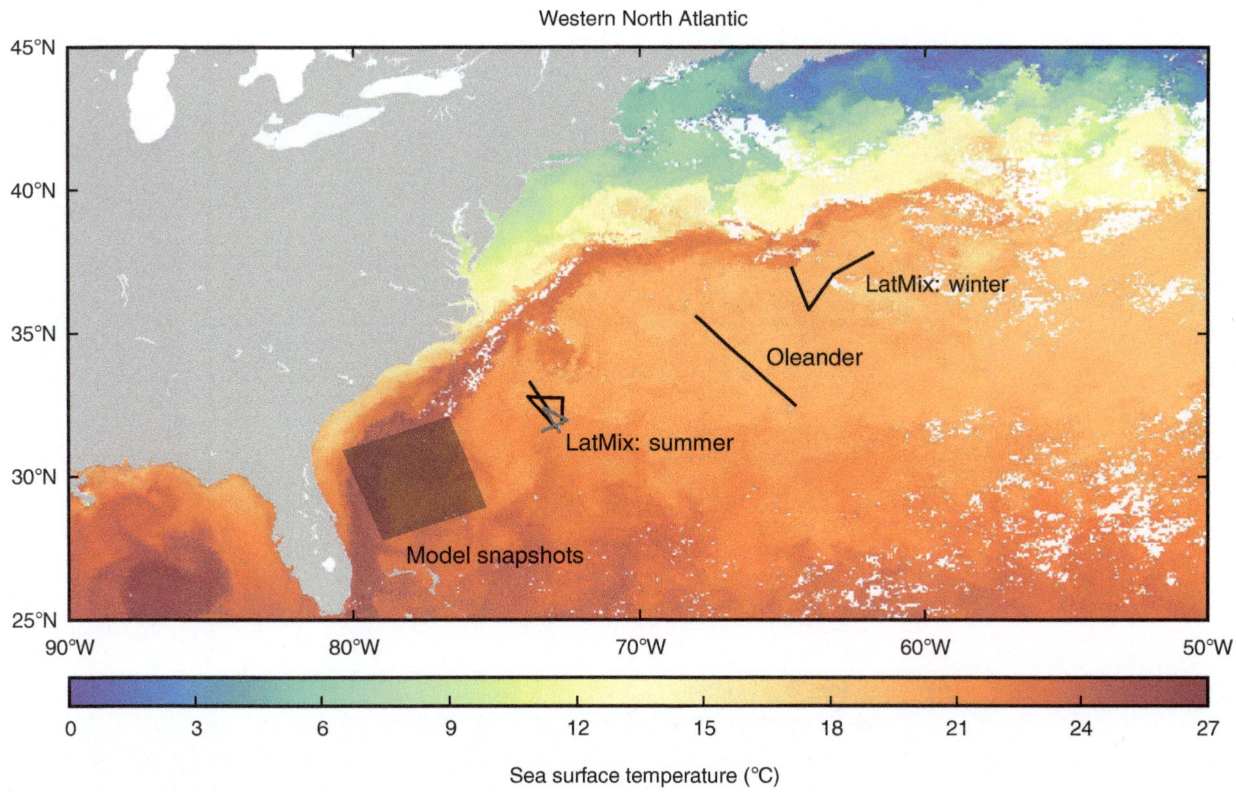

Figure 2 | Measurement locations. Locations of velocity transects (black lines), of additional buoyancy transects (dark gray lines) and of the model snapshots shown in Fig. 1 (transparent shading). The colour shading shows

are deepest[28]) and summer ones (June through August). In the Oleander region, mixed layers are on average 150–200 m deep in winter and shoal to order 10 m in summer[28]. In the LatMix winter experiment, the mixed layer was about $h = 220$ m deep and had an average stratification of $N = 2 \times 10^{-3}\,\mathrm{s}^{-1}$, giving an Eady instability scale $4Nh/f = 20$ km. In the LatMix summer experiment, the mixed layer was no deeper than $h = 5$ m and although there were no shallow measurements of stratification, the Eady instability scale was most likely no larger than 100 m.

Power spectra are computed at constant depths for the longitudinal (along-track) velocities u, the transverse (across-track) velocities v (both obtained with acoustic Doppler current profilers, ADCPs) and the buoyancy b (obtained from towed conductivity–temperature–depth sensors, available only for the LatMix data):

$$S^u(k) = \langle |\hat{u}(k)|^2 \rangle, \quad S^v(k) = \langle |\hat{v}(k)|^2 \rangle, \quad S^b(k) = \langle |\hat{b}(k)|^2 \rangle / N^2, \tag{1}$$

where k is the along-track wavenumber, the caret denotes a Fourier transform and $\langle \cdot \rangle$ denotes the average over transects. More detail on the observations and the computation of these spectra is provided in the Methods section. Departing from recent literature, we do not focus solely on spectral slopes, because they do not uniquely determine the dynamics: different dynamics can produce the same spectral slopes. We instead use power spectra primarily to diagnose submesoscale energy levels at different times and depths. Together with the relationships between the spectra of u, v and b, this allows a more nuanced assessment of submesoscale dynamics.

Seasonality. The near-surface spectra of kinetic and potential energy,

$$K(k) = \tfrac{1}{2}[S^u(k) + S^v(k)], \quad P(k) = \tfrac{1}{2}S^b(k), \tag{2}$$

at 50 m depth for Oleander and 20 m depth for LatMix are shown in Fig. 3. The kinetic energy spectra from Oleander and LatMix have similar shapes, suggesting that the dynamics are similar across the different geographical locations. All spectra exhibit marked differences between summer and winter. While the energy levels at the mesoscale (scales larger than 100 km) remain fairly constant across seasons, the submesoscale energy levels (scales of order 10 km) undergo a seasonal transition: they are significantly higher in winter than in summer (Fig. 3). This seasonal difference in submesoscale energy levels is reflected in how rapidly the energy falls off with wavenumber in the submesoscale range. In winter, the spectra are relatively flat and approximately follow a k^{-2} power law (Fig. 3a,b). In summer, the spectra in the range 20–100 km fall off more rapidly and approximately follow a k^{-3} power law (Fig. 3c,d); at scales smaller than 20 km, the spectra flatten out and roughly match the Garrett–Munk (GM) empirical model spectrum of internal waves[30]. Potential and kinetic energy spectra are approximately equal in the LatMix winter experiment (Fig. 3b). In the LatMix summer experiment,

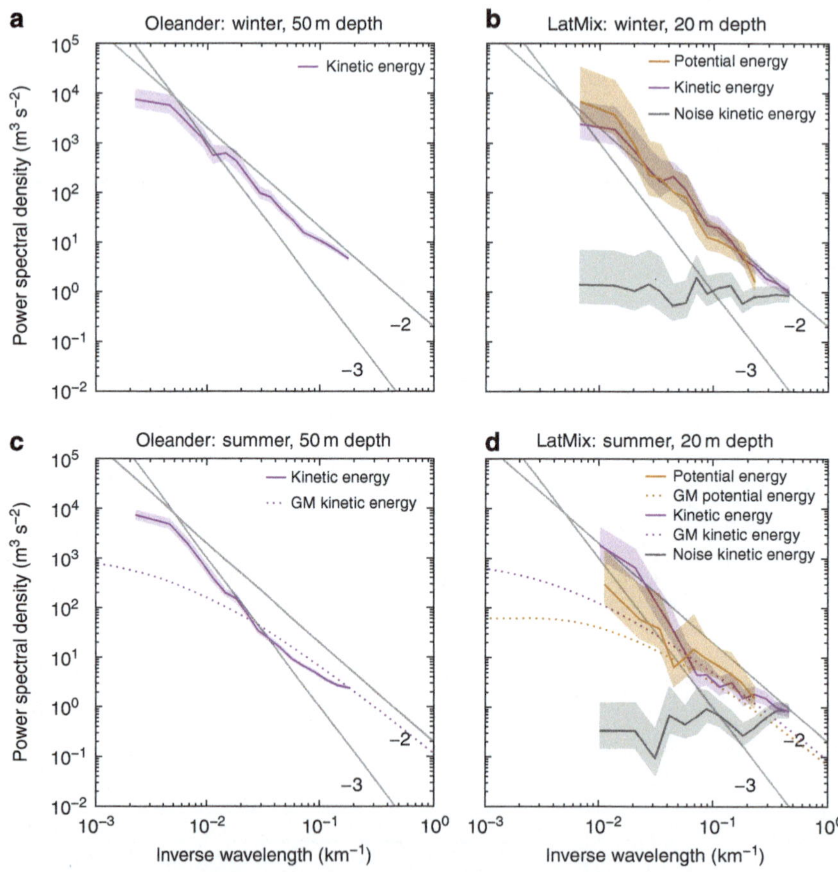

Figure 3 | Seasonality in observations. (**a**) Kinetic energy spectrum at 50 m depth for the Oleander winter data. (**b**) Potential and kinetic energy spectra at 20 m depth for the LatMix winter experiment. (**c**) Kinetic energy spectrum at 50 m depth for the Oleander summer data. (**d**) Potential and kinetic energy spectra at 20 m depth for the LatMix summer experiment. The light shadings are 95% confidence intervals. Also shown are the GM model spectra for internal waves in the seasonal thermocline (with parameters from ref. 30), estimates for the noise level of the LatMix velocity data and reference lines with slopes −2 and −3.

the kinetic energy spectrum is considerably larger than the potential energy spectrum at scales larger than 20 km (Fig. 3d). At smaller scales, they are roughly equal and consistent with the GM spectrum for internal waves.

These spectra confirm that submesoscale turbulence is more energetic in winter than in summer. The seasonal transition in submesoscale potential energy levels is consistent with the seasonal transition in frontal strength seen in numerical simulations (Fig. 1). In winter, the observed equipartition between kinetic and potential energy, $K(k)/P(k) \simeq 1$, and the k^{-2} spectral slopes are likely the result of the turbulent dynamics induced by mixed-layer instabilities and their interaction with the mesoscale straining field, but how these dynamics shape the spectra is currently not understood. The wintertime spectral slopes are also consistent with the mesoscale-driven surface frontogenesis mechanism, but the seasonal transition is not. In summer, the steep energy spectra are consistent with interior quasi-geostrophic turbulence[2]. Eddies generated through baroclinic instability in the thermocline transfer very little energy to submesoscales. The observation that the kinetic energy spectrum is substantially larger than the potential energy spectrum in this range is also consistent with the ratio $K(k)/P(k) \simeq 4$ predicted by interior quasi-geostrophic turbulence[2,31]. The emergence of the GM spectrum at scales below 20 km confirms that internal waves dominate these scales and mask an even more dramatic drop-off of geostrophic submesoscale motions in summer, as pointed out in recent literature[31,32].

Wintertime submesoscale flows. We examine the vertical structure of kinetic energy in the LatMix winter experiment, which provides further evidence for the mixed layer instability mechanism. We use the LatMix data, because the accompanying buoyancy data provide information on the mixed-layer depth. It is roughly constant at 220 m across the transects, which facilitates the interpretation of the vertical structure in observed kinetic energy. The submesoscale kinetic energy is fairly constant throughout the mixed layer at scales larger than 20 km (Fig. 4a). At smaller scales, the energy is slightly enhanced at the surface and at the base of the mixed layer, as expected from Eady dynamics at scales below the deformation radius. The spectra uniformly scale like k^{-2} in the mixed layer. Below the base of the mixed layer, the energy decays rapidly; the spectra steepen to k^{-3} in the permanent thermocline at scales larger than 20 km (Fig. 4b). At smaller scales, the spectra in the thermocline follow the GM model and internal waves mask a steep drop-off. This vertical structure is consistent with submesoscale turbulence induced by baroclinic instabilities in the mixed layer, but inconsistent with mesoscale-driven surface frontogenesis, for which the vertical decay of submesoscale turbulence would start in the mixed layer[31].

The observations further suggest that the energetic submesoscale flows in winter are to leading order in geostrophic balance, as required for an inverse energy cascade. Horizontally non-divergent isotropic flows have a distinct signature in the power spectra of transverse and longitudinal velocities: for a flow with a k^{-2} kinetic energy spectrum, the ratio $S^v(k)/S^u(k) \simeq 2$ is expected[2,31–33]. Error bars are substantial, but a clear trend across scales is evident, which suggests that the observed spectra are roughly consistent with this relation. In the Oleander data, where the large number of transects results in smaller error bars, the relation is roughly satisfied over the entire submesoscale range (Fig. 5a). In the LatMix winter experiment, the relation is roughly satisfied throughout the submesoscale range (Fig. 5b), except for scales smaller than 5 km, where noise may affect the spectra significantly (*cf.* Fig. 3b). This is consistent with a flow that is to

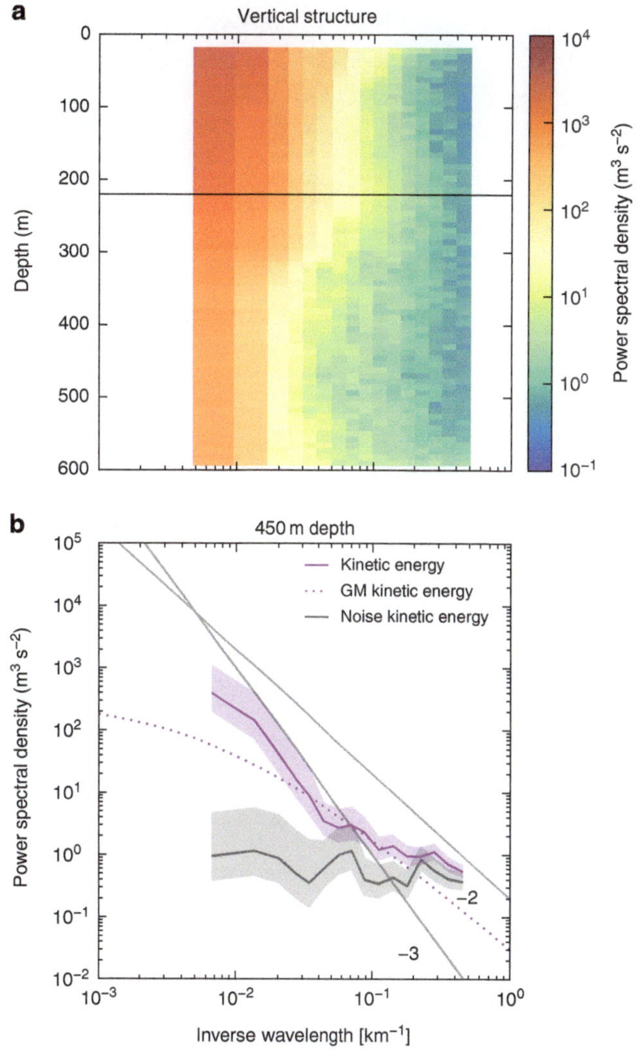

Figure 4 | Vertical structure in the LatMix winter experiment. (**a**) Kinetic energy spectrum as a function of depth. The depth of the mixed layer $h = 220$ m is indicated by the black horizontal line. (**b**) Kinetic energy spectrum at a depth of 450 m. The light shadings are 95% confidence intervals. Also shown is the GM model spectrum for internal waves in the thermocline (with parameters from ref. 30), an estimate of the noise level of the velocity data and reference lines with slopes −2 and −3.

dynamics, on the other hand, have a leading-order horizontal divergence and thus a different $S^v(k)/S^u(k)$ ratio. Near-inertial waves forced by strong winter winds or wind- and buoyancy-driven three-dimensional turbulence can thus be ruled out to dominate the submesoscale observations in winter.

Calculation of the Rossby and Froude numbers as functions of scale, $\mathrm{Ro}(k) = [k^3 K(k)]^{1/2}/f$ and $\mathrm{Fr}(k) = [k^3 P(k)]^{1/2}/f$, further suggests that the observed wintertime submesoscale flows follow quasi-geostrophic dynamics. Quasi-geostrophic dynamics describe flows with $\mathrm{Fr}(k) \sim \mathrm{Ro}(k) \ll 1$ (for example, see ref. 34). For k^{-2} kinetic energy spectra, as observed in winter, $\mathrm{Ro}(k)$ increases moderately with k, approximately like $k^{1/2}$. As $\mathrm{Ro}(k)$ is small in the mesoscale, it remains fairly small throughout the observed submesoscale range (Fig. 5c,d). The condition $\mathrm{Fr}(k) \sim \mathrm{Ro}(k)$ is satisfied because of the approximate equipartition between kinetic and potential energy (Figs 3b and 5d).

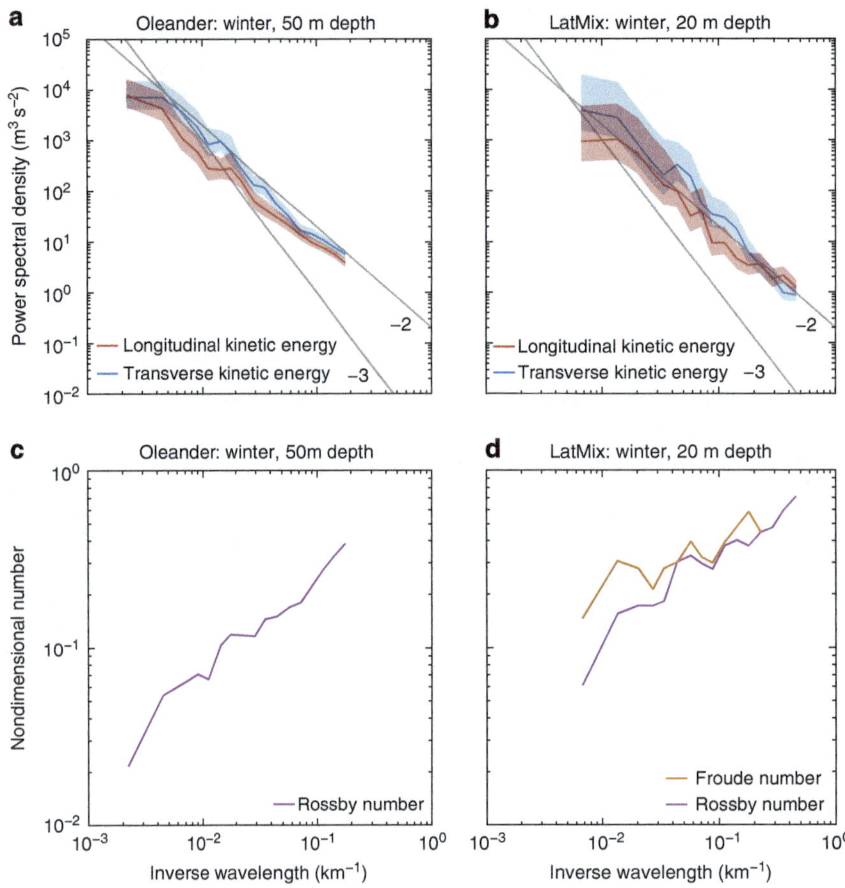

Figure 5 | Testing whether the wintertime submesoscale flow is quasi-geostrophic. (**a**) Power spectra of transverse and longitudinal velocities at 50 m depth for the Oleander winter data. (**b**) Power spectra of transverse and longitudinal velocities at 20 m depth for the LatMix winter experiment. (**c**) Rossby number at 50 m depth for the Oleander winter data. (**d**) Rossby and Froude numbers at 20 m depth for the LatMix winter experiment. The light shadings are 95% confidence intervals.

to become strongly ageostrophic—allowing effective transfer of energy to dissipation scales[24,25]—at scales smaller than the observed range.

Discussion

The near-surface observations are characterized by a pronounced seasonal cycle in kinetic and potential energy: submesoscale flows are much more energetic in deep winter mixed layers than in the seasonal thermocline in summer. In contrast, the mesoscale energy remains fairly constant throughout the year. The near-surface submesoscale spectral roll-off changes from a rapid k^{-3} in summer to a frontal k^{-2} in winter. In winter, submesoscale energies are fairly constant with depth throughout the mixed layer and decay rapidly below its base, transitioning to a k^{-3} regime in the main thermocline.

We can rationalize these results in terms of quasi-geostrophic turbulence theory. A very weak forward transfer of energy below the scale of the deep mesoscale instability prevails in the seasonal thermocline in summer and in the main thermocline throughout the year. In winter, deep mixed layers allow a secondary baroclinic instability in the mixed layer that converts potential to kinetic energy in the submesoscale, over a range of scales around the mixed-layer deformation radius between 1 and 10 km. This injection of kinetic energy throughout the depth of the mixed layer drives an inverse cascade of kinetic energy to larger scales and energizes the entire range between the mesoscale and the mixed-layer deformation radius.

The observations are consistent with realistic high-resolution numerical simulations[6,7] and thereby suggest that these simulations capture the leading-order dynamics of submesoscale turbulence. This consistency further builds confidence in the instability mechanisms, because in the simulations one can directly diagnose a seasonal cycle in the release of available potential energy through mixed-layer instabilities and an inverse cascade of kinetic energy thus created[6,7]. Although this inverse cascade is not directly diagnosed from data, two crucial ingredients are. First, the lack of a spectral peak at the instability scale indicates that the instability has grown to finite amplitude and become fully turbulent. Second, the diagnosed horizontal non-divergence indicates that the flow is to leading order in geostrophic balance, which is necessary for turbulent scale interactions to preferentially transfer energy to large scales.

The observed seasonal cycle of submesoscale energy levels, the vertical structure of the flow and its horizontal non-divergence in winter are all consistent with the instability mechanism. In conjunction with the results of numerical simulations, the observations therefore strongly favour the hypothesis that the submesoscale is energized through a baroclinic instability in the mixed layer. This consistency, of course, does not constitute a proof of the instability mechanism and further progress can be made by deriving additional falsifiable predictions or by obtaining additional observations. But at the moment, the instability mechanism appears to be the only available explanation for the observations on hand.

Previous studies suggested that mesoscale-driven surface frontogenesis generates most of the submesoscale kinetic energy in the upper ocean[3]. This process appears to be weak in the western subtropical North Atlantic. In summer, when the mixed-layer instability is not active, the submesoscale spectra roll off like k^{-3}—surface frontogenesis, which can occur year round, would generate steps in the velocity field and thus produce k^{-2} spectra. It appears that in this region the summertime flow is dominated by interior quasi-geostrophic turbulence, even close to the surface.

The mixed-layer instability mechanism is likely to dominate the generation of submesoscale kinetic energy in all parts of the ocean that are characterized by deep mixed layers. For example, the mechanism appears to be at work in the entire Gulf Stream and Kuroshio regions in winter, as found in high-resolution numerical simulations[6,7]. We also expect this mechanism to be at work in parts of the Southern Ocean, where winter mixed layers can reach as deep as 500 m (ref. 35).

Our results further suggest that a strong seasonal cycle in submesoscale energy is probably characteristic of regions with relatively strong interior potential vorticity gradients reversing sign at depth, such as the western subtropical North Atlantic analysed here[36]. With such potential vorticity profiles, the mesoscale field is generated through a Phillips-type baroclinic instability[37], which energizes low-mode eddies and transfers little energy to scales smaller than the mesoscale instability scale. Surface buoyancy gradients are relatively weak, rendering any mesoscale-driven surface frontogenesis ineffective. In other regions, where surface buoyancy gradients are larger and deep potential vorticity gradients weaker, Charney-type baroclinic instabilities and mesoscale-driven surface frontogenesis may become a leading-order driver of submesoscale turbulence in the upper ocean[38].

Submesoscale turbulence is associated with strong flows along steep isopycnals. These generate large vertical fluxes of physical and chemical tracers[4] and may have an impact on the exchange of these tracers between the mixed layer and the interior ocean. The submesoscale enhancement of fluxes of heat and salt has been suggested to affect the water mass properties and circulation of the permanent thermocline[39]. Furthermore, submesoscale fluxes of nutrients between the nutrient-depleted mixed layer and the nutrient-rich thermocline are believed to play an important role in maintaining primary production in subtropical gyres[40]. It remains to be studied, however, how the seasonality in submesoscale turbulence affects these exchanges. Stommel[41] argued that the properties of waters subducted from the mixed layer into the permanent thermocline are set in winter, when mixed layers are deepest. Our study has shown that winter is also the time when submesoscale turbulence is most vigorous—just when Stommel's demon opens the trapdoor. An impact of submesoscale turbulence on the physical and chemical properties of interior waters appears plausible.

Methods

Numerical simulation. The snapshots of buoyancy gradients shown in Fig. 1 are taken from a numerical simulation of the Gulf Stream region performed with the Regional Ocean Modeling System[42]. This simulation has a horizontal resolution of 750 m and 50 vertical levels. The model domain spans 1,050 km by 900 km, which provides generous padding to the domain shown in the snapshots. Boundary conditions are supplied by a lower-resolution simulation that spans the Atlantic basin[43]. The simulation is forced by daily winds and diurnally modulated surface fluxes. The modelling approach is described in detail in Gula *et al.*[44].

***In-situ* observations.** The Oleander data were collected in 2005–2013 with a 75-kHz ADCP, averaged over 8-m depth bins and 4.5- to 6-min intervals, which at a ship speed of about 8 m s^{-1} results in an average spacing of 2.0–2.5 km. Data west of 68°W and north of

Gulf Stream extension do not affect the results (Fig. 2). Subsequently, transects with fewer than 190 data points are discarded. The measured velocities are transformed into a coordinate system aligned with the ship track and interpolated with cubic splines onto a regular grid with a spacing of 2.6 km. After selecting for the season, a Hann window is applied and the Fourier transforms are averaged over the different transects to form the spectra. The summer spectrum is an average over 46 transects and the winter spectrum is an average over 11 transects. The spectra are further averaged over ten wavenumber bins per decade. The spectral amplitudes at the smallest resolved scales are affected by the averaging and interpolation procedures, but an assessment of these effects using synthetic signals suggests that the impacts are small and that they do not affect our conclusions, which depend mostly on larger scales.

A comparison with the Oleander data collected with a 150-kHz ADCP in 1994–2004 reveals that although the two data sets are consistent in summer, the 1994–2004 data set exhibits less submesoscale energy in winter than the 2005–2013 data set. More accurate navigational data can be used in the processing of the 2005–2013 data, which makes it more reliable and leads us to suspect that the submesoscale energy is artificially reduced in the 1994–2004 data set during winter.

The LatMix velocity data were collected with 75 kHz ADCPs aboard the *RV Endeavor* (summer) and *RV Atlantis* (winter). They are averaged over 8-m depth bins and 5-min intervals, which at a ship speed of about 3 m s^{-1} results in an average spacing of about 1 km. Four straight large-scale transects are used from the summer experiment and three from the winter experiment (Fig. 2). The winter data were previously analysed by Shcherbina *et al.*[29]. The measured velocities are transformed into a coordinate system aligned with the ship track and interpolated onto a regular grid with a spacing of 1 km. The spectra are obtained following the same procedure as described above for the Oleander data. Estimates of the noise level are calculated as in Shcherbina *et al.*[29].

The LatMix buoyancy data were collected using Moving Vessel Profilers, sampling the water column 5–100 m in summer and 10–250 m in winter. We use three transects occupied by the *RV Endeavor* in summer (Fig. 2) and three transects occupied by the *RV Atlantis* in winter (same as for velocity). Data from the nearly vertical downcasts are averaged over 1-m depth bins; data from upcasts are discarded. The data are further interpolated onto a regular along-track grid with a spacing of 2 km. The stratification is obtained by fitting a second-order polynomial to the horizontally averaged buoyancy profile in the range 10–100 m in summer and 20–200 m in winter. The spectra are obtained following the same procedure as described above for the velocities.

References

1. Stammer, D. Global characteristics of ocean variability estimated from regional TOPEX/POSEIDON altimeter measurements. *J. Phys. Oceanogr.* **27,** 1743–1769 (1997).
2. Charney, J. G. Geostrophic turbulence. *J. Atmos. Sci.* **28,** 1087–1095 (1971).
3. Lapeyre, G. & Klein, P. Dynamics of the upper oceanic layers in terms of surface quasigeostrophy theory. *J. Phys. Oceanogr.* **36,** 165–176 (2006).
4. Capet, X., McWilliams, J. C., Molemaker, M. J. & Shchepetkin, A. F. Mesoscale to submesoscale transition in the California Current System. Part I: flow structure, eddy flux, and observational tests. *J. Phys. Oceanogr.* **38,** 29–43 (2008).
5. Capet, X., Campos, E. J. & Paiva, A. M. Submesoscale activity over the Argentinian shelf. *Geophys. Res. Lett.* **35,** L15605 (2008).
6. Mensa, J. A. *et al.* Seasonality of the submesoscale dynamics in the Gulf Stream region. *Ocean Dyn.* **63,** 923–941 (2013).
7. Sasaki, H., Klein, P., Qiu, B. & Sasai, Y. Impact of oceanic-scale interactions on the seasonal modulation of ocean dynamics by the atmosphere. *Nat. Commun.* **5,** 5636 (2014).
8. Stone, P. H. Frontogenesis by horizontal wind deformation fields. *J. Atmos. Sci.* **23,** 455–465 (1966).
9. Lapeyre, G., Klein, P. & Hua, B. L. Oceanic restratification forced by surface frontogenesis. *J. Phys. Oceanogr.* **36,** 1577–1590 (2006).
10. Held, I. M., Pierrehumbert, R. T., Garner, S. T. & Swanson, K. L. Surface quasi-geostrophic dynamics. *J. Fluid Mech.* **282,** 1–20 (1995).
11. Blumen, W. Uniform potential vorticity flow. Part I: theory of wave interactions and two-dimensional turbulence. *J. Atmos. Sci.* **35,** 774–783 (1978).
12. Hoskins, B. J. & Bretherton, F. P. Atmospheric frontogenesis models: mathematical formulation and solution. *J. Atmos. Sci.* **29,** 11–37 (1972).
13. Boyd, J. P. The energy spectrum of fronts: time evolution of shocks in Burgers' equation. *J. Atmos. Sci.* **49,** 128–139 (1992).
14. Haine, T. W. N. & Marshall, J. Gravitational, symmetric, and baroclinic instability of the ocean mixed layer. *J. Phys. Oceanogr.* **28,** 634–658 (1998).
15. Boccaletti, G., Ferrari, R. & Fox-Kemper, B. Mixed layer instabilities and restratification. *J. Phys. Oceanogr.* **37,** 2228–2250 (2007).
16. Eady, E. T. Long waves and cyclone waves. *Tellus* **1,** 33–52 (1949).
17. Blumen, W. On short-wave baroclinic instability. *J. Atmos. Sci.* **36,** 1925–1933 (1979).
18. Rivest, C., Davis, C. A. & Farrell, B. F.

19. Juckes, M. Quasigeostrophic dynamics of the tropopause. *J. Atmos. Sci.* **51,** 2756–2768 (1994).
20. Vallis, G. K. *Atmospheric and Oceanic Fluid Dynamics* (Cambridge University Press, 2006).
21. Fjørtoft, R. On the changes in the spectral distribution of kinetic energy for two-dimensional, nondivergent flow. *Tellus* **5,** 225–230 (1953).
22. Kraichnan, R. H. Inertial ranges in two-dimensional turbulence. *Phys. Fluids* **10,** 1417 (1967).
23. Marshall, J. & Schott, F. Open-ocean convection: observations, theory, and models. *Rev. Geophys.* **37,** 1–64 (1999).
24. McWilliams, J. C., Molemaker, M. J. & Yavneh, I. *From Stirring to Mixing Momentum: Cascades from Balanced Flows to Dissipation* in *Proc. 12th 'Aha Huliko'a Hawaiian Winter Work* 59–66 (University of Hawaii, 2001).
25. Capet, X., McWilliams, J. C., Molemaker, M. J. & Shchepetkin, A. F. Mesoscale to submesoscale transition in the California Current System. Part III: energy balance and flux. *J. Phys. Oceanogr.* **38,** 2256–2269 (2008).
26. Armstrong, E. M., Wagner, G., Vazquez-Cuervo, J. & Chin, T. M. Comparisons of regional satellite sea surface temperature gradients derived from MODIS and AVHRR sensors. *Int. J. Remote Sens.* **33,** 6639–6651 (2012).
27. Ferrari, R. & Rudnick, D. L. Thermohaline variability in the upper ocean. *J. Geophys. Res.* **105,** 16857–16883 (2000).
28. de Boyer Montégut, C., Madec, G., Fischer, A. S., Lazar, A. & Iudicone, D. Mixed layer depth over the global ocean: an examination of profile data and a profile-based climatology. *J. Geophys. Res.* **109,** C12003 (2004).
29. Shcherbina, A. Y. *et al.* Statistics of vertical vorticity, divergence, and strain in a developed submesoscale turbulence field. *Geophys. Res. Lett.* **40,** 4706–4711 (2013).
30. Munk, W. Internal waves and small-scale processes in *Evolution of Physical Oceanography.* (eds Warren, B. A. & Wunsch, C.) 264–291 (The MIT Press, 1981).
31. Callies, J. & Ferrari, R. Interpreting energy and tracer spectra of upper-ocean turbulence in the submesoscale range (1–200 km). *J. Phys. Oceanogr.* **43,** 2456–2474 (2013).
32. Bühler, O., Callies, J. & Ferrari, R. Wave–vortex decomposition of one-dimensional ship-track data. *J. Fluid Mech.* **756,** 1007–1026 (2014).
33. Leith, C. E. Atmospheric predictability and two-dimensional turbulence. *J. Atmos. Sci.* **28,** 145–161 (1971).
34. Pedlosky, J. *Geophysical Fluid Dynamics* 2nd edn (Springer, 1987).
35. Dong, S., Sprintall, J., Gille, S. T. & Talley, L. Southern Ocean mixed-layer depth from Argo float profiles. *J. Geophys. Res.* **113,** C06013 (2008).
36. Tulloch, R., Marshall, J., Hill, C. & Smith, K. S. Scales, growth rates, and spectral fluxes of baroclinic instability in the ocean. *J. Phys. Oceanogr.* **41,** 1057–1076 (2011).
37. Phillips, N. A. Energy transformations and meridional circulations associated with simple baroclinic waves in a two-level, quasi-geostrophic model. *Tellus* **6,** 273–286 (1954).
38. Roullet, G., McWilliams, J. C., Capet, X. & Molemaker, M. J. Properties of steady geostrophic turbulence with isopycnal outcropping. *J. Phys. Oceanogr.* **42,** 18–38 (2012).
39. Lévy, M. *et al.* Modifications of gyre circulation by sub-mesoscale physics. *Ocean Model.* **34,** 1–15 (2010).
40. Klein, P. & Lapeyre, G. The oceanic vertical pump induced by mesoscale and sub mesoscale turbulence. *Ann. Rev. Mar. Sci.* **1,** 351–375 (2009).
41. Stommel, H. Determination of water mass properties of water pumped down from the Ekman layer to the geostrophic flow below. *Proc. Natl Acad. Sci. USA* **76,** 3051–3055 (1979).
42. Shchepetkin, A. F. & McWilliams, J. C. The Regional Oceanic Modeling System (ROMS): a split-explicit, free-surface, topography-following-coordinate oceanic model. *Ocean Model.* **9,** 347–404 (2005).
43. Mason, E. *et al.* Procedures for offline grid nesting in regional ocean models. *Ocean Model.* **35,** 1–15 (2010).
44. Gula, J., Molemaker, M. J. & McWilliams, J. C. Gulf Stream dynamics along the southeastern U.S. seaboard. *J. Phys. Oceanogr.* **45,** 690–715 (2015).

Acknowledgements

We thank the Office of Naval Research for funding and the entire LatMix group for conducting the experiment. We thank the Oleander group for collecting and making available the long-term ADCP data. In particular, we thank Kathleen A. Donohue for help with understanding the imprint of processing on the spectra derived from the first set of Oleander data (1994–2004). J.C. and R.F. acknowledge financial support under grants ONR-N00014-09-1-0458 and NSF-OCE-1233832; J.M.K. under grants ONR-N00014-11-1-0165 and NSERC-327920-2006; J.G. under grant ONR-N00014-12-1-0105.

Author contributions

J.C. and R.F. planned the research and wrote the manuscript. J.C. analysed the data. J.M. K. planned, collected and processed the Moving Vessel Profiler observations. J.G. performed the model simulations shown in Fig. 1.

Additional information

Competing financial interests: The authors declare no competing financial interests.

2

Detecting anthropogenic footprints in sea level rise

Sönke Dangendorf[1], Marta Marcos[2], Alfred Müller[3], Eduardo Zorita[4], Riccardo Riva[5], Kevin Berk[3] & Jürgen Jensen[1]

While there is scientific consensus that global and local mean sea level (GMSL and LMSL) has risen since the late nineteenth century, the relative contribution of natural and anthropogenic forcing remains unclear. Here we provide a probabilistic upper range of long-term persistent natural GMSL/LMSL variability ($P = 0.99$), which in turn, determines the minimum/maximum anthropogenic contribution since 1900. To account for different spectral characteristics of various contributing processes, we separate LMSL into two components: a slowly varying volumetric component and a more rapidly changing atmospheric component. We find that the persistence of slow natural volumetric changes is underestimated in records where transient atmospheric processes dominate the spectrum. This leads to a local underestimation of possible natural trends of up to ~1 mm per year erroneously enhancing the significance of anthropogenic footprints. The GMSL, however, remains unaffected by such biases. On the basis of a model assessment of the separate components, we conclude that it is virtually certain ($P = 0.99$) that at least 45% of the observed increase in GMSL is of anthropogenic origin.

[1] Department of Civil Engineering, Research Institute for Water and Environment, University of Siegen, Paul-Bonatz-Strasse 9-11, 57076 Siegen, Germany. [2] IMEDEA (UIB-CSIC), Miquel Marquès, 21, E-07190 Esporles, Spain. [3] Department of Mathematics, University of Siegen, Walter-Flex-Strasse 3, 57072 Siegen, Germany. [4] Helmholtz-Centre Geesthacht, Max-Planck-Strasse 1, 21502 Geesthacht, Germany. [5] Departement of
Delft University of Technology, Stevinweg 1, 2628 Delft

Tide gauge records along coastlines provide evidence that mean sea levels (MSLs) have risen since the late nineteenth century with globally averaged rates of 1.33–1.98 mm per year[1–4]. These rates derived from instrumental records are larger than the value reported for the past 2,000 years using indirect proxy estimators[5,6]. A question that remains is how much of this rise is related to past and current natural climate variations or to anthropogenic forcing. Answering this question is challenging, since our measurements available over the past two centuries represent only a very narrow window of the Earth's history, limiting our knowledge about natural low-frequency signals in the climate system.

Understanding the persistence of natural variations is a fundamental prerequisite to distinguish between natural and anthropogenic causes of the contemporary global MSL (GMSL) and local MSL (LMSL) rise. Knowing the persistence of naturally forced GMSL and LMSL change, in turn, enables a statistical assessment of the probability that the recent rise is outside the range of natural variability, and more robust estimates of the uncertainty in (and therefore the significance of) observed trends. While in most sea level investigations it has been assumed that the persistence of GMSL and LMSL change diminishes after a few months to years[1–4], recent works demonstrate that there is also persistence on timescales of decades[7,8]. This persistence, which can be accurately described by the Hurst exponent α (see Methods), leads naturally to multi-decadal excursions from the mean state and can therefore produce relatively strong multi-decadal to centennial trends unrelated to any external forcing[7,8]. On the basis of the assumption that the natural variability can be accurately modelled by a purely long-term correlated process, it was recently suggested that more than 50% of the observed twentieth-century LMSL rise (and more than 58% of the GMSL rise) can be attributed to external causes unrelated to natural climate variations[8] (thus including an anthropogenic contribution and local human interventions or vertical land motion). However, this assumption still needs to be verified considering that MSL is an integrative measure representing the cumulative response of the ocean to multiple physical processes[9] with diverse spectral properties[7].

While the two main contributors to the recent GMSL rise (thermal expansion and the melting of land-locked ice[9]) are, beside the possible anthropogenic long-term trend, governed by slow natural variations with Hurst exponents α larger than 0.8 (that is, strong long-term correlations)[7], the major drivers of LMSL change (ocean redistribution of heat, salt and water mass in response to changing winds, ocean circulation and air–sea fluxes[10–12]) were found to decay exponentially, thus rather resembling a short-term correlated process[7]. Therefore, the processes contributing to LMSL changes act on various timescales, leading to different spectral properties. This further raises the question whether one statistical model is able to reproduce these different properties.

Here we revisit the characteristics of natural variability in LMSL and GMSL by assessing the spectral properties of different components. We demonstrate that in some regions the high-frequency variability due to atmospheric forcing masks intrinsic low-frequency variability related to volumetric MSL changes (density and barystatic mass changes and their redistribution). These masking effects lead locally and regionally to an under-estimation of the true natural variability and, consequently, to an erroneously enhanced significance of the observed twentieth-century LMSL rise. This finding is underpinned by Monte-Carlo experiments with time series where the spectral properties are *a priori* known. Since the masking effect is mainly related to wind-driven redistribution processes, the GMSL, however, remains unaffected. We provide even with an improved estimate of natural GMSL variability unambiguous evidence for a significant anthropogenic footprint in GMSL rise since 1900.

Results

Spectral properties of LMSL. We base our approach on established techniques[8,13–16] and estimate the upper range of possible naturally forced centennial trends in LMSL and GMSL for a given confidence level ($P = 0.99$) and the resulting minimum contribution related to external causes (here we concentrate only on the linear contribution that can be produced by external/natural forcing). This is achieved by computing the probability of the occurrence of naturally forced centennial trends with the statistical parameters derived from a second-order detrended fluctuation analysis (DFA2) or standard autoregressive models of the order one (AR1; see Methods). From the distribution of natural trends, one can directly obtain the significance S of an observed trend as well as the minimum external contribution unrelated to any natural fluctuation (for a given level of confidence). This minimum external contribution is defined as the difference between the observed trend and the 99th percentile of the distribution of natural trends[8,13–15] (Methods and

Table 1 | Results from the estimation of natural and external trends in SLR.

Station	Availability	Hurst exponent α			LAG1 autocorrelation c			Observed trend (mm per year) and significance			Natural trend (mm per year)		External trend (mm per year)	
		OBS	ATM	RES	OBS	ATM	RES	OBS	Case 1*	Case 2†	Case 1	Case 2	Case 1	Case 2
Vlissingen	1900-2011	0.72	0.52	0.86	0.14	0.01	0.51	2.18	(1.00)	(1.00)	0.66	0.80	1.52	1.38
Hoek van Holland	1900-2011	0.71	0.51	0.86	0.17	0.00	0.45	2.24	(1.00)	(1.00)	0.69	0.81	1.54	1.42
Ijmuiden	1900-2011	0.73	0.52	0.90	0.16	0.01	0.51	2.04	(1.00)	(1.00)	0.85	1.05	1.19	0.99
Den Helder	1900-2011	0.67	0.54	0.89	0.13	0.06	0.45	1.25	(1.00)	(0.99)	0.64	0.89	0.61	0.36
Delfzijl	1900-2011	0.62	0.51	0.86	0.13	0.04	0.40	1.55	(1.00)	(0.99)	0.59	0.98	0.96	0.57
Norderney	1900-2011	0.61	0.56	0.84	0.15	0.11	0.50	1.84	(1.00)	(1.00)	0.60	0.91	1.25	0.94
Cuxhaven	1900-2011	0.60	0.56	0.82	0.16	0.11	0.50	1.76	(1.00)	(0.99)	0.65	0.85	1.11	0.91
Esbjerg	1900-2011	0.57	0.56	0.80	0.15	0.11	0.43	0.71	(1.00)	(0.81)	0.58	0.84	0.13	–
Hirtshals	1900-2011	0.65	0.56	0.86	0.20	0.14	0.37	1.17	(1.00)	(0.99)	0.69	1.04	0.48	0.13
Aberdeen	1900-2011	0.73	0.50	0.88	0.21	0.05	0.39	1.79	(1.00)	(1.00)	0.66	0.89	1.13	0.89
North Shields	1900-2011	0.88	0.52	0.96	0.35	0.04	0.53	2.29	(1.00)	(1.00)	1.25	1.43	1.05	0.86

The table provides an overview over the main values derived from an assessment of each tide gauge record in the North Sea. The Hurst exponents α are derived from a DFA2. All trends are calculated for the common period from 1900 to 2011 using a confidence level of 99% ($P = 0.99$). The significance S of each trend is provided in brackets.
*Case 1: integrated assessment.
†Case 2: separate assessment.

Supplementary Fig. 1). These rather conservative confidence bounds are used, since we search for the upper bounds of natural variability. Results for the respective 95% confidence intervals can be found in Supplementary Figs 5–6 and Supplementary Table 1. Unlike in previous works[8,13–15] we categorize the observed LMSL (OBS) in two components and assess their statistical properties separately: a quickly acting atmospheric component (ATM) and a more slowly varying residual component (RES; difference between OBS and ATM), comprising the volumetric contributions from density and mass changes and their redistribution. Focusing on 11 centennial (seasonally adjusted) tide gauge records (see Methods and Table 1) from the North Sea (Fig. 1b), we estimate the ATM at each location using the multiple linear regression model (LRM) from ref. 11. The LRM identifies relevant predictors from a pool of atmospheric reanalysis data (zonal/meridional winds and sea level pressure (SLP) from the twentieth century reanalysis[17] in an area of $\pm 2°$ around the respective tide gauge) using a stepwise algorithm with forward selection[10,11]. The model has been successfully validated against a state-of-the art barotropic hydrodynamical model (Hamburg Shelf Ocean Model: HAMSOM[18]) for the second half of the twentieth century[11] (see also Methods and Supplementary Fig. 2) and is used here to model the ATM over the period 1871–2011.

To better understand the temporal characteristics of ATM and RES compared with OBS, all contributions (Fig. 1a) are assessed for the tide gauge of Cuxhaven in the non-stationary (Fig. 1c,d) and stationary spectral power space (Fig. 1e). The two components ATM and RES show different timescale characteristics. While the ATM displays its largest spectral power on timescales below a few years, the spectral power of the RES increases with frequency becoming largest at low frequencies (Fig. 1e). This leads to a dominant coherence between ATM and OBS at timescales shorter than a decade (Fig. 1c), whereas longer variations are closely linked to RES (Fig. 1d). In other words, rapid MSL variations in Cuxhaven (and in the North Sea in general, see Supplementary Fig. 3) are mostly associated with barotropic processes in response to the atmospheric forcing[11,19], while decadal fluctuations are mostly related to persistent volumetric changes and their redistribution[7].

Given their different spectral characteristics, DFA2 is applied separately to OBS, its ATM and RES at each site. Figure 1f shows the fluctuation functions for Cuxhaven (those for the other sites can be found in the Supplementary Fig. 4). The slope α of the fluctuation function (equivalent to the Hurst exponent α) was computed between 13 and 423 months ($\sim$35 years; Fig. 1e). The fluctuation function of the ATM steadily increases with α close to 0.5, an indicator that these processes have characteristics similar to white or AR1 noise. In contrast, the analysis of OBS reveals the presence of long-term correlations, with Hurst exponents α of 0.6. Once the ATM is removed from observations α significantly increases at all stations (0.8–0.9 for RES, Table 1). This is in line with ref. 7 who noted that semi-enclosed basins and continental shelves display smaller Hurst exponents than large oceanic basins, where the contribution of atmospheric forcing is smaller and damped by the advection of heat leading to a reddening of the spectra[20]. We thus conclude that the presence of rapidly varying short-term processes, such as those generated by the atmospheric forcing over shallow shelves, mask the long-term correlation in MSL records driven by slow and persistent processes, such as gradual MSL rise due to ocean thermal expansion or mass exchange and its redistribution[9,10].

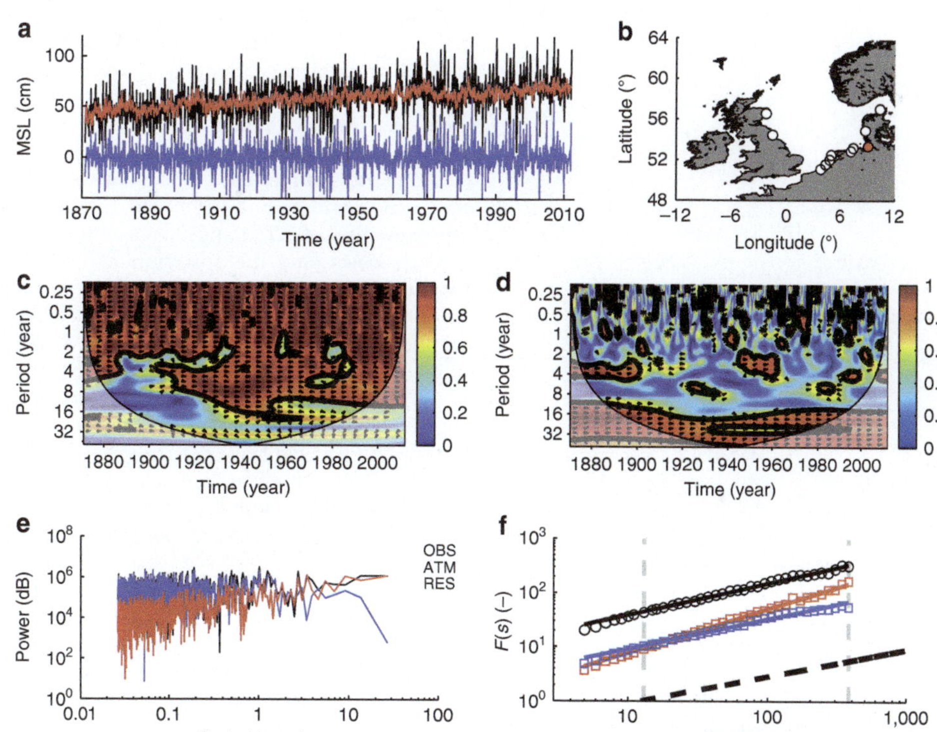

Figure 1 | Spectral properties of LMSL and its components. (**a**) Time series of OBS (black), ATM (blue) and RES (red) at the tide gauge of Cuxhaven over the period from 1871 to 2011. The location of the tide gauges is shown in (**b**) with Cuxhaven marked in red. Wavelet coherence plots (Morlet wavelet) of ATM and RES versus OBS are shown in (**c**,**d**). The black lines show the 5% significance level using the red noise model. The corresponding global power spectra for OBS, ATM and RES inferred from a fast Fourier transformation are shown in (**e**,**f**). Fluctuation functions estimated with DFA2 f... OBS, ATM and RES, respectively. The grey dotted lines mark the time window (13...
The black dotted line marks a H...

Short-term noise and the estimation of natural LMSL trends. The inability to estimate the actual long-term correlation in presence of short-term correlated processes implies that the natural variability of a given time series may not be properly described[21]. Consequently, the distribution of trends associated with natural variability can be biased, with large implications for the determination of the statistical significance of observed trends as well as their minimum external contribution. This is underpinned by a Monte-Carlo experiment (Fig. 2a–e), where the distributions of natural trends in a set of long-term correlated data are compared in cases with and without additive AR1 noise. Different degrees of long-term correlations have been tested, with Hurst exponents between 0.6 and 1.0, as well as different AR1 noise weights having a s.d. 1–10 times larger than the purely long-term correlated time series (note that the weight between ATM and RES ranges in the North Sea from 0.6 to 3 depending on the site). It is obvious that for small Hurst exponents (Fig. 2a, $\alpha = 0.6$), which are close to the theoretical value of white or AR1 noise ($\alpha = 0.5$), the distribution of the associated trends does not change when AR1 noise (with such small lag-1 autocorrelations as obtained from ATM) is added. However, with increasing α, the presence of noise induces an underestimation of natural trends reaching 0.4–0.5 mm per year for centennial records with Hurst exponents comparable to those inferred from the RES in the North Sea (Table 1, Fig. 1c,d).

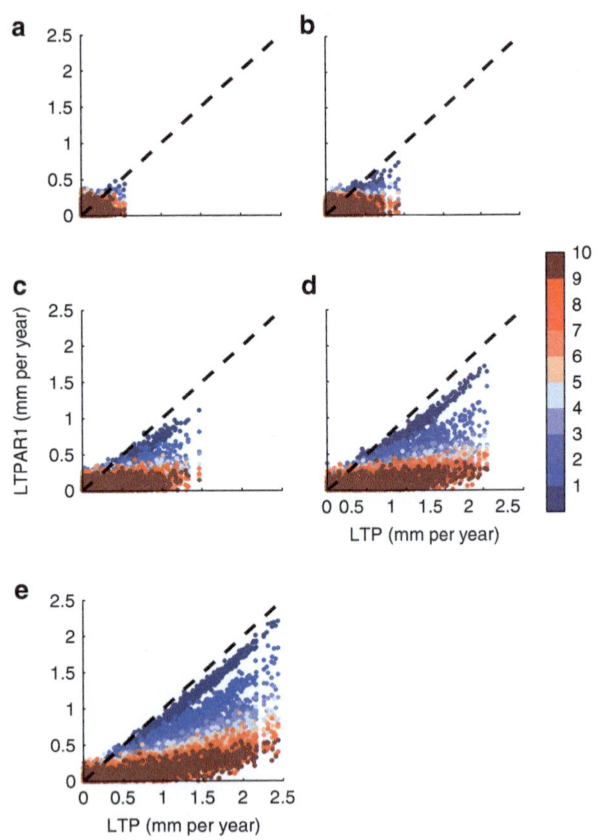

Figure 2 | The effect of AR1 noise on modelling natural and external trends in LMSL. (**a–e**) Natural trends in long-term correlated Gaussian data (LTP; 1,000 time series with a length of 1,692 months) versus natural trends in long-term correlated Gaussian data with additive AR1 noise (LTPAR1). The different colour shades correspond to the weight with which the AR1 noise was added to the long-term correlated data (1–10 times larger than the s.d. of the purely long-term correlated Gaussian data; in the North Sea the s.d. of the ATM exceeds that of the RES by 1–3 times). The natural trends correspond to a time series with a unit s.d. of 10 cm in the combined time series (which is roughly the median in the analysed North Sea records). Simulation results are shown for long-term correlated Gaussian data with (**a**) $\alpha = 0.6$, (**b**) $\alpha = 0.7$, (**c**) $\alpha = 0.8$, (**d**) $\alpha = 0.9$ and (**e**) $\alpha = 1.0$. For the lag-1 autocorrelation a value of $c = 0.1$, which is the maximum value obtained from the ATM, has been applied.

Applying the separate calculation of natural and minimum external trends to the 11 tide gauge records in the North Sea basin (Fig. 3a,b) leads to estimates that are in close agreement to those inferred from the Monte-Carlo experiment (Fig. 2a–e). The sum (here, the root sum of squares) of natural trends derived by a separate assessment of ATM and RES exceeds those applying the integrated approach[8] by up to 0.4 mm per year on a centennial timescale from 1900 to 2011 (Fig. 3a), which corresponds to roughly one-fourth of the entire twentieth-century LMSL rise. When the ATM is subtracted, the long-term correlation of the LMSL record is better captured (increase of α) and the magnitude of the trend associated to natural variations is increased by an amount of up to 0.4 mm per year. In consequence, the significance of the observed trends is reduced (Table 1) and the minimum fraction of the observed local trends related to external causes is smaller (Fig. 3b), implying that previous attempts[8] have likely overestimated the minimum amount of anthropogenic forcing in LMSL rise. It is also important to notice that the resulting confidence bounds of the observed trends are even up to four times larger than in the classical approach where AR1 noise has been assumed (Fig. 3c and Supplementary Fig. 5)[1–4].

An assessment on a global scale is currently hampered by the absence of reliable long-term estimates of the ATM contribution. Instead, we use synthetically modelled LMSL (LMSLsyn) fields as a combination of the dynamic sea-surface height (SSH) from a centennial ocean reanalysis[22], the barystatic contribution resulting from the mass loss of 18 glacier regions[23], the Greenland ice sheet[24] and their corresponding fingerprints[25] as well as variations in land hydrology[26] (see Methods). We separate the SSH from the ocean reanalysis into the contributions from the local steric height and ocean bottom pressure (OBPsyn) variations and combine the local steric height with the fingerprints from glaciers and ice sheets as well as hydrology to a volumetric component RESsyn. Although OBP still contains non-local steric variations[22], the evaluations provide a first approximation of ATM and therefore for the bias induced by not properly describing the spectrum of LMSL. We calculated the Hurst exponents α (Fig. 4a–c) and the resulting natural trends for the LMSLsyn, the OBPsyn and the RESsyn (Fig. 4e,f and Supplementary Fig. 6). In large parts of the world oceans the long-term correlations are confined to the RESsyn (Fig. 4b), while the OBPsyn signal exhibits Hurst coefficients close to AR1 noise (Fig. 4c). This leads to an underestimation of the Hurst exponent α present in RESsyn (Fig. 4d). As expected, significant differences are found between natural trends calculated within an integrated and a separate assessment (Fig. 4h). These differences are largest in areas where OBP changes are known to dominate the LMSL variations[27–29] (for example, South Pacific, Mediterranean Sea and Arctic Ocean). In such dynamically active regions naturally forced centennial trends of LMSLsyn tend to be underrated by an amount of up to 1.0 mm per year or even larger (Arctic Ocean), demonstrating the importance of accounting for the spectral diversity of different processes involved into LMSL variations.

Natural and external trends in GMSL. Since the differences found in the naturally forced centennial trends are mostly related to the masking effect of wind-driven redistribution processes, they do not have any effect on GMSL trends and their significance. GMSL variations are driven by barystatic mass and density changes[1,9], both of which are strongly long-term

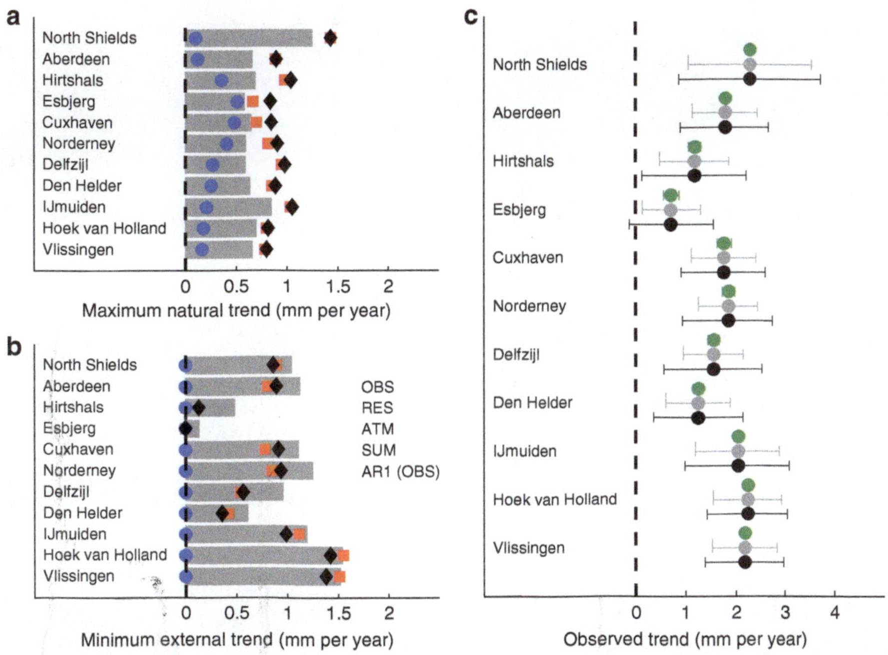

Figure 3 | Maximum natural and minimum/maximum external trends in LMSL at tide gauges in the North Sea. (**a**) Maximum natural trends ($P = 0.99$) derived by the method from refs 9,13,15 for Hurst exponents α estimated with OBS (grey bars), ATM (blue dots), RES (red squares) and the root sum of squares of ATM and RES (SUM, black diamonds) over the period from 1900 to 2011. The corresponding minimum external trends, calculated as the difference between the observed and maximum natural trends, are shown in (**b**). In (**c**) the observed trends are shown in its classical expression with their lower and upper 99% confidence bounds (that is, the minimum and maximum external contribution, see also Methods) obtained from OBS (grey) and SUM (black). For comparison also the results for a classical AR1 model are shown (green). Note that the observed trends still contain vertical land motions, which are responsible for the vast majority of the differences obtained between the different stations.

correlated[7]. Hence, naturally forced centennial trends can be assessed for the GMSL using the integrated assessment by applying the DFA2 for the estimation of the Hurst exponent α to the original GMSL record. However, a major problem of recent GMSL reconstructions[2–4] is that while they all agree in the sign of their long-term trends[1–4,26], their temporal variability remains highly uncertain with respect to the true global mean[26]. For instance, the variability in the Church and White[2] (CW11) reconstruction mostly reflects LMSL variations measured by tide gauges along the coast rather than the true GMSL variations[26], whereas the Hay *et al.*[4] (H15) reconstruction only represents a smoothed signal without any inter-annual variability. The temporal variability in the Jevrejeva *et al.*[3] (J14) reconstruction, by contrast, is significantly overestimated compared with the true global mean (Supplementary Fig. 7) and becomes increasingly uncertain back in time due to the sparse tide gauge network in the early twentieth century[3]. Consequently, an accurate description of naturally forced centennial trends with these time series[8] is not possible. To overcome the sampling problem of sparse tide gauge records in the earlier part of the twentieth century, we form a new synthetic GMSL curve (GMSLsyn) from the spatially homogeneous LSMLsyn fields. On inter-annual and longer timescales, this GMSLsyn curve is significantly correlated with the observed GMSL from altimetry (~ 0.5) and shows a considerable smaller s.d. (although still slightly overestimated) than the other reconstructions[2–3] over the altimetry period since 1993 (Supplementary Fig. 7).

We search for the presence of long-term correlations in the GMSLsyn curve using DFA2 (Fig. 5a,b), which yields a Hurst exponent α in the order of 1.28 (Fig. 5b). In combination with a s.d. of 9.1 mm over the entire twentieth century this leads to maximum naturally forced centennial trends of 0.73 mm per year ($P = 0.99$), 0.52 mm per year ($P = 0.95$) and 0.43 ($P = 0.90$) for the period 1900–2009. These values are significantly larger than previous estimates[1–4], where the presence of long-term correlations was neglected. However, even with this improved estimate all twentieth-century GMSL reconstructions[2–4] point towards a significant ($S > 0.99$) long-term trend, which cannot be solely explained by natural variability.

Discussion

The results demonstrate that the presence of short-term processes impacts significantly on the determination of the trends associated with natural variations in LMSL time series (up to 1 mm per year on a centennial scale). The removal of the ATM is therefore essential for a correct description of the natural variability of long-term processes, especially in those areas where the ATM contribution accounts for a large part of the total LMSL variance, as is the case in most shallow continental shelf seas[19]. When this natural variability is properly characterized, it permits a more sophisticated statistical judgment of the significance of observed trends and therefore a quantification of the minimum fraction that can be attributed to an external forcing unrelated to natural climate long-term variations. In the North Sea, 10 out of 11 stations show a significant long-term trend ($S > 0.99$) leading to minimum external trends ($P = 0.99$) that range between 0.13 and 1.42 mm per year (Fig. 3b and Table 1). These values should not be interpreted simply as the contribution of a common external forcing because an external trend in a single record may well be due to local effects such as measurement errors (as it is probably the case in North Shields (Fig. 3a,b)) or vertical land movements due to glacial isostatic adjustment or local subsidence/uplift, and they refer to the minimum value, meaning that the actual

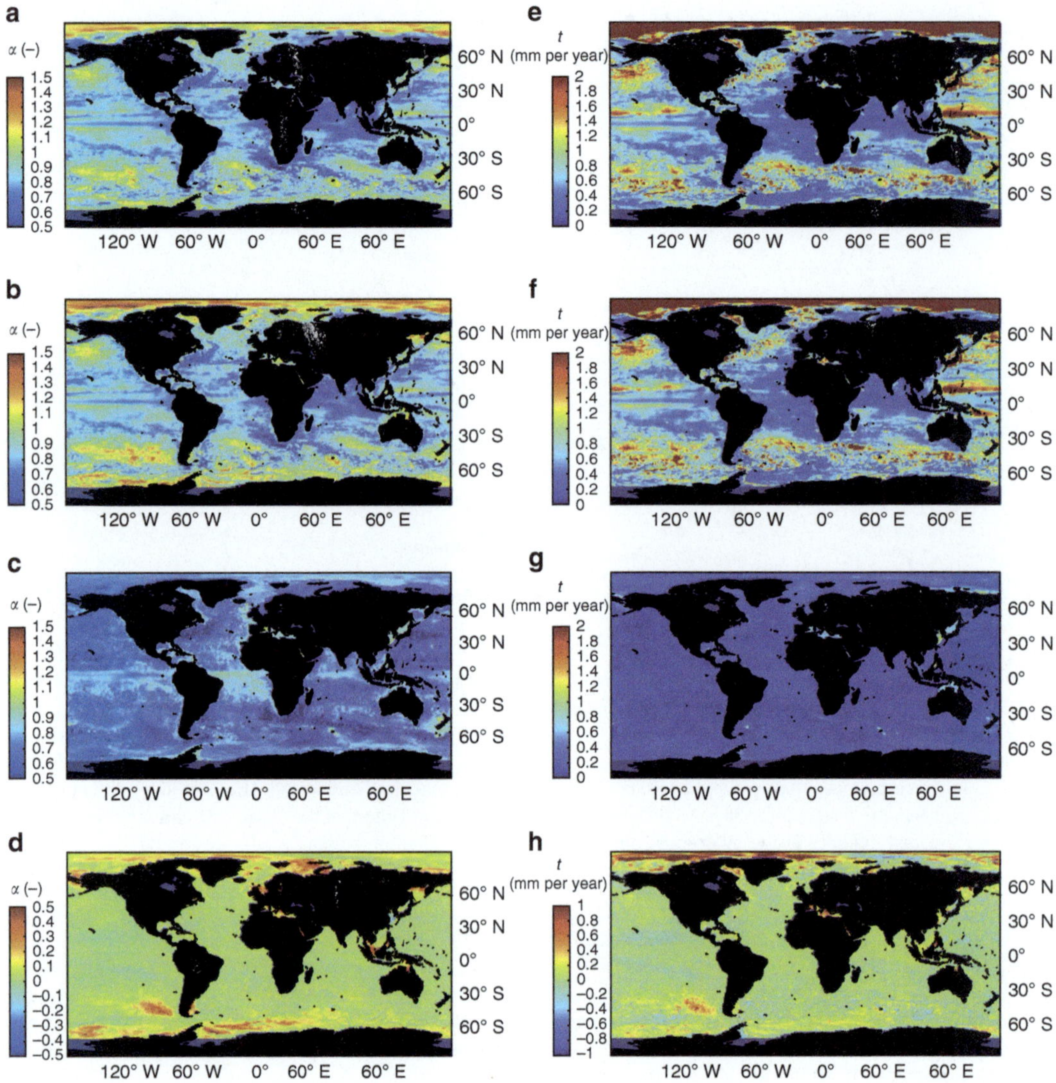

Figure 4 | Hurst exponents α and maximum natural centennial trends in modelled LMSLsyn. (a–c) α values as calculated for different components of LMSLsyn, RESsyn and OBPsyn over the period from 1899 to 2008. **(d)** Differences between the α values from LMSLsyn and RESsyn. **(e–g)** Maximum natural trends ($P = 0.99$) for LMSLsyn, RESsyn and OBPsyn fields under the assumption of a short-term ($\alpha = 0.5$) or a long-term correlated ($\alpha > 0.5$) process. **(h)** Differences between maximum natural trends in sea level derived from an integrated assessment of LMSLsyn minus the root sum of squares of natural trends calculated for OBPsyn and RESsyn separately.

On a global scale the short-term fluctuations of wind-driven redistribution processes cancel out and the resulting GMSL changes are suggested to represent a hybrid of long-term correlated natural variability (Hurst exponent α of 1.28) and relatively smooth anthropogenic forcing[7,8,30]. Accounting for the long-term correlated structure of GMSL variations yields an upper range of possible natural centennial trends in the order of 0.73 mm per year since 1900. Glacial isostatic adjustment corrected local and regional late Holocene proxy evidence shows rates that do not exceed this value[5,6,31]. Considering that the GMSL variability in observations and model simulations is at least an order of magnitude less than the local/regional variability[30], our findings are therefore consistent with the paleo evidence from periods in which the anthropogenic contribution was absent. Comparing our model-based estimate of natural GMSL variability with the observed twentieth-century GMSL rise of 1.33–1.98 mm per year[1–4] therefore suggests that it is virtually certain ($P = 0.99$) that at least 45% (1.33–0.73 mm per year) of the observed twentieth-century GMSL rise is of anthropogenic origin, extremely likely ($P = 0.95$) that it is at least 61% (1.33–0.52 mm per year) and very likely ($P = 0.90$) that it is at least 68% (1.33–0.43 mm per year). Likewise, our estimate of possible naturally forced centennial trends enhances the upper bound of the possible anthropogenic contribution. Hence, it is also virtually certain ($P = 0.99$) that the anthropogenic contribution does not exceed a value of 2.71 mm per year (1.98 + 0.73 mm per year), extremely likely ($P = 0.95$) that it does not exceed 2.50 mm per year (1.98 + 0.52 mm per year) and very likely ($P = 0.90$) that it does not exceed 2.41 mm per year (1.98 + 0.43 mm per year). The majority of the external contribution is probably related to the two dominant contributors to global MSL rise over the twentieth century[1], that is, glacier melting and thermosteric MSL, both of which became increasingly determined by anthropogenic forcing during the second half of the twentieth century[32–34].

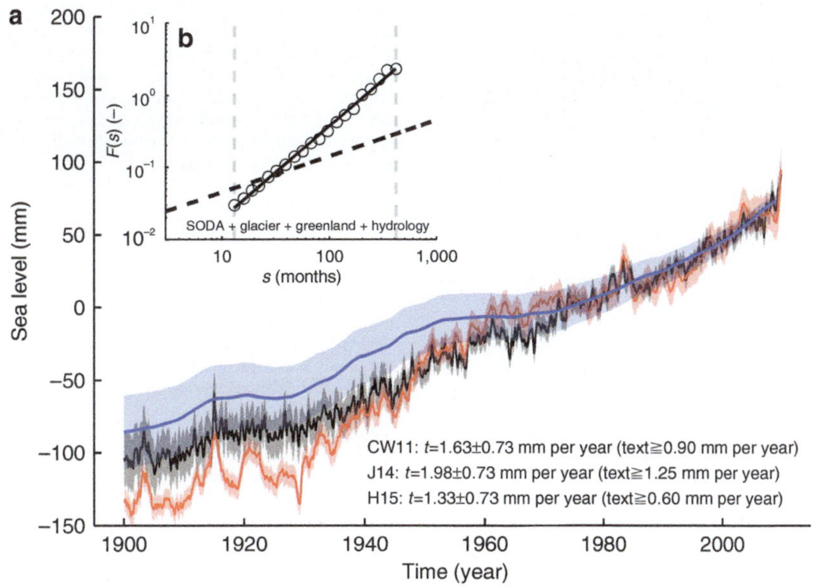

Figure 5 | Maximum natural and minimum external trends in GMSL. (**a**) Shown are three recent reconstructions of GMSL based on CW11 (ref. 2), J14 (ref. 3) and H15 (ref. 4) over the period 1900–2009. The shadings show their 1σ uncertainties. Their linear trends are given in the legend. All reconstructions suffer, with respect to their temporal variability, from sampling problems related to the temporally and spatially unevenly distributed location of tide gauge measurements[23]. Therefore, the temporal GMSL variability and the resulting naturally forced centennial trends ($P = 0.99$) have been assessed using spatial average of the LMSLsyn fields (SSH, glacier, Greenland ice sheet and hydrology), which are available over the entire global ocean over the period from 1871 to 2008. The resulting fluctuation function derived from a DFA2 is shown in (**b**). The fluctuation function yields an α value of 1.28. Following the approach described in refs 8,13–15 this implies an upper bound of naturally forced centennial trends (1900–2009) of 0.73 mm per year ($P = 0.99$). This suggests that the observed twentieth century GMSL rise is already outside the range of natural variability with a minimum external contribution (dependent on the reconstruction) of 0.60–1.25 mm per year ($P = 0.99$).

Methods

Tide gauge data. Eleven centennial tide gauge records of monthly resolution from the North Sea basin were selected for this study (Supplementary Table 1 and Fig. 1b). Nine of the records were downloaded from the website of the Permanent Service for Mean Sea Level[35], while the other two records (Cuxhaven, Norderney) stem from the database published by ref. 36. The reason for limiting the analyses of tide gauge data to this particular region is that reliable long-term estimates for the ATM are freely available only here[11]. For our assessment we remove the mean seasonal cycle from each record by subtracting the average of each calendar month taken over the entirely available period. No corrections for the effects of vertical land motion have been applied, since these values are still very scarce[37] and models for glacial isostatic adjustment very uncertain[3].

Modelled MSL data. In addition to the tide gauge records, we built model-based LMSLsyn fields on a global scale by adding together recent reconstructions of the dynamic SSH, the glacier contribution from 18 major mass loss regions and the Greenland ice sheet combined with the respective fingerprints, and temporally varying land hydrology. The contributions from the Antarctic ice sheet is expected to be of minor importance over the twentieth century[38,39] and therefore neglected in the present study. For the dynamic SSH the Simple Ocean Data Assimilation (SODA) reanalysis[22] over the period from 1871–2008 is used. SODA is a global baroclinic ocean reanalysis forced with winds from the twentieth-century reanalysis[17] that assimilates temperature and salinity observations[22]. Since SODA is based on the Boussinesq approximation, we accounted for missing steric effects following the approach introduced by ref. 40: that is, adding a globally uniform average of the steric height at each grid point and time step. For completeness we also added the missing long-term trend to SODA using the CMIP5 historical ensemble trend of the global mean thermosteric component over the period 1901–2000 (0.47 mm per year) as published by ref. 38. The good performance of SODA with respect to inter-annual and decadal LMSL variability has already been demonstrated[7,26] and is therefore suggested to be of sufficient quality for our purposes.

The contribution of glacier melting and the Greenland ice sheet is incorporated at each grid point by multiplying the sea level contribution as reconstructed by refs 23,24 with the respective fingerprints[25]. Glacier fingerprints are computed by solving the sea level equation[41], which accounts for the gravitational coupling between surface redistribution of water masses and deformation of the solid earth, on a compressible elastic earth and including the effect of changes in earth rotation[42].

Despite these two factors, which determined the majority of the long-term trends in twentieth-century GMSL[1,38,39], hydrology is known to be responsible for a large fraction of the observed inter-annual GMSL variability[43]. We follow ref. 26 and produce an estimate of this contribution under the assumption that it acts as a barotropic load on the ocean, that is, monthly soil moisture fields from the Climate Prediction Centre[44] are used to model hydrology changes as a globally uniform but temporally varying signal in the form of:

$$\eta_{\mathrm{h}} = -\frac{1}{A_{\mathrm{Ocean}}} \int_{A_{\mathrm{Land}}} F \mathrm{d}A \qquad (1)$$

where A_{Ocean} and A_{Land} are the surface area of the ocean and the continents, respectively, and F, the detrended soil moisture[26]. Unfortunately, the soil moisture component is only available from 1948 on. To artificially extend this time series back in time to 1871 we apply a phase-randomized Fourier algorithm[45]. The algorithm is based on a Fourier transformation of the original time series and a randomization of the phases at different frequencies in the range of 0 and 2π. This allows for an artificial extension of the existing record back to 1871 showing a similar spectrum and autocorrelation function as the observations. Although this time series does not provide the truly observed signal before 1948, it gives a reasonable estimate of the underlying processes with a similar temporal characteristic[45].

To separate the contributions from local steric and OBP variations from the dynamic SSH in SODA the common nonlinear equation of the state of the ocean has been applied to estimate the local steric height. The OBPsyn component was derived by simply removing the local steric height from the model SSH at each grid point. The modelled volumetric component RESsyn is consequently derived as the sum of the steric height from SODA, the glacier and ice sheet contribution and hydrology at each time step and grid point, while OBPsyn serves as a proxy for the ATM.

Determination of the ATM. The determination of the ATM is based on a multiple stepwise LRM introduced by refs 11,46. For the LRM, SLP and zonal and meridional wind stress time series showing the highest bivariate correlation with the observed MSL from an area of $\pm 4°$ around the tide gauge location are selected as independent forcing factors. Monthly detrended MSL anomalies from the respective tide gauge location are then used as the dependent variable, while the monthly detrended SLP and wind stress anomalies are taken as independent predictors. The stepwise algorithm chosen in the LRM is when one of the obtained regression coefficients gets insignificant (95% confidence level using common F-statistics). To estimate the contribution of the ATM to long-term trends, we assume that the relationship found for the detrended records also holds on longer timescales and simply combine the estimated regression coefficients with the non-detrended atmospheric forcing variables. By definition, ATM and RES are uncorrelated.

The performance of this model has been intensively tested in ref. 11. However, to provide an overview about the performance for the 11 selected stations from this study, we validate the model against a state-of-the art barotropic hydrodynamical HAMSOM model of the North Sea introduced by ref. 18. The HAMSOM model is driven by the eight main tidal constituents at the model boundaries and 10 m surface winds from reanalysis data sets and has a spatial resolution of 3 km[18]. For the validation both models, that is, the LRM as well as HAMSOM, have been forced with atmospheric fields from the National Centers for Environmental Prediction reanalysis[47]. The HAMSOM runs are available over the period 1953–2003. Hence, we limit the validation to this period. The results are shown in Supplementary Fig. 2. For 9 out of 11 tide gauge locations the estimated ATMs show reasonable agreement between the two different models as underpinned by correlation coefficients larger than 0.9. Only two stations display somewhat smaller correlations coefficients in the order of 0.6–0.8. These smaller values are probably related to local bathymetric inaccuracies in the numerical HAMSOM model rather than a poor performance of the LRM as pointed out in ref. 11. Overall, the validation demonstrates the ability of the LRM in capturing the barotropic ATM in the North Sea with sufficient accuracy, which underpins the applicability for the purposes of this study.

Estimation of natural and external trends. To estimate a relative MSL trend over a specific time period L, we apply a common linear least-squares-fit to the monthly MSL anomalies, which gives us an absolute change Δ (that is, the difference between the last and the first value of the regression line) and the s.d. σ of the residuals around the regression line. Following refs 8,13–15 we consider the relative trend ratio $x = \Delta/\sigma$ as the quantity of interest. A central question regarding the observed MSL change is now whether x is natural or anthropogenic in origin. Therefore, we require the probability $P(x; L)dx$ that in time series with a similar autocorrelation structure as the observed data, a relative trend between x an $x + dx$ occurs. From P we can derive the exceedance probability

$$W(x;\,L) = \int_{x}^{\infty} P(x';\,L)\mathrm{d}x' \tag{2}$$

and the respective significance

$$S(x,\,L) = \int_{-x}^{x} P(x';\,L)\mathrm{d}x' = 1 - 2W(x;\,L) \tag{3}$$

of x. By definition, S is the probability that the relative trend in the record lies between $-x$ and x (refs 8,13–15).

Here we assume that an observed trend x cannot be explained by natural variability alone, if S is above 0.99 ($P = 0.99$). Consequently, relative trends x between $-x_{99}(L)$ and $x_{99}(L)$ are considered as natural. If x_{99} is exceeded in an observational record, the part $x - x_{99}$ cannot be explained by natural variations alone. Thus, $x - x_{99}$ must be externally forced and can be considered as the minimum external trend. Furthermore, we also know that the external trend cannot exceed a value of $x + x_{99}$, which therefore appears to be the maximum possible external trend[15].

It has been shown in extensive Monte-Carlo experiments[15] that for records of sufficient length L the probability density function $P(x; L)$ is distributed in many models as Student's t, which can be written as

$$P(x;\,L) = \frac{\Gamma\left(\frac{l(L)+1}{2}\right)}{\Gamma\left(\frac{l(L)}{2}\right)\sqrt{\pi l(L)}a}\left(1 + \frac{\left(\frac{x}{a}\right)^2}{l(L)}\right)^{-\frac{l(L)+1}{2}}. \tag{4}$$

$P(x; L)$ is therefore a function of the scaling parameter a and the effective length l, both of which depend on the autocorrelation structure of the observed record. For instance, if a record is only characterized by short-term correlations with a specific lag-1 autocorrelation c one can easily obtain l and a by following the approach of ref. 48: that is,

$$l = L\frac{1-c}{1+c} - 2 \tag{5}$$

and

$$a = \frac{\sqrt{12}}{\sqrt{l}}. \tag{6}$$

However, as recently shown[7,8], this assumption might not be appropriate with respect to the fact that MSL records also exhibit long-term correlations characterized by a Hurst exponent $\alpha > 0.5$. In these cases tables for a and l based on extensive Monte-Carlo simulations have been provided in the appendix of ref. 15. To decide which case is appropriate in our assessment, we calculate the lag-1 autocorrelation as well as the Hurst exponent α using DFA2 (ref. 49). DFA2 has been established especially for the case of non-stationary time series and is very robust against external trends in a given record[49]. DFA2 integrates a time series to a random walk series $Y(t)$ (by calculating the cumulative sum), separates the random walk series in non-overlapping segments, each of them containing s points, removes quadratic trends from each segment, calculates the variance Y_s for each segment, takes the average over all segments, and finally computes the root mean square to obtain a fluctuation function $F(s)$:

$$F(s) = \left[\frac{1}{2N_s}\sum_{\nu=1}^{2N_s}\left(\frac{1}{s}\sum_{i=1}^{s} Y_s^2[(\nu-1)s + i]\right)\right]^{\frac{1}{2}} \tag{7}$$

The Hurst exponent α can be estimated as the asymptotic slope of $F(s)$ in the double logarithmic space. A time series is long-term correlated if α is significantly different from 0.5 and short-term correlated or uncorrelated if α is equal to 0.5. If we find significant long-term correlations we consequently derive l and a from the tables in ref. 15, whereby in case of a Hurst exponent α smaller or equal 0.5, l and a are calculated following ref. 48. Further information about the general mathematical concept of estimating natural and external trends can be found in the literature[13–15].

When dividing the observed sea level signal into the two separate components ATM and RES we estimate the quantiles of naturally forced centennial trends based on two separate t-distributions with different scaling parameters a and effective length l. In case of a normal distribution and uncorrelated records of ATM and RES one can easily obtain approximations of the combined quantiles via a common root sum of squares:

$$x_{99,\mathrm{SUM}} = \sqrt{x_{99,\mathrm{ATM}}^2 + x_{99,\mathrm{RES}}^2}, \tag{8}$$

while in case of significant correlations between RES and ATM the quantiles can be calculated as:

$$x_{99,\mathrm{SUM}} = \sqrt{x_{99,\mathrm{ATM}}^2 + x_{99,\mathrm{RES}}^2 + 2Kx_{99,\mathrm{ATM}}x_{99,\mathrm{RES}}}, \tag{9}$$

where K represents the linear correlation between ATM and RES. However, the natural trends are obtained from Student's t-distributions for which the accuracy of the root sum of squares approximation is not known. Therefore, we have tested whether the approximations from equations (8) and (9) also hold for Student's t-distributions with similar characteristics as obtained from RES and ATM. As shown in Supplementary Fig. 8 this is still the case. Note that significant correlations between RES (RESsyn) and ATM (OBPsyn) have been only found in the Arctic Ocean and some smaller sub-areas with minor influences on the calculation of the naturally forced trends (Supplementary Fig. 9). For the North Sea tide gauges all correlations between ATM and RES are zero (not shown).

References

1. Church, J. A. *et al.* in *Climate Change 2013: The Physical Science Basis. Contribution of Working Group I to the Fifth Assessment Report of the Intergovernmental Panel on Climate Change* (Cambridge University Press, 2013).
2. Church, J. A. & White, N. J. Sea-level rise from the late 19th to the early 21st century. *Surv. Geophys.* **32,** 585–602 (2011).
3. Jevrejeva, S., Moore, J. C., Grinsted, A. & Woodworth, P. L. Recent global sea level acceleration started over 200 years ago? *Geophys. Res. Lett.* **35,** L08715 (2008).
4. Hay, C. C., Morrow, E., Kopp, R. E. & Mitrovica, J. X. Probabilistic reanalysis of twentieth-century sea-level rise. *Nature* **517,** 481–484 (2015).
5. Kemp, A. C. *et al.* Climate related sea-level variations over the past two millennia. *Proc. Natl Acad. Sci. USA* **105,** 13252–13257 (2011).
6. Gehrels, R. W. & Woodworth, P. L. When did modern rates of sea-level rise start? *Global Planet. Change* **100,** 263–277 (2013).
7. Dangendorf, S. *et al.* Evidence for long-term memory in sea level. *Geophys. Res. Lett.* **41,** 5564–5571 (2014).
8. Becker, M., Karpytchev, M. & Lennartz-Sassinek, S. Long-term sea level trends: natural or anthropogenic? *Geophys. Res. Lett.* **41,** 5571–5580 (2014).
9. Milne, G. W., Gehrels, R., Hughes, C. W. & Tamisiea, M. E. Identifying the causes of sea level change. *Nat. Geosci.* **2,** 471–478 (2009).
10. Calafat, F. M. & Chambers, D. P. Quantifying recent acceleration in sea level unrelated to internal climate variability. *Geophys. Res. Lett.* **40,** 3661–3666 (2013).
11. Dangendorf, S. *et al.* Mean sea level variability in the North Sea: processes and implications. *J. Geophys. Res.* **119,** 6820–6841 (2014).
12. Thompson, P. R. & Merrifield, M. A. A unique asymmetry in the pattern of recent sea level change. *Geophys. Res. Lett.* **41,** 7675–7683 (2014).
13. Lennartz, S. & Bunde, A. Trend evaluation in records with long-term memory: application to global warming. *Geophys. Res. Lett.* **36,** L16706 (2009).
14. Lennartz, S. & Bunde, A. Distribution of trends in long-term correlated records: a scaling approach. *Phys. Rev. E* **84,** 021129 (2011).
15. Tamazian, A., Ludescher, J. & Bunde, A. The significance of trends in long-term correlated records. *Phys. Rev. E* **91,** 032806 (2015).
16. Zorita, E., Stocker, T. F. & von Storch, H. How unusual is the recent series of warm years? *Geophys. Res. Lett.* **35,** L24706 (2008).

17. Compo, G. P. *et al.* The twentieth century reanalysis project. *Q. J. Roy. Meteorol. Soc.* **137,** 1–28 (2011).
18. Chen, X. *et al.* On sea level change in the North Sea influenced by the North Atlantic Oscillation: local and remote steric effects. *Estuar. Coast. Shelf Sci.* **151,** 186–195 (2014).
19. Marcos, M. & Tsimplis, M. N. Forcing of coastal sea level rise patterns in the North Atlantic and the Mediterranean Sea. *Geophys. Res. Lett.* **34,** L18604 (2007).
20. Marshall, J., Johnson, H. & Goodman, J. A study of the interaction of the North Atlantic Oscillation with Ocean Circulation. *J. Climate* **14,** 1399–1421 (2001).
21. Ludescher, J., Bogachev, M. I., Kantelhardt, J. W., Schumann, A. Y. & Bunde, A. On spurious and corrupted multifractality: the effects of additive white noise, short-term memory and periodic trends. *Physica A* **390,** 2480–2490 (2011).
22. Carton, F. M. & Giese, B. S. A reanalysis of ocean climate using Simple Ocean Data Assimilation (SODA). *Mon. Weather Rev.* **136,** 2999–3017 (2008).
23. Marzeion, B., Jarosch, A. H. & Hofer, M. Past and future sea-level change from the surface mass balance of glaciers. *Cryosphere* **6,** 1295–1322 (2012).
24. Box, J. E. & Colgan, W. Greenland ice sheet mass balance reconstruction. Part III: marine ice loss and total mass balance (1840-2010). *J. Climate* **26,** 6990–7002 (2013).
25. Bamber, J. & Riva, R. The fingerprint of recent ice mass flux. *Cryosphere* **4,** 621–627 (2010).
26. Calafat, F. M., Chamber, D. P. & Tsimplis, M. N. On the ability of global sea level reconstructions to determine trends and variability. *J. Geophys. Res.* **119,** 1572–1592 (2014).
27. Boening, C., Lee, T. & Zlotnicki, V. A record-high ocean bottom pressure in the South Pacific observed by GRACE. *Geophys. Res. Lett.* **38,** L04602 (2011).
28. Piecuch, C. G., Quinn, K. J. & Ponte, R. Satellite-derived interannual ocean bottom pressure variability and its relation to sea level. *Geophys. Res. Lett.* **40,** 3106–3110 (2013).
29. Marcos, M. Ocean bottom pressure variability in the Mediterranean Sea and its relationship with sea level from a numerical mode. *Global Planet. Change* **124,** 10–21 (2015).
30. Gregory, J. M., Lowe, J. A. & Tett, S. F. B. Simulated global-mean sea level changes over the last half-millennium. *J. Climate* **19,** 4576–4591 (2006).
31. Saher, M. H. *et al.* Sea-level changes in Iceland and the influence of the North Atlantic Oscillation during the last half millennium. *Quat. Sci. Rev.* **108,** 23–36 (2015).
32. Marzeion, B., Cogley, J. G., Richter, K. & Parkes, D. Attribution of global glacier mass loss to anthropogenic and natural causes. *Science* **345,** 919–921 (2014).
33. Marcos, M. & Amores, A. Quantifying anthropogenic and natural contributions to thermosteric sea level rise. *Geophys. Res. Lett.* **41,** 2502–2507 (2014).
34. Slangen, A .B. A., Church, J. A., Zhang, X. & Monselesan, D. Detection and attribution of global mean thermosteric sea level change. *Geophys. Res. Lett.* **41,** 5951–5959 (2014).
35. Holgate, S. *et al.* New data systems and products at the permanent service for mean sea level. *J. Coast. Res.* **29,** 493–504 (2013).
36. Wahl, T., Haigh, I. D., Frank, T. & Jensen, J. Improved estimates of mean sea level changes in the German Bight over the last 166 years. *Ocean Dynamics* **61,** 701–715 (2011).
37. Wahl, T. *et al.* Observed mean sea level changes around the North Sea coastline from 1800 to present. *Earth Sci. Rev.* **124,** 51–67 (2013).
38. Gregory, J. *et al.* Twentieth-century global-mean sea-level rise: is the whole greater than the sum of the parts? *J. Climate* **26,** 4476–4499 (2013).
39. Church, J. A., Monselesan, D., Gregory, J. M. & Marzeion, B. Evaluating the ability of process based models to project sea-level change. *Environ. Res. Lett.* **8,** 014051 (2013).
40. Greatbach, R. J. A note on the representation of steric sea level in models that conserve volume rather than mass. *J. Geophys. Res.* **99,** 12767–12771 (1994).
41. Farrell, W. E. & Clark, J. A. Postglacial sea-level. *Geophys. J. R. Astr. Soc.* **46,** 647–667 (1976).
42. Milne, G. A. & Mitrovica, J. X. Postglacial sea-level change on a rotating Earth. *Geophys. J. Int.* **133,** 1–19 (1998).
43. Fasullo, J. T., Boening, C., Landerer, F. W. & Nerem, S. R. Australia'a unique influence on global sea level in 2010-2011. *Geophys. Res. Lett.* **40,** 4368–4373 (2013).
44. Fan, Y. & van den Dool, H. Climate Prediction Centre global monthly soil moisture data set at 0.5° resolution for 1948 to present. *J. Geophys. Res.* **109,** D10102 (2004).
45. Pricard, D. & Theiler, J. Generating surrogate data for time series with several simultaneously measured variables. *Phys. Rev. Lett.* **73,** 951–954 (1994).
46. Dangendorf, S., Mudersbach, C., Wahl, T. & Jensen, J. Characteristics of intra-, inter-annual and decadal sea level variability and the role of meteorological forcing: the long record of Cuxhaven. *Ocean Dynamics* **63,** 209–224 (2013).
47. Kalnay, E. *et al.* The NCEP/NCAR 40-year reanalysis project. Bull. *Am. Meteorol. Soc.* **77,** 437–471 (1996).
48. Santer, B. D. *et al.* Statistical significance of trends and trend differences in layer-average atmospheric temperature time series. *J. Geophys. Res.* **105,** 7337–7356 (2000).
49. Kantelhardt, J. W., Koscielny-Bunde, E., Rego, H. H. A., Havlin, S. & Bunde, A. Detecting long-range correlations with detrended fluctuation analysis. *Physica A* **295,** 441–454 (2001).

Acknowledgements

S.D. was funded by an internal fellowship at the University of Siegen. M.M. acknowledges a 'Ramon y Cajal' contract funded by the Spanish Ministry of Economy; R.R. acknowledges VIDI project no. 864.12.012 funded by the Netherlands Organization for Scientific Research; we are further grateful to Jan Even Øie Nilsen, Joaquim Pinto, Holger Steffen and Roland Gehrels for providing valuable comments on an earlier version of the manuscript; Ben Marzeion and Jonathan Gregory are acknowledged for the discussion of the anthropogenic contribution to twentieth century glacier melting and thermosteric sea level rise. We further thank Ben Marzeion and Jason Box for providing their estimates for the glacier and ice sheet contribution to sea level rise. The work by E.Z, is part of CLISAP (Germany). K.B. is funded by the Ministry of Innovation, Science, Research and Technology of North Rhine-Westphalia, Germany and the European Commission in the context of the Ziel 2 framework (No. 280411902).

Author contributions

S.D. conceived the idea of the study, performed the analyses and wrote the paper together with M.M.; M.M. further participated in many technical discussions; A.M. and K.B. provided a MATLAB file for the statistical assessment of trend significance; R.R. calculated fingerprints for the 18 melting regions; E.Z., A.M. and J.J. participated in technical discussions at an early stage when the idea for the study evolved and co-wrote the paper.

Additional information

Competing financial interests: The authors declare no competing financial interests.

3

Anthropogenic effects are associated with a lower persistence of marine food webs

Luis J. Gilarranz[1], Camilo Mora[2] & Jordi Bascompte[1,†]

Marine coastal ecosystems are among the most exposed to global environmental change, with reported effects on species biomass, species richness and length of trophic chains. By combining a biologically informed food-web model with information on anthropogenic influences in 701 sites across the Caribbean region, we show that fishing effort, human density and thermal stress anomaly are associated with a decrease in local food-web persistence. The conservation status of the site, in turn, is associated with an increase in food-web persistence. Some of these associations are explained through effects on food-web structure and total community biomass. Our results unveil a hidden footprint of human activities. Even when food webs may seem healthy in terms of the presence and abundance of their constituent species, they may be losing the capacity to withstand further environmental degradation.

[1] Department of Integrative Ecology, Estación Biológica de Doñana, (EBD-CSIC), Calle Américo Vespucio s/n, E-41092 Sevilla, Spain. [2] Department of Geography, University of Hawaii, Honolulu, Hawaii 96822, USA. † Present address: Department of Evolutionary Biology and Environmental Studies, University of Zurich, Winterthurerstrasse 190, CH-8057 Zurich, Switzerland. Correspondence and requests for materials should be addressed to J.B. (email: jordi.bascompte@ieu.uzh.ch).

Anthropogenic influences are affecting species richness, abundance and genetic composition, with far reaching consequences for ecosystem functioning[1–4]. The majority of published studies, however, tend to treat species as independent units. In contrast, recent research is putting the emphasis on the consequences of global environmental change on the network of interactions among species[5,6]. Indeed, network persistence can be eroded before there is a change in community composition, with potential implications for systemic collapse[7,8].

Previous work on trophic cascades has already identified anthropogenic signatures in food-web dynamics following the removal of key species, mainly top predators[9–15]. Here we address whether changes in food-web persistence can precede changes in species composition following anthropogenic influences. We tackle this problem by integrating several layers of information. Our starting point is a previously compiled regional marine food web encompassing 249 species/trophic groups and 3,313 interactions[16], originated from reports on trophic interactions[17] and stomach content analyses[18]. Next, we use a spatial dataset with species' demographic information from 701 locations across the Caribbean Sea, Gulf of Mexico and North Atlantic Ocean reefs (Fig. 1). Combining both data sets, we obtain not only the potential food web at every location, but also the body mass of the individuals, and the strength of the interactions.

Each local food web determines the skeleton of a dynamic model parametrized with empirical data (Methods). We estimate persistence as the fraction of species surviving at the end of a simulation[19–21]. We proceed by comparing the persistence of a local community with indicators of human influence at such a location, specifically fishing effort, nearby human density, thermal stress anomaly (TSA)—related with climate change, and the protected status of the area—protected versus non-protected as well as the size and age of the protected area. We also take in consideration recorded properties of the local communities, such as the total biomass or the architecture of each local food web—in terms of its tendency to be organized in compartments, where species within a compartment tend to interact among themselves while showing much fewer trophic links with species from other compartments. This later property seems to be pervasive across ecological networks[22–24] and has been suggested to increase their robustness through the buffering of perturbations within specific compartments[21,25].

Our results show that anthropogenic influences such as fishing affect the persistence of food webs beyond their more evident effects on species richness or total biomass. This suggests that human activities may have a previously unnoticed effect on natural communities by eroding their resilience in the face of further perturbations.

Results

Ecological associations. Figure 1 represents a regional persistence map for the marine food web. This map provides the opportunity to look for signatures of food-web erosion along the wide range of human pressures existing in this region. First, and to fully disentangle the anthropogenic effects, we describe basic associations between ecological variables. Thus, we find the already known negative relationship between the number of species in a network and the number of surviving species at the end of the numerical simulations (Fig. 2; Supplementary Table 1). The larger the number of species, the narrower the parameter space in which species can coexist. This results in a lower persistence in larger food webs. We also find a significant relationship between the observed total biomass in one place and the number of species in that place (Fig. 2; Supplementary Table 2), probably mediated by trait complementarity[26,27]. These associations have to be taken

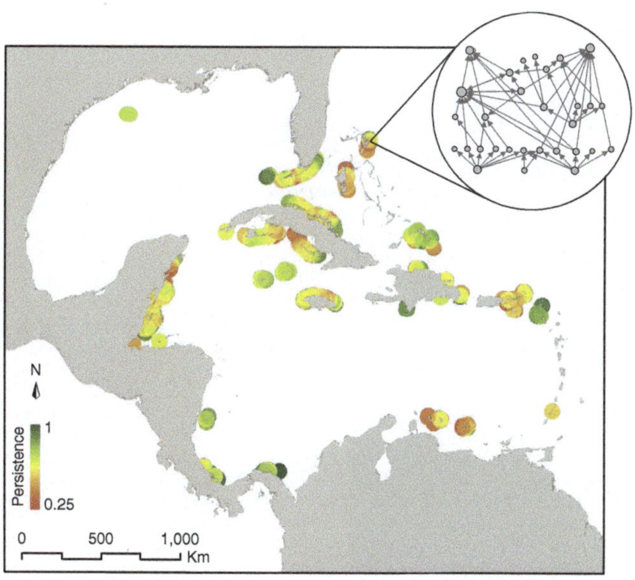

Figure 1 | Persistence map of the Caribbean coral reef food webs. Each point indicates a sampling site and the colour represents the average persistence value of its simulated food web. Red and green colours represent low and high persistence, respectively. This map was made creating a 50 km buffer around each sampling site. The value of persistence of the sampling site is assigned to the area inside the buffer. Then, for representation purposes, each pixel in the map represents the average of the values of the buffers that overlap in that pixel. The insert shows one of the local food webs, where node size is proportional to its number of trophic interactions.

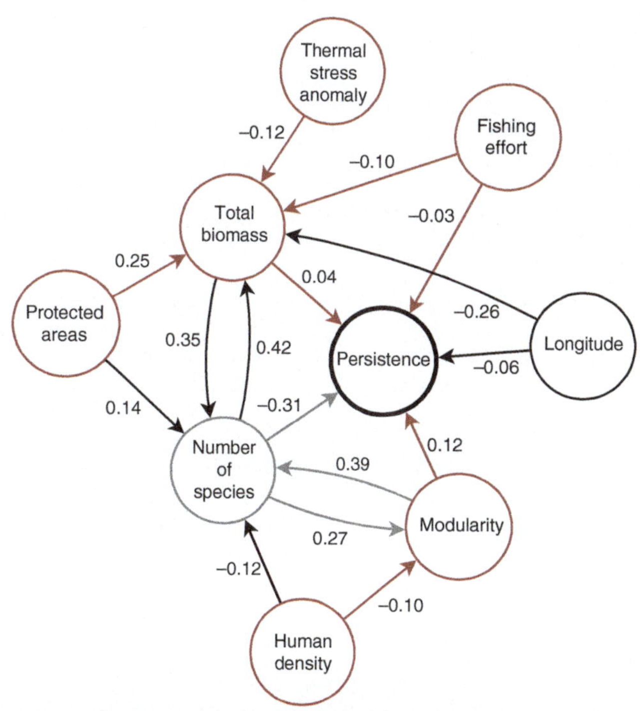

Figure 2 | Network of statistical associations between variables. Each link represents a statistically significant effect of one variable on another. The numbers next to each link are the effect sizes. The associations between anthropogenic effects and food-web persistence are represented in red. Light grey represent into account when assessing the anthropogenic footprints on marine food webs. Next, we look at such anthropogenic effects across the regional persistence map.

Anthropogenic associations with species richness and biomass. We observe that while fishing shows no association with the number of species present at a site, human density is negatively associated with species richness (Fig. 2; Supplementary Table 3). Arguably, this is because while fishing reduces species biomass, human density has an effect on the habitat inhabited by the local community, specially in water quality. This leads to a different species assembly near urban areas. However, locations inside protected areas tend to have greater biomass (Fig. 2; Supplementary Table 2).

Anthropogenic associations with food-web persistence. Beyond the above effects of human activity on the size of the food webs, we now turn to the anthropogenic associations with the persistence of the food web as colour coded in Fig. 1. Specifically, our results show that fishing effort, human density and TSA are associated with a decrease in community persistence (Fig. 2). The conservation status of a site, on the other hand, is positively associated with food-web persistence (Fig. 2). Importantly, these correlations between food-web persistence and anthropogenic influences remain significant even when total biomass is accounted for (Fig. 2). This shows that anthropogenic effects on food-web persistence are beyond any signal in total biomass.

Role of food-web structure. Some of the above associations between anthropogenic effects and food-web persistence are partly explained by human-induced changes in food-web structure and biomass. For example, we found that communities surrounded by higher human densities have a lower tendency to be organized in compartments (Fig. 2; Supplementary Table 4). In turn, consistent with theory[21], our results show that food-web compartmentalization is associated with an increase in persistence (Supplementary Table 1). Therefore, the negative association between human density and food-web persistence can be explained by a decrease in food-web compartmentalization. Since we control for the number of species, the change in food-web structure is driven by a change in the identity of the constituent species.

Role of food-web biomass. Part of the association between fishing effort and food-web persistence is explained by the negative association between fishing effort and total community biomass (Supplementary Table 2). More intensively fished communities tend to have a significantly lower fish biomass. However, it is not the reduction of the overall community biomass, but the decrease in the biomass of certain species targeted by fishing[2], that provides a highly plausible mechanism explaining the relationship between fishing effort and community persistence[16]. The effect of fishing effort on community biomass reveals that the reduction of the targeted species' biomass is not compensated by an increase in some other species biomass. Therefore, fishing is correlated with an overall decrease in community biomass.

Similarly, those areas suffering higher TSAs tend to have lower community biomass (Supplementary Table 2). A lower community biomass, in turn, is related to a lower food-web persistence (Fig. 2). Since TSA is related with climate change, this represents another pathway in which anthropogenic stressors are negatively associated with a decrease in food-web persistence.

Last, our analysis shows that sites located inside protected areas harbour a significant larger community biomass (Supplementary Table 2) than those outside (as reported by refs 28,29), which in turn is positively correlated with food-web persistence (Fig. 2). This results on the observed association between the conservation status of the site and food-web persistence. Further analyses show that while there is no association between marine protected area (MPA) age and food-web persistence, there is a significant positive association of MPA size with persistence (effect $= 0.068$; s.e. $= 0.020$; $df = 245$; $t = 3.425$; $P < 0.001$).

Discussion

We have mapped community persistence along a wide regional scale similar to how ecologists plot maps of species richness. While the local number of species is informative of the ecological value of a site, our approach goes a step further by looking at the ability of these communities to persist and withstand future perturbations. Although we cannot infer causality from our statistical analysis, we provide strong evidence revealing a new dimension of the wide-scale impact of human activities in the biosphere. Thus, human activity may not only be affecting population abundances of independent species, but it may also be eroding the robustness of these food webs and therefore their ability to persist in the future.

Availability of spatially distributed information on food-web structure will open the door towards more functional, dynamically based resource management and conservation. If we want to conserve not only species but the persistence of entire communities and the services they provide, we need to ensure the maintenance of trophic structures.

Methods

Food webs. The Caribbean food web[16] describes the trophic relationships between 249 species/trophic groups in a wide geographic region of the Caribbean sea, comprising $\sim 1{,}000\,km^2$ and stretching from the surface to 100-m depth. Obtained through gut content analysis[17,18], it is the most complete and accurate description of the feeding relationships in the region. One limitation that should be noted is that the level of resolution of the food web is not homogeneous through all ecological groups. Thus, while fishes are resolved at the species level, other species such as zooplankton and invertebrates are highly aggregated. It should also be acknowledged that sampling surveys may fail to account for cryptic species and/or small species difficult to identify while in the field across a vast region (see Supplementary Information of ref. 16 for a detailed description of the strengths and limitation of data). Regardless of the nature of these limitations, they do not systematically vary among sites so as to be likely to generate spurious results. Another limitation is represented by our lack of information on diet shits. It has been shown theoretically[30,31] that interaction plasticity increases persistence. We do not know, however, if interaction plasticity is constant across sites. Therefore, we decided not to consider diet shifts and to use a model that is more grounded in the food web modelling tradition. Also, although we could have included diet shifts from a theoretical perspective (assigning a certain probability), the number of free parameters would have increased enormously, precluding any general conclusion. From all possible scenarios, our results would only become weaker if less persistent food webs could become more plastic and therefore experience a higher increase in persistence.

The Atlantic and Gulf Rapid Reef Assessment (AGRRA)[32] is a multinational consortium of scientists in the Caribbean Region monitoring the status of coral reefs in a standardized manner. Under this methodology, a highly precise database has been compiled. All scientists involved have higher education degrees and specializations in the study of coral reefs as to ensure data consistency and reductions in human errors implementing the sampling and identifying species. It contains presence–absence information, along with the biomass ($g\,m^{-2}$), size and number of individuals—per square metre of sampling area—of the fish species found at 701 different locations across the region (Fig. 1). The sampled units were underwater visual censuses over an area of $50\,m^2$. The number of sampling units ranged from 6 to 10 per site. Additional details can be found at www.agrra.org.

By combining both databases, we obtain not only the potential food web at every location (Supplementary Fig. 1), but also the body mass of the individuals, and the strength of the interactions. The survey does not provide data on primary resources. However, we know which type of primary producer is targeted by each herbivore species. To 'feed' the herbivores, we introduce three primary producers in the networks: benthic autotrophs, plankton and colonial invertebrates. Regarding secondary consumers, it may happen that a few of them have been sampled in a site, but none of their prey has been found in that site. In those cases, we add a node called 'alloctonous input' that simulates the pool of prey for those secondary consumers[16] (Supplementary Fig. 1). The result of this procedure is a set of 701 local networks of sizes ranging from 15 to 37 species, and connectances from 0.04 to 0.16.

Dynamic bioenergetic model. To simulate species dynamics, we use the bioenergetic model proposed by Yodzis and Innes[33] and widely used in recent years[16,19,20,34]. This model simulates the biomass B of each species over time. The change in biomass is a function of growth, respiration, and biomass gain and loss trough predation. Mathematically, for species i, the model can be written as:

$$\frac{dB_i}{dt}=r_iG_iB_i - B_ix_i + B_ix_i\sum_{j=prey}^{n} y_iF_{ij} - \sum_{k=pred}^{n}\frac{x_ky_kB_kF_{ki}}{e_{ki}}, \quad (1)$$

where G_i is the normalized growth rate of basal species; r_i is the mass-specific maximum growth rate; x_i is the mass-specific metabolic rate; y_i is the species maximum consumption relative to its metabolic rate; e_{ki} is the fraction of the biomass of species i lost due to consumption by species k that is actually metabolized; F_{ij} is the type II functional response described as:

$$F_{ij}=\frac{\omega_{ij}B_i}{B_0+\sum_{k=prey}^{n}\omega_{ik}B_k}, \quad (2)$$

where ω_{ij} is the interaction preference between species i and j measured as the relative fraction of prey species i in the gut content of predator j, and B_0 is the half-saturation density. For basal species, $x_i=0$, while for consumers $G_i=0$. Interaction strength depends on preference and on the biomass of the species involved in the interaction[16].

The growth rate G_i of basal species is calculated as a competition model:

$$G_i=1-\frac{\sum_{j=prod}^{n}B_j}{K}, \quad (3)$$

where K is the carrying capacity, and for every location it is equal to the biomass of the most abundant species in that location $prod$ indicates the total number of primary producers.

The timescale of the system is defined by normalizing the mass-specific growth rate of a basal species used as reference to one. Then, we also normalize the mass-specific metabolic rate x_i and the maximum consumption rate relative to its metabolic rate y_i by the metabolic rates. Therefore, the mass-specific maximum growth rate r_i is equal to one, and x_i is defined as follows:

$$x_{ij}=\frac{a_x}{a_r}\left(\frac{M_b}{M_j}\right)^{\frac{1}{4}}, \quad (4)$$

where a_r is the maximal production-to-biomass ratio for ectotermic vertebrates, with $a_r=a_y-a_x$. The allometric constants a_y and a_x are, respectively, defined as the maximal ingestion and respiration rates for ectothermic vertebrates. M_j is the average body mass of the predator and M_b is the body mass of a reference basal species (set to 1). Finally, the species maximum consumption rate relative to its metabolic rate is defined as $y_i=\frac{a_y}{a_x}$.

Average body mass of prey and predators are obtained empirically trough the AGRRA survey and might be different at every location. The allometric constants for ectotermic vertebrates take the values $a_y=8.9$ and $a_x=2.3$. The half-saturation density takes the value $B_0=0.75$. The fraction of the biomass of prey lost to consumption by the predator that is actually metabolized (e_{ij}) is 0.85. The initial biomasses at each location are those described by the AGRRA survey and are measured in $\mathrm{g\,m^{-2}}$.

The unknown preferences of the interactions between primary producers and herbivores, and between the alloctonous input and the fishes that feed on it are sampled from a probability distribution. To obtain this distribution, we pool together all the known interaction preferences in the local food webs and then fit every possible distribution to the data. The best fit is to a log-normal distribution of $\mu=-4.14\pm0.10$ and $\sigma=1.09\pm0.07$.

The unknown initial biomass of the basal species and the alloctonous input are also taken at random from a distribution. To parametrize the distribution, we have used the biomass of each species at each location. The best fit is to a log-normal distribution of $\mu=2.72\pm0.02$ and $\sigma=2.27\pm0.01$. With this procedure, even those parameters that are not directly measured in the field are taken from distributions fitted to what is observed in the field.

To solve the system of equations, we use the Runge–Kutta method implemented in Matlab under function ode45. To avoid numerical inconsistencies, we do not allow the solution to have negative biomass values; and to improve accuracy, we manually decrease the relative error tolerance down to 10^{-6}. We let the model run for a thousand time steps—enough time for the model to reach the steady state. Since some parameters are not directly observed in the field and are taken from the parametrized distributions, we run a hundred replicates taking those parameters at random from their distributions. Although most studies of this kind run a larger number of replicates, the precise parametrization of the local food webs allows us to obtain a convergent value of average persistence even after 20 replicates.

Compartmentalization analysis. The modularity M of a network is defined as the fraction of links within compartments minus the expected fraction of such links[35]. The expected fraction is estimated f

nodes have the same number of interactions but they are rewired to other nodes. In a directed random network, the probability of finding a link between two vertices i and j is $k_i^{in}k_j^{out}/L$, where k_i^{in} is the in-degree of the node i, k_j^{out} is the out-degree of node j, and L is the number of links in the network. Therefore, the modularity of directed networks can be defined as:

$$M=\frac{1}{2L}\sum_{ij}\left[A_{ij}-\frac{k_i^{in}k_j^{out}}{L}\right]\delta_{m_i,m_j}, \quad (5)$$

where A is the adjacency matrix of the network, δ_{ij} is the Kronecker delta symbol, and m_i is the label of the module to which the node i is assigned. The maximum value of M, obtained with the algorithm described in ref. 36, is the modularity of the network.

Socioeconomic impacts and Geographic Information Systems analysis. Fishing effort is measured as tons of fish per $\mathrm{km^2}$ per year, as taken from Halpern *et al.*[37] It is an aggregated measure of different fishing practices (artisanal, demersal and pelagic). The fishing effort at each location is therefore calculated as the intersection between the sampling site coordinates and the fishing effort raster layer.

The number of human inhabitants near the reef is obtained from the Gridded Population of the World Database[38]. To calculate the human population size around every sampling site, we establish a 25-km radius[39] around the coordinates of the sampling site and then count the number of humans within the radius. This choice was made by Mora *et al.*[39] based on a tradeoff between the resolution of the data and the buffer over which human populations affect coral reefs directly.

The delimitation of the MPAs in the Caribbean Sea is obtained from a global database of MPAs (WDPA; www.protectedplanet.net). The overlap between the protected areas and the sampling site coordinates determines whether a sampling site is inside a MPA.

TSA is defined as the weekly sea surface temperature (SST) minus the maximum weekly climatological SST. It is obtained from remote sensing data downloaded from the NOAA (National Oceanic and Atmospheric Administration). In particular, we used data from the Coral Reef Temperature Anomaly Database. The database uses Pathfinder Version 5.2 SSTs to quantify thermal stress patterns on the world's coral reefs between November 1981 and 2010.

Statistical analysis. From each of the 701 sampling sites, we have information on the following variables: simulated persistence, fishing effort, human density, species richness, network connectance, network modularity, geographical coordinates, total recorded biomass, TSA, and finally whether that location is located within a MPA. However, it is unlikely that these variables explain the totality of the observed variance. To incorporate unknown variables that may affect network structure and dynamics, we use a linear mixed effect model that assigns a different intercept for each sampling site.

To model logit(persistence), we use as a random factor the sampling site's ID, acknowledging in this way intrinsic differences within sites. As fixed effects, we use the log of fishing effort, the log of human density, species richness, network modularity, latitude, longitude, protected areas as a categorical factor with two levels (protected or unprotected), and log of recorded total biomass. We excluded other variables—such as phylogenetic diversity and network connectance—that were strongly correlated with any of the variables used as fixed effects. All variables are scaled so variance is of the order of one. The model is fitted by maximizing the log-likelihood. The output of the linear mixed effect model used to analyse the data are given in Supplementary Table 1.

To construct a linear mixed effect model for community biomass, we use as a random factor the different sites and as fixed effects the log fishing effort, the log of human density, food-web modularity, geographical coordinates, the number of species in a certain site, TSA at the site and the conservation status of the site—as a categorical factor with two levels (protected or unprotected). All variables are scaled so variance is of the order of one. The model is fitted by maximizing the log-likelihood. Details of the statistical results are given in Supplementary Table 2.

We also explored the variables that have an effect on the current biodiversity of the sampling sites. To model species richness, we use as a random factor the different sites, and as fixed effects the log fishing effort, the log of human density, modularity, geographical coordinates, the TSA, community biomass, the conservation status of the site as a categorical factor with two levels and log of recorded total biomass in a certain site. All variables are scaled so variance is of the order of one. The model is fitted by maximizing the log-likelihood. Details of the output provided by the linear mixed effect model after model selection are given in Supplementary Table 3.

To study the factors that affect food-web structure, quantified as modularity, we use as a random factor the different sites, and as fixed effects the log fishing effort, the log of human density, number of species, geographical coordinates, the TSA, community biomass, the conservation status of the site as a categorical factor with two levels and log of recorded total biomass in a certain site. All variables are scaled so variance is of the order of one.

So far, we were considering whether a site is inside or outside a MPA. But we could also analyse which properties of the MPA might be driving the observed effect. To do so, we performed a linear mixed effect model analysis similar to the one underlying Fig. 2 of the main text—with persistence as the response variable—but considering only the sample sites located within a protected area. This allows us to add MPA size and age as predictive variables, and not just whether a site is inside or outside a MPA.

In all cases, we performed a model selection to obtain the simplest model with the strongest explanatory power. We first run a linear mixed effect model with all the possible variables and obtain the effect sizes and the *P* values for each of the variables. From this first model, we remove the variable with the highest *P* value. Then, we run another linear mixed effect model without that variable. We iterate this process until all remaining variables are significant. We follow this procedure for all the statistical models involved.

References

1. Jackson, J. B. *et al.* Historical overfishing and the recent collapse of coastal ecosystems. *Science* **293,** 629–637 (2001).
2. Pauly, D. Fishing down marine food webs. *Science* **279,** 860–863 (1998).
3. Myers, R. & Worm, B. Rapid worldwide depletion of predatory fish communities. *Nature* **423,** 280–283 (2003).
4. Anderson, C. N. K. *et al.* Why fishing magnifies fluctuations in fish abundance. *Nature* **452,** 835–839 (2008).
5. Tylianakis, J. M., Tscharntke, T. & Lewis, O. T. Habitat modification alters the structure of tropical host-parasitoid food webs. *Nature* **445,** 202–205 (2007).
6. Tylianakis, J. M., Didham, R. K., Bascompte, J. & Wardle, D. A. Global change and species interactions in terrestrial ecosystems. *Ecol. Lett.* **11,** 1351–1363 (2008).
7. Scheffer, M. *et al.* Anticipating critical transitions. *Science* **338,** 344–348 (2012).
8. Barnosky, A. D. *et al.* Approaching a state shift in Earth's biosphere. *Nature* **486,** 52–58 (2012).
9. Paine, R. T. Food webs: linkage, interaction strength and community infrastructure. *J. Anim. Ecol.* **49,** 667–685 (1980).
10. Carpenter, S. R. & Kitchell, J. F. *The Trophic Cascade in Lake Ecosystems* (Cambridge Univ. Press, 2003).
11. Pauly, D., Christensen, V., Dalsgaard, J., Froese, R. & Torres, F. J. Fishing down marine food webs. *Science* **279,** 860–863 (1998).
12. Pace, M. L., Cole, J. J., Carpenter, S. R. & Kitchell, J. F. Trophic cascades revealed in diverse ecosystems. *Trends Ecol. Evol.* **14,** 483–488 (1999).
13. Shurin, J. B. *et al.* A cross-ecosystem comparison of the strength of trophic cascades. *Ecol. Lett.* **5,** 785–791 (2002).
14. Myers, R. A. & Worm, B. Rapid worldwide depletion of predatory fish communities. *Nature* **423,** 280–283 (2003).
15. Culp, J. M., Glozier, N. E., Cash, K. J. & Baird, D. J. in *Dynamic Food Webs: Multi-species Assemblages, Ecosystem Development and Environmental Change.* (eds de Ruiter, P., Wolters, V. & Moore, J.) (Academic Press, 2005).
16. Bascompte, J., Melián, C. J. & Sala, E. Interaction strength combinations and the overfishing of a marine food web. *Proc. Natl Acad. Sci. USA* **102,** 5443–5447 (2005).
17. Opitz, S. *Trophic Interactions in Caribbean Coral Reefs* (International Center for Living Aquatic Resources Management, 1996).
18. Randall, J. Food habits of reef fishes of the west indies. *Stud. Trop. Oceanogr* **5,** 665–847 (1976).
19. Williams, R. J. Effects of network and dynamical model structure on species persistence in large model food webs. *Theor. Ecol.* **1,** 141–151 (2008).
20. Stouffer, D. B. & Bascompte, J. Understanding food-web persistence from local to global scales. *Ecol. Lett.* **13,** 154–161 (2010).
21. Stouffer, D. B. & Bascompte, J. Compartmentalization increases food-web persistence. *Proc. Natl Acad. Sci. USA* **108,** 3648–3652 (2011).
22. Melián, C. J. & Bascompte, J. Food web cohesion. *Ecology* **85,** 352–358 (2004).
23. Olesen, J. M., Bascompte, J., Dupont, Y. L. & Jordano, P. The modularity of pollination networks. *Proc. Natl Acad. Sci. USA* **104,** 19891–19896 (2007).
24. Guimerà, R. *et al.* Origin of compartmentalization in food webs. *Ecology* **91,** 2941–2951 (2010).
25. May, R. M. Will a large complex system be stable? *Nature* **238,** 413–414 (1972).
26. Petchey, O. L. & Gaston, K. J. Functional diversity (FD), species richness and community composition. *Ecol. Lett.* **5,** 402411 (2002).
27. Maherali, H. & Klironomos, J. N. Influence of phylogeny on fungal community assembly and ecosystem functioning. *Science* **316,** 1746–1748 (2007).
28. Lester, S. E. *et al.* Biological effects within no-take marine reserves: a global synthesis. *Mar. Ecol. Prog. Ser.* **384,** 3346 (2009).
29. Halpern, B. S. The impact of marine reserves: do reserves work and does reserve size matter? *Ecol. Appl.* **13,** S117S137 (2003).
30. Valdovinos, F. S., Ramos-Jiliberto, R., Garay-Narvaaez, L., Urbani, P. & Dunne, J. A. Consequences of adaptive behaviour for the structure and dynamics of food webs. *Ecol. Lett.* **13,** 1546–1559 (2010).
31. Ramos-Jiliberto, R., Valdovinos, F. S., Moisset de Espanaes, P. & Flores, J. D. Topological plasticity increases robustness of mutualistic networks. *J. Anim. Ecol.* **81,** 896–904 (2012).
32. Marks, K. W. AGRRA Database, version (10/2007). Available at http:www.agrra.org/ (2007).
33. Yodzis, P. & Innes, S. Body size and consumer-resource dynamics. *Am. Nat.* **139,** 1151–1175 (1992).
34. Williams, R. J., Brose, U. & Martinez, N. D. in *From Energetics to Ecosystems: The Dynamics and Structure of Ecological Systems.* (eds Rooney, N., McCann, K. & Noakes, D.) Ch. 2, 37–51 (Springer, 2007).
35. Leicht, E. & Newman, M. Community Structure in Directed Networks. *Phys. Rev. Lett.* **100,** 118703 (2008).
36. Guimerà, R., Sales-Pardo, M. & Amaral, L. A. N. Module identification in bipartite and directed networks. *Phys. Rev. E* **76,** 036102 (2007).
37. Halpern, B. S. *et al.* A global map of human impact on marine ecosystems. *Science* **319,** 948–952 (2008).
38. Center for International Earth Science Information Network (CIESIN), Centro Internacional de Agricultura Tropical (CIAT), Gridded Population of the World Version 3 (GPWv3) (2005).
39. Mora, C. *et al.* Global human footprint on the linkage between biodiversity and ecosystem functioning in reef fishes. *PLoS Biol.* **9,** e1000606 (2011).

Acknowledgements

We thank Jonathan Shurin for an influential discussion that lead to this project, Daniel B. Stouffer and Miguel A. Fortuna for helpful comments, and Rudolf P. Rohr and Serguei Saavedra for statistical advice. Ben Halpern provided the data on fishing effort. Raúl Ortega and the LAST-EBD provided technical assistance. We also thank the Supercomputing Center of Galicia (CESGA) for computing capabilities. Funding was provided by FPU PhD fellowship to L.J.G. and by an ERC's Advanced Grant to J.B.

Author contributions

L.J.G. and J.B. designed the research; C.M. provided the data; L.J.G. performed the simulations and analysed the data; L.J.G. and J.B. wrote a first draft of the paper and C.M. contributed to the final version of the manuscript.

Additional information

Competing financial interests: The authors declare no competing financial interests.

4

Decadal predictions of the North Atlantic CO_2 uptake

Hongmei Li[1], Tatiana Ilyina[1], Wolfgang A. Müller[1] & Frank Sienz[1]

As a major CO_2 sink, the North Atlantic, especially its subpolar gyre region, is essential for the global carbon cycle. Decadal fluctuations of CO_2 uptake in the North Atlantic subpolar gyre region are associated with the evolution of the North Atlantic Oscillation, the Atlantic meridional overturning circulation, ocean mixing and sea surface temperature anomalies. While variations in the physical state of the ocean can be predicted several years in advance by initialization of Earth system models, predictability of CO_2 uptake has remained unexplored. Here we investigate the predictability of CO_2 uptake variations by initialization of the MPI-ESM decadal prediction system. We find large multi-year variability in oceanic CO_2 uptake and demonstrate that its potential predictive skill in the western subpolar gyre region is up to 4–7 years. The predictive skill is mainly maintained in winter and is attributed to the improved physical state of the ocean.

[1] Max Planck Institute for Meteorology, Bundesstraße 53, 20146 Hamburg, Germany. Correspond
H.L. (email: hongmei.li@mpimet.mpg.de) or to T.I.

The world's oceans currently take over about 25–30% of anthropogenic CO_2 from the atmosphere, and the North Atlantic is a key component of this oceanic carbon sink[1,2]. Steered by deep convection, most of the North Atlantic CO_2 uptake takes place in the subpolar gyre (SPG) region, a region also characterized by pronounced variability in CO_2 fluxes[3–5]. The variability of carbon uptake in the subpolar North Atlantic ocean was found to be largely related to variations of the North Atlantic Oscillation (NAO), the Atlantic meridional overturning circulation (AMOC), sea surface temperature (SST) and ocean-mixing strength, and all these processes are intrinsically related[3,5–9]. The SPG region experienced several abrupt warming and cooling events in the twentieth century, with SST increasing or decreasing by more than 1 °C in only a few years[10–13]. These changes are related to variations of the NAO and the associated AMOC, heat transports[11,14], SPG circulation and Labrador Sea convection[15,16]. The key elements such as AMOC and SST have been shown to be predictable several years ahead by initialization of Earth system models (ESMs)[10–12,17]. However, no such investigation of predictability has been performed for the CO_2 uptake so far. Are variations in the North Atlantic CO_2 uptake represented in decadal prediction simulations with an ESM? And if so, to what extent is the CO_2 uptake predictable?

We address these questions by using the Max Planck Institute-ESI (MPI-ESM) decadal prediction system[18], which includes an assimilation simulation by nudging observations into the atmospheric and oceanic components of the ESM (see the Methods section) and a set of initialized retrospective prediction simulations (that is, starting from the assimilation run). For comparison, a set of uninitialized simulations is performed without any observations nudged. No carbon cycle observations are used in the assimilation run, so that the ocean biogeochemical processes perform as a passive component. Such an assimilation run, including both anthropogenic trends and natural fluctuations, allows us to investigate the spatial and temporal variability of the oceanic carbon uptake and its connection to local and large-scale physical processes. The initialized retrospective prediction simulations enable us to assess the predictability of carbon uptake. Here we mainly focus on interannual and longer timescale and if not stated otherwise use multi-year and annual mean data set. In this study, we focus on the SPG region because first it plays a prominent role in the oceanic carbon budget. Second, both the physical and the biogeochemical states of the ocean in the SPG region exhibit pronounced natural variability, which is governed by processes acting on decadal scales. Third, this region demonstrates robust predictive skill for the physical fields offering promise for predictions of the oceanic carbon uptake. Our analysis indicates that the North Atlantic CO_2 uptake shows large spatial and temporal variations on decadal scale. Furthermore, these variations are better represented in the initialized simulations than in the uninitialized simulations. Direct comparison to observations indicates improved representation of the ocean surface pCO_2 by initialization. We find the potential prediction skill of the western SPG CO_2 uptake until lead time of 4–7 years. This predictability of CO_2 uptake is related to the initialization of SST and AMOC by observations.

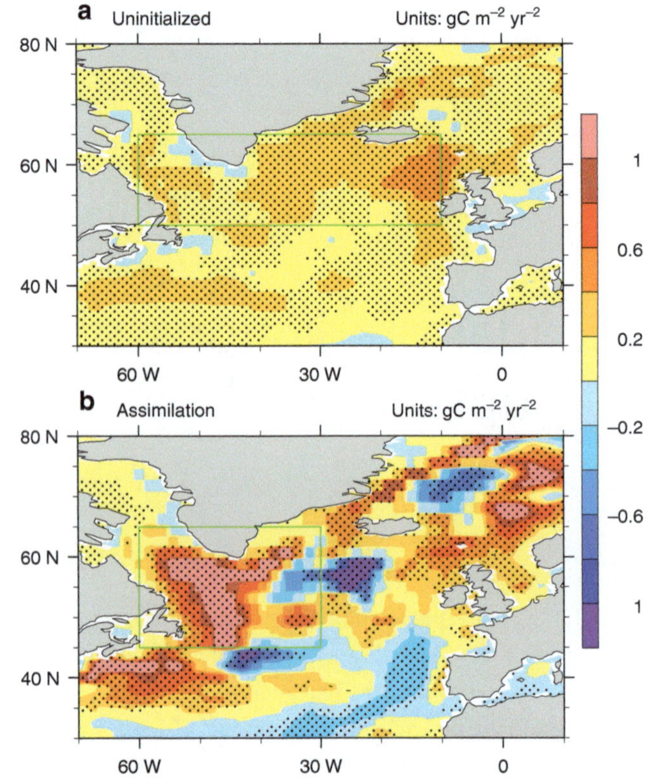

Figure 1 | Trends of annual mean CO_2 flux into the ocean from 1970 to 1995. Ensemble mean of uninitialized simulations (**a**) and assimilation simulation (**b**) (units: $gC\,m^{-2}$ per year[2]). The trends are calculated based on linear regression. The dots show grid points where the trends are significant at 95% level, based on a two-sided *t*-test. The green box in **a** denotes the SPG region where the mid-1990s abrupt warming occurred[11], and the green box in **b** denotes the western SPG region where the CO_2 flux increases the most.

Results

Spatial variations of CO_2 flux in the North Atlantic. Spatial distributions of trends in CO_2 flux during the 25 years before the mid-1990s SPG abrupt warming (that is, from 1970 to 1995) show a remarkably different pattern between the uninitialized and assimilation runs (Fig. 1). The oceanic CO_2 uptake in the uninitialized simulations increases at about 0.2 $gC\,m^{-2}$ per year[2], following the increase in atmospheric CO_2 due to rising fossil fuel carbon emissions, with a somewhat larger increase in the eastern part of the North Atlantic. The CO_2 flux in the assimilation run does not increase uniformly in the North Atlantic; the largest increase is found in the western SPG region (Fig. 1b), and decreasing trends are found in the eastern SPG region. A zonal dipole distribution in delta pCO_2 anomalies was also found in a previous study based on the simulations with an ocean only model[6]. This dipole trend pattern of CO_2 uptake can be attributed to the NAO-related western–eastern heat loss gradient[19] and the related ocean-mixing strength changes (Supplementary Fig. 1 and Supplementary Note 1). Regarding the mid-1990s warming, further decomposition of the SPG region by a previous study[20] suggests a longer predictive skill in the western SPG region, which involves AMOC and meridional heat transport changes in response to persistent positive NAO phase from 1988 to 1995, and a shorter predictive skill in the eastern SPG region, which involves gyre circulation adjustment in response to NAO phase switch from positive to negative in 1995. Here we focus on the western SPG region with the largest trend of CO_2 uptake and with longer prediction skill of the ocean physical state.

Temporal evolution of CO_2 uptake and related processes. To further explore the temporal variability of the oceanic carbon uptake and how it is related to prominent local hydrodynamic processes in the SPG region, we compare the CO_2 flux calculated in the assimilation run with the observed NAO index[21], observed

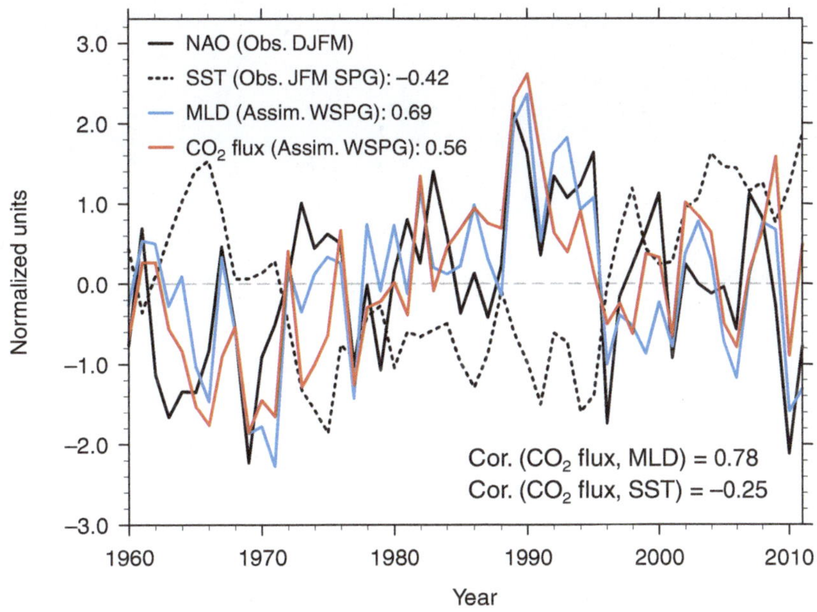

Figure 2 | Normalized time series of CO_2 flux and related physical variables. The observed (Obs.) DJFM NAO[21], observed JFM SST[22] in SPG region, western SPG (WSPG) MLD and CO_2 flux calculated with the assimilation run are shown with black solid line, black dashed line, blue solid line and red solid line, respectively. The location of SPG and WSPG refers to green box in Fig. 1a,b, respectively. The correlation (Cor.) coefficients between time series and NAO (CO_2 flux) are shown on the top (bottom) of the figure. The normalized time series are calculated by dividing individual variable anomalies with their respective standard deviation.

SST[22] and mixed layer depth (MLD) in the assimilation run (Fig. 2). The western SPG MLD and CO_2 flux from our model results are highly correlated (with a correlation coefficient of 0.78), and both are highly correlated (with correlation coefficients of 0.69 and 0.56, respectively) with the observed December–January–February–March (DJFM) NAO index. The correlations are significantly different from zero with P values below 0.01. Consistent with previous model studies[6], the close connection between the NAO and the North Atlantic CO_2 uptake at interannual and decadal timescales is also reproduced in the assimilation run. On the one hand, the heat loss related to a positive NAO leads to an enhancement of the ocean mixing and deep-water formation in the North Atlantic[19]; this is indicated by the relatively high correlation (0.69) between NAO and MLD. Moreover, the SST anomalies in the SPG region with deep convection recur from winter to winter through a re-emergence process[23,24], which works as follows. The winter thermal anomalies related to positive NAO remain at depth below the shallow summer mixed layer, these anomalies persist through summer and are partially re-entrained into the following winter mixed layer. On the other hand, the North Atlantic atmospheric circulation changes can also affect CO_2 uptake through the associated changes in AMOC and the corresponding meridional heat transport changes. Enhanced northward heat transport leads to increased SST thereby affecting the oceanic carbon uptake in the SPG. These processes lag the NAO changes by several years[14]. Accordingly, given all the processes, the correlation between CO_2 flux and SST is not significant in our simulations.

Potential predictive skill of CO_2 flux and SST. The state-of-the-art decadal prediction systems consistently identify the North Atlantic as a key region with pronounced forecast skill for different parameters of the climate system[25–31]. Several abrupt climate events in the twentieth century have been shown to be predictable several years ahead[10–13]. The SPG abrupt warming and cooling events are also well

simulations, whereas the uninitialized simulations only capture the long-term warming trend and display large spread among ensemble members (Supplementary Fig. 2 and Supplementary Note 2). Given the robust predictive skill of the physical state of the ocean, can variations in the North Atlantic CO_2 uptake be predicted as well? A previous study explored multi-year predictability of tropical marine productivity and found a predictive skill of 3 years, whereas the predictive skill of SST there is only 1 year[32]. In our analysis, as indicated by the initialized time series at different lead times (Supplementary Fig. 3 and Supplementary Note 3), the variability of CO_2 flux, which is absent in the uninitialized simulations, can be reproduced by the model with initialization several years in advance and implies predictability of CO_2 uptake.

Owing to lack of observational data, we first use model fields calculated in the assimilation run as a proxy to quantify the potential predictive skill of SST and of the CO_2 flux into the ocean (Fig. 3, see the Methods section). The correlation coefficients of the initialized 4-year mean SST exceed the uninitialized correlations for lead times of 1–4 and 2–5 years. The correlation goes down in the intermediate lead time, however, it recovers from a lead time of 5–8 years. The results suggest improved potential predictive skills of the background ocean physical fields. The correlation coefficient of initialized CO_2 flux is significantly higher than the uninitialized correlation until a lead time of 4–7 years. The P values are lower than 0.05 until the lead time of 4–7 years suggesting significant improvement of potential predictive skill due to initialization of the physical states constrained by observations.

Further investigation of the potential predictive skill with seasonal and monthly time series (Fig. 3c,d) reveals that the high predictive skill of CO_2 flux in the initialized simulations is mainly maintained during winter months. This is related to the seasonal shift of processes in regulating the CO_2 uptake and the fact of assimilation for the decadal prediction system. In winter, the CO_2 uptake in

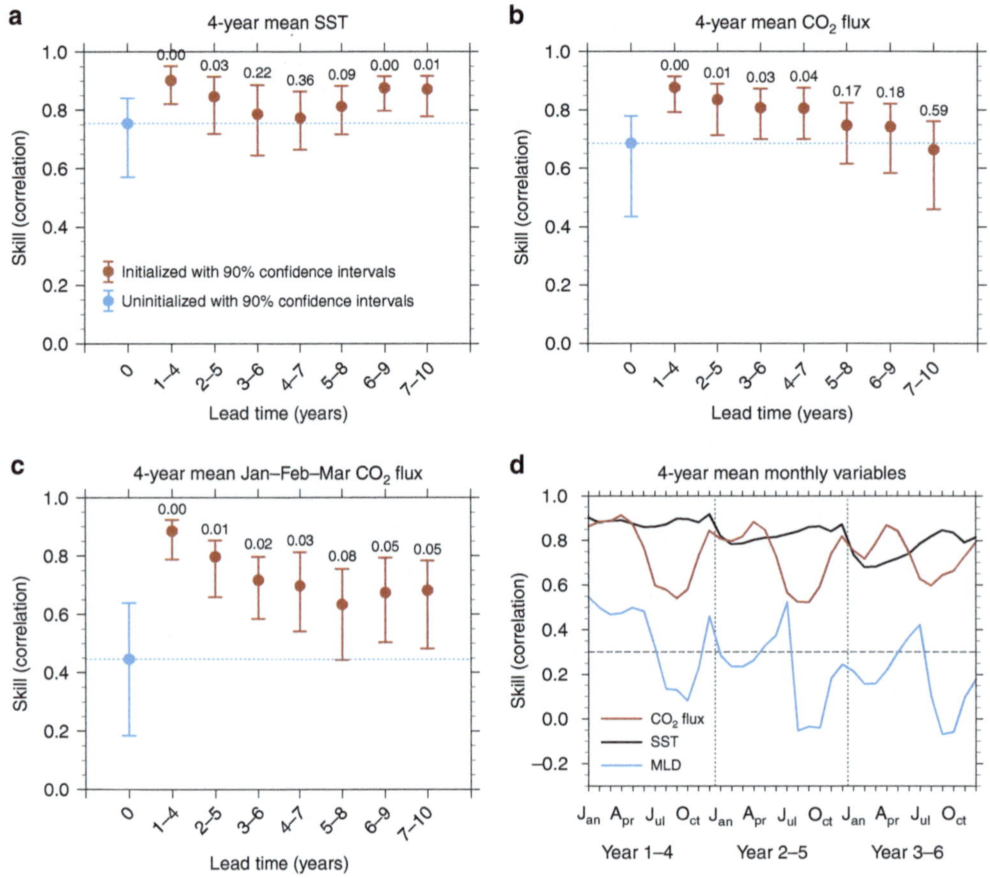

Figure 3 | Potential predictive skill of CO_2 flux and related physical variables. Correlation skill of ensemble mean of 4-year mean SST (**a**), 4-year mean CO_2 flux into the ocean (**b**), seasonally stratified 4-year mean CO_2 flux into the ocean (late winter, that is, January-February-March) (**c**) and monthly stratified 4-year mean CO_2 flux, SST and MLD (**d**) in the western SPG region. Shown are uninitialized (blue dot in **a-c**) and initialized (red dots in **a-c**) simulations at different lead time verified against assimilation. The correlations are calculated from 4-year mean predictions (that is, years 1-4, 2-5 and so on) of initialized simulations, and 4-year running mean of the corresponding assimilation and uninitialized time series. To ensure that the number of validation years is the same for both initialized and uninitialized simulations, we use the common time period from 1967-1970 mean to 2008-2011 mean. The blue dashed line in **a-c** extends the uninitialized correlation for easy comparison. The vertical lines in **a-c** provide 90% confidence intervals based on a bootstrap approach[46]. The numbers on the top of the bars in **a-c** show the *P* values based on the hypothesis that the difference of correlations between the initialized and uninitialized simulations is smaller or equal to zero based on 1,000 bootstrapped resamples. The potential predictive skills of monthly stratified 4-year mean CO_2 flux, SST and MLD in **d** are shown with red, black and blue curves, respectively. Three-month running mean is applied before monthly correlation estimation in **d**. The horizontal dashed grey line in **d** shows the 95% level of significance based on a two-sided *t*-test. The vertical dashed grey lines in **d** are added to better distinguish different lead year results.

circulation[33–35]. The variations of ocean physical fields are well represented in the initialized simulations, as the corresponding initial states are constrained by observations through assimilation. However, in spring and early summer, when the ocean warms, the biological primary production draws down seawater pCO_2 and regulates the oceanic CO_2 uptake in the western SPG region[33–35]; the correlation of the initialized simulations goes down from May, as no ocean biological observations are assimilated into the system.

We further explore possible causes of the potential predictive skill of CO_2 uptake in comparison with that of the SST and MLD (Fig. 3d). The SST correlations peak in December and go down from late winter, which is coherent with that of the MLD and the CO_2 uptake. It suggests that both the evolution of the ocean thermal state and the local mixing strength contribute to the predictive skill of CO_2 uptake in winter. Although the correlation skill of MLD is generally lower than that of CO_2 uptake, both share similar seasonal cycle of predictive skill, indicating effects of MLD changes in maintaining the predictive skill of CO_2 uptake. Moreover, the potential predictive skill of CO_2 uptake is related to the AMOC variability. As revealed by previous studies, the predictive skill of SPG SST is assured by initialization of the AMOC variability[27], and the AMOC shows predictive skill up to 4 years[17]. The potential predictive skill of AMOC in our system is up to 2–5 years (figure omitted). The AMOC at lead time of 1 year is highly correlated with CO_2 flux at lead time of 1 year and onwards (Supplementary Fig. 4 and Supplementary Note 4). An observational study also suggested a close connection between the AMOC and the CO_2 uptake in SPG region[3].

From these our results suggest that the potential predictive skill of CO_2 uptake in the western SPG is up to 4–7 years. The predictive skill is mainly attributed to the combination of improved ocean physical states and circulation variability, primarily in winter.

Evaluation of model predictions against observations. The high potential predictability of CO_2 uptake provides a basis for assessing our predictions against observations using the surface ocean CO_2 atlas (SOCAT) measurement[36]. Although observational data from SOCAT are sparse in the SPG region,

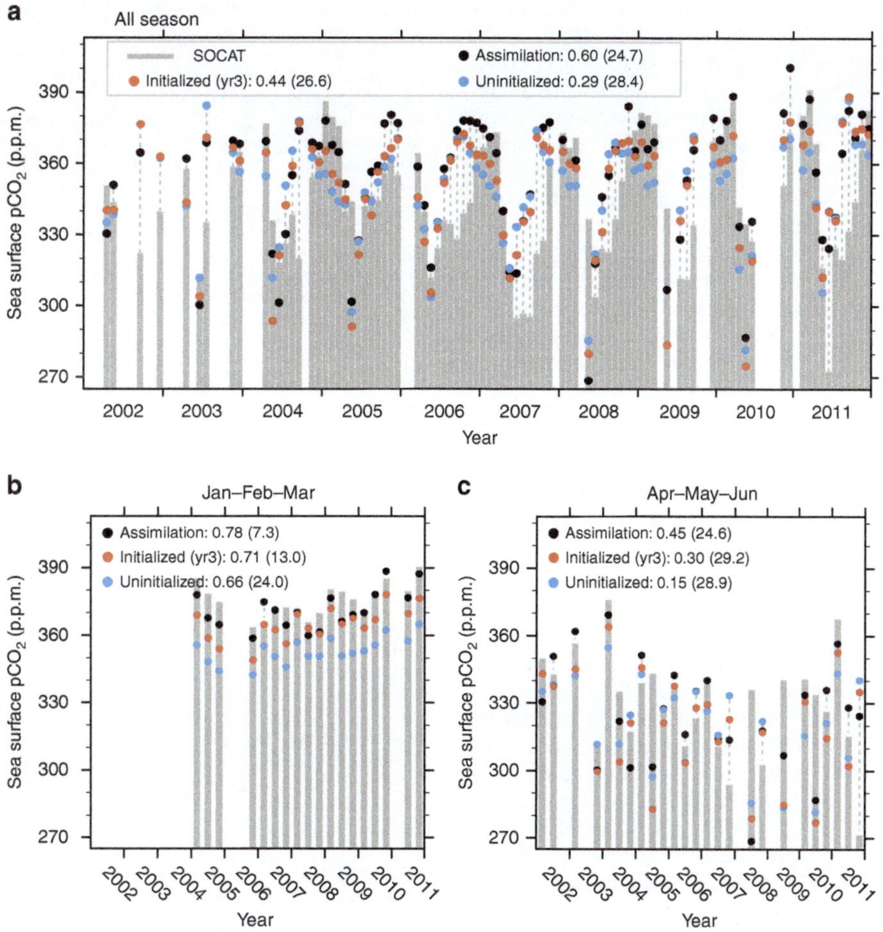

Figure 4 | Observed and model simulated monthly mean ocean surface pCO_2 in western SPG region. (**a**) All season time series, (**b**) only January-February-March are shown and (**c**) only April-May-June are shown. The SOCAT observations[36] are shown with grey bars, and model output of assimilation, uninitialized simulations and initialized simulations at a lead time of 3 years (yr3) are shown in black, blue and red dots, respectively. The grey dashed lines connect outlier dots with the grey bars to make the comparison clearer. The numbers in legend are correlation coefficients and root mean squared error (in brackets) between model simulations and SOCAT observations. For the all season time series in **a** the correlations of the assimilation and initialized simulations at lead time of 3 years are significantly larger than that of the uninitialized simulations at 95% significant level based on a bootstrap test. Note the SOCAT observational data are not continuously distributed in space and time; the model simulations are averaged over the grid points where SOCAT data are available.

these are the best ocean surface observations we can get for this region. The ocean surface pCO_2 in the SPG region peaks in winter as a result of the enhanced vertical supply of carbon from the intermediate waters by deep convection; surface pCO_2 values reach a minimum in summer due to biological draw down[33-35] (Fig. 4). The initialized predictions produce ocean surface pCO_2 closer to SOCAT observation than the uninitialized simulations as indicated by the correlations and root mean squared errors. The correlations of assimilation (0.60) and initialized simulation at a lead time of 3 years (0.44) are significantly larger than the correlation of initialized simulation (0.29) at 95% significance level. The root mean squared error is lower in the assimilation and initialized simulations than in the uninitialized simulation. As we use monthly data due to lack of continuous observations, the better performances of initialized simulations are partially due to better representation of the seasonal cycle in the initialized simulations. We further separate the time series seasonally, and find that in addition to the seasonal cycle there is an improvement of the initialized run against the uninitialized run particularly in the winter months when the pCO_2 is high. The root mean squared error of ocean surface pCO_2 is much smaller in the assimilation (7.3 p.p.m.) and initi... lead time of 3 years (13.0 p.p.m.) than in the uninitialized simulations (24.0 p.p.m.; Fig. 4b). The coherences between model simulations and observations are generally lower in spring months (Fig. 4c). The improvement of prediction in individual seasons further demonstrates that the interannual variations of oceanic carbon cycle are improved in the initialized simulations. The higher correlations between SOCAT and initialized simulations and the lower root mean squared error of the initialized simulations against SOCAT confirm that the oceanic carbon cycle can be predicted several years ahead by initialization of the ESM (Supplementary Fig. 5 and Supplementary Note 5). However, owing to temporal and spatial gaps in observations, the precise prediction skill with respect to observations cannot be estimated. Comparison of simulations against SOCAT observations only confirm that prediction of the state of the oceanic carbon cycle is improved by initialization of the physical ocean state with observations.

Discussion

We have focused here on th...

uptake. The drawdown of pCO_2 by biological processes such as carbon fixation during seasonal phytoplankton growth, with consequent export of carbon into the deep ocean is also critical for the air–sea carbon exchange in the nutrient-rich SPG region[37]. These marine biogeochemical processes themselves are influenced by variability in climate and circulation. For instance, an increase in ocean temperature leads to a decrease of CO_2 uptake through decrease in solubility. Biological production starts in spring when the ocean warms up. This draws down the ocean surface pCO_2 and enhances CO_2 uptake. However, the growth of phytoplankton in the North Atlantic is more limited by nutrient supply than by temperature. The nutrient supply, in turn, is constrained by the enhanced thermal stratification associated with the ocean warming, thus resulting in a decrease of primary production[38]. Therefore, while the biological CO_2 drawdown regulates the seasonal cycle of CO_2 flux, the physical regulation is more important for interannual and decadal variations of CO_2 flux in the SPG region[39]. Our findings show that there is an improved representation of the oceanic carbon flux because the background oceanic circulation and thermal state are well reproduced. Physical processes such as the ocean thermal state, local mixing and large-scale circulation contribute to the predictability of CO_2 uptake. It requires further large ensembles of multi-model decadal prediction simulations and sensitivity experiments to disentangle the relative role of an individual physical mechanism on the predictability of CO_2 uptake.

We find that beside the trend due to CO_2 emissions, predictions of the oceanic uptake and storage of carbon show considerable decadal variations that are not fully captured in the uninitialized simulations with modern ESMs. We demonstrate that variations in the oceanic carbon cycle can be predicted several years ahead. Hence, predictions of the North Atlantic carbon sink considering both anthropogenic change and natural fluctuations are important to understand the evolution of climate and ocean acidification and to reduce uncertainties in CO_2 uptake estimates. Predictions of the oceanic carbon sink are also necessary to provide information to monitoring programs aimed at the present and future oceanic carbon sink.

Methods

Model description. We use the MPI-ESM[40] to conduct historical (uninitialized) experiments and retrospective decadal prediction (initialized) experiments for the period 1961–2013. A low-resolution configuration (MPI-ESM-LR) is used with the resolution of the ocean model being 1.5° on average with 40 vertical levels. The ocean component of MPI-ESM-LR use bipolar configuration, with one pole over Greenland and another over Antarctica, so the resolution in the North Atlantic SPG region is from ~20 km in the north (65° N) to ~70 km in the south (45° N). MPI-ESM-LR is capable of producing North Atlantic deep convection confined to the Labrador Sea, which is close to observation[41]. The ocean biogeochemistry component of MPI-ESM is represented by the Hamburg Ocean Carbon Cycle Model (HAMOCC)[42].

Numerical simulations. An ensemble of 10 uninitialized historical simulations and RCP4.5 scenario simulations is performed for the periods 1850–2005 and 2006–2013, respectively. The uninitialized experiments are started from a preindustrial control simulation and forced with historical greenhouse gas and aerosol concentrations together with solar variability and volcanic eruptions. The initial conditions of the uninitialized run ensemble members differ based on the preindustrial control simulation starting from every 50 years. On the basis of an assimilation run, an ensemble of 10 initialized simulations is started on 1 January in every year over the period 1960–2012. The ensemble members of the initialized simulations are generated with lagged 1-day initialization, that is, the runs start from 10 consecutive days centred on 1 January. Each initialized simulation has a length of 10 years[18]. The assimilation experiment is also performed with MPI-ESM-LR. For the oceanic component, monthly temperature and salinity anomalies from the ECMWF ocean reanalysis system 4 (ORAS4)[43] are added to the model climatology and nudged into the model state at every time step. For the atmospheric component, globally full-field temperature, vorticity, divergence and surface pressure data from the ECMWF ERA40[44] and ERA-Interim[45] reanalyses are nudged.

Predictive skill assessment. We mainly show ensemble mean results in this study. For the uninitialized experiments, the anomalies are calculated by removing the climatology of the ensemble mean. For the initialized experiments, the anomalies are calculated in terms of the climatology of the ensemble member and the lead time. The observed temporal evolution of the DJFM NAO index (station based)[21] and the SST[22] in SPG is investigated together with the mixing and CO_2 flux around the SPG region in the assimilation run. We use the fields in the assimilation run as a proxy to quantify the historical variability and to estimate the potential predictive skill of CO_2 uptake. The potential predictions are verified with skill scores based on anomaly correlation coefficients against the assimilation run. As short-term/small-scale variability may add noise and hence reduce predictive skill, temporal and spatial averaging was recommended[46] and has been used in most decadal prediction studies[25,31,47]. A set of 4-year mean (that is, 1–4, 2–5, ..., 7–10 years) predictions is verified in this study against corresponding running mean of the assimilation and uninitialized simulations.

Bootstrap approach for significance test. Statistical tests and confidence intervals of the correlations in Figs 3 and 4 are calculated through a bootstrapping approach[46]. The resampling scheme considers uncertainty arising from both ensemble size and hindcast period. A block size of 2 is used to account for autocorrelation in the SOCAT time series (Fig. 4). Note that the significance of correlation differences in Fig. 3 remains unaltered for the same block size.

Conversion of ocean surface CO_2 observation data. The gridded observations of the SOCAT[36] surface ocean fugacity of CO_2 (fCO_2) from 2002 to 2011 are used to verify model simulations. The fugacity is converted to partial pressure[48]. As observational data are not continuously distributed in space and time, for the model evaluation presented in Fig. 4 and Supplementary Fig. 5 we only consider those model grid points where SOCAT data are available.

Code availability. The MPI-ESM model is freely available to the scientific community. The code can be accessed with a license agreement on the Max Planck Institute for Meteorology model distribution website: http://www.mpimet.mpg.de/en/science/models/license/

References

1. Sabine, C. L. *et al.* The oceanic sink for anthropogenic CO_2. *Science* **305**, 367–371 (2004).
2. Le Quéré, C. *et al.* Global carbon budget 2015. *Earth Syst. Sci. Data* **7**, 349–396 (2015).
3. Pérez, F. F. *et al.* Atlantic Ocean CO_2 uptake reduced by weakening of the meridional overturning circulation. *Nat. Geosci.* **6**, 146–152 (2013).
4. McKinley, G. A., Fay, A. R., Takahashi, T. & Metzl, N. Convergence of atmospheric and North Atlantic carbon dioxide trends on multidecadal timescales. *Nat. Geosci.* **4**, 606–610 (2011).
5. Corbière, A., Metzl, N., Reverdin, G., Brunet, C. & Takahashi, A. Interannual and decadal variability of the oceanic carbon sink in the North Atlantic subpolar gyre. *Tellus Ser. B* **59**, 168–178 (2007).
6. Thomas, H. *et al.* Changes in the North Atlantic Oscillation influence CO_2 uptake in the North Atlantic over the past 2 decades. *Global Biogeochem. Cycles* **22**, GB4027 (2008).
7. Metzl, N. *et al.* Recent acceleration of the sea surface fCO_2 growth rate in the North Atlantic subpolar gyre (1993-2008) revealed by winter observations. *Global Biogeochem. Cycles* **24**, GB4004 (2010).
8. Tjiputra, J. F., Olsen, A., Assmann, K., Pfeil, B. & Heinze, C. A model study of the seasonal and long-term North Atlantic surface pCO_2 variability. *Biogeosciences* **9**, 907–923 (2012).
9. Halloran, P. R. *et al.* The mechanisms of North Atlantic CO_2 uptake in a large Earth System Model ensemble. *Biogeosciences* **12**, 4497–4508 (2015).
10. Robson, J., Sutton, R. & Smith, D. Decadal predictions of the cooling and freshening of the North Atlantic in the 1960s and the role of ocean circulation. *Clim. Dyn.* **42**, 2353–2365 (2014).
11. Robson, J. I., Sutton, R. T. & Smith, D. M. Initialized decadal predictions of the rapid warming of the North Atlantic Ocean in the mid 1990s. *Geophys. Res. Lett.* **39**, L19713 (2012).
12. Müller, W. A., Pohlmann, H., Sienz, F. & Smith, D. Decadal climate prediction for the period 1901-2010 with a coupled climate model. *Geophys. Res. Lett.* **41**, 2100–2107 (2014).
13. Yeager, S., Karspeck, A., Danabasoglu, G., Tribbia, J. & Teng, H. A decadal prediction case study: Late twentieth-century North Atlantic Ocean heat content. *J. Clim.* **25**, 5173–5189 (2012).
14. Robson, J., Sutton, R., Lohmann, K., Smith, D. & Palmer, M. D. Causes of the rapid warming of the North Atlantic Ocean in the mid-1990s. *J. Clim.* **25**, 4116–4134 (2012).
15. Hakkinen, S. & Rhines, P. B. Decline of subpolar North Atlantic circulation during the 1990s. *Science* **304**, 555–559 (2004).

16. Hátún, H., Sando, A. B., Drange, H., Hansen, B. & Valdimarsson, H. Influence of the Atlantic subpolar gyre on the thermohaline circulation. *Science* **309,** 1841–1844 (2005).
17. Matei, D. *et al.* Multiyear prediction of monthly mean atlantic meridional overturning circulation at 26.5°N. *Science* **335,** 76–79 (2012).
18. Pohlmann, H. *et al.* Improved forecast skill in the tropics in the new MiKlip decadal climate predictions. *Geophys. Res. Lett.* **40,** 5798–5802 (2013).
19. Eden, C. & Willebrand, J. Mechanism of interannual to decadal variability of the North Atlantic circulation. *J. Clim.* **14,** 2266–2280 (2001).
20. Barrier, N., Deshayes, J., Treguier, A.-M. & Cassou, C. Heat budget in the North Atlantic subpolar gyre: impacts of atmospheric weather regimes on the 1995 warming event. *Prog. Oceanogr.* **130,** 75–90 (2015).
21. Hurrell, J. W. Decadal trends in the North Atlantic oscillation: regional temperatures and precipitation. *Science* **269,** 676–679 (1995).
22. Rayner, N. A. *et al.* Global analyses of sea surface temperature, sea ice, and night marine air temperature since the late nineteenth century. *J. Geophy. Res. Atmos.* **108,** 4407 (2003).
23. Polkova, I., Koehl, A. & Stammer, D. Impact of initialization procedures on the predictive skill of a coupled ocean-atmosphere model. *Clim. Dyn.* **42,** 3151–3169 (2014).
24. Deser, C., Alexander, M. A. & Timlin, M. S. Understanding the persistence of sea surface temperature anomalies in midlatitudes. *J. Clim.* **16,** 57–72 (2003).
25. Meehl, G. A. *et al.* Decadal climate prediction: an update from the trenches. *Bull. Am. Meteorol. Soc.* **95,** 243–267 (2014).
26. Smith, D. M. *et al.* Skilful multi-year predictions of Atlantic hurricane frequency. *Nat. Geosci.* **3,** 846–849 (2010).
27. Matei, D. *et al.* Two tales of initializing decadal climate prediction experiments with the ECHAM5/MPI-OM model. *J. Clim.* **25,** 8502–8523 (2012).
28. Müller, W. A. *et al.* Forecast skill of multi-year seasonal means in the decadal prediction system of the Max Planck Institute for Meteorology. *Geophys. Res. Lett.* **39,** L22707 (2012).
29. Pohlmann, H., Jungclaus, J. H., Koehl, A., Stammer, D. & Marotzke, J. Initializing decadal climate predictions with the GECCO Oceanic synthesis: effects on the North Atlantic. *J. Clim.* **22,** 3926–3938 (2009).
30. Doblas-Reyes, F. J. *et al.* Initialized near-term regional climate change prediction. *Nat. Commun.* **4,** 1715 (2013).
31. Keenlyside, N. S., Latif, M., Jungclaus, J., Kornblueh, L. & Roeckner, E. Advancing decadal-scale climate prediction in the North Atlantic sector. *Nature* **453,** 84–88 (2008).
32. Séférian, R. *et al.* Multiyear predictability of tropical marine productivity. *Proc. Natl Acad. Sci. USA* **111,** 11646–11651 (2014).
33. Körtzinger, A., Send, U., Wallace, D. W. R., Karstensen, J. & DeGrandpre, M. Seasonal cycle of O_2 and pCO_2 in the central Labrador Sea: atmospheric, biological, and physical implications. *Global Biogeochem. Cycles* **22,** GB1014 (2008).
34. Takahashi, T. *et al.* Global sea-air CO_2 flux based on climatological surface ocean pCO_2, and seasonal biological and temperature effects. *Deep Sea Res. Part II* **49,** 1601–1622 (2002).
35. Ullman, D. J., McKinley, G. A., Bennington, V. & Dutkiewicz, S. Trends in the North Atlantic carbon sink: 1992-2006. *Global Biogeochem. Cycles* **23,** GB4011 (2009).
36. Bakker, D. C. E. *et al.* An update to the Surface Ocean CO_2 Atlas (SOCAT version 2). *Earth Syst. Sci. Data* **6,** 69–90 (2014).
37. Takahashi, T. *et al.* Climatological mean and decadal change in surface ocean pCO_2, and net sea-air CO_2 flux over the global oceans. *Deep Sea Res. Part I* **56,** 554–577 (2009).
38. Behrenfeld, M. J. *et al.* Climate-driven trends in contemporary ocean productivity. *Nature* **444,** 752–755 (2006).
39. Bennington, V., McKinley, G. A., Dutkiewicz, S. & Ulman, D. What does chlorophyll variability tell us about export and air-sea CO_2 flux variability in the North Atlantic? *Global Biogeochem. Cycles* **23,** GB3002 (2009).
40. Giorgetta, M. A. *et al.* Climate and carbon cycle changes from 1850 to 2100 in MPI-ESM simulations for the Coupled Model Intercomparison Project phase 5. *J. Adv. Model. Earth Syst.* **5,** 572–597 (2013).
41. Jungclaus, J. H. *et al.* Characteristics of the ocean simulations in the Max Planck Institute Ocean Model (MPIOM) the ocean component of the MPI-Earth system model. *J. Adv. Model. Earth Syst.* **5,** 422–446 (2013).
42. Ilyina, T. *et al.* The global ocean biogeochemistry model HAMOCC: model architecture and performance as component of the MPI-Earth System Model in different CMIP5 experimental realizations. *J. Adv. Model. Earth Syst.* **5,** 287–315 (2013).
43. Balmaseda, M. A., Mogensen, K. & Weaver, A. T. Evaluation of the ECMWF ocean reanalysis system ORAS4. *Q. J. R. Meteorol. Soc.* **139,** 1132–1161 (2012).
44. Uppala, S. M. *et al.* The ERA-40 re-analysis. *Q. J. R. Meteorol. Soc.* **131,** 2961–3012 (2005).
45. Dee, D. P. *et al.* The ERA-Interim reanalysis: Configuration and performanceof the data assimilation system. *Q. J. R. Meteorol. Soc.* **137,** 553–597 (2011).
46. Goddard, L. *et al.* A verification framework for interannual-to-decadal predictions experiments. *Clim. Dyn.* **40,** 245–272 (2013).
47. Smith, D. M. *et al.* Improved surface temperature prediction for the coming decade from a global climate model. *Science* **317,** 796–799 (2007).
48. Zeebe, R. E. & Wolf-Gladrow, D. *CO2 in Seawater: Equilibrium, Kinetics, Isotopes* Vol. 65 (Elsevier, 2001).

Acknowledgements

We thank J. Marotzke, H. Pohlmann and C. Li for their helpful comments and suggestions on the manuscript. We acknowledge funding from the Federal Ministry of Education and Research in Germany (BMBF) through the research programme 'MiKlip' (FKZ: 01LP1144A; 01LP1145A; 01LP1519A; 01LP1157B; and 01LP1517B). This study has also received funding from EU FP7 project CARBOCHANGE (grant 264879) and the European Union's Horizon 2020 research and innovation programme (grant 641816). The simulations were performed at the German Climate Computing Center (DKRZ).

Author contributions

H.L., T.I. and W.M. jointly designed the study and wrote the manuscript; H.L. analysed the data; F.S. provided suggestions and comprehensive support on statistical analyses; all authors discussed the results and commented on the manuscript.

Additional information

Competing financial interests: The authors declare no competing financial interests.

5

No inter-gyre pathway for sea-surface temperature anomalies in the North Atlantic

Nicholas P. Foukal[1] & M. Susan Lozier[1]

Recent Lagrangian analyses of surface drifters have questioned the existence of a surface current connecting the Gulf Stream (GS) to the subpolar gyre (SPG) and have cast doubt on the mechanism underlying an apparent pathway for sea-surface temperature (SST) anomalies between the two regions. Here we use modelled Lagrangian trajectories to determine the fate of surface GS water and satellite SST data to analyse pathways of GS SST anomalies. Our results show that only a small fraction of the surface GS water reaches the SPG, the water that does so mainly travels below the surface mixed layer, and GS SST anomalies do not propagate into the SPG on interannual timescales. Instead, the inter-gyre heat transport as part of the Atlantic Meridional Overturning Circulation must be accomplished via subsurface pathways. We conclude that the SST in the SPG cannot be predicted by tracking SST anomalies along the GS.

[1] Nicholas School of the Environment, Duke University, Durham, North Carolina 27708, USA. Correspondence and requests for materials should be addressed to N.P.F. (email: Nicholas.Foukal@Duke.edu).

The surface waters of the eastern subpolar gyre (50° N–60° N and 30° W–15° W) in the North Atlantic are exceptionally warm and salty compared with the surface waters at similar latitudes, leading to the supposition that the subpolar gyre (SPG) is primarily supplied by the similarly warm and salty waters of the subtropical gyre (STG)[1–3]. This connection has been of particular interest because the sea-surface characteristics of the North Atlantic SPG provide the environment for density-driven overturning—the warm surface water loses heat to the atmosphere, becomes denser and sinks to depth as it flows around the SPG. This water mass transformation is one component of the Atlantic Meridional Overturning Circulation (AMOC), which moves warm water northwards in its upper layers and colder water southwards at depth, affecting regional and global climate by redistributing heat[4]. Another impetus for understanding the structure and pathways of the upper-layer AMOC is the potential for predicting climate on timescales of months to years in advance[5–7].

Eulerian-based analyses of surface drifter velocities—in which drifter velocities within a spatially defined bin are averaged over time—indicate that surface waters in the Cape Hatteras region are advected north-eastward into the eastern SPG[3,8]. In contrast, Lagrangian analyses of the pathways of individual trajectories from the same surface drifter data show little to no surface connection between the Gulf Stream (GS) and the eastern SPG[9,10]. Only 1 out of 273 (0.37%) surface drifters deployed south of 45° N in the North Atlantic between 1990 and 2002 made it to the SPG[9]. Though this surface throughput has increased since 2002, the largest exchange observed during any given time period is only 3.3% (ref. 10). A subsequent modelling study demonstrates that Lagrangian trajectories initialized in the STG at depth shoal as they move northward along isopycnal surfaces and are more likely to reach the SPG than those initialized at the surface in the STG[11]. When coupled with recent findings on the formation and fate of subtropical mode water (STMW)[12,13], these Lagrangian results yield a new circulation pattern of North Atlantic upper-layer currents: surface GS water recirculates in the STG, becomes part of the STMW, re-enters the GS at depth and is then exported to the eastern SPG[14].

This proposed paradigm, with subsurface STG waters supplying the eastern SPG, begs the question as to whether winter sea-surface temperature (SST) anomalies are transmitted to the SPG via the advective pathway described in previous studies[5,15,16] in which SST anomalies re-emerge during subsequent winters as they progress along the GS/North Atlantic Current (GS/NAC). The goal of this paper is to answer that question by first carefully analysing and quantifying the fate of surface GS waters and second, determining whether winter SST anomalies in the modern satellite SST record propagate from the STG into the SPG.

Here we find that only a small fraction of surface GS water reaches the SPG, and that the water that does reach the SPG does so at depths below the winter-mixed layer and is thus out of contact with the SST. We also find that the SST anomaly pathway previously reported is most likely an artefact of a smoothing filter on the same timescale as SST changes associated with the North Atlantic Oscillation (NAO), and not an actual physical pathway for SST anomalies. We conclude that interannual changes in the SST of the SPG cannot, as once thought, be predicted from information on the GS SST and that inter-gyre heat transport of the AMOC must occur below the surface layers.

Results

Modelled Lagrangian pathways. To trace the pathways and fate of surface GS waters, we initialize Lagrangian ...

eddy-resolving ocean circulation model (FLAME)[17,18]. FLAME is an ocean-only, fully dynamic model with 1/12° horizontal resolution and 45 depth layers varying from 5-m resolution at the surface to 250-m resolution at depth. The model is forced by NCEP-NCAR reanalysis wind and buoyancy forcing for 15 years (1990–2004). We find Lagrangian particle positions using the model's 3-day velocity fields, and record those positions every 15 days. Particles are initialized at the surface in the vicinity of Cape Hatteras (VCH) in a spatial box defined as in a previous study[5] (31.5° N–38.5° N and 80° W–60° W). We also require that the particles' initial positions be located south of the GS north wall—defined to be the 15 °C isotherm at 200-m depth[19]—so that all trajectories are of subtropical origin. To calculate particle pathways from the model's velocity field, we initially use only horizontal velocities—that is, we restrict the simulated floats to the surface to mimic the constraint placed on the observational surface drifters (Fig. 1a). In a subsequent Lagrangian experiment, we use the three-dimensional flow field (Fig. 1b) for the trajectory calculations. In both experiments, water particles are tagged at the surface in the VCH box every 15 days for the first 10 years of the model run (1990–1999) and tracked for 5 years. In the surface-constrained run, the trajectories are heavily influenced by the winds, accumulating in the Ekman convergence zone at 30° N.

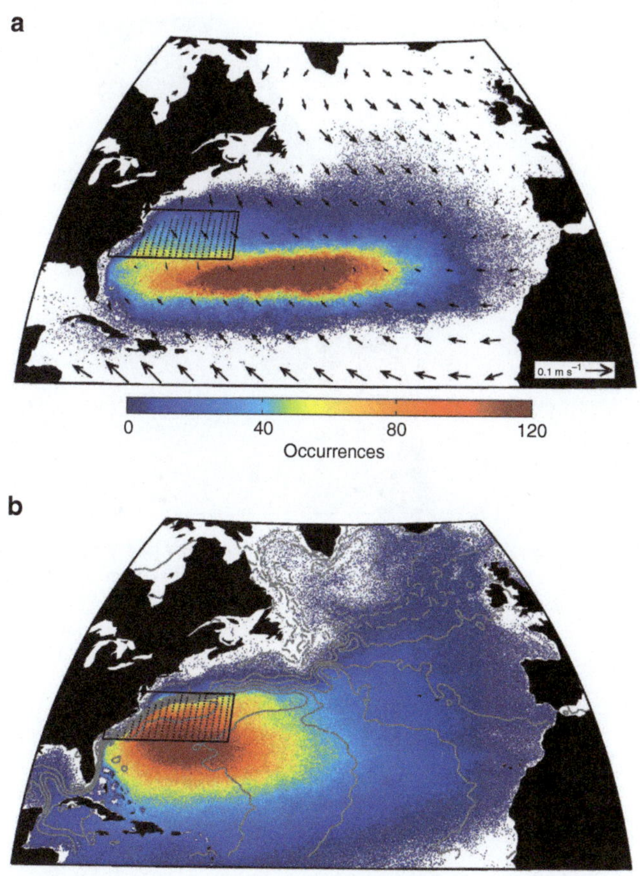

Figure 1 | Distributions of modelled particle positions. Floats are initialized from the surface of the VCH box (black rectangle) in FLAME and are constrained to follow either the two-dimensional surface flow field (**a**) or the three-dimensional flow field (**b**) for 5 years. The colour bar refers to the total number of trajectories that pass through a given location (that is, occurrences) over the 15 years. Arrows in **a** are the time-mean Ekman velocities from the Center for Topographic Studies of the Ocean and Hydrosphere[28] and conto...

Only 5 out of 60,906 modelled trajectories (<0.01%) cross north of 53° N, the latitude at which we consider the floats to be subpolar. When allowed to follow the three-dimensional flow field, the vast majority of the trajectories (97.5%) trace the wind-driven, western-intensified geostrophic STG, accumulating in its convergent centre. Only 1,526 out of the 60,906 total floats (2.51%) reach the SPG in this experiment. Along with vertical mixing, Ekman pumping from negative wind-stress curl forcing yields a net downward transport in the STG[20], removing particles from the surface layer and creating the disparity between a and b in Fig. 1. These results agree with the observed and modelled Lagrangian experiments on the fate of surface water in the GS[9–11,21], and we elaborate on those previous studies by exploring the pathways of the Lagrangian particles from the GS region to the SPG.

An examination of those trajectories that reach the SPG within their 5-year lifetime reveals the canonical GS/NAC pathway (Fig. 2). However, only 0.5% of the floats remain above the 125-m climatological winter-mixed layer in the GS/NAC[22] on their way to the SPG (Fig. 2, inset). In addition, the small number of particles that travel to the eastern SPG vary in their transit time (median = 3.4 years, s.d. = 1.2 year); the spread demonstrates that waters from the VCH do not travel coherently to the eastern SPG and instead mix with waters of other ages (Fig. 3). In comparison, the Lagrangian integral timescale of temperature anomalies for the trajectories that reach the SPG (~48 days) is much shorter than the advective timescale, thus individual trajectories are not expected to carry their temperature anomaly to the SPG.

Satellite SST anomaly pathways. Given the lack of a surface advective pathway from the STG to the SPG, we revisit the question of whether winter SST anomalies propagate along the GS/NAC[5,15,16]. Our investigation uses a lagged correlation analysis with the modern satellite SST record—31 years (1981–2012) of AVHRR Pathfinder SST data[23]. We switch to observational data for this analysis because a reliable data set exists with an observational period long enough for this analysis. A number of steps are followed to provide consistency with previous studies: (1) we use winter (November–April mean) SST anomalies at 1° × 1° spatial resolution; (2) the VCH box is defined as 80° W–60° W and 31.5° N–38.5° N; and (3) we apply a 5-year running mean filter to the SST anomalies before calculating correlations. A time series of winter SST anomalies averaged over the VCH box is extracted and then correlated with

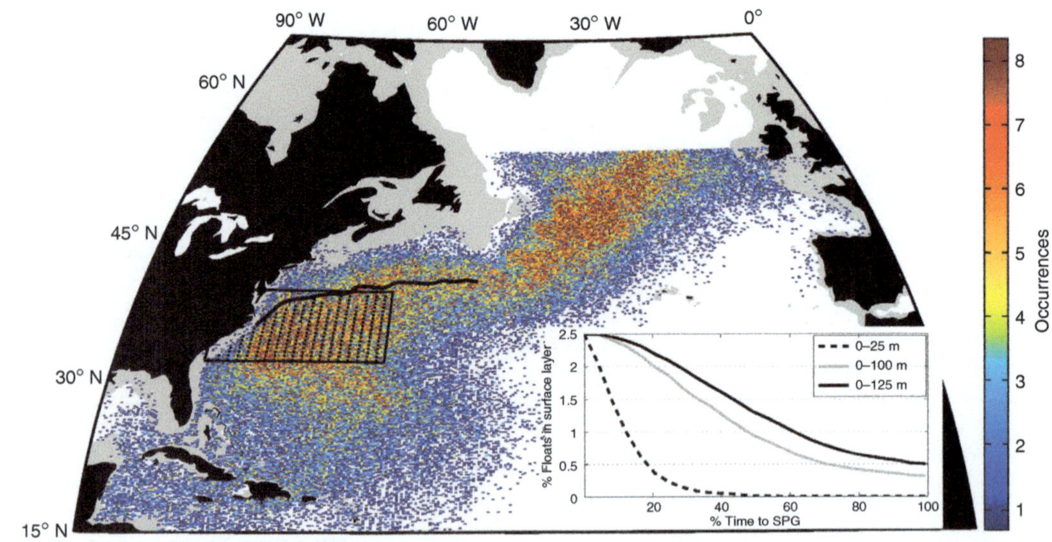

Figure 2 | Distribution of modelled particle positions that are SPG-bound. Particles that reach 53° N within 5 years in the three-dimensional flow field (2.51% of total). The colour bar refers to the total number of trajectories that passed through a given location (that is, occurrences) over the 15 years. Particle locations north of 53° N are not displayed. Solid black line indicates the average position of the GS north wall in FLAME. Inset panel shows the percent of trajectories that remain in the surface layer as a function of time, normalized by the total time required to reach the SPG. Three definitions of the surface layer are provided (0–25, 0–100 and 0–125 m).

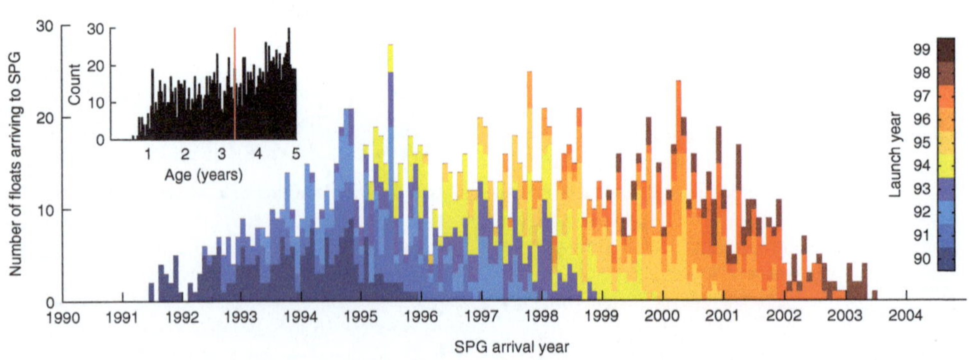

Figure 3 | Number of modelled particles crossing 53° N every 30 days. Colours denote the year of initialization in the VCH box. Inset panel shows the distribution of float ages when crossing 53° N, with a median (red line) at 3.37 years.

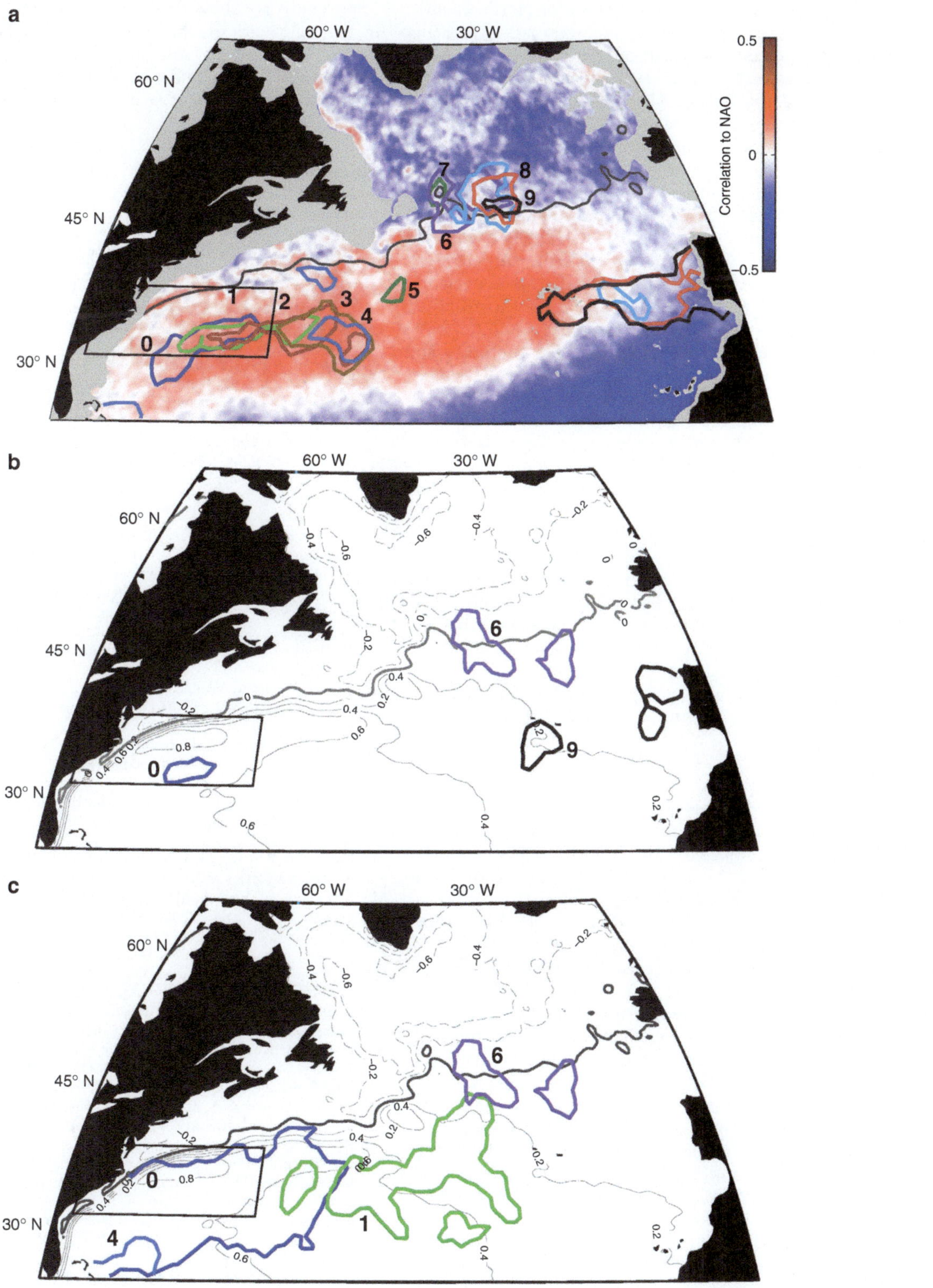

Figure 4 | Lagged correlation analysis with the modern satellite SST record. Correlations are calculated between the local SST and the time series from the spatially averaged VCH box (black rectangle). (**a**) Correlations that exceed subjectively chosen thresholds that decrease as the lag increases (lag 0 = 0.8, lag 1 = 0.75, lag 2 = 0.65, lag 3 = 0.55, lag 4 = 0.50, lag 5 = 0.40, lag 6 = 0.40, lag 7 = 0.40, lag 8 = 0.35 and lag 9 = 0.35). The thresholds used in lags 7 through 9 are not statistically significant ($P > 0.05$). Winter SST anomalies are smoothed with a 5-year running mean before calculating correlations. Correlations between the NAO and North Atlantic SST from 1981-2012 are shown in the background. Bathymetry above 1,000 m is shaded light grey, and the dark grey line represents the zero SSH contour. (**b**) Correlation contours are calculated as in **a** but the 5-year running mean filter is removed. (**c**) Objectively chosen, statistically significant ($P < 0.05$) correlation contour thresholds with no running mean filter. In **b** and **c**, time-mean AVISO SSH (1993-2014) contours (metres) are shown in grey. An identical analysis with March SST winter SST anomalies for each $1^\circ \times 1^\circ$ grid in the North Atlantic. Correlations are calculated for zero to nine-year lags. Areas that exceed subjectively chosen threshold correlations that decrease as the lag increases so as to maximize a propagation pattern are outlined for each lag in Fig. 4a. The resulting contours reveal a propagation pattern similar to that in Fig. 2 of Sutton and Allen[5]. However, if the 5-year running mean filter is removed before calculating the lagged correlations (Fig. 4b), the propagation pattern disappears. This implies that the apparent pathway in Fig. 4a results solely from the use of temporal smoothing. When we instead outline objectively chosen, statistically significant correlation thresholds ($P < 0.05$) and do not use a running mean filter (Fig. 4c), years 0 and 1 encompass most of the western and central STG (STMW formation regions). STMW temperature anomalies have been hypothesized to resurface in the following winter due to summer stratification capping the anomalies and protecting them from air–sea exchanges[13,24]. Our results support this hypothesis with the caveat that our cutoff for significance at 0- and 1-year lag was relatively low ($r \geq 0.36$). Though the individual particles have Lagrangian integral timescales of < 50 days, the STMW as a whole exhibits a 1-year SST memory due to particles entering and exiting the STMW during their lifetimes[12,13]. We also note that the STMW re-emergence region (Fig. 4c, green) is south of the GS/NAC indicated by either the GS north wall (Fig. 2) or the zero SSH contour (Fig. 4), both of which can be used to track the northern extent of the STG[25]. Thus, not only are the thermal anomalies short-lived, they are also primarily constrained to the STG. We conclude that SST anomalies are not advected or propagated along the GS/NAC from the STG to the SPG.

To determine why an inter-gyre pathway for SST anomalies appears with the 5-year running mean filter, we correlate the monthly time series of the NAO, an index of the atmospheric pressure gradient between Iceland and the Azores, with the AVHRR satellite SST (Fig. 4a). The spatial expression of the NAO forms a tri-pole pattern across the North Atlantic, with the SPG in phase with the tropical Atlantic and out of phase with the STG. This spatial pattern, combined with NAO timescales of roughly 2–7 years, gives context to the lagged correlation contours in Fig. 4a. With the 5-year running mean filter, the correlation contours of lags 0 through 5 years occur in the STG, where NAO-induced SST variability is in phase with the SST of the VCH box. For lags of 6 through 9 years, the correlations shift into the SPG where the NAO forcing is out of phase with the VCH SST. This shift into the SPG results not from the advection of water from the STG, but from the changing phase of the NAO and its opposing effect on the two gyres[26]. In addition, the 5-year running mean filter artificially increases the magnitude and prolongs the duration of the correlations that appear without the smoothing; the original 1-year memory of STMW SST in Fig. 4c is extended to 5 years by the temporal smoothing in Fig. 4a. Taken together, these results imply that the appearance of a SST anomaly pathway from the GS to the SPG in Fig. 4a is an artefact of the 5-year running mean filter introducing memory into NAO-induced SST oscillations rather than a signal of SST anomalies propagating along the GS/NAC[16,27]. Though other studies have pointed to the divergences and convergences in oceanic heat transport due to NAO forcing as a means to generate the SST anomalies in the GS region[16], we propose that the apparent propagation pathway results from a smoothing window (5 years) that is on the same timescale as the NAO forcing (2–7 years).

Though for decades it has been assumed that waters from the surface GS flow northward as part of the upper limb of the AMOC, building on a recent study[11] we demonstrate here that this throughput is not achieved by a surface connection. Importantly, we show that surface thermal anomalies generated in the subtropics are not transmitted to the surface waters at higher latitudes on interannual timescales. This work has implications for our understanding of how the AMOC achieves its poleward transport of heat.

References

1. Maury, M. F. *The Physical Geography of the Sea and its Meteorology* 13th edn (Sampson Low, 1868).
2. Broecker, W. S. The Great Ocean Conveyor. *Oceanography* **4,** 79–89 (1991).
3. Fratantoni, D. M. North Atlantic surface circulation during the 1990s observed with satellite-tracked drifters. *J. Geophys. Res.* **106,** 22067–22093 (2001).
4. Kostov, Y., Armour, K. C. & Marshall, J. Impact of the Atlantic meridional overturning circulation on ocean heat storage and transient climate change. *Geophys. Res. Lett.* **41,** 2108–2116 (2014).
5. Sutton, R. T. & Allen, M. R. Decadal predictability of North Atlantic sea surface temperature and climate. *Nature* **388,** 563–567 (1997).
6. Latif, M., Collins, M., Pohlmann, H. & Keenlyside, N. A review of predictability studies of Atlantic sector climate on decadal time scales. *J. Clim.* **19,** 5971–5987 (2006).
7. Msadek, R., Dixon, K. W., Delworth, T. L. & Hurlin, W. Assessing the predictability of the Atlantic meridional overturning circulation and associated fingerprints. *Geophys. Res. Lett.* **37,** L19608 (2010).
8. Lumpkin, R. & Johnson, G. C. Global ocean surface velocities from drifters: mean, variance, El-Niño-Southern Oscillation response, and seasonal cycle. *J. Geophys. Res.* **118,** 2992–3006 (2013).
9. Brambilla, E. & Talley, L. D. Surface drifter exchange between the North Atlantic subtropical and subpolar gyres. *J. Geophys. Res.* **111,** C07026 (2006).
10. Häkkinen, S. & Rhines, P. B. Shifting surface currents in the northern North Atlantic Ocean. *J. Geophys. Res.* **114,** C0405 (2009).
11. Burkholder, K. C. & Lozier, M. S. Subtropical to subpolar pathways in the North Atlantic: deductions from Lagrangian trajectories. *J. Geophys. Res.* **116,** C07017 (2011).
12. Gary, S. F., Lozier, M. S., Kwon, Y.-O. & Park, J. J. The fate of North Atlantic subtropical mode water in the FLAME model. *J. Phys. Oceanogr.* **44,** 1354–1371 (2014).
13. Kwon, Y.-O., Park, J. J., Gary, S. F. & Lozier, M. S. Year-to-year reoutcropping of eighteen degree water in an eddy-resolving ocean simulation. *J. Phys. Oceanogr.* **45,** 1189–1204 (2015).
14. Burkholder, K. C. & Lozier, M. S. Tracing the pathways of the upper limb of the North Atlantic Meridional Overturning Circulation. *Geophys. Res. Lett.* **41,** 4254–4260 (2014).
15. Hansen, D. V. & Bezdek, H. F. On the nature of decadal anomalies in the North Atlantic sea surface temperature. *J. Geophys. Res.* **101,** 8749–8758 (1996).
16. Krahmann, G., Visbeck, M. & Reverdin, G. Formation and propagation of temperature anomalies along the North Atlantic current. *J. Phys. Oceanogr.* **31,** 1287–1303 (2001).
17. Böning, C. W., Scheinert, M., Dengg, J., Biastoch, A. & Funk, A. Decadal variability of subpolar gyre transport and its reverberation in the North Atlantic overturning. *Geophys. Res. Lett.* **33,** L21S01 (2006).
18. Biastoch, A., Böning, C. W., Getzlaff, J., Molines, J.-J. & Madec, G. Causes of interannual-decadal variability in the meridional overturning circulation of the midlatitude North Atlantic Ocean. *J. Clim.* **21,** 6599–6615 (2008).
19. Joyce, T. M. & Zhang, R. On the path of the Gulf Stream and the Atlantic Meridional overturning circulation. *J. Clim.* **23,** 3146–3154 (2010).
20. Palter, J., Lozier, M. S. & Barber, R. T. The effect of advection on the nutrient reservoir in the North Atlantic subtropical gyre. *Nature* **437,** 687–692 (2005).
21. Rypina, I. I., Pratt, L. J. & Lozier, M. S. Near-surface transport pathways in the North Atlantic Ocean: looking for throughput from the subtropical to the subpolar gyre. *J. Phys. Oceanogr.* **41,** 911–925 (2011).
22. de Boyer Montégut, C., Fischer, A. S., Lazar, A. & Iudicone, D. Mixed layer depth over the global ocean: an examination of profile data and a profile-based climatology. *J. Geophys. Res.* **109,** C12003 (2004).
23. Casey, K. S., Brandon, T. B., Cornillon, P. & Evans, R. in *Oceanography from Space: Revisited* (eds Barale, V., Gower, J. F. R. & Alberotanza, L.) 273–287 (Springer, 2010).
24. Alexander, M. A. & Deser, C. A mechanism for the recurrence of wintertime midlatitude SST anomalies. *J. Phys. Oceanogr.* **25,** 122–137 (1995).
25. Frankignoul, C., de Coëtlogon, G., Joyce, T. & Dong, S. Gulf Stream variability and ocean-atmosphere interactions. *J. Phys. Oceanogr.* **31,** 3516–3529 (2001).
26. Lozier, M. S. *et al.* The spatial pattern and mechanisms of heat content change in the North Atlantic. *Science* **319,** 800–803 (2008).
27. de Coëtlogon, G. & Frankignoul, C. The persistence of winter sea surface temperature in the North Atlantic. *J. Clim.* **16,** 1364–1377 (2003).
28. Sudre, J. & Morrow, R. A. Global surface currents: a high-resolution product for investigating ocean dynamics. *Ocean Dyn.* **58,** 101–118 (2008).

29. Ducet, N., Traon, P.-Y. & Reverdin, G. Global high resolution mapping of ocean circulation from TOPEX/Poseidon and ERS-1 and -2. *J. Geophys. Res.* **105,** 19477–19498 (2000).

Acknowledgements

We thank Stefan Gary (SAMS) for developing the platform to run the Lagrangian trajectories in the FLAME model, Apurva Dave (EPA) for his code to process the Ekman velocities and Claus Böning and Arne Biastoch (IFM-GEOMAR) for access to the FLAME model output. We are also grateful for the publicly available data from NOAA (AVHRR SST), CNES (AVISO SSH) and CTOH (Ekman velocities). This work was funded by the NSF (OCE-1259102) and the NASA Earth and Space Science Fellowship (NNX13AO21H).

Author contributions

N.P.F. and M.S.L. devised the study and wrote the manuscript.

Additional information

Competing financial interests: The authors declare no competing financial interests.

6

Drying of Indian subcontinent by rapid Indian Ocean warming and a weakening land-sea thermal gradient

Mathew Koll Roxy[1], Kapoor Ritika[1,2], Pascal Terray[3,4], Raghu Murtugudde[5], Karumuri Ashok[1,6] & B.N. Goswami[1,7]

There are large uncertainties looming over the status and fate of the South Asian summer monsoon, with several studies debating whether the monsoon is weakening or strengthening in a changing climate. Our analysis using multiple observed datasets demonstrates a significant weakening trend in summer rainfall during 1901–2012 over the central-east and northern regions of India, along the Ganges-Brahmaputra-Meghna basins and the Himalayan foothills, where agriculture is still largely rain-fed. Earlier studies have suggested an increase in moisture availability and land-sea thermal gradient in the tropics due to anthropogenic warming, favouring an increase in tropical rainfall. Here we show that the land-sea thermal gradient over South Asia has been decreasing, due to rapid warming in the Indian Ocean and a relatively subdued warming over the subcontinent. Using long-term observations and coupled model experiments, we provide compelling evidence that the enhanced Indian Ocean warming potentially weakens the land-sea thermal contrast, dampens the summer monsoon Hadley circulation, and thereby reduces the rainfall over parts of South Asia.

[1] Centre for Climate Change Research, Indian Institute of Tropical Meteorology, Pashan, Pune 411008, India. [2] Department of Environmental Sciences, Fergusson College, University of Pune, Pune 411004, India. [3] Sorbonne Universites (UPMC, Univ Paris 06)-CNRS-IRD-MNHN, LOCEAN Laboratory, 4 place Jussieu, F-75005 Paris, France. [4] Indo-French Cell for Water Sciences, IISc-IITM-NIO-IRD Joint International Laboratory, IITM, Pune 411008, India. [5] ESSIC, University of Maryland, College Park, Maryland 20740, USA. [6] Centre for Earth and Space Sciences, University of Hyderabad, Hyderabad 500046, India. [7] Earth and Climate Science Department, Indian Institute of Science Education and Research, Pune 411008, India. Correspondence and requests for materials should be addressed to M.K.R. (email: roxy@tropmet.res.in)

The South Asian summer monsoon rainfall during June to September plays a significantly large role in the 'roti and dal' of roughly one-half of the world population[1]. The variability of the monsoon makes the region one of the most susceptible areas around the world to the impacts of climate-related natural disasters such as droughts and floods. Through observations and modelling, the scientific community has advanced its understanding of the past and future changes in the monsoons. With regard to the northern summer monsoon, some studies suggest that the monsoon circulation and rainfall have intensified in a changing climate[2]. Wang *et al.*[2] show that the rainfall over the northern hemisphere summer monsoon domain, as well as the Hadley and Walker circulations, have all undergone substantial intensification during the last three decades (1979–2011), with a striking increase of monsoon rainfall by 9.5% per degree of global warming. A few recent studies have also pointed out the increasing frequency of extreme rainfall events over the South Asian monsoon domain, due to a warming climate[3]. However, quite a few other studies indicate that the monsoon rainfall[4–6] and the Hadley and Walker circulations[7–9] are weakening over the South Asian domain during the past half century (since 1950s). Some of these studies suggest that, though the extreme rainfall events have increased over some regions[3], the frequency of moderate-to-heavy rainfall events has decreased over the subcontinent[8]. In addition, several recent studies based on the Coupled Model Inter-comparison Project phase 5 (CMIP5) indicate a significant weakening of the large-scale Asian summer monsoon circulation, especially in the middle to upper atmosphere[10–12]. Monsoonal changes are found to be different when the region and time periods considered are different, and so are the implications[13]. In fact, a recent study[14] using proxy records for the Asian monsoon over the past two millennia suggests that the monsoon is currently in the decreasing phase of a multi-decadal oscillation. This study also cautions that monsoonal changes due to anthropogenic forcing will be difficult to detect against a backdrop of large natural variability. The Intergovernmental Panel on Climate Change (IPCC) Fifth Assessment Report (AR5) is also suggestive of the uncertainties looming over the status and fate of the monsoons[15]. Considering these uncertainties, there is an urgent need to provide robust information through the use of long-term observed datasets and climate models that could resolve the spatial and temporal lacuna associated with the monsoon variability in a changing climate.

Climate models consistently show that, under increasing greenhouse gas forcing, the land-sea temperature contrast increases, possibly due to enhanced evaporation (latent heat loss) over the ocean[16]. An increase in the land-sea thermal contrast should ideally increase the strength of the monsoon[17–19]. Also, the sea surface temperatures (SST) over the Indian Ocean have shown a warming trend over the last several decades[20–22], potentially enhancing the supply of moisture to the monsoon regions and increasing the rainfall amounts[23]. In a recent review of climate change and the South Asian monsoon, Turner and Annamalai[13] point out that though the land-sea contrast and the increased moisture availability should favour an increased monsoon rainfall, the observations do not provide any convincing evidence for such a positive trend. On the contrary, the South Asian monsoon rainfall has decreased significantly during the past several decades and this time evolution is not well reproduced by the CMIP5 models used in the IPCC AR5 (refs 4,11,12,24–26). This is indeed intriguing and escalates the uncertainty associated with the monsoon in a changing climate.

In this study, we find that the summer monsoon rainfall during 1901–2012 shows a weakening trend over parts of South Asia. The reduction in rainfall is significant over the central-east and northern regions of India, along the Ganges-Brahmaputra basins and the Himalayan foothills (hereafter referred to as central Indian subcontinent), parts of India that are heavily populated and where agriculture is still mostly rain-fed. Using observations and climate model experiments, we demonstrate that this reduction in rainfall is linked to the rapid warming of the Indian Ocean, especially its western part, during the past century. The Indian Ocean warming, along with a relatively subdued warming of the Indian subcontinent, has played a key role in weakening the land-sea thermal contrast, a major driver of the South Asian monsoon. The weakened land-sea thermal contrast dampens the summer monsoon Hadley circulation, reducing the rainfall over the central Indian subcontinent.

Results

Rainfall trends during the past century. Figure 1a,b shows the trend in precipitation over the South Asian region, during summer monsoon for the years 1901–2012, using gridded datasets from the India Meteorological Department (IMD) and the Climate Research Unit (CRU). The trends are similar in both the datasets, with a weakening trend over the central-east and

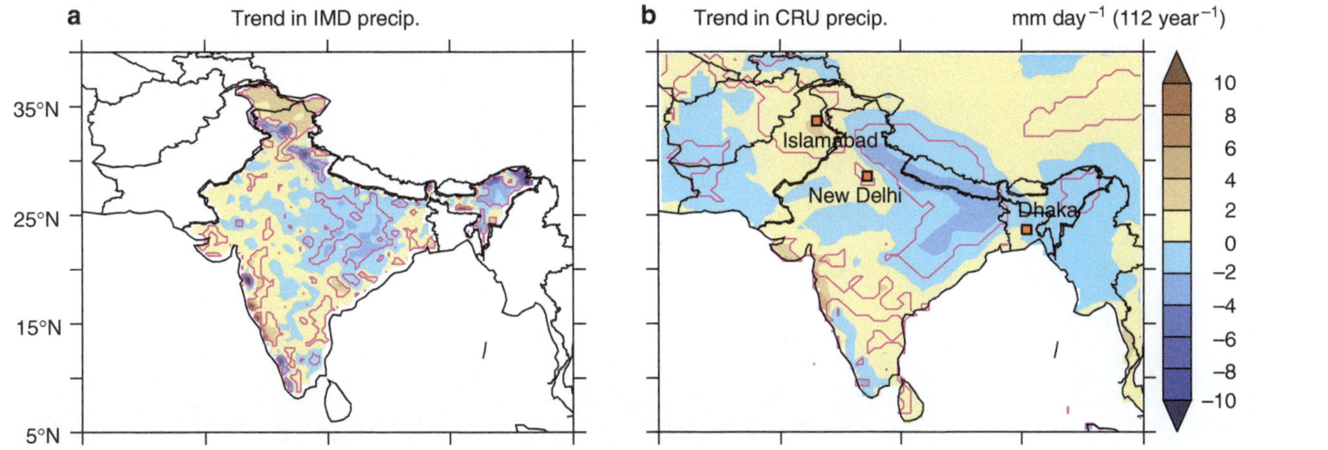

Figure 1 | Summer monsoon precipitation trends for the years 1901–2012. Observed trend in precipitation (mm day−1 112 ...
CRU datasets, during June–September ...

northern regions of Indian subcontinent, and south of the Western Ghats region. The CRU dataset shows that the negative trend over the central subcontinent extends west to east, from south of Pakistan to central-east India, with a prominent horseshoe pattern with one of its arms placed on the foothills of the Himalayas, and the other obliquely along the central-east coast of India. The reduction in summer rainfall over central-east India during the past century is about 10–20%. This could have socioeconomic consequences as the agriculture in this region is still largely rain-fed. Significant positive trends in precipitation are consistently observed to the north of the Western Ghats region, in both the datasets. However, the positive rainfall trends are confined to a small domain along the west coast.

Warming Indian Ocean. Previous studies using atmospheric general circulation models (AGCM) forced by observed SST variations suggest that a decrease in global land monsoon rainfall can be caused by warming trends over the tropical oceans, especially the central-eastern Pacific and the western Indian Ocean[27]. Among these tropical oceans, the Pacific SST anomalies do not show any long-term significant trend, while the western Indian Ocean exhibited rapid warming throughout the past century[28] (Fig. 2). In comparison with the rest of the Indian Ocean, the western Indian Ocean generally has cooler mean SSTs in summer, owing to the strong monsoon winds and the resultant upwelling over the west. Meanwhile, the central-east Indian Ocean is characterized by a warm pool with SSTs greater than 28.0 °C. Though previous studies demonstrate a basin-wide warming in the Indian Ocean during the past half century[21,22], our analysis using long-term SST datasets show that over an extended period of 112 years, the western Indian Ocean (WIO, 50–65°E, 5°S-10°N) has experienced anomalous warming of 1.2 °C (Fig. 2), which is 0.5 °C greater than the warming over the warm pool region[28]. Even though a warming trend over the Indian Ocean may result in an increased supply of moisture locally over the ocean, it is not necessary that the surplus moisture is carried on to the peninsula[29]. The sustained Indian Ocean warming in the west has led to a greater westward extension of this Indian Ocean warm pool in recent decades[21,28]. The expansion of the Indian Ocean warm pool can in turn modify the land-sea thermal gradient, which could modulate the strength and flow of the monsoon circulation and the moisture laden winds towards the South Asian subcontinent. Hence, deciphering the relationship between the Indian Ocean warming and the monsoon precipitation might give a clue to the cause and effect of the observed rainfall trends over the region.

To evaluate the association of the nearly monotonic warming over the western Indian Ocean on the observed monsoon trends over the subcontinent, we carried out a correlation analysis between SST and precipitation anomalies over the two regions. Figure 3a,b show the correlation between the western Indian Ocean SST anomalies and precipitation anomalies over the South Asian region, using multiple datasets. The spatial distribution of the correlation coefficients is consistent among the different data sets. The fascinating aspect is that, the correlation of precipitation anomalies with the Indian Ocean SST anomalies and the trend in these precipitation anomalies are strikingly similar. A pattern correlation computed between the trends in Fig. 1b and the correlation coefficients in Fig. 3b (for the region 60°–100°E, 5°–35°N at 0.5° horizontal grid resolution) indicates that they are strongly correlated ($r = 0.73$, significant at 95% confidence level). The most robust common feature among the trend and correlation patterns is a horseshoe pattern over central India. Apart from this, a dipolar pattern in the rainfall trends, with positive in the north and negative in the south is also visible along the Western Ghats of the Indian subcontinent (Fig. 1). A recent study[30] suggests that enhanced warming to the north of western Indian Ocean in recent decades has also led to a poleward shift of the core of low-level monsoon winds, increasing the precipitation to the north of the Western Ghats, while weakening it over the south. This dipolar rainfall pattern between the south and north of the Western Ghats is also found to be associated with the western Indian Ocean warming at both the interannual and low frequency (for example, trend) timescales (Fig. 3).

For a time series analysis, we consider the precipitation over the central Indian subcontinent (80–90°E, 20–30°N, inset box in Fig. 3b) along with the western Indian Ocean SST anomalies (Fig. 3c). The rainfall region under consideration is chosen because the decreasing trend in rainfall is significant over this region, and also consistent with other studies utilizing station-wise analysis of rain gauge data[6]. Indeed, the SST and precipitation time series in Fig. 3c are negatively correlated ($r = -0.35$ at $df = 110$, for the unsmoothed time series), significant at 95% confidence level. Analysis using detrended time series data also shows significant negative correlation over the same region (Supplementary Fig. 1). This means that similar processes may operate at interannual and lower frequency timescales to generate a secular downward monsoon rainfall trend in response to the warming of the western Indian Ocean, or alternatively, that a part of the Indian Ocean warming is due to the weakening of the monsoon[22]. The negative correlation is not observed during the first half of the 20th century, but becomes statistically significant post-1950s, probably due to the accelerated warming observed during the latter half. This implies that the relationship between a warm ocean and a weak monsoon is not linear. Since the ocean warming is a slow process, the heat has to build-up over several decades to make a substantial difference and reach the warm pool SST values (28.0 °C)[21,28]. The sustained western Indian Ocean warming has

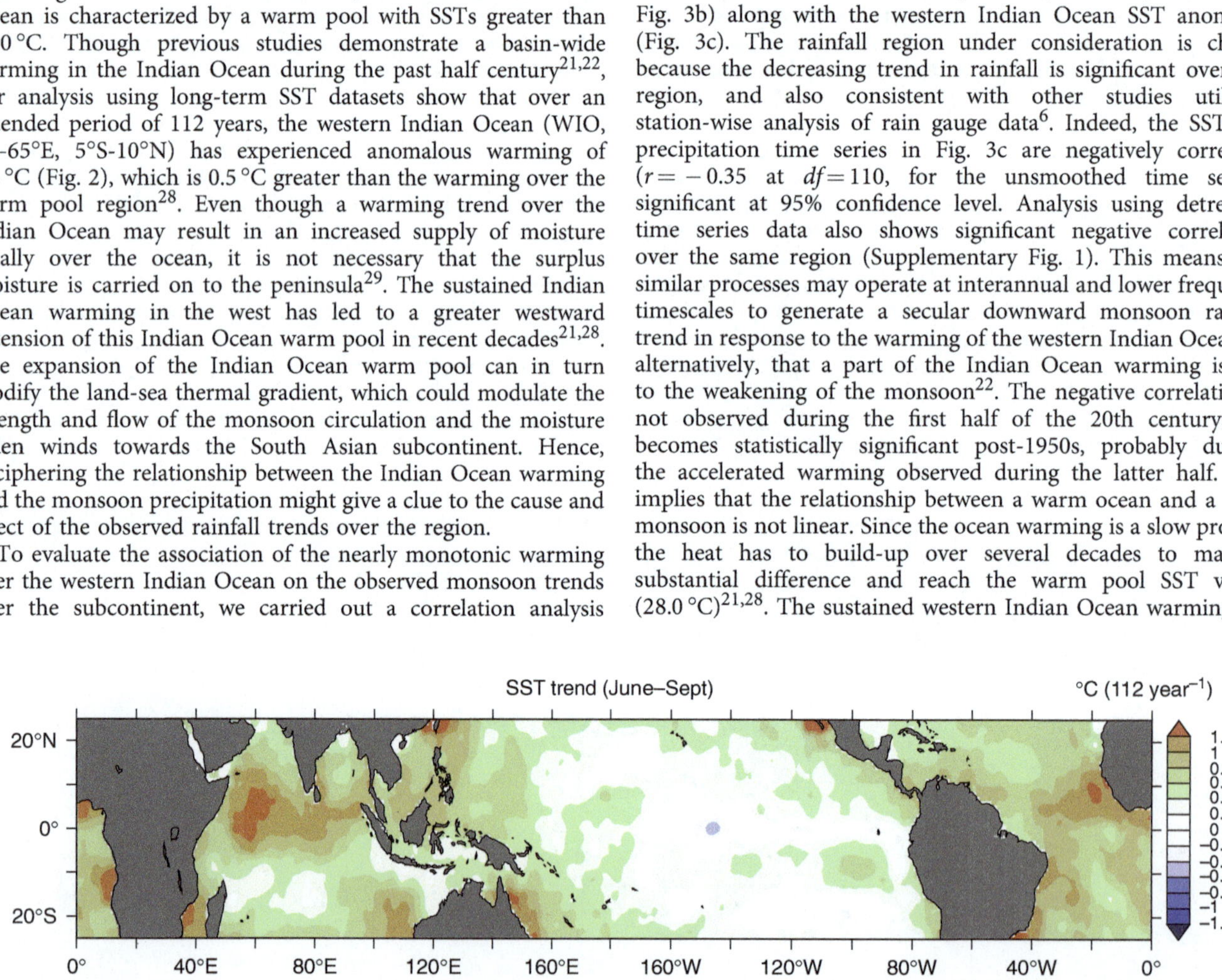

Figure 2 | Summer sea surface temperature trends for the years 1901-2012. Observed trend in mean summer (June-September) SST (°C 112 year^{-1}) over the global tropics during 1901-2012.

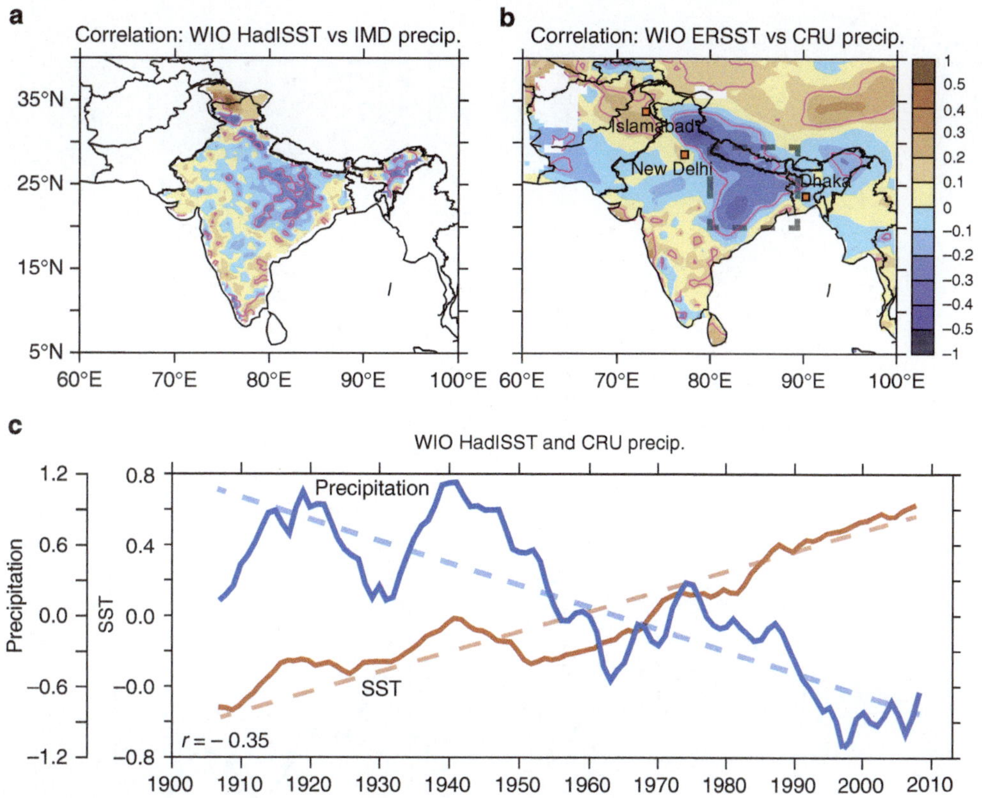

Figure 3 | Correlation between western Indian Ocean sea surface temperatures and monsoon precipitation. Correlation between SST over the western Indian Ocean (WIO, 50-65°E, 5°S-10°N) and precipitation over the South Asian subcontinent, for (**a**) HadISST and IMD precipitation, and (**b**) ERSST and CRU precipitation, for June-September 1901-2012. Contours denote regions significant at the 95% confidence level. The inset box includes parts of the central Indian subcontinent where the weakening trend in precipitation is significant (**c**) Time series of SST anomalies (°C, *red*) over western Indian Ocean along with CRU (*blue*) precipitation (mm day^{-1}) over central Indian subcontinent (80-90°E, 20-30°N, inset box in ***b***), smoothed with a 10 year moving average. Note that the correlation coefficient ($r = -0.34$) between HadISST and CRU precipitation is estimated using non-smoothed time series. Kendall's rank correlation test for the two variables provided a tau coefficient of − 0.3 ($P<0.01$, two tailed). Mann-Kendall test for the trend in the time series provided a tau coefficient of 0.6 for SST and − 0.2 for precipitation, both significant at 95% confidence level.

led to a greater spatial extension of the Indian Ocean warm pool in the recent decades only[21].

Changing land-sea thermal contrast. As pointed out earlier, changes in the land-sea thermal contrast impact the strength of the monsoon[16,17–19]. Studies show that over the northern hemisphere, the surface temperatures over land increase more rapidly than over sea, under increasing greenhouse gases[19]. Ideally, such a scenario should strengthen the South Asian monsoon[16,17–19]. Figure 4a,b shows the climatological mean and trends of surface temperatures over the Indian Ocean and the South Asian monsoon domain, during summer. Obviously, the monotonic warming trend over the Indian Ocean is prominent, with the strongest warming over the west. At the same time the Indian subcontinent exhibits suppressed warming during the past century. In fact, the surface temperatures in the northern Indian peninsula show a cooling trend. This implies a weakening meridional thermal gradient over the South Asian domain, in contrast with earlier studies that suggest a strengthening of the thermal gradient under a globally warming scenario. The weakening land-sea thermal gradient is reproduced in other datasets also, indicating the robustness of the trend (Supplementary Fig. 2).

The land-sea contrast in surface temperatures, though important may not provid

factors modulating the strength of the monsoon circulation. Once the monsoon rains set in on the South Asian subcontinent, the land surface cools down considerably. However, the troposphere above the land remains warm due to the latent heat release from the convective activity, keeping the thermal contrast functional[31]. This means that the tropospheric temperature is an ideal parameter for examining the thermodynamic forcing related to the monsoon[31]. However, unlike the surface temperature, robust records of air temperature at different levels of the troposphere are not available before the 1950s, which makes it difficult to examine the role of long-term tropospheric temperature trends on the monsoon. Nevertheless, to confirm that our results are valid for the recent decades also, we analyse the tropospheric temperatures over the monsoon domain. The trend in the thermal contrast is well reflected in the upper tropospheric (200 hPa) temperature anomalies also, as demonstrated in Fig. 4c. Along with a cooling trend over the larger Indo-Tibetan landmass, increasing positive temperature anomalies are observed over the western Indian Ocean, indicating its potential role in weakening the monsoon flow towards the Indian peninsula. The warming of the upper troposphere over the Indian Ocean (Fig. 4c) is co-located with that over the sea surface (Fig. 4b). This may be attributed to the fact that the tropical Indian Ocean is a highly convective region during the northern summer, as the mean

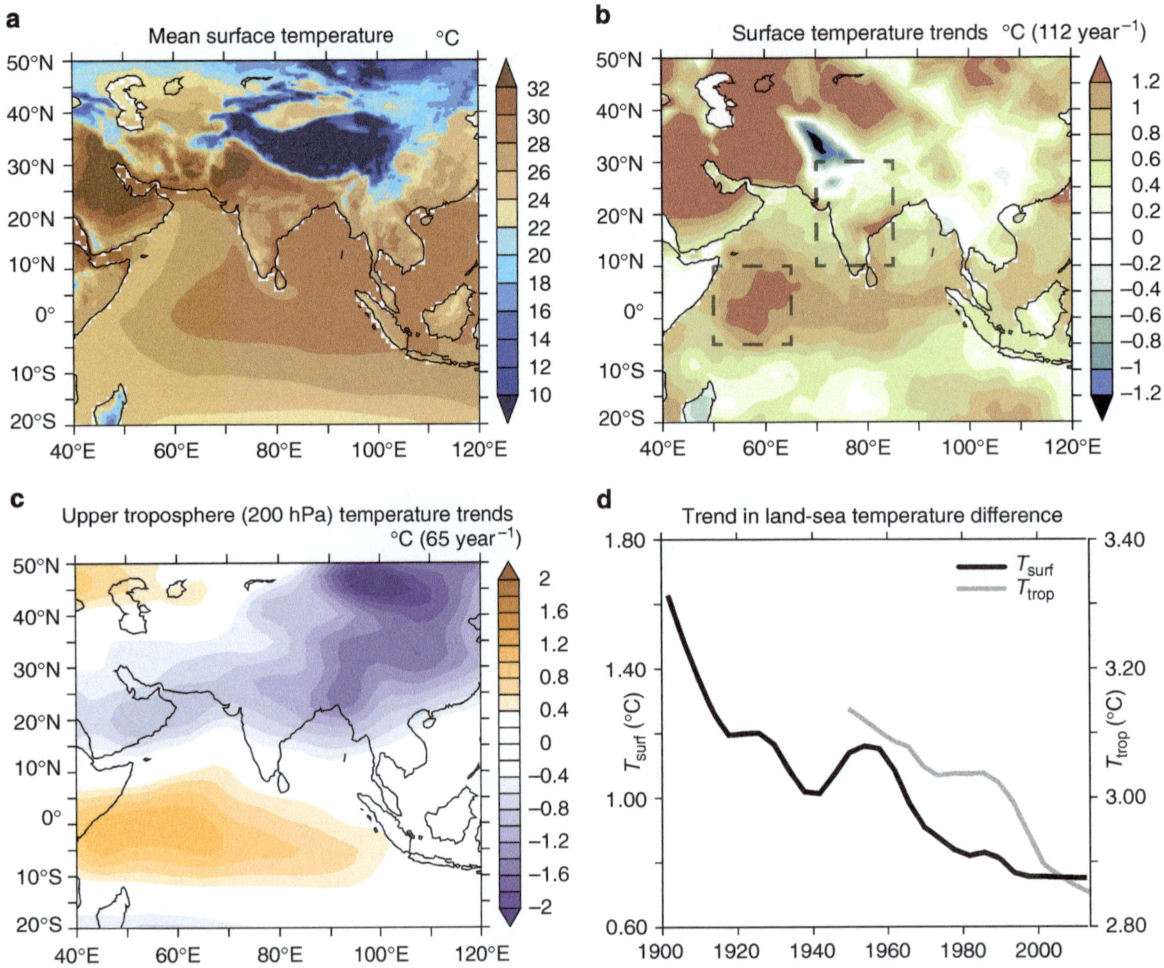

Figure 4 | Summer mean surface temperatures and land-sea temperature gradients for the years 1901-2012. Observed (**a**) mean (°C) and (**b**) trend (°C 112 year^{-1}) of surface temperatures during June-September, 1901-2012. For the trend, the colour shades represent regions significant at the 95% confidence level. (**c**) Observed trend (°C 65 year^{-1}) of upper tropospheric (200 hPa) temperatures during June-September, 1948-2012. (**d**) Time series of the trend in land-sea contrast in the surface (T_{surf}) and tropospheric (T_{surf}, 850-200hPa average) temperatures (°C). The land-sea contrast is estimated as the difference in the values between the boxes over the South Asian subcontinent (70-85°E, 10-30°N) and western Indian Ocean (50-65°E, 5°S-10°N), and the trend is estimated over 31 year sliding periods. The land surface temperature is from HadCRUT, SST from HadISST and tropospheric temperature from NCEP reanalysis.

conducive for enhanced moisture convergence[32]. Further warming of the Indian Ocean enhances the convection[23], transferring surplus heat to the upper troposphere[33-35].

To evaluate the trends observed in land-sea thermal contrast in Fig. 4b,c, we analyse the difference between the summer tropospheric temperatures over western Indian Ocean and the Indian subcontinent, at 850 and 200 hPa levels (Fig. 4d). The surface temperature gradient (T_{surf}) shows a sharp decline throughout the past century, largely contributed by an increased warming over the western Indian Ocean. The tropospheric temperature gradient (T_{trop}) post-1950s also indicates a trend similar to that in the surface temperatures. This is suggestive of a warm Indian Ocean having a strong hold on the whole troposphere, from the surface to the top, possibly due to enhanced convective mixing over the ocean[33-35]. Indeed, a separate analysis of the trend at each of these levels gave consistent results of a warming troposphere over the Indian Ocean, and a weakening thermal contrast between the subcontinent and the ocean. All these results indicate a reduced land-sea thermal contrast during the past century, partly due to increased warming over the Indian Ocean, especially over the western region, and partly due to a relatively weaker warming over the Indian peninsula.

Weakening of the summer monsoon Hadley circulation. Many studies have shown that warm SST anomalies are accompanied by large variations in the lower and upper troposphere due to enhanced latent heating aloft from convection over the ocean[33-35]. Furthermore, these changes are highly correlated with the strength of the monsoon circulation[31,34]. A more recent study using a coupled model framework also points out that a spatial extension of Indian Ocean warm pool could enhance the convection over the ocean while introducing a dry bias over land, by modulating the meridional Hadley circulation[36]. It is therefore likely that the impact of the Indian Ocean warming on the monsoon has become more prominent in recent decades. Consistently, an examination of the vertical wind velocity over the South Asian domain (50–100°E) during the years 1948–2012 indicates largescale upward motion over the equatorial ocean (10°S-10°N), extending up to the upper troposphere and favouring intense local convection (Fig. 5a). Furthermore, this

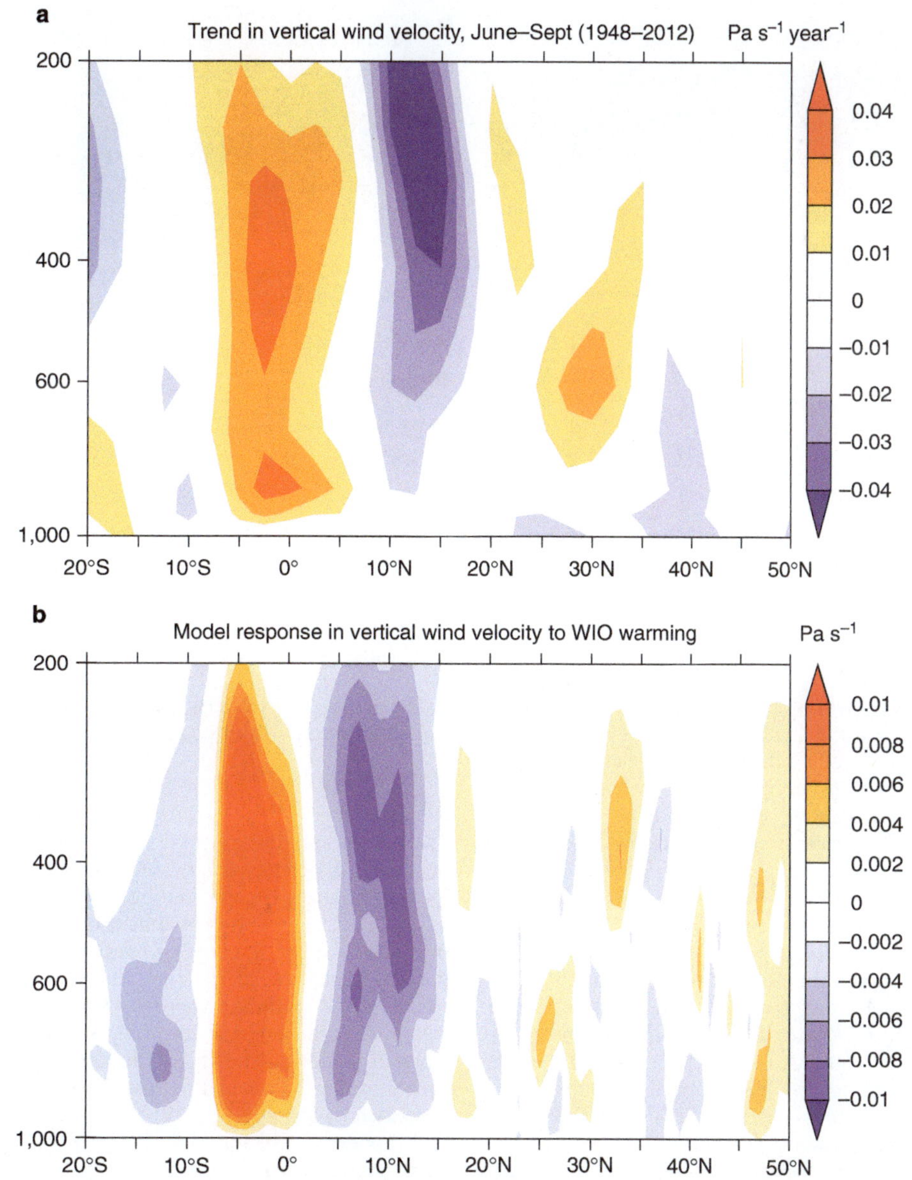

Figure 5 | Observed trends and model simulated changes in summer monsoon Hadley circulation in response to Indian Ocean warming. (**a**) Trend from NCEP re-analysis ($Pa\,s^{-1} year^{-1}$) and (**b**) model simulated response ($Pa\,s^{-1}$) to western Indian Ocean warming in vertical velocity (omega) during northern summer (June-September) along the South Asian domain (50-100°E). Positive (red) values indicate upward motion.

enhanced upward motion over the ocean is compensated by subsidence of air over the subcontinent (10–20°N), inhibiting convection over the landmass and drying the region, through a modulation of the local Hadley cell (Fig. 5a). This suggests that though the warming ocean engenders enhanced local rainfall due to increased moisture availability, it weakens the monsoon Hadley circulation and reduces the rainfall over the land, ultimately building up a competition among the land and ocean rainfall in the South Asian monsoon domain.

Climate model response to Indian Ocean warming. Attribution of the decreasing monsoon rainfall to the Indian Ocean warming from observations alone is very difficult and probably impossible. Thus, to delineate and examine the role of the Indian Ocean warming on the monsoon circulation and rainfall, model sensitivity experiments using an ocean-atmosphere coupled climate model were carried out. In the

SST anomalies similar to those in the observations were added to the western Indian Ocean region (Supplementary Fig. 3). Figure 5b shows the model response in the summer monsoon Hadley circulation, and features a weakening of the local Hadley cell, much similar to the observed trends. The fact that the warming has resulted in enhanced convection over the ocean and subsidence over the landmass serves as the evidence that the observed warming trend in the Indian Ocean has the potential to weaken the monsoon circulation.

Figure 6 compares the observed trends in precipitation and surface winds with the model-simulated anomalies in response to warming over the western Indian Ocean. The interesting, robust element in the model simulations (Fig. 6b) is the horseshoe pattern in the negative precipitation anomalies, with one arm placed over the Himalayan foothills and the other over the central Indian landmass, similar to the one in the precipitation trends (Figs 1 and 6a) and the

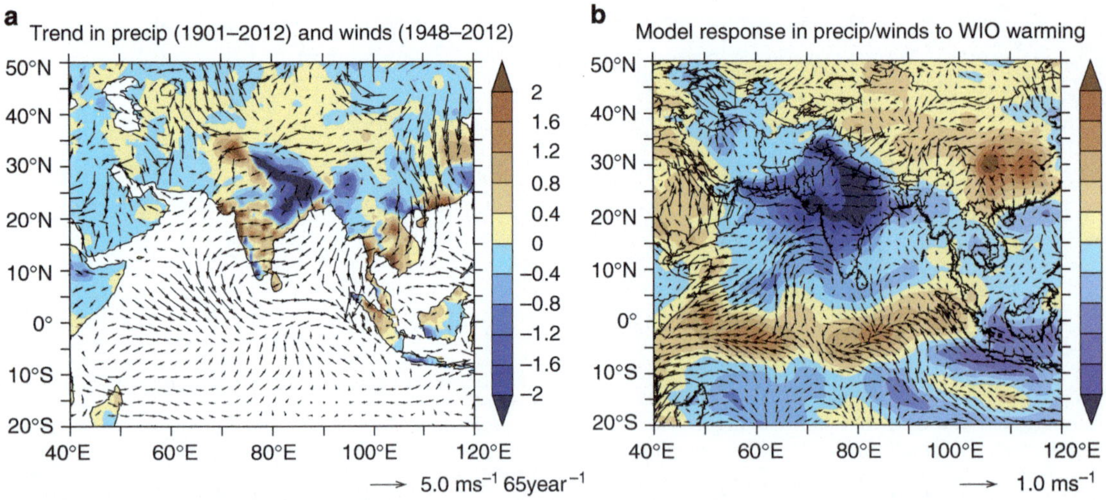

Figure 6 | Observed trends and model simulated changes in summer monsoon precipitation in response to Indian Ocean warming. (**a**) Observed trend in precipitation (CRU data, 1901-2012, mm day^{-1}) and near-surface winds (NCEP, 1948-2012, m s^{-1}), for June-September. (**b**) Model-simulated mean precipitation (mm day^{-1}) and wind (10 m) anomalies (m s^{-1}) in response to warming over the western Indian Ocean. The model-simulated anomalies are estimated from the sensitivity run where SST anomalies of the order of 1.5 °C is introduced over the western Indian Ocean (CFSv2$_{WIO}$), with respect to a model control run (CFSv2$_{CTL}$).

precipitation, the surface winds also suggest weakened mean southwesterly winds. The trend analysis (Fig. 6a) shows a Rossby wave-like response[37] to the Indian Ocean warming, with northeasterly wind anomalies over most of the Arabian Sea, converging towards the central Indian Ocean. The model simulation (Fig. 6b) also shows similar results, with respect to the Rossby wave response near the subcontinent and over the Arabian Sea. However, the zonal direction of the wind anomalies near the equator appears different in the observations and the model. This is not an unexpected result as the periods under consideration and the model simulations are different, since the observed winds are available only for 1948–2012, a period during which the Indian Ocean experienced a basin-wide warming[21,22]. In fact previous studies using AGCM experiments driven by recent basin-wide warming in the tropical Indian Ocean reproduced the observed zonal wind changes in the recent decades[38]. In the model sensitivity experiment, however, the positive SST offset is prescribed only over the western Indian Ocean region, where the long-term warming trend is observed in the SST records during 1901–2012. This suggests that while western Indian Ocean warming plays a significant role in the long-term change in winds, basin-wide warming over the tropical Indian Ocean has emerged as a co-conspirator in the change in land-sea thermal contrast and winds in recent decades.

A further comparison of the model response to the warming is provided in Supplementary Fig. 4. The thermal response (surface and upper troposphere) is similar to those in the observed trends over the ocean, but exhibit some discrepancies over land. More specifically, for the model response in temperature at 200 hPa, the results reproduce the upper tropospheric warming over the ocean, but do not replicate the cooling over the Indo-Tibetan landmass. This is possibly because of the role of other processes on the upper tropospheric temperatures, such as aerosol forcing[39,40] and stratosphere–troposphere interactions[41], which are not included in the model experiment. It may also be noted that previous studies[41] have suggested that the NCEP upper tropospheric data overestimate the cooling trend over the Indo-Tibetan landmass. Nevertheless, the similarity between the rainfall and local Hadley responses between the model experiments and observed fields clearly demonstrates the importance of western Indian Ocean warming in determining the observed monsoon response.

Discussion

The above analysis using observations and climate model sensitivity experiments indicates the potential role of the Indian Ocean warming in weakening the monsoon circulation and rainfall over central Indian subcontinent. The Indian Ocean and monsoon exhibit a coupled relationship[32,42,43], and it is also possible that a feedback from a weak monsoon accelerates the warming in the Indian Ocean[22]. Some studies have suggested a role for anthropogenic aerosols in weakening the monsoon circulation during the recent decades[40,44–48]. It is possible that such a weakened circulation amplifies the SST warming due to a decreased evaporation. In contrast, recent studies show that the anthropogenic aerosols tend to have a cooling effect and compete with the greenhouse warming over the Indian Ocean[49,50]. Despite the fact that aerosols are a competing factor, the observed SST trends do not show any change during the post-1980s where aerosols have increased (Fig. 3c). In fact, a study[50] points out that a large share of the CMIP5 models with aerosol forcing is not able to reproduce the observed SST warming over the Indian Ocean. They attribute this issue to the large uncertainties in representing the direct and indirect effects of aerosols in these models. Meanwhile, a recent study demonstrates that apart from the warming due to greenhouse gas forcing[51], the long-term warming trend in the western Indian Ocean is significantly associated with the asymmetry in the ENSO forcing[28]. That is, while the quasi-periodic El Niño events induce a warming over the Indian Ocean via the Walker circulation, the La Niña events fail to do the inverse. Along with this, an increase in the magnitude and frequency of El Niños in the recent decades has accelerated this warming[28]. The heat pile-up due to this ENSO forcing[52] persists in the land-locked Indian Ocean that has a restricted thermohaline circulation, sustained by a feedback from the local air-sea interaction[53,54] and ocean dynamics[55,56]. Considering the strong seasonal variability in this region and short-term persistence in atmospheric changes, it is rather self-evident that, on longer time scales, the warmer ocean plays a larger role in weakening the monsoon than vice-versa.

Though the observed increase in anthropogenic aerosols may not have a direct effect in warming the Indian Ocean, they may have a potential role in suppressing (or even cooling) the surface warming over the South Asian domain in the recent decades[39,40]. Aerosols could be considered as the perfect partner in reducing the land-sea thermal contrast, along with a warming Indian Ocean. An evaluation of the CMIP5 models[40] suggests that aerosols could potentially reduce the incoming solar radiation over the Indo-Gangetic plain, thereby cooling the surface. However, it may be pointed out that majority of these models could not simulate the spatial distribution of aerosols or related processes over the South Asian subcontinent[40]. Also, ambiguity still remains on the extent of cooling (or warming) induced by different types of aerosols[45–47]. Meanwhile, other studies link the upper tropospheric cooling over the larger Indo-Tibetan landmass during summer to stratospheric temperature changes that are channelled through stratosphere–troposphere interactions[41]. Further research is required to differentiate the individual contributions of these components on the observed trends in surface and upper tropospheric temperatures over land.

The current study discusses the role of Indian Ocean warming in weakening of the monsoon, but does not rule out the possibility that similar impacts may also arise due to SST changes in the tropical Pacific and the Atlantic[9,54,57–61]. Recent studies show that the Pacific warming trend during the past century is not prominent as compared with the Indian Ocean warming while the tropical and extratropical Atlantic show strong warming trends[28]. However, Prodhomme *et al.*[61] pointed out that a warming in the tropical Atlantic does not change the tropospheric temperature gradient between the Indian Ocean and the subcontinent as the temperature changes due to the teleconnection are uniform over this domain. Meanwhile, Goswami *et al.*[59] suggest that a North Atlantic warming should ideally strengthen the Asian monsoon by increasing the tropospheric temperatures over northern India and southern Eurasia. However, temperature and rainfall trends during recent decades suggest that any such role of the Atlantic is offset by the tropospheric temperature trends induced by a warming Indian Ocean. Deconvolving the non-stationary relations between the monsoon and the warming over these remote regions is beyond the scope of this study. But considering the robustness of the Indian Ocean warming and its direct role in monsoon dynamics in comparison with the other tropical oceans, it is justified to say that the Indian Ocean SST trends play a relatively larger role on the observed rainfall trends in the South Asian domain.

A recent study comparing present observations and future regional climate model projections of the summer monsoon suggests a further weakening of the monsoon circulation and peninsular rainfall in the future[62]. Meanwhile, the CMIP5 future projections suggest an increase in rainfall over the central Indian subcontinent[26], though one must note the ongoing debate about whether CMIP5 models are in fact capturing the historic monsoon trends accurately[4,11,12,24] and whether the monsoon projections are reliable in these models[25]. The overall weakening trend of monsoon rainfall over South Asia is a matter of grave concern since the socio-economic livelihood in this region, including agriculture, water resources and power generation are irrevocably dependent on it. The role of Indian Ocean warming might not be solely limited to weakening the Asian monsoon. Research based on climate proxy records and models point out that the climate signals forced by warm SST anomalies over the tropical Indian and western Pacific oceans synergistically contribute to widespread drying over South Asia and mid-latitudes, an ideal scenario for a large-scale drought[63,64]. In addition, a recent study points out that the Indian Ocean warming may also be playing an important role in modulating the Pacific climate

favouring stronger trade winds in the Pacific, thereby contributing to the recent global warming hiatus[65]. Hence, the critical role of the warm Indian Ocean deserves special attention not only for its decisive effect on the food security of a large fraction of the world's population, but also due to its disproportionately large role as a partner to the warm pool Pacific in inducing a drought, as well as its potential role in the global warming pause.

Methods

Rainfall and SST historical data analysis. Two long-term precipitation datasets for the northern summer during the period 1901–2012, obtained from the India Meteorological Department (IMD) and Climate Research Unit (CRU) are used in this study. The recent IMD precipitation data use daily rainfall data from all the rain gauge stations over India, gridded at 0.25° horizontal resolution[66]. This includes rainfall records from 6,955 stations, which is the highest number of stations used by any gridded datasets currently available. The data density varies from year to year from about 1,500 in the first couple of years to about 4,000 by the end of the century. The trend in precipitation over the central Indian region under consideration is compared with studies utilizing station wise analysis of rain gauge data[6]. SST data for the same period are obtained from the HadISST1 dataset provided by the Met Office Hadley Centre, and also the Extended Reconstructed Sea Surface Temperature (ERSST) provided by NOAA. To examine the role of land-sea temperature gradient, surface temperatures from HadCRUT (1901–2012) and tropospheric temperature values from NCEP reanalysis (1948–2012) are used. Observed trends in the monsoon southwesterlies are inferred from the wind data based on NCEP reanalysis during 1948–2012. The significance of the trends and correlations is examined using standard two-tailed Student's *t*-tests. The correlation analysis in Fig. 3c has also been tested using Kendall's rank correlation that is non-parametric and therefore makes no assumptions about the distribution and at the same time determine the direction and significance of the relation between the two variables[67]. The correlated variables are said to be concordant if their ranks vary together (-1) and discordant if they vary differently ($+1$). The significance of the rainfall and SST linear trend values are further assessed with Mann–Kendall tests[68] with block bootstrap to improve the significance test when a time series shows auto-correlation.

Model and experimental design. For the numerical model experiments, we use the standard configuration of a global coupled ocean-atmosphere model, the Climate Forecast System version 2 (CFSv2). The CFSv2 model simulates the mean monsoon and its interannual variability reasonably well, and hence used in this study[42]. The oceanic component has a 0.25–0.5° horizontal resolution, 40 vertical levels and includes an ice model. The atmospheric component is at T126 ($\sim$0.9°) horizontal resolution and 64 sigma-pressure hybrid levels. The atmosphere and ocean exchange quantities such as heat and momentum fluxes every half an hour, with no flux adjustment or correction. See Saha *et al.*[69] for further details on the model and its components. The coupled configuration of CFSv2 is time integrated over a period of 100 years, and utilized as the reference run ($CFSv2_{CTL}$). To examine the role of a warming Indian Ocean, a sensitivity experiment ($CFSv2_{WIO}$) using ensembles (20 members) of short integrations for the summer monsoon season during June–September were performed by adding temperature anomalies to the SSTs passed over the western Indian Ocean, to the atmosphere. Positive anomalies of the order of 1.5 °C were added over the region, in such a way that it tapers out by the limits of the domain (50–65°E, 5°S–10°N). The positive offset added to the simulated SSTs over western Indian Ocean in the sensitivity experiments is similar to the amplitude of the observed trends (Fig. 2, Supplementary Fig. 3). The difference between $CFSv2_{WIO}$ and $CFSv2_{CTL}$ is taken as the model response to the summer warming over the western Indian Ocean.

References

1. Gadgil, S. & Rupa Kumar, K. *The Asian Monsoon* 651-683 (Springer, 2006).
2. Wang, B. *et al.* Northern Hemisphere summer monsoon intensified by mega-El Niño/southern oscillation and Atlantic multidecadal oscillation. *Proc. Natl Acad. Sci. USA* **110**, 5347–5352 (2013).
3. Goswami, B. N., Venugopal, V., Sengupta, D., Madhusoodanan, M. & Xavier, P. K. Increasing trend of extreme rain events over India in a warming environment. *Science* **314**, 1442–1445 (2006).
4. Kitoh, A. *et al.* Monsoons in a changing world: a regional perspective in a global context. *J. Geophys. Res.* **118**, 3053–3065 (2013).
5. Zhou, T., Zhang, L. & Li, H. Changes in global land monsoon area and total rainfall accumulation over the last half century. *Geophys. Res. Lett.* **35**, L16707 (2008).
6. Guhathakurta, P. & Rajeevan, M. Trends in the rainfall pattern over India. *Int. J. Climatol.* **28**, 1453–1469 (2008).
7. Tokinaga, H., Xie, S.-P., Deser, C., Kosaka, Y. & Okumura, Y. M.

8. Krishnan, R. *et al.* Will the South Asian monsoon overturning circulation stabilize any further? *Clim. Dyn.* **40,** 187–211 (2013).
9. Vecchi, G. A. *et al.* Weakening of tropical Pacific atmospheric circulation due to anthropogenic forcing. *Nature* **441,** 73–76 (2006).
10. Ma, J. & Yu, J. Y. Paradox in South Asian summer monsoon circulation change: Lower tropospheric strengthening and upper tropospheric weakening. *Geophys. Res. Lett.* **41,** 2934–2940 (2014).
11. Ogata, T. *et al.* Projected future changes of the Asian Monsoon: A Comparison of CMIP3 and CMIP5 model results. *J. Meteorol. Soc. Japn* **92,** 207–225 (2014).
12. Sooraj, K. P., Terray, P. & Mujumdar, M. Global warming and the weakening of the Asian summer Monsoon circulation: assessments from the CMIP5 models. *Clim. Dyn.* **45,** 1–20 (2014).
13. Turner, A. G. & Annamalai, H. Climate change and the South Asian summer monsoon. *Nat. Clim. Change* **2,** 587–595 (2012).
14. Sinha, A. *et al.* Trends and oscillations in the Indian summer monsoon rainfall over the last two millennia. *Nat. Commun.* **6,** 6309 (2015).
15. Christensen, J. H., Kanikicharla, K. K., Marshall, G. & Turner, J. *Climate Phenomena and Their Relevance for Future Regional Climate Change* (Cambridge University Press, 2013).
16. Sutton, R. T., Dong, B. & Gregory, J. M. Land/sea warming ratio in response to climate change: IPCC AR4 model results and comparison with observations. *Geophys. Res. Lett.* **34,** L02701 (2007).
17. Chou, C. Land–sea heating contrast in an idealized Asian summer monsoon. *Clim. Dyn.* **21,** 11–25 (2003).
18. Wu, G. *et al.* Thermal controls on the Asian summer monsoon. *Scientific Rep.* **2,** 404 (2012).
19. Kamae, Y., Watanabe, M., Kimoto, M. & Shiogama, H. Summertime land–sea thermal contrast and atmospheric circulation over East Asia in a warming climate—Part I: Past changes and future projections. *Clim. Dyn.* **43,** 2553–2568 (2014).
20. Alory, G., Wijffels, S. & Meyers, G. Observed temperature trends in the Indian Ocean over 1960–1999 and associated mechanisms. *Geophys. Res. Lett.* **34,** L02606 (2007).
21. Rao, S. A. *et al.* Why is Indian Ocean warming consistently? *Clim. Change* **110,** 709–719 (2012).
22. Swapna, P., Krishnan, R. & Wallace, J. Indian Ocean and monsoon coupled interactions in a warming environment. *Clim. Dyn.* **42,** 2439–2454 (2014).
23. Roxy, M. Sensitivity of precipitation to sea surface temperature over the tropical summer monsoon region—and its quantification. *Clim. Dyn.* **43,** 1159–1169 (2013).
24. Saha, A., Ghosh, S., Sahana, A. & Rao, E. Failure of CMIP5 climate models in simulating post-1950 decreasing trend of Indian monsoon. *Geophys. Res. Lett.* **41,** 7323–7330 (2014).
25. Sabeerali, C., Rao, S. A., Dhakate, A., Salunke, K. & Goswami, B. Why ensemble mean projection of south Asian monsoon rainfall by CMIP5 models is not reliable? *Clim. Dyn.* **45,** 1–14 (2014).
26. Sharmila, S., Joseph, S., Sahai, A., Abhilash, S. & Chattopadhyay, R. Future projection of Indian summer monsoon variability under climate change scenario: An assessment from CMIP5 climate models. *Global Planet. Change* **124,** 62–78 (2015).
27. Zhou, T., Yu, R., Li, H. & Wang, B. Ocean forcing to changes in global monsoon precipitation over the recent half-century. *J. Climate* **21,** 3833–3852 (2008).
28. Roxy, M. K., Ritika, K., Terray, P. & Masson, S. The curious case of Indian Ocean warming. *J. Climate* **27,** 8501–8509 (2014).
29. Prodhomme, C. *et al.* Impacts of Indian Ocean SST biases on the Indian Monsoon: as simulated in a global coupled model. *Clim. Dyn.* **42,** 271–290 (2014).
30. Sandeep, S. & Ajayamohan, R. Poleward shift in Indian summer monsoon low level jetstream under global warming. *Clim. Dyn.* **45,** 337–351 (2014).
31. Xavier, P. K., Marzin, C. & Goswami, B. N. An objective definition of the Indian summer monsoon season and a new perspective on the ENSO–monsoon relationship. *Q. J. Roy. Meteorol. Soc.* **133,** 749–764 (2007).
32. Gadgil, S., Joshi, N. V. & Joseph, P. V. Ocean-atmosphere coupling over monsoon regions. *Nature* **312,** 141–143 (1984).
33. Danielsen, E. F. In situ evidence of rapid, vertical, irreversible transport of lower tropospheric air into the lower tropical stratosphere by convective cloud turrets and by larger-scale upwelling in tropical cyclones. *J. Geophys. Res.* **98,** 8665–8681 (1993).
34. Dai, A. *et al.* The relative roles of upper and lower tropospheric thermal contrasts and tropical influences in driving Asian summer monsoons. *J. Geophys. Res.* **118,** 7024–7045 (2013).
35. Su, H., Neelin, J. D. & Meyerson, J. E. Sensitivity of tropical tropospheric temperature to sea surface temperature forcing. *J. Climate* **16,** 1283–1301 (2003).
36. Bollasina, M. A. & Ming, Y. The general circulation model precipitation bias over the southwestern equatorial Indian Ocean and its implications for simulating the South Asian monsoon. *Clim. Dyn.* **40,** 823–838 (2013).
37. Gill, A. E. *Some Simple Solutions for Heat-Induced Tropical Circulation* vol. 106 (John Wiley & Sons, Ltd, 1980).
38. Zhou, T. *et al.* Why the western Pacific subtropical high has extended westward since the late 1970s. *J. Climate* **22,** 2199–2215 (2009).
39. Krishnan, R. & Ramanathan, V. Evidence of surface cooling from absorbing aerosols. *Geophys. Res. Lett.* **29,** 54-51–54-54 (2002).
40. Sanap, S., Pandithurai, G. & Manoj, M. On the response of Indian summer monsoon to aerosol forcing in CMIP5 model simulations. *Clim. Dyn.* **2,** 1–13 (2015).
41. Yu, R., Wang, B. & Zhou, T. Tropospheric cooling and summer monsoon weakening trend over East Asia. *Geophys. Res. Lett.* **31,** L22212 (2004).
42. Roxy, M., Tanimoto, Y., Preethi, B., Pascal, T. & Krishnan, R. Intraseasonal SST-precipitation relationship and its spatial variability over the tropical summer monsoon region. *Clim. Dyn.* **41,** 45–61 (2012).
43. Webster, P. J. *The Asian Monsoon* 3-66 (Springer-Verlag, 2006).
44. Bollasina, M. A., Ming, Y. & Ramaswamy, V. Anthropogenic aerosols and the weakening of the South Asian summer monsoon. *Science* **334,** 502–505 (2011).
45. Ramanathan, V. *et al.* Atmospheric brown clouds: Impacts on South Asian climate and hydrological cycle. *Proc. Natl Acad. Sci. USA* **102,** 5326–5333 (2005).
46. Meehl, G. A., Arblaster, J. M. & Collins, W. D. Effects of black carbon aerosols on the Indian monsoon. *J.Climate* **21,** 2869–2882 (2008).
47. Menon, S., Hansen, J., Nazarenko, L. & Luo, Y. Climate effects of black carbon aerosols in China and India. *Science* **297,** 2250–2253 (2002).
48. Lau, K. M. & Kim, K. M. Observational relationships between aerosol and Asian monsoon rainfall, and circulation. *Geophys. Res. Lett.* **33,** L21810 (2006).
49. Dong, L., Zhou, T. & Wu, B. Indian Ocean warming during 1958–2004 simulated by a climate system model and its mechanism. *Clim. Dyn.* **42,** 203–217 (2014).
50. Ning, H., Li-Juan, L. & Bin, W. The role of the aerosol indirect effect in the northern Indian Ocean warming simulated by CMIP5 models. *Atmos. Oceanic Sci. Lett.* **7,** 411–416 (2014).
51. Du, Y. & Xie, S. P. Role of atmospheric adjustments in the tropical Indian Ocean warming during the 20th century in climate models. *Geophys. Res. Lett.* **35,** L08712 (2008).
52. Compo, G. P. & Sardeshmukh, P. D. Removing ENSO-related variations from the climate record. *J. Climate* **23,** 1957–1978 (2010).
53. Du, Y., Xie, S.-P., Huang, G. & Hu, K. Role of air-sea interaction in the long persistence of El Niño-induced North Indian Ocean warming. *J. Climate* **22,** 2023–2038 (2009).
54. Lau, N. C. & Nath, M. J. Impact of ENSO on the variability of the Asian-Australian monsoons as simulated in GCM experiments. *J. Climate* **13,** 4287–4309 (2000).
55. Chowdary, J. & Gnanaseelan, C. Basin-wide warming of the Indian Ocean during El Niño and Indian Ocean dipole years. *Int. J. Climatol.* **27,** 1421–1438 (2007).
56. Rahul, S. & Gnanaseelan, C. Net heat flux over the Indian Ocean: trends, driving mechanisms, and uncertainties. *Geosci. Remote Sensing Lett.* **10,** 776–780 (2013).
57. Ashok, K., Sabin, T., Swapna, P. & Murtugudde, R. Is a global warming signature emerging in the tropical Pacific? *Geophys. Res. Lett.* **39,** 02701 (2012).
58. Kumar, A., Jha, B. & L'Heureux, M. Are tropical SST trends changing the global teleconnection during La Niña? *Geophys. Res. Lett.* **37,** L12702 (2010).
59. Goswami, B. N., Madhusoodanan, M., Neema, C. & Sengupta, D. A physical mechanism for North Atlantic SST influence on the Indian summer monsoon. *Geophys. Res. Lett.* **33,** L02706 (2006).
60. Kucharski, F., Bracco, A., Yoo, J. & Molteni, F. Atlantic forced component of the Indian monsoon interannual variability. *Geophys. Res. Lett.* **35,** L04706 (2008).
61. Prodhomme, C., Terray, P., Masson, S., Boschat, G. & Izumo, T. Oceanic factors controlling the Indian Summer Monsoon onset in a coupled model. *Clim. Dyn.* **44,** 977–1002 (2015).
62. Dash, S. K. *et al.* Projected seasonal mean summer monsoon over India and adjoining regions for the twenty-first century. *Theor. Appl. Climatol.* 1–13, doi:10.1007/s00704-014-1310-0 (2014).
63. Hoerling, M. & Kumar, A. The perfect ocean for drought. *Science* **299,** 691–694 (2003).
64. Ummenhofer, C. C., D'Arrigo, R. D., Anchukaitis, K. J., Buckley, B. M. & Cook, E. R. Links between Indo-Pacific climate variability and drought in the Monsoon Asia Drought Atlas. *Clim. Dyn.* **40,** 1319–1334 (2013).
65. Luo, J.-J., Sasaki, W. & Masumoto, Y. Indian Ocean warming modulates Pacific climate change. *Proc. Natl Acad. Sci. USA* **109,** 18701–18706 (2012).
66. Pai, D., Sridhar, L., Badwaik, M. & Rajeevan, M. Analysis of the daily rainfall events over India using a new long period (1901–2010) high resolution (0.25×0.25) gridded rainfall data set. *Clim. Dyn.* 1–22, doi:10.1007/s00382-014-2307-1 (2014).
67. Kendall, M. G. *Rank Correlation Methods* 1 edn (ed. Griffin, C.) (Oxford, 1948).

68. Wang, B. & Ding, Q. Changes in global monsoon precipitation over the past 56 years. *Geophys. Res. Lett.* **33,** L0671 (2006).
69. Saha, S. *et al.* The NCEP climate forecast system version 2. *J. Climate* **27,** 2185–2208 (2013).

Acknowledgements

We gratefully acknowledge the financial support given by the Earth System Science Organization, Ministry of Earth Sciences, Government of India, to conduct this research under the National Monsoon Mission (Grant #MM/SERP/CNRS/2013/INT-10/002, Contribution #MM/PASCAL/RP/03).

Author contributions

M.K.R. designed the study and the model experiment. M.K.R. and K.R. performed the model sensitivity experiments and the analysis. All authors contributed ideas in developing the research, discussed the results and wrote the paper.

Additional information

Accession codes: The climate model code utilized in this study, the NCEP Climate Forecast System version 2, is available at the 'Monsoon Desk' at NOAA's National Centres for Environmental Prediction (NCEP), as part of the National Monsoon Mission setup by the Ministry of Earth Sciences, Government of India [url: http://www.emc.ncep.noaa.gov/index.php?branch=monsoondesk].

Competing financial interests: The authors declare no competing financial interests.

7

Constraints on soluble aerosol iron flux to the Southern Ocean at the Last Glacial Maximum

T.M. Conway[1,2,†], E.W. Wolff[1,2], R. Röthlisberger[2,†], R. Mulvaney[2] & H.E. Elderfield[1]

Relief of iron (Fe) limitation in the Southern Ocean during ice ages, with potentially increased carbon storage in the ocean, has been invoked as one driver of glacial-interglacial atmospheric CO_2 cycles. Ice and marine sediment records demonstrate that atmospheric dust supply to the oceans increased by up to an order of magnitude during glacial intervals. However, poor constraints on soluble atmospheric Fe fluxes to the oceans limit assessment of the role of Fe in glacial-interglacial change. Here, using novel techniques, we present estimates of water- and seawater-soluble Fe solubility in Last Glacial Maximum (LGM) atmospheric dust from the European Project for Ice Coring in Antarctica (EPICA) Dome C and Berkner Island ice cores. Fe solubility was very variable (1–42%) during the interval, and frequently higher than typically assumed by models. Soluble aerosol Fe fluxes to Dome C at the LGM (0.01–0.84 mg m^{-2} per year) suggest that soluble Fe deposition to the Southern Ocean would have been $\geq$10 $\times$ modern deposition, rivalling upwelling supply.

[1] Department of Earth Sciences, University of Cambridge, Cambridge CB2 3EQ, UK. [2] British Antarctic Survey, High Cross, Madingley Road, Cambridge CB3 0ET, UK. † Present addresses: Department of Earth Sciences, Institute for Geochemistry and Petrology, ETH Zürich, 8092 Zürich, Switzerland (T.M.C.); Federal Office for the Environment, Bern, Switzerland (R.R.). Correspondence and requests for materials should be addressed to T.M.C. (email: conway.tm@gmail.com).

Changes in atmospheric iron (Fe) supply to the oceans (described here as aerosol Fe) during glacial intervals have been suggested as one of a number of possible drivers for changes observed in atmospheric CO_2 between glacial and interglacial climate states (The Iron Hypothesis)[1–3]. Fe fertilization of the presently Fe-limited Southern Ocean during glacial times could be responsible for up to 40 p.p.m.v.[3–5] of the regular 80–120 p.p.m.v. cycles in atmospheric CO_2 that coincide with glacial–interglacial cycles over at least the last 800,000 years[6]. Increased atmospheric Fe supply to the glacial Southern Ocean, where today low dissolved Fe concentrations limit phytoplankton growth[4,7], may have enhanced the efficiency of the 'biological pump', retaining more carbon in the deep ocean and driving a reduction in pCO_2 (refs 1,2). While changes to ocean circulation and deep ocean carbon storage may play a dominant role in glacial–interglacial cycling[3,8–10], atmospheric Fe may also strongly influence the efficiency of the biological pump on both glacial and millennial timescales[3,11,12], with potentially large effects on the carbon cycle. Indeed, a recent study showed that increased nutrient utilization and greater productivity in the Sub-Antarctic are linked to periods of increased dust (and Fe) supply over the past 160,000 years[13], highlighting atmospheric Fe fertilization in the Sub-Antarctic surface ocean as a candidate for explaining millennial-scale fluctuations in atmospheric CO_2. Through this mechanism, Fe fertilization may play an important role in glacial–interglacial cycling, in combination with other processes[9,13,14]. Evidence for higher dust fluxes comes from Antarctic ice cores, which record marked increases in atmospheric dust deposition to Antarctica during colder intervals over the past 800,000 years[15], most likely due to expanded dust source areas in South America and enhanced atmospheric lifetime and transport[16]. Sedimentary records also show that the temporal pattern of Antarctic ice-core dust fluxes is representative of dust reaching the Southern Ocean, with the dust–climate coupling extending back at least 4 million years[5]. However, modelling efforts to understand the sensitivity of the oceans and climate to Fe (for example, ref. 17) are still limited by poor constraints on aerosol Fe solubility and the fluxes of 'bioavailable' (the Fe that can be taken up for growth by phytoplankton) aerosol Fe to oceans during glacial times.

Converting ice and sedimentary dust records to bioavailable aerosol Fe fluxes is not straightforward. Previous studies have used dust flux records with fixed composition and Fe solubility[18], or non-seasalt Ca (nss-Ca) as proxies for dissolved Fe (ref. 11). Recently, measurements of Fe concentration and fluxes in ice cores have become available[19–23]; however, these were made under acidic conditions (pH 1–2), and are thus far-removed from dissolution in seawater (pH ~8) where Fe solubility can be $<1\%$ (ref. 18). Fe solubility is strongly pH dependent (Supplementary Discussion), and acid leaches may access a variable fraction of the total Fe present[21,24]. Equally, dissolved Fe concentrations within melted Antarctic ice of glacial age (for example, 1–70 ng g^{-1} in this study) are typically well above the low-equilibrium saturation concentration of Fe(III) (<0.1 ng g^{-1}; Supplementary Fig. 1), meaning that Fe loss due to precipitation/adsorption of Fe hydroxides is likely under weakly acidic pH conditions (Supplementary Discussion). While Fe(II) is more soluble than Fe(III) under these conditions[25], a recent study suggests that a large proportion of Fe within ice-core material is present as Fe(III) (ref. 26). Any loss of Fe(III) via precipitation or adsorption to plastic equipment during processing that is not recovered during acidification will lead to underestimation of soluble Fe, which might affect studies analysing Fe after melting using discrete bottled samples or flow-through systems (for example, refs 19,20,23,26). Thus, while previous ice-core records provide valuable informatio

fluxes, they provide incomplete information about soluble or bioavailable aerosol Fe supplied to the oceans. We were therefore motivated to seek techniques for more representative estimates of soluble Fe flux to the Southern Ocean during glacial intervals to allow models to more accurately assess the effect of changing atmospheric Fe supply on the carbon cycle.

Here we present high-resolution fluxes of total and soluble atmospheric Fe, as well as calculated aerosol Fe solubility, from the Last Glacial Maximum (LGM; 21–26 kyr before present (BP)) of the EPICA Dome C ice core[15,20] from East Antarctica (EDC; 75° 06′ S 123° 21′ E, 3,233 m.a.s.l.; Fig. 1) and from Marine Isotope Stage (MIS) 2-3 (22–52 kyr BP) from an ice core from Berkner Island (79° 32.9′ S 45° 32.9′ E; Fig. 1). Soluble Fe was determined using a novel rapid-filtration technique, in meltwater at pH ~5.3 (the natural pH of the melted ice) or following leaching of sublimated dry dust with natural seawater, analogous to instantaneous techniques used with modern aerosols[27]. Our data are much more representative of dissolution of atmospheric dust in the ocean than previous ice-core studies, avoiding both unrepresentative leaching by strong acids[19,20,22,23] or precipitation of insoluble Fe(III) hydroxides during melting. These methods are relevant for the possible modes of deposition of dust into the ocean, either deposition within rain, where dust has already been exposed to liquid at pH ~5.3, or deposition of dry dust or solid ice/snow containing dust to the ocean where the first liquid encountered is seawater. We find that both total aerosol Fe fluxes and aerosol Fe solubility were very variable throughout the LGM in EDC ice (1–42% soluble; pH ~5.3 mean 10%, median 6%) leading to highly variable soluble aerosol Fe fluxes of 0.01–0.84 mg m^{-2} per year at Dome C throughout the interval. At Berkner Island, although aerosol Fe solubility was lower than Dome C (mean ~3%), higher dust fluxes resulted in similar mean soluble aerosol Fe fluxes to those at Dome C during the LGM. While these calculated soluble Fe fluxes are not fully representative of bioavailable Fe fluxes to the Southern Ocean, they enable an improved estimate of the atmospheric Fe that would have been deposited and dissolved in seawater at the LGM compared with previous studies. Our measured aerosol Fe solubilities and soluble aerosol Fe fluxes are higher than typically used in models of ocean biogeochemistry, and thus we suggest that models should consider higher and more spatially and temporally variable soluble atmospheric Fe fluxes to the Southern Ocean during glacial intervals.

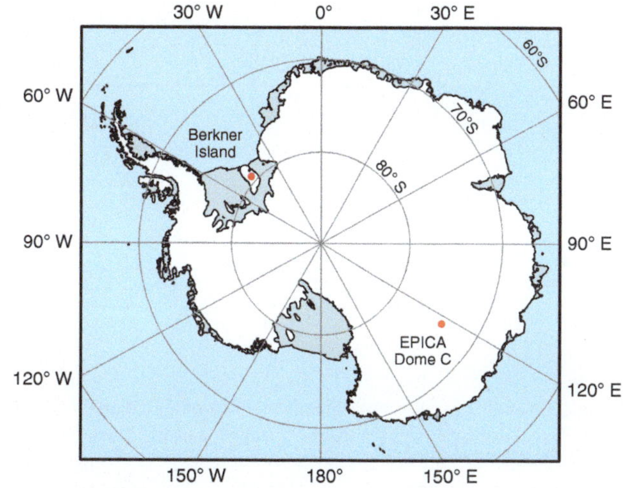

Figure 1 | Th

Results

Figure summary. Aerosol Fe solubility and calculated total and soluble Fe fluxes from EDC and Berkner ice cores are shown compared with other ice-core parameters in Figs 2–4, with a comparison of Fe solubility in EDC ice of LGM age at pH ~5.3 and in seawater shown in Fig. 2b. The data are also available in Supplementary Data.

Aerosol Fe solubility in EDC and Berkner ice cores. Fe solubility in atmospheric dust deposited at Dome C across the LGM had a mean of 10% ($n=23$) and 17% ($n=11$) at pH ~5.3 and in seawater, respectively, and a median of 6% at pH ~5.3 (Fig. 2 and Supplementary Data). These aerosol Fe solubilities are higher than solubility of loess and desert dust (<0.5%; refs 18,28) or glacial flour (2–3%; ref. 28), but consistent with high-end solubility estimates for modern atmospheric dust, where transport-induced size/compositional fractionation and chemistry enhance solubility away from source regions[29,30]. Indeed, dust dry-deposited[20] at Dome C during the LGM represents the distal fine-fraction of dust leaving the source regions on Patagonia[31]. EDC dust is enriched in both Al and Fe compared with both the crust and Patagonian topsoil (Supplementary Data), suggesting that high Fe solubility at Dome C could be driven by a compositional bias towards smaller clay particles containing more readily soluble Fe[32] during atmospheric transport[30,33]. By contrast, mean Fe solubility at Berkner Island, closer to South American dust sources[18], was only 3% during 23–50 kyr BP (Fig. 2c) and 1–5% at the LGM.

Fe solubility in EDC samples was very variable across the interval (Fig. 2a; 1–42%). At several points within the LGM, we measured samples of very similar ages (Fig. 3), showing that variability sometimes also occurs at annual–decadal scales (Figs 2a and 3); for example, Fe solubility in samples at 23,350 and 23,352 kyr BP were 4 and 14%, and samples at 23,391 and 23,396 kyr BP were 17 and 6%, respectively (Supplementary Data and Fig. 3). Variability in Fe solubility through the LGM could be driven[24,29] by seasonal changes in dust supply or discrete dust events, but is notable in ice samples, which by their nature represent combined dust deposition over several years. The observed variability in Fe solubility has no clear relationship with total Fe, dust loading or particle size (Fig. 4), suggesting that, if real, it is driven either by particle composition or transport-induced changes in Fe speciation. Similar EDC temperature (Fig. 2a) and assumed atmospheric transport pathways[16], combined with a homogenous grain-size distribution, throughout the interval (Fig. 4), suggest that variability is more likely the result of compositional and Fe-speciation changes than transport-induced changes in Fe chemistry. Mineralogy exerts a strong control on Fe solubility, with Fe in clay and silicate minerals much more soluble than Fe hydroxides[32], and glacial flour enriched in both Fe(II) and clays compared with loess[28]. While higher solubility may be driven by relative enrichment of clay minerals with transport[29,33], variable solubility could be the result of changing contributions from different types of South American dust (for example, loess or glacial material[34]). Contributions from local dust sources is unlikely at Dome C,

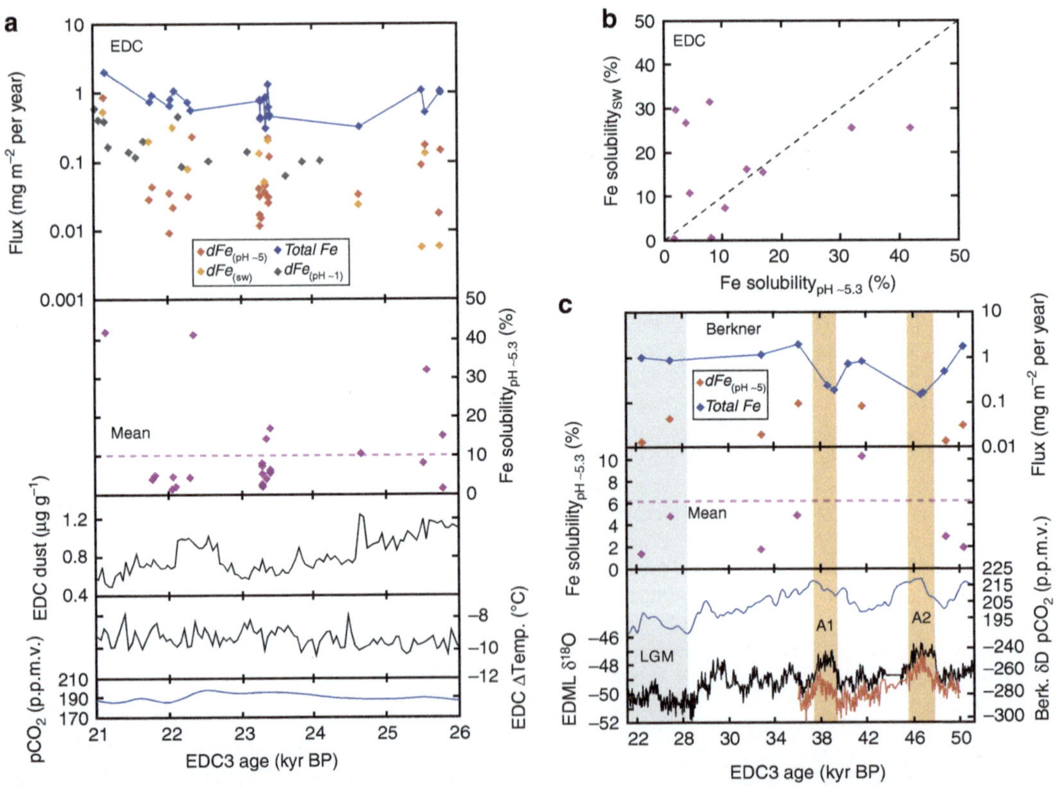

Figure 2 | Variability in fluxes and chemistry of aerosol Fe deposited to Antarctica shown with other climatic parameters. (**a**) Fluxes of total and soluble (pH ~5.3, seawater, pH ~1 (ref. 20)) aerosol Fe, and Fe solubility across the LGM interval from the EPICA Dome C core. (**b**) Comparison of aerosol Fe solubility at pH ~5.3 and in natural seawater (pH ~8) from EDC ice of identical LGM age. Dashed line indicates 1:1. (**c**) Fluxes of total and soluble (pH ~5.3) aerosol Fe and Fe solubility from MIS 2-3 of the Berkner Island Core. The grey bar denotes the LGM and the orange bars denote Antarctic warm events. Data points represent two measurements of a single sample, and errors are smaller than the size of points. All data are shown on the EDC3 age scale[47]. See Methods section for details of other climatic parameters, and details of conversion of Berkner age to EDC3.

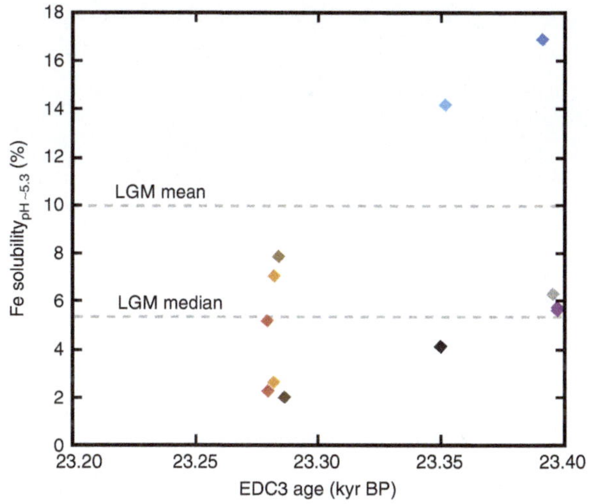

Figure 3 | Fe solubility at pH ~5.3 in EPICA Dome C ice of LGM age at high resolution. Ice of identical/similar age was subsampled (Supplementary Data) to investigate variability in Fe solubility. Ice samples of different age are shown in different colours, with three ice samples sampled twice (red 23.280 kyr BP, orange 23.282 kyr BP and purple 23.398 kyr BP; Supplementary Data). Data points represent two measurements of a single sample, and errors are smaller than the size of points. Dashed lines denote the mean and median of all EDC LGM measurements from Fig. 2.

where most glacial-age dust is considered to be from Patagonia[31], but they could play a role at Berkner Island.

Comparison of solubility at pH ~5.3 and in seawater (Fig. 2b) also shows scatter about a 1:1 line, similar to that observed in modern studies comparing ultrapure water and seawater rapid-leaching techniques[27]. This scatter may reflect heterogeneous particle distribution and chemistry or could partly be the result of differences between the two methods, such as temperature, number of leaches, or solution chemistry enhancing dissolution in seawater. Similarly, while comparison of pH ~5.3 leaching between three samples of identical age showed broadly similar solubility, there was also some variability (6 and 6%, 2 and 5% and 3 and 7%; Fig. 3). It is therefore possible that the full observed variability between samples in this study is not real; however, if anything, we would expect methodological concerns to result in underestimation of Fe solubility due to incomplete leaching or precipitation, which does not change our primary observation of high aerosol Fe solubility at times within LGM ice.

LGM aerosol Fe fluxes at Dome C and Berkner Island. Decadal to centennial variability in both Fe solubility and total Fe fluxes throughout the LGM at Dome C (Fig. 2a; 0.3–2 mg m^{-2} per year; mean 0.76 ± 0.4 mg m^{-2} per year) combines to mean that soluble Fe fluxes show great variability (Fig. 2a; 0.01–0.84 mg m^{3} per year), with mean values of 0.09 ± 0.17 mg m^{-2} per year (1 s.d.) for pH ~5.3 soluble Fe and 0.15 ± 0.13 mg m^{-2} per year (1 s.d.) for seawater-soluble Fe. Despite the lower Fe solubility at Berkner Island compared with Dome C, higher atmospheric Fe fluxes at the LGM (0.95 ± 0.07 mg m^{-2} per year), probably attributable to greater dust deposition owing to closer proximity to South American dust sources[18], result in similar mean soluble Fe fluxes (0.03 ± 0.02 mg m^{-2} per year) compared with those at Dome C. Total Fe fluxes at Berkner Island were almost an order of magnitude lower during Antarctic Warm Events A1 and A2 than during colder periods (Fig. 2c), reflecting lower regional dust

Discussion

The aerosol Fe flux data presented here can be compared with previous ice-core estimates of Fe fluxes to Antarctica and the nearby Southern Ocean, during the LGM. We find that our total EDC aerosol Fe fluxes (5.5–36 μmol m^{-2} per year; mean 13.7 μmol m^{-2} per year) are considerably higher than previous estimates based on acidic (pH ~1–2) leaching or mineral dust proxies at several Antarctic locations (4.3–8 μmol m^{-2} per year; refs 22,23), including those based on acidic leaches at EDC with a conversion to total Fe (refs 19,22; 6.6 μmol m^{-2} per year). The difference may partially reflect the high-resolution variability in Fe flux at EDC, but also the fact that, similar to modern dust[24], an acidic leach represents a variable portion of the total Fe, dependent on dust composition and leaching timescale. While pH ~1 Fe EDC fluxes are generally higher than estimates at pH 5–8 (Fig. 2a), the two ranges overlap, meaning that acidic leaches of ice core do not represent either total or soluble aerosol Fe, although they may capture broad-scale trends. This offset between acidic leaches and total digested Fe may be partially to do with methodology; the pH 1 EDC Fe records were measured in samples that were melted, acidified and only left to dissolve for >24 h (refs 19,20), while other studies have shown that Fe continues to dissolve at pH 1–2 for up to 3 months[22,35]. As a result, these short timescale acidic leaches in EDC ice samples may not have fully captured the pH 1 leachable fraction that is present. Previous comparison of pH 1 leached methods with digestion procedures on EDC ice suggested that 30–65% of total Fe was recovered by pH 1 leaching methods[19,21], but it seems likely that the proportion accessed by these short-term acidic leaches would vary with dust composition and mineralogy. Also, even with 1–3 month leaching times, acidic leaches are unlikely to fully dissolve dust particles.

Our data also allow evaluation of the use of other proxies for ice core Fe (Fig. 5). We find that both nss-Ca[11] (Ca with Na-calculated seasalt component removed) and total Al are good proxies for the flux of total Fe (Fig. 5). However, taking the EDC dust flux and calculating total Fe concentrations, assuming a constant crustal Fe percentage of ~4% in dust[18] underestimates total Fe for EDC samples, which are enriched in Fe (mean 8 wt%; Supplementary Data) compared with crustal values of 2–3% (ref. 36) or Patagonian topsoil (2–6%; ref. 37). EDC samples are similarly enriched for Al (mean 15 wt%; Supplementary Data), compared with crustal values of 8% (ref. 36) and Patagonian topsoil of 6–10% (ref. 37). This difference in the dust composition presumably relates to compositional-fractionation processes during transport, such as relative enrichment in silicate minerals, such as clays and feldspars[33]. As such, changes in composition and mineralogy may be spatially variable and site specific for ice-coring locations, complicating efforts to use dust, Ca or Al fluxes as proxies for total Fe. In addition, despite the possible usefulness of these proxies for providing information about timing and magnitude of changes in atmospheric Fe[11], the lack of relationship between total and soluble Fe (Fig. 5c) means that they do not approximate soluble or bioavailable Fe. Instead, direct measurements of soluble Fe, together with biological experiments, are needed for constraining changes in soluble and bioavailable Fe supply to the oceans during climatic change and especially within stable climate intervals. However, we do note the strong covariance between ice-core dust and Fe fluxes and both Sub-Antarctic productivity and heavier Sub-Antarctic surface nitrogen isotope ratios (indicative of increased nutrient utilization) during the last 160 kyr (ref. 13), as well as a similar covariance between ice-core dust and (pH 1) Fe fluxes, sediment Fe mass accumulation rates and Sub-Antarctic export productivity over the last eight

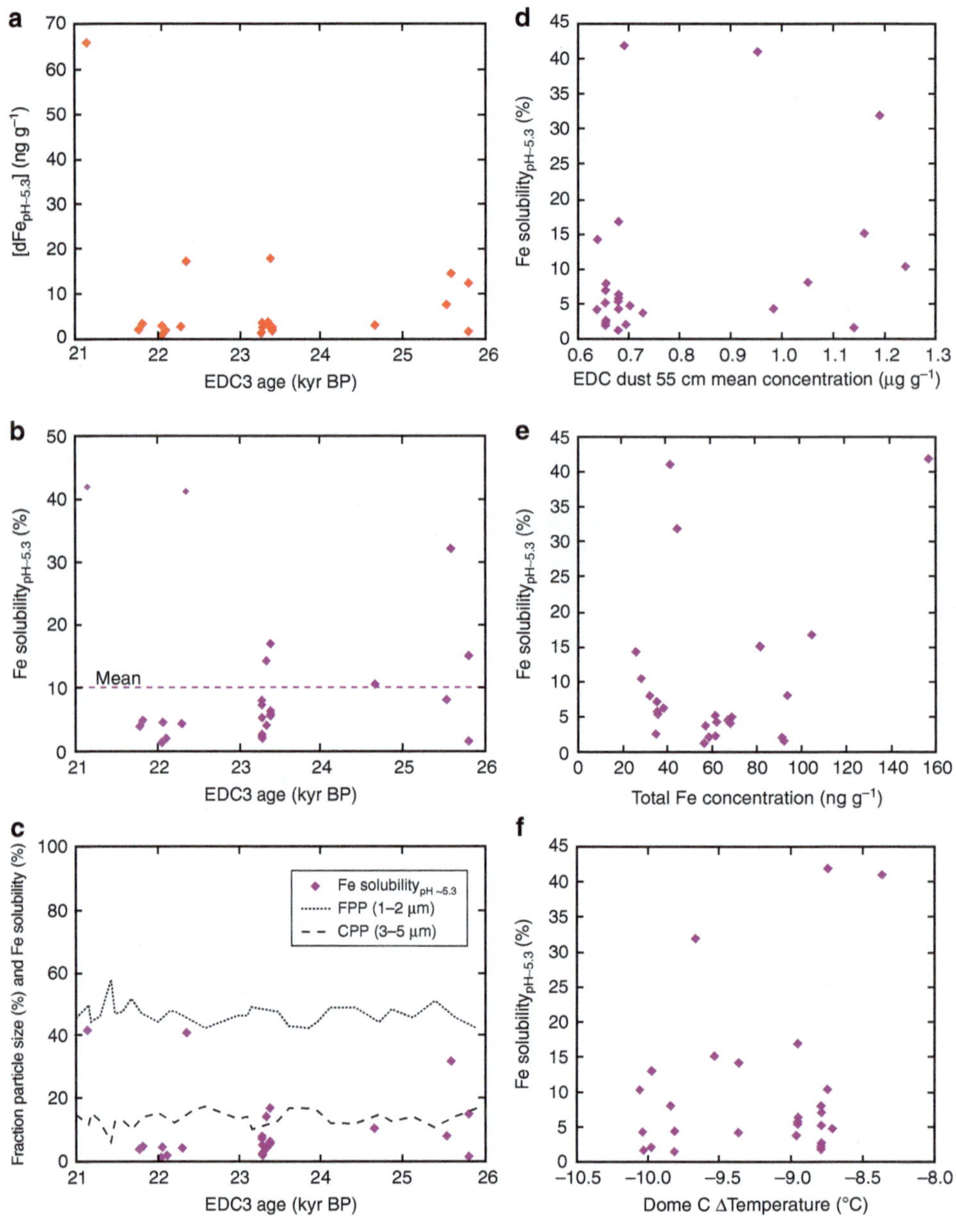

Figure 4 | Variability in EPICA Dome C over the LGM. (**a**) pH ~5.3 dissolved Fe concentration, (**b**) pH ~5.3 Fe solubility and (**c**) pH ~5.3 Fe solubility shown with grain size (fine particle, FPP; coarse particle, CPP) from coulter-counter analysis[15,70]. The relationship between pH ~5.3 Fe solubility and (**d**) dust concentration[15], (**e**) Total Fe concentration (this study), (**f**) Dome CΔ air temperature[68]. Data points represent two measurements of a single sample, and errors are smaller than the size of points.

ice-core Fe fluxes do capture the order of magnitude changes in Fe supply to the Southern Ocean that are important for understanding millennial-scale climate and pCO_2 variability.

By comparison with estimates of Holocene Fe fluxes to Antarctica[22], our mean total Fe flux estimates (13.7 μmol m^{-2} per year) represent a ~30–50× increase over EDC (0.3–0.4 μmol m^{-2} per year), and a ~10–20× increase over coastal sites such as Law Dome (ref. 22; 0.8–1.1 μmol m^{-2} per year). The modern gradient in dust deposition between coastal sites and EDC is thought to be largely influenced by wet deposition, which dominates at coastal sites and is a minor factor inland, and thus might be somewhat smoothed out during drier glacial intervals[22]. In terms of soluble Fe, the mean EDC seawater-soluble Fe flux of 2.7 μmol m^{-2} per year represents a >10-fold increase over modern soluble Fe fluxes (even assuming a similar high solubility in modern dust), and is ~10× higher than modern calculated Southern Ocean atmospheric Fe fluxes (0.2 μmol m^{-2} per year; ref. 4). This suggests that the soluble atmospheric Fe flux to the Southern Ocean increased by at least a factor of 10 at the LGM. Such an Fe flux would be broadly equivalent to the expected upwelling supply of deep Fe, which would be reduced from calculated modern values of 8–16 μmol m^{-2} per year due to decreased circulation and stratification due to increased sea ice[2,4]. The range in soluble Fe fluxes observed in this study (0.2–15 μmol m^{-2} per year) is also similar to modelled modern upwelling Fe flux values for entrainment (deep winter mixing; 9.1–33 μmol m^{-2} per year) and vertical diffusion (1.2–3.6 μmol m^{-2} per year), which have been suggested to sustain surface water Fe concentrations in the modern Southern Ocean[39].

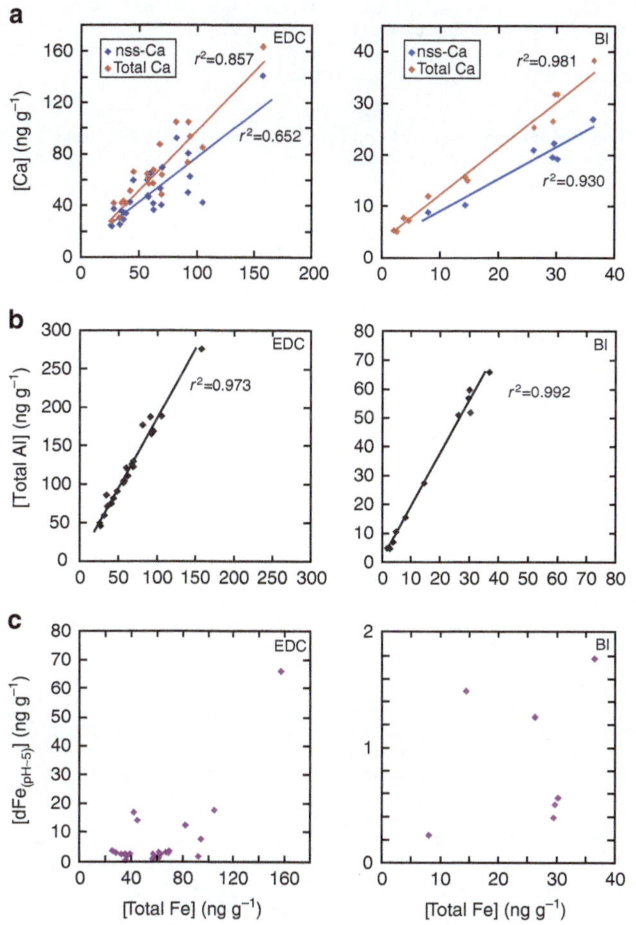

Figure 5 | Direct comparison of different parameters as proxies for aerosol Fe in ice cores. Total Fe concentration, in EDC ice of LGM age (left panel) and Berkner Island (BI) ice from MIS 2-3 (right panel), is shown plotted against (**a**) nss-Ca and total Ca concentration, (**b**) total Al concentration and (**c**) pH ~5.3 dissolved Fe concentration. Best-fit linear regression lines with r^2 values are shown. Data points represent two measurements of a single sample, and errors are smaller than the size of points.

Of course, upwelling of deep water supplies both Fe and carbon, and it is the proportion of this carbon that is utilized by primary productivity in surface waters, and thus retained in the ocean, which determines the surface Southern Ocean's role in setting pCO_2. While the surface ocean is Fe-limited (insufficient bioavailable Fe to fully utilize all the upwelled nutrients), surface waters can act as a source for pCO_2. Deposition of atmospheric dust only supplies Fe and thus provides the means to increase utilization of upwelled carbon and nutrients in surface waters. A full understanding of the role of atmospheric Fe supply in carbon cycling of the past thus requires not only knowledge of upwelling and dust deposition Fe fluxes but also the resulting change in the carbon to bioavailable Fe concentration ratio in surface waters, as well as how both Fe and carbon are recycled and/or taken to depth. However, models do suggest that an increase in bioavailable atmospheric Fe supply of 5–10 × could double export production[40], meaning that the LGM EDC soluble Fe flux could be large enough to drive big changes in primary productivity, nutrient utilization and carbon sequestration, especially in the Sub-Antarctic zone of the Southern Ocean[9,13]. Although a recent ocean Fe biogeochemical modelling study[17]

globally, lower than the up to 40 p.p.m.v. suggested by recent Sub-Antarctic productivity/nutrient utilization records[13], the model used a Fe solubility of only 2% (ref. 17), highlighting the need for constraints on soluble and bioavailable fluxes during glacial intervals.

In conclusion, our EDC soluble Fe fluxes suggest that soluble Fe fluxes to the Southern Ocean during the LGM and glacial times were likely higher than previously considered. Both Fe content (8 wt%) and mean Fe solubility (pH ~5.3 10%; seawater 17%) in EDC dust are greater than typically assumed by models (~1–10%; refs 17,18,41–44), suggesting that models may underestimate soluble (and therefore bioavailable) Fe fluxes to the Southern Ocean at the LGM. This might especially be the case in modelling studies where 0.5–2% Fe solubility is used[17,41–43]. To address this, the median LGM pH ~5.3 Fe solubility from this study (6%) is simplest for consideration in models, but the rapid variability in Fe fluxes on short timescales and higher solubilities should also be considered. We attribute variability in aerosol Fe solubility to changes in the composition and fluxes of atmospheric dust arriving at EDC, even on decadal timescales. Therefore, to fully capture the influence of atmospheric Fe on glacial–interglacial cycling, models must take into account both temporal and spatial variability in dust composition and Fe solubility. Closer to source regions, higher dust fluxes may offset lower Fe solubility, and short-term changes in the contribution of different dust types could have important effects on surface ocean biogeochemistry. We anticipate that these better general constraints on the soluble Fe flux to the Southern Ocean will be useful for modellers determining Fe's role in marine biogeochemical cycles over glacial–interglacial changes, as well as providing insight into the response of the oceans to anthropogenic CO_2 release.

Methods

General laboratory procedures and equipment cleaning. All ice-preparation was carried out in a cold room (−20 °C) at the British Antarctic Survey, Cambridge, and all sample processing and analysis were carried out in clean laboratories at the University of Cambridge. Seawater analysis was carried out at Old Dominion University, USA. All clean work was carried out within a Class 1,000 HEPA-filtered clean laboratory and/or under Class 100 laminar flow air.

All water used was ultrapure (18.2 MOhm cm) and acids were either quartz-distilled (QD) to high purity from reagent grade or purchased as ultrapure (Fisher Scientific). Strict trace metal protocols were used, including wearing polyethylene (PE) gloves to minimize metal contamination during cleaning, processing and sampling. All plastic equipment was rigorously acid-cleaned following the procedures outlined below, which were based on previously published work[45] or adapted from routine cleaning procedures in the Department of Earth Sciences at Cambridge.

New perfluroalkoxy (PFA) or fluorinated ethylene propylene (FEP) Teflon filter rigs, vials, jars, bottles and micro centrifuge tubes were cleaned by sequential immersion in weak (~5%) Decon detergent (24 h), warm 50% (v/v) reagent-grade HNO_3 (24–48 h) and warm 50% (v/v) reagent-grade HCl (24–48 h). Teflon equipment was re-cleaned by immersion in either warm 50% (v/v) HNO_3 or HCl (24 h). All Teflon equipment was thoroughly rinsed with ultrapure water following each step and then left to dry on a clean drying rack under laminar flow. For any Teflon equipment that was to be in contact with ice or samples at near-neutral pH, the plastic was filled with ultrapure water to condition the surface. Nalgene low-density PE and polycarbonate (PC) bottles were cleaned following previously published cleaning methods for preparation of equipment for seawater sampling[45]: bottles were sequentially immersed in weak (~5%) Decon detergent (48 h), 1 M reagent-grade HCl (48 h), 6 M reagent-grade HCl (48 h) and ~1% (v/v) QD-HCl (>3 months), with extensive rinsing with ultrapure water between steps. Bottles were conditioned to neutral pH by filling with ultrapure water for 1 week before use. Whatman Nucleopore PC membranes for filtration (0.2 μm) were cleaned by immersion in 6 M QD-HCl (48 h) and then rinsed extensively with ultrapure water. Membranes were always handled with clean PFA tweezers to minimize contamination. Membranes were stored in and flushed with ultrapure water before use to condition the plastic to near-neutral pH for contact with ice samples. Pipette tips were either acid-cleaned[45] with 50% (v/v) reagent-grade HCl (24 h) and rinsed with ultrapure water (seawater samples), or rinsed twice with 10% (v/v) QD-HNO_3 and twice with ultrapure water before use (non-seawater samples).

sequential immersion in 6 M reagent-grade HCl (24 h) and 1 M reagent-grade HCl (24 h), followed by extensive rinsing with ultrapure water. PTFE boards were cleaned by immersion in 6 M reagent-grade HCl and then rinsed extensively with ultrapure water. Arm-length PE gloves for handling ice were cleaned by immersion in weak (~0.01% v/v) QD-HCl (24 h) and rinsed thoroughly with ultrapure water to ensure no transfer of acid to samples.

Ice samples and cutting and decontamination. Detailed information about the age of ice sampled in this study is shown in Supplementary Data. EDC samples from 21 to 26 kyr BP were chosen to cover the very high dust portion of the LGM (~0.6–1.2 mg kg^{-1} dust; ref. 15) based on previous work[15,46]. 'A cuts' (or ~$\frac{1}{4}$ of the core) from 15 separate 'bags' of ice (55 cm vertical section; ~50 years) from the EDC ice core were supplied by EPICA (see EPICA cutting plan in Supplementary Fig. 2a). For EDC sampling, ~3 cm pieces were typically cut from either the bottom or top of each bag, using a steel bandsaw within the cold room (−20 °C) at the British Antarctic Survey. Two vertical cuts were then made to create three subsamples of identical age (Supplementary Fig. 2b), which were then taken for decontamination and processing to generate subsamples of identical age for total or soluble analysis. The sampling resolution was therefore typically 50 years, although each sample represents only ~3 years. Four bags (959, 984, 986 and 987) were sampled at higher resolution (two to four 3 cm subsamples per 50 years) to investigate higher-resolution variability as well as reproducibility between ice subsamples of identical age. Precise age for each 3 cm sample was calculated from depth within each bag and the EDC3 age scale[47]. Berkner samples were chosen to cover MIS 2–3, with two samples within the LGM (23 and 26 kyr BP) and 10 samples across the 'A' events[48] of MIS 3 (36–50 kyr BP) based on the Berkner δD record (Mulvaney, unpublished data). Ice was supplied as $\frac{1}{4}$ pieces of Berkner core (diameter 50 mm), with 3 cm vertical pieces cut and subsampled analogous to EDC samples.

The outside of ice core material is typically unavoidably contaminated with trace metals from drilling, cutting with bandsaws and handling with gloves[49]. Thus, the outside of the ice must be decontaminated to remove outer layers where contamination overwhelms the natural signal. Previously, this has been carried out using a melt-head[50], polyethylene lathe[51], stainless steel blade[52], ceramic blade or chisels[53]. Decontamination may require removal of several cm of core[22,51,52,54] during low-dust or less-compacted intervals, or when elements of interest are only present at pg g^{-1} levels. Preliminary testing in this study demonstrated that, as expected, ultrapure water ice was severely contaminated following cutting with a bandsaw (20 ng g^{-1} Ca, 4 ng g^{-1} Fe and 8 ng g^{-1} Al). The small size of ice samples used in this study prevented the use of techniques that remove several cm of core, and so we developed a technique using Kyocera ceramic blades that sequentially removed contaminated ice-core material. Each piece of ice was placed on a PTFE board and held with a clean PE glove, while the outside few mm were scraped off with a clean ceramic blade. This was repeated three times using separate clean equipment. To achieve the lowest contamination levels, all ice sampling equipment and ice were only handled with clean PE gloves, and all ice decontamination was carried out under Class 100 laminar flow air within the cold room (−20 °C) at the British Antarctic Survey.

Decontamination blanks were 47 pg g^{-1} for Fe and 116 pg g^{-1} for Al (<0.1% of natural conc.), and testing established that using acid-cleaned equipment had no effect on the natural pH of ice (pH ~5.3). Our simpler technique did not achieve the full decontamination of more rigorous chiselling methods[22] (46 pg g^{-1} c.f. 4 pg g^{-1} Fe). However, our method can be applied to smaller pieces of ice, and the procedure reduces Fe contamination by factor of ~100×, resulting in insignificant contamination that represents <0.1% of typical natural Fe concentrations within ice from the high dust portion of the LGM.

Total digestion and pH ~5.3 rapid-filtration procedures. Following decontamination, ice samples for total digestion were allowed to melt within PFA vials and acidified with 50 µl of QD-HNO_3 (conc.) per 5 ml. A 5-ml aliquot was evaporated to dryness, refluxed overnight at ~80 °C with 100 µl QD-HNO_3 (conc.) and 200 µl Teflon-distilled conc. HF, evaporated to dryness, redissolved in 100 µl of 6 mol l^{-1} QD-HCl, refluxed at ~80 °C overnight and evaporated to dryness. Finally, samples were redissolved in 250–500 µl of 0.1 mol l^{-1} QD-HNO_3 for analysis by inductively coupled plasma optical emission spectrometry (ICP-OES). Procedural blanks (n = 26) were 1.9 ng g^{-1} (Fe), 5.7 ng g^{-1} (Al) and 2.8 ng g^{-1} (Ca), all <0.5% of natural concentrations.

Following decontamination, ice samples for soluble analysis were processed by either 'rapid-filtration' (melting at natural pH, referred to as 'pH ~5.3) or 'seawater-leaching' methods (see later sections). Ice samples for rapid filtration were allowed to melt at room temperature (~2 h) on a 0.2-µm Whatman polycarbonate membrane inside a 47-mm PFA Savillex filter rig, under vacuum. Rigs were covered to prevent contamination. Liquid that formed was quickly filtered and collected in a PFA reservoir containing an appropriate volume of conc. HNO_3 to acidify samples to pH 0. On completion of melting, dust was leached with three 25 ml aliquots of ultrapure water, similarly acidified. This procedure, analogous to instantaneous leaches of modern aerosols[27], was designed to minimize any $FeOH_3$ precipitation or wall adsorption, which are concerns with flow-through and long leaching procedures (Supplementary Discussion). Acidified samples were evaporated to dryness and redissolved in 300 µl of 0.1 mol l^{-1} QD-HNO_3 for analysis by ICP-OES. Fe concentration from the meltwater and subsequent ultrapure leaches were combined in final calculations. Procedural blanks (n = 78) were 17 pg g^{-1} (Fe), 244 pg g^{-1} (Ca) and 83 pg g^{-1} (Na), all corresponding to 0–2% of natural concentrations.

Elemental analysis. Samples from total digestion or rapid filtration were analysed for dissolved Al, Ca, Fe and other cations using a Varian Vista ICP-OES at the University of Cambridge. Each sample was analysed six times and a mean concentration was determined by reference to a quadratic calibration curve based on five standards. Single-session accuracy on a 250-ng per gram internal multi-element reference standard for Al, Ca and Fe was >99.6% with precision of 0.41% (1 s.d.; n = 10). Long-term reproducibility of the standard was 100 ± 1% for all elements (n = 60). Detection limits for Fe were 0.1–0.4 ng g^{-1}. We assign 5% uncertainty on data to account for weighing, pipetting and %RSD of six ICP-OES analyses (typically <5%).

Sublimation. Sublimation allows the extraction of gases or dust from ice-core material while preventing melting, which could compromise the sample fidelity, and is achieved by lowering the pressure and temperature of ice below the triple point (0.01 °C, 6.1173 mbar). Sublimation has been previously used to investigate concentrations and isotope ratios of gases (for example, CO_2 and CH_4) released from bubbles trapped in ice-core material[55–58], as well as having been utilized to cross-check CO_2 measurements obtained using other techniques in Bern[59]. In the present study, we were motivated to develop a sublimation technique to extract dust from ice cores without exposing the dust to liquid water at a pH lower than that of seawater (pH ~8), which may alter the speciation or solubility of Fe within dust before analysis (Supplementary Discussion). The major mode of deposition of dust at Dome C during the LGM is thought to be dry[20], meaning that EDC dust extracted by sublimation should be representative of dry dust deposition at Dome C during the LGM.

We designed a sublimation apparatus based on previous studies on gases[58], adapted to cleanly extract large masses of dust[57] from ice without melting. The apparatus consists of a Schott borosilicate glass desiccator, sealed with Apiezon N vaccum grease, a vaccum pump with oil cleaner and pressure gauge and a −100 °C freeze trap for removal of water vapour (Supplementary Fig. 3). To allow flexibility depending on the mass of dust/ice required by experiments, two configurations of the system were designed (Supplementary Fig. 3). In the present study, ice samples were placed directly on membranes within filter rigs within a Savillex PFA Teflon Jar with top ports open (25–50 g; Supplementary Fig. 3a). A second configuration, not used in the present study, allows for the collection of large dust masses, with larger pieces of ice (>50 g) placed within 1 l bottles within a 4-l bottle with holes drilled in the lid (Supplementary Fig. 3b). Although ice sublimes under any pressures and temperatures below the triple point, previous studies have shown that quasi- and surface-melting can occur on ice or ice–glass surfaces at temperatures (−1 to −4 °C) and pressures below the triple point[60–62]. To avoid this, it is necessary to keep the ice cold until the pressure is sufficiently lowered, and to operate at pressures much lower than the triple point (<2 mbar). To achieve this, first the glass desiccator was taken into a Class 100 laminar flow bench within the cold room and allowed to cool before being cleaned with propanol. Jars or bottles were then placed within the cleaned desiccator and the desiccator was closed and sealed before being transferred to the warm laboratory and connected to the pumps. This ensured that samples remained clean and that the ice remained cold for ~10 min while the pressure was reduced below the triple point to the typical running pressure of 0.8–1.6 mbar.

Sublimation was allowed to proceed until all ice was gone, which was assessed visually or when the pressure had dropped to the background of the empty chamber (0.26 mbar). Rates of sublimation were relatively slow, dependent on ice mass and volume, with a maximum rate of ~250 g per 24 h. Small samples (5–10 g) sublimed in several hours, but were typically left overnight to ensure all ice was gone.

Once all ice had sublimed, the desiccator was sealed, disconnected and returned to the laminar flow bench to slowly allow clean air to enter the chamber; this prevented both dust being blown away by a sudden increase in pressure and the entry of dirty air from the pump. The bottles containing the dry dust samples were then carefully bagged and taken for 'seawater-leaching' procedures (see next section). Blank contamination for the whole procedure over 24 hours was found to be acceptably low (~40 pg Fe per ice sample).

Seawater-leaching procedures and Fe analysis. Ice samples for seawater-soluble Fe analysis ('seawater-leaching') were placed on a membrane on a PFA rig within a 1 l PFA jar within a glass desiccator at −20 °C, and quickly placed cleanly under vacuum (before moisture could form) at ~18 °C. Samples remained under vacuum until all the ice had sublimed, leaving dry dust on the membrane (see 'Sublimation' section). The dust on the membrane was then leached with four 60-ml aliquots of low-Fe Ross Sea seawater at 2 °C. The Ross Sea seawater used in leaching was previously collected[63] from 10 m depth in the Southern Ross Sea (76° 02′ S, 169° 53′ E) under clean conditions in December 2003. The water was filtered with 0.2 µm polypropylene cartridges and stored in 125 l polyethylene barrels, before being subsampled for this work using clean procedures[63]. We determined the background dissolved Fe concentration in this seawater to be 4.7 pg g^{-1} (0.11 ± 0.02 nmol l^{-1}, 2 s.d.; $n = 4$), consistent with previous determination[63,64] of 0.07 ± 0.02 nmol l^{-1}. Each aliquot of seawater was allowed to leach the dust momentarily, vacuum was applied and then the liquid was pulled through and acidified to pH 1.6 with 120 µl of conc. ultrapure HCl. Samples were left acidified for >2 weeks before being analysed for dissolved Fe by FIA-8HQ, following published techniques[45]. This method has low detection limits (<5 pg g^{-1}) and analytical precision is typically 5–10% during routine analysis; we therefore applied 10% uncertainty on seawater Fe concentration. Fe concentrations from four leaches were combined in final calculations.

Berkner Deuterium and parameters and age scales. The ice core at Berkner Island was drilled to bedrock, and completed in 2005 (ref. 65 and Fig. 2). Deuterium (δD) ratios (expressed relative to the international standard Vienna Standard Mean Ocean Water) were measured at the NERC Isotope Geosciences Laboratory using standard methods[66] and have a typical precision of 1.0‰. In Fig. 2b, the Berkner δD record across A events A1 and A2 is shown overlain on the EPICA Dronning Maud Land (EDML) $\delta^{18}O$ record[67], which is shown on the EDML1 age model that has been synchronized to EDC3 (refs 47,67). In the absence of an official age model for this part of the Berkner Island ice core, Berkner δD is approximately visually tuned to the EDML $\delta^{18}O$ record so that EDML $\delta^{18}O$, Berkner δD and Berkner Fe data are all shown on approximately the EDC3 age scale. Dust and temperature in Fig. 2a are from the EDC core[15,68], while atmospheric CO_2 concentration (pCO_2) in both Fig. 2a,b is taken from the composite CO_2 record of Schilt *et al.*[69]

Fe solubility and flux calculation. Elemental fluxes were calculated:

$$\text{Fe flux}(\text{mg m}^{-2}\text{ per year}) = [\text{Fe}] \times A_R \quad (1)$$

where [Fe] is the concentration of Fe (mg kg^{-1}) in the ice, A_R is the calculated ice-accumulation rate (kg m^{-2} per year; ref. 47, Mulvaney, unpublished data).
Fe solubility was defined:

$$\text{Fe solubility }(\%) = 100 \times ([\text{pH} \sim 5.3 \text{ or seawater soluble Fe}]/[\text{total Fe}]) \quad (2)$$

where [pH ~5.3 or seawater-soluble Fe] and [total Fe] are the concentration in ice meltwater (ng g^{-1}), combined from multiple leaches.

Calculating nss-Ca. nss-Ca in each ice-core sample was calculated in this study from measured dissolved Ca and Na concentration data using paired simultaneous equations[50]:

$$[\text{ss-Na}^+] = [\text{dNa}] - ([\text{nss-Ca}^{2+}]/R_t) \quad (3)$$

$$[\text{nss-Ca}^{2+}] = [\text{dCa}] - R_m[\text{ss-Na}^+] \quad (4)$$

where ss denotes from seasalt, nss from non-seasalt sources, [dNa] dissolved Na concentration at pH ~5.3, [dCa] dissolved Ca concentration at pH ~5.3, R_t and R_m are the Ca/Na ratios in crustal material (1.78) and seasalt (0.038), respectively[50].

References

1. Martin, J. H. Glacial-interglacial CO_2 change: The Iron Hypothesis. *Paleoceanography* **5,** 1–13 (1990).
2. Sigman, D. M. & Boyle, E. A. Glacial/interglacial variations in atmospheric carbon dioxide. *Nature* **407,** 859–869 (2000).
3. Kohfeld, K. E. & Ridgwell, A. J. in *Surface Ocean - Lower Atmosphere Processes*. (eds le Quéré, C. & Saltzman, E. S.) 251–286 (AGU, 2009).
4. Watson, A. J. *et al.* Effect of iron supply on Southern Ocean CO_2 uptake and implications for glacial atmospheric CO_2. *Nature* **407,** 730–733 (2000).
5. Martínez-Garcia, A. *et al.* Southern Ocean dust-climate coupling over the past four million years. *Nature* **476,** 312–315 (2011).
6. Lüthi, D. *et al.* High-resolution carbon dioxide concentration record 650,000-800,000 years before present. *Nature* **453,** 379–382 (2008).
7. Boyd, P. W. *et al.* Mesoscale iron enrichment experiments 1993-2005: synthesis and future directions. *Science* **315,** 612–617 (2007).
8. Anderson, R. F. *et al.* Wind-driven upwelling in the Southern Ocean and the deglacial rise in atmospheric CO_2. *Science* **323,** 1443–1448 (2009).
9. Sigman, D. M., Hain, M. P. & Haug, G. H. The polar ocean and glacial cycles in atmospheric CO_2 concentration. *Nature* **466,** 47–55 (2010).
10. Adkins, J. F. The role of deep ocean circulation in setting glacial climates. *Paleoceanography* **28,** 539–561 (2013).
11. Röthlisberger, R. *et al.* Ice core evidence for the extent of past atmospheric CO_2 change due to iron fertilisation. *Geophys. Res. Lett.* **31,** L16207 (2004).
12. Ziegler, M., Diz, P., Hall, I. R. & Zahn, R. Millennial-scale changes in atmospheric CO_2 levels linked to the Southern Ocean carbon isotope gradient and dust flux. *Nat. Geosci.* **6,** 457–461 (2013).
13. Martínez-García, A. *et al.* Iron fertilization of the Subantarctic ocean during the last ice age. *Science* [illegible]
14. Jaccard, S. L. *et al.* Two modes of change in Southern Ocean productivity over the past million years. *Science* **339,** 1419–1423 (2013).
15. Lambert, F. *et al.* Dust-climate couplings over the past 800,000 years from the EPICA Dome C ice core. *Nature* **452,** 616–619 (2008).
16. Fischer, H. *et al.* Reconstruction of millennial changes in dust emission, transport and regional sea ice coverage using the deep EPICA ice cores from the Atlantic and Indian Ocean sector of Antarctica. *Earth Planet. Sci. Lett.* **260,** 340–354 (2007).
17. Tagliabue, A., Aumont, O. & Bopp, L. The impact of different external sources of iron on the global carbon cycle. *Geophys. Res. Lett.* **41,** 920–926 (2014).
18. Mahowald, N. M. *et al.* Atmospheric global dust cycle and iron inputs to the ocean. *Global Biogeochem. Cycles* **19,** GB4025 (2005).
19. Gaspari, V. *et al.* Atmospheric iron fluxes over the last deglaciation: climatic implications. *Geophys. Res. Lett.* **33,** L03704 (2006).
20. Wolff, E. W. *et al.* Southern Ocean sea-ice extent, productivity and iron flux over the past eight glacial cycles. *Nature* **440,** 491–496 (2006).
21. Ruth, U. *et al.* Proxies and measurement techniques for mineral dust in Antarctic ice cores. *Environ. Sci. Technol.* **42,** 5675–5681 (2008).
22. Edwards, R., Sedwick, P. N., Morgan, V. & Boutron, C. Iron in ice cores from Law Dome: a record of atmospheric iron deposition for maritime East Antarctica during the Holocene and Last Glacial Maximum. *Geochem. Geophys. Geosyst.* **7,** Q12Q01 (2006).
23. Vallelonga, P. *et al.* Iron fluxes to Talos Dome, Antarctica, over the past 200 kyr. *Clim. Past* **9,** 597–604 (2013).
24. Tian, Z., Ollivier, P., Véron, A. & Church, T. M. Atmospheric Fe deposition modes at Bermuda and the adjacent Sargasso Sea. *Geochem. Geophys. Geosyst.* **9,** Q08007 (2008).
25. Morgan, B. & Lahav, O. The effect of pH on the kinetics of spontaneous Fe(II) oxidation by O_2 in aqueous solution - basic principles and a simple heuristic description. *Chemosphere* **68,** 2080–2084 (2007).
26. Spolaor, A. *et al.* Iron speciation in aerosol dust influences iron bioavailability over glacial-interglacial timescales. *Geophys. Res. Lett.* **40,** 1618–1623 (2013).
27. Buck, C. S., Landing, W. M., Resing, J. A. & Lebon, G. T. Aerosol iron and aluminum solubility in the northwest Pacific Ocean: results from the 2002 IOC cruise. *Geochem. Geophys. Geosyst.* **7,** 21 (2006).
28. Schroth, A. W., Crusius, J., Sholkovitz, E. R. & Bostick, B. C. Iron solubility driven by speciation in dust sources to the ocean. *Nat. Geosci.* **2,** 337–340 (2009).
29. Trapp, J. M., Millero, F. J. & Prospero, J. M. Trends in the solubility of iron in dust-dominated aerosols in the equatorial Atlantic trade winds: Importance of iron speciation and sources. *Geochem. Geophys. Geosyst.* **11,** Q03014 (2010).
30. Baker, A. R. & Croot, P. L. Atmospheric and marine controls on aerosol iron solubility in seawater. *Mar. Chem.* **120,** 4–13 (2010).
31. Basile, I. *et al.* Patagonian origin of glacial dust deposited in East Antarctica (Vostok and Dome C) during glacial stages 2, 4 and 6. *Earth Planet. Sci. Lett.* **146,** 573–589 (1997).
32. Journet, E., Desboeufs, K. V., Caquineau, S. & Colin, J.-L. Mineralogy as a critical factor of dust iron solubility. *Geophys. Res. Lett.* **35,** L07805 (2008).
33. Glaccum, R. A. & Prospero, J. M. Saharan aerosols over the tropical North Atlantic - Mineralogy. *Mar. Geol.* **37,** 295–321 (1980).
34. Sugden, D. E., McCulloch, R. D., Bory, A. J.-M. & Hein, A. S. Influence of Patagonian glaciers on Antarctic dust deposition during the last glacial period. *Nat. Geosci.* **2,** 281–285 (2009).
35. Koffman, B. G., Handley, M. J., Osterberg, E. C., Wells, M. L. & Kreutz, K. J. Dependence of ice-core relative trace-element concentration on acidification. *J. Glaciol.* **60,** 103–112 (2014).
36. Taylor, S. R. & McLennan, S. M. *The Continental Crust: its Composition and Evolution* (Blackwell Scientific Publishing, 1985).
37. Gaiero, D. M. *et al.* Iron and other transition metals in Patagonian riverborne and windborne materials: geochemical control and transport to the southern South Atlantic Ocean. *Geochim. Cosmochim. Acta* **67,** 3603–3623 (2003).
38. Martínez-Garcia, A. *et al.* Links between iron supply, marine productivity, sea surface temperature, and CO_2 over the last 1.1 Ma. *Paleoceanography* **24,** PA1207 (2009).
39. Tagliabue, A. *et al.* Surface-water iron supplies in the Southern Ocean sustained by deep winter mixing. *Nat. Geosci.* **7,** 314–320 (2014).
40. Ridgwell, A. J. & Watson, A. J. Feedback between aeolian dust, climate, and atmospheric CO_2 in glacial time. *Paleoceanography* **17,** 1059 (2002).
41. Moore, J. K., Doney, S. C. & Lindsay, K. Upper ocean ecosystem dynamics and iron cycling in a global three-dimensional model. *Global Biogeochem. Cycles* **18,** GB4028 (2004).
42. Fung, I. Y. *et al.* Iron supply and demand in the upper ocean. *Global Biogeochem. Cycles* **14,** 281–295 (2000).
43. Tagliabue, A., Bopp, L. & Aumont, O. Evaluating the importance of atmospheric and sedim[illegible]

44. Zhang, Y. *et al.* Modeling the global emission, transport and deposition of trace elements associated with mineral dust. *Biogeosci. Discuss.* **11,** 17491–17541 (2014).
45. Sedwick, P. N., DiTullio, G. R. & Mackey, D. J. Iron and manganese in the Ross Sea, Antarctica: seasonal iron limitation in Antarctic shelf waters. *J. Geophys. Res.* **105,** 11321 (2000).
46. Delmonte, B. *et al.* Comparing the Epica and Vostok dust records during the last 220,000 years: stratigraphical correlation and provenance in glacial periods. *Earth-Sci. Rev.* **66,** 63–87 (2004).
47. Parrenin, F. *et al.* The EDC3 chronology for the EPICA Dome C ice core. *Clim. Past* **3,** 485–497 (2007).
48. Blunier, T. & Brook, E. J. Timing of millennial-scale climate change in Antarctica and Greenland during the last glacial period. *Science* **291,** 109–112 (2001).
49. Boutron, C. F. & Batifol, F. M. Assessing laboratory procedures for the decontamination of Polar snow or ice samples for the analysis of toxic metals and metalloids. *Ann. Glaciol.* **7,** 7–11 (1985).
50. Röthlisberger, R. *et al.* Dust and sea salt variability in central East Antarctica (Dome C) over the last 45 kyrs and its implications for southern high-latitude climate. *Geophys. Res. Lett.* **29,** 1963 (2002).
51. Candelone, J. P., Hong, S. & Boutron, F. C. An improved method for decontaminating polar snow or ice cores for heavy metal analysis. *Anal. Chim. Acta* **299,** 9–16 (1994).
52. Gabrielli, P. *et al.* Determination of Ir and Pt down to the sub-femtogram per gram level in polar ice by ICP-SFMS using preconcentration and a desolvation system. *J. Anal. At. Spectrom.* **19,** 831–837 (2004).
53. Chambers, R. Trace elements in an Antarctic ice core as potential indicators of marine biogenic activity (PhD Thesis, Univ. East Anglia, 2006).
54. Gabrielli, P. *et al.* Variations in atmospheric trace elements in Dome C (East Antarctica) ice over the last two climatic cycles. *Atmos. Environ.* **39,** 6420–6429 (2005).
55. Wilson, A. T. & Long, A. New Approaches to CO_2 Analysis in polar ice cores. *J. Geophys. Res.* **102,** 26601–26606 (1997).
56. Wilson, A. T. Application of AMS ^{14}C dating to ice core research. *Radiocarbon* **37,** 637–641 (1995).
57. Wilson, A. T. & Donahue, D. J. AMS carbon-14 dating of ice: progress and future prospects. *Nucl. Instrum. Methods Phys. Res.* **52,** 473–476 (1990).
58. Güllük, T., Slemr, F. & Stauffer, B. Simultaneous measurements of CO_2, CH_4, and N_2O in air extracted by sublimation from Antarctic ice cores: confirmation of the data obtained using other extraction techniques. *J. Geophys. Res.* **103,** 15971–15978 (1998).
59. Siegenthaler, U. CO_2 konzentrationsmessungen an polaren eisbohrkernen und tests mit einer neuen extraktionsmethode. *Dip. Thesis* (Univ. Bern, 2002).
60. Elbaum, M., Lipson, S. G. & Dash, J. G. Optical study of surface melting on ice. *J. Cryst. Growth* **129,** 491–505 (1993).
61. Furukawa, Y. & Ishikawa, I. Direct evidence for melting transition at interface between ice crystal and glass substrate. *J. Cryst. Growth* **128,** 1137–1142 (1993).
62. Furukawa, Y., Yamamoto, M. & Kuroda, T. Ellipsometric study of the transition layer on the surface of an ice crystal. *J. Cryst. Growth* **82,** 665–677 (1987).
63. Garcia, N. S., Sedwick, P. N. & DiTullio, G. R. Influence of irradiance and iron on the growth of colonial Phaeocystis antarctica: implications for seasonal bloom dynamics in the Ross Sea, Antarctica. *Aquat. Microb. Ecol.* **57,** 203–220 (2009).
64. Sedwick, P. N., Garcia, N. S., Riseman, S. F., Marsay, C. M. & DiTullio, G. R. Evidence for high iron requirements of colonial *Phaeocystis antarctica* at low irradiance. *Biogeochemistry* **83,** 83–97 (2007).
65. Mulvaney, R., Alemany, O. & Possenti, P. The Berkner Island (Antarctica) ice-core drilling project. *Ann. Glaciol.* **47,** 115–124 (2007).
66. Abram, N. J., Mulvaney, R. & Arrowsmith, C. Environmental signals in a highly resolved ice core from James Ross Island, Antarctica. *J. Geophys. Res.* **116,** D20116 (2011).
67. Ruth, U. *et al.* 'EDML1': a chronology for the EPICA deep ice core from Dronning Maud Land, Antarctica, over the last 150 000 years. *Clim. Past* **3,** 475–484 (2007).
68. Jouzel, J. *et al.* Orbital and millennial Antarctic climate variability over the past 800,000 years. *Science* **317,** 793–796 (2007).
69. Schilt, A. *et al.* Atmospheric nitrous oxide during the last 140,000 years. *Earth Planet. Sci. Lett.* **300,** 33–43 (2010).
70. Delmonte, B. *et al.* Aeolian dust in East Antarctica (EPICA-Dome C and Vostok): Provenance during glacial ages over the last 800 kyr. *Geophys. Res. Lett.* **35,** L07703 (2008).

Acknowledgements

We thank Mervyn Greaves, Jason Day, Susan Foord (deceased), Hazel Chapman, Paul Geissler and Jack Triest for their technical assistance with sampling and analysis, and Peter Sedwick for providing Ross Sea seawater and facilities for analysis of dissolved Fe in seawater samples. This work is a contribution to the European Project for Ice Coring in Antarctica (EPICA), a joint European Science Foundation/European Commission (EC) scientific programme. This study was funded by a Natural Environment Research Council (NERC) studentship to T.M.C., and E.W.W. is funded by a Royal Society professorship.

Author contributions

All authors devised the study; T.M.C performed sample processing and analysis, R.M. supplied samples and data, and the paper was written by T.M.C. and E.W.W. with contributions from other authors.

Additional information

Competing financial interests: The authors declare no competing financial interests.

8

Evidence for ice-free summers in the late Miocene central Arctic Ocean

Ruediger Stein[1,2], Kirsten Fahl[1], Michael Schreck[3], Gregor Knorr[1], Frank Niessen[1], Matthias Forwick[4], Catalina Gebhardt[1], Laura Jensen[1], Michael Kaminski[5], Achim Kopf[2,6], Jens Matthiessen[1], Wilfried Jokat[1,2] & Gerrit Lohmann[1,6]

Although the permanently to seasonally ice-covered Arctic Ocean is a unique and sensitive component in the Earth's climate system, the knowledge of its long-term climate history remains very limited due to the restricted number of pre-Quaternary sedimentary records. During *Polarstern* Expedition PS87/2014, we discovered multiple submarine landslides along Lomonosov Ridge. Removal of younger sediments from steep headwalls has led to exhumation of Miocene sediments close to the seafloor. Here we document the presence of IP_{25} as a proxy for spring sea-ice cover and alkenone-based summer sea-surface temperatures $>4\,°C$ that support a seasonal sea-ice cover with an ice-free summer season being predominant during the late Miocene in the central Arctic Ocean. A comparison of our proxy data with Miocene climate simulations seems to favour either relatively high late Miocene atmospheric CO_2 concentrations and/or a weak sensitivity of the model to simulate the magnitude of high-latitude warming in a warmer than modern climate.

[1] Alfred Wegener Institute (AWI) Helmholtz Centre for Polar and Marine Research, Am Alten Hafen 26, Bremerhaven 27568, Germany. [2] Department of Geosciences (FB5), University of Bremen, Klagenfurter Strasse 4, Bremen 28359, Germany. [3] Arctic Research Centre, Korea Polar Research Institute, 26 Songdomirae-ro, Yeonsu-gu, Incheon 406-840, Korea. [4] Institute of Geology, University of Tromsø—The Arctic University of Norway, P O Box 6050 Langnes, Tromsø 9037, Norway. [5] Geosciences Department, College of Petroleum Engineering & Geosciences, King Fahd University of Petroleum & Minerals, Dhahran 31261, Saudi Arabia. [6] MARUM-Center for Marine Environmental Sciences

There is a general consensus that the polar regions—and in particular the Arctic Ocean and surrounding areas—are at present, and were over historic and geologic time scales, subject to rapid and dramatic environmental changes. Owing to complex feedback processes, collectively known as 'polar amplification', the Arctic is both a contributor to climate change and a region that will be most affected by global warming[1-3]. Despite the importance of the Arctic Ocean in the global climate

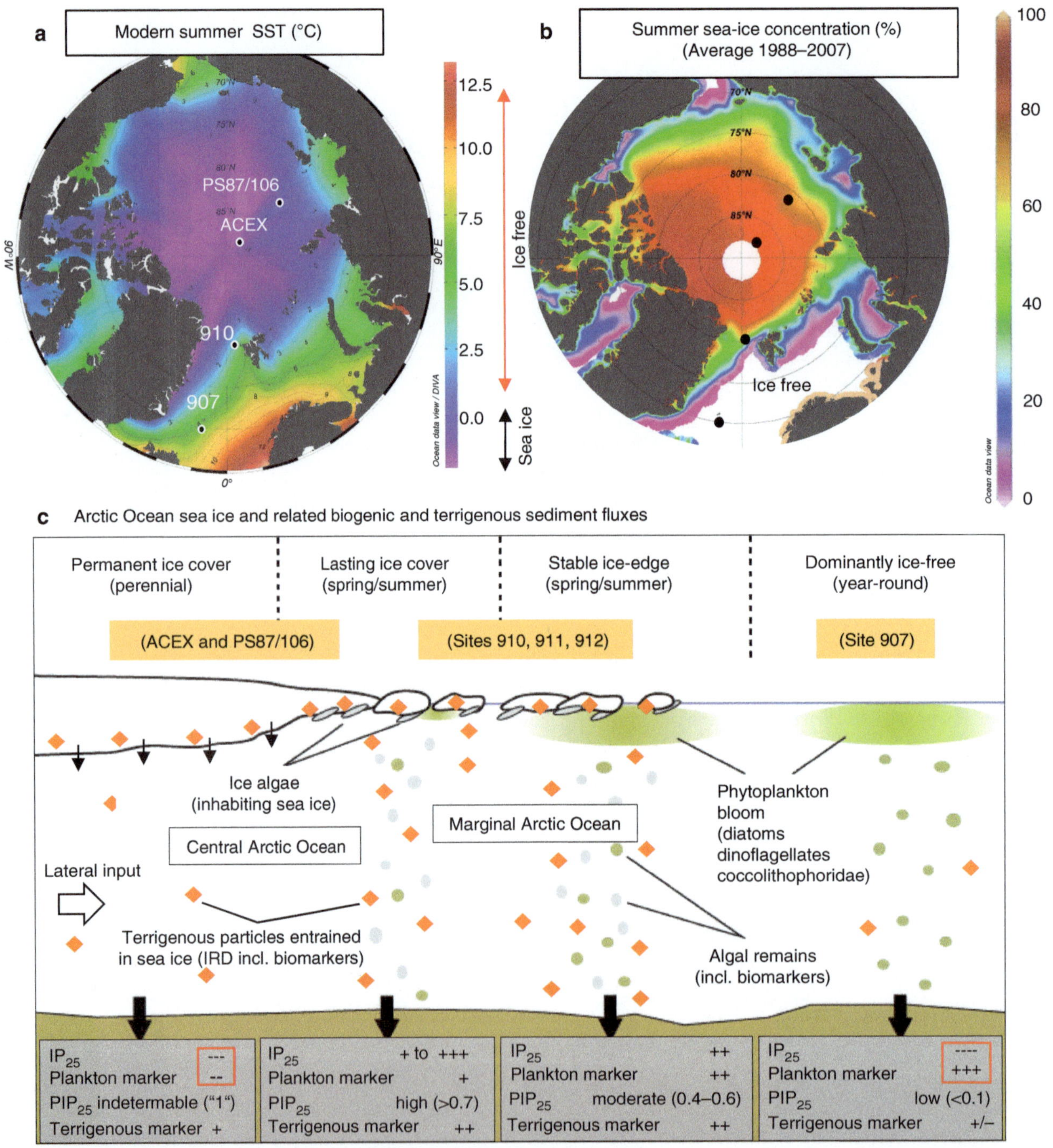

Figure 1 | Modern Arctic Ocean SST and sea-ice conditions. (**a**) Map of modern August SST with locations of ODP Site 907, ODP Site 910 and IODP Expedition 302 (ACEX) Site, as well as Core PS87/106 (August SST data—average of 1955-2012—from World Ocean Atlas; http://odv.awi.de/en/data/ocean/world_ocean_atlas_2013/). (**b**) Map of summer sea-ice concentration—average of 1988-2007—with locations of studied sites (black dots); data source: http://nsidc.org. Maps **a** and **b** produced with Ocean Data View software (source: http://odv.awi.de/). (**c**) Generalized scheme illustrating (1) sea surface conditions and respective (spring/summer) productivities of ice algae and phytoplankton, and (2) sedimentary contents of IP_{25}, terrigenous biomarkers and IRD, phytoplankton-derived biomarkers and PIP_{25} index for different settings in the modern Arctic Ocean (after ref. 14, supplemented). Owing to ice melting and related nutrient and sediment release, a stable ice-edge situation is characterized by high concentrations of IP_{25} and phytoplanktion biomarkers, but also by high concentrations of terrigenous biomarkers and IRD. Modern situations at locations of ACEX and PS87/106, ODP Site 910 (and close-by ODP sites 911 and 912) and ODP Site 907 are indicated.

system, this permanently to seasonally ice-covered region (Fig. 1) is one of the last major physiographic provinces on Earth, whose climate history and its transition from early Cenozoic Greenhouse to late Cenozoic Icehouse conditions remain still poorly known. Only one drill site recovered from the central Arctic Ocean during the Arctic Coring Expedition (ACEX)—the Integrated Ocean Drilling Program (IODP) Expedition 302 in 2004—gives some insight into the early Cenozoic climate[4–6].

Concerning recent climate change, the most prominent example is the dramatic decrease of the extent and thickness of the Arctic sea-ice cover the last decades, a decrease that seems to be by far more rapid than predicted by climate models[1,3]. The scientific community recognized this drastic change with major concern as the Arctic sea ice is a critical component in the global climate system, which contributes to changes in the Earth's albedo, primary productivity and deep-water formation, a driving mechanism for global thermohaline circulation[7]. The causes of these recent changes, that is, natural versus anthropogenic forcings, and their relevance within the global climate system, however, are subject of intense scientific and societal debate. Thus, understanding the processes controlling Arctic sea-ice variability is of overall interest and significance[8,9]. In this context, records of past climate and sea-ice conditions going beyond instrumental records and representing times of different boundary conditions are of major value: such records can be used to assess the sensitivity of the Earth's climate system to changes of different forcing parameters, for example, level of CO_2, and to test the reliability of climate models by evaluating their simulations for conditions very different from the modern climate. This type of records giving detailed information about past Arctic sea-ice conditions are still very rare, especially due to the lack of precise proxies for sea-ice reconstructions (see review in ref. 9).

The ability to (semi-)quantitatively reconstruct paleo-sea-ice distributions has been significantly improved by a biomarker approach based on the determination of a highly branched isoprenoid (HBI) with 25 carbons (C_{25} HBI monoene $= IP_{25}$)[10]. This biomarker is only biosynthesized by specific diatoms living in the Arctic ice[11] and appears to be a specific, sensitive and stable proxy for Arctic sea ice in sedimentary sections representing Pliocene to Pleistocene times[12,13]. When using this proxy, one has to consider that IP_{25} is absent under a permanent sea-ice cover limiting light penetration and, in consequence, sea-ice algal growth (that is, $IP_{25} = 0$). The same consequence applies to totally ice-free conditions (Fig. 1). Müller *et al.*[14,15] overcame this difficulty in interpreting IP_{25} data by the additional use of phytoplankton-derived open-water biomarkers such as brassicasterol or dinosterol (Fig. 1). Furthermore, Müller *et al.*[14] combined the environmental information carried by IP_{25} and phytoplankton biomarkers in a phytoplankton-IP_{25} index, the so-called 'PIP_{25} index' (for calculation of PIP_{25}, see 'Methods'). In general, the PIP_{25} values determined in Arctic Ocean surface sediments correlate reasonably well with the modern sea-ice distribution based on satellite data, although regional differences in the correlation patterns seem to be obvious[14,16]. In the modern central Arctic Ocean characterized by a mostly perennial sea-ice cover, the PIP_{25} values of most of the surface sediments are between 0.65 and 1 (ref. 16). These data suggest that the PIP_{25} index seems to be a promising proxy for more semi-quantitative estimates of the present and past sea-ice extent, although certainly further groundtruth data are needed (see Methods for some more details)[8,16].

Here we apply the new sea-ice biomarker approach together with alkenone-based sea-surface temperatures (SSTs) to sediment cores most recently recovered during *Polarstern* Expedition PS87 and SST conditions. The proxy data are combined with climate model simulations using a coupled atmosphere-ocean general circulation model (AOGCM; see Methods) with focus on seasonal changes in the high northern latitudes. Based on our new proxy records, we demonstrate that only a seasonal sea-ice cover has been predominant in the central Arctic Ocean during (most of) the Late Miocene time interval. Furthermore, our combined data/modelling approach seems to indicate either relatively high atmospheric CO_2 concentrations and/or an overly weak sensitivity of the model to simulate the magnitude of warm polar temperatures in the late Miocene. These new findings from the Arctic region provide new benchmarks for groundtruthing global climate reconstructions and modelling.

Results

Large submarine slide scars along Lomonosov Ridge. *Polarstern* Expedition PS87 was scheduled for August–September 2014 to explore the Lomonosov Ridge area (Fig. 2) with the objective of collecting seismic data and sediment cores to reconstruct the short- and long-term climate history and the tectonic evolution of the central Arctic Ocean[17]. More than 3,000 km of high-quality multi-channel seismic (MCS) profiles and ~10,000 km of high-quality multibeam bathymetry and sub-bottom sediment-echosounding (PARASOUND) profiles were acquired along the ship's track and numerous sediment cores were recovered (Fig. 2). One major finding of the expedition was the discovery of numerous submarine slide scars that occurred on both sides of the crest of Lomonosov Ridge over a distance of *ca.* 350 km between 81°07' N and 84°14' N in water depths from ~800 to 1,500 m (Fig. 3a). Single scars are up to several kilometres wide and long, and their head walls are 100–500 m high. Swath-bathymetry data indicate that different processes probably triggered slope failures, that various processes of sediment evacuation took place, and that failures occurred at various times. Slide scars were earlier described from a restricted area on Lomonosov Ridge near 88°N[18]. However, the wide lateral distribution of mass wasting as presented here is a new discovery.

On top of the southern Lomonosov Ridge in areas between the slide scars, we discovered SE–NW oriented, streamlined landforms over distances of >100 km at water depths between 800 and 1,000 m (Fig. 3a). These features are interpreted to be glacial lineations that formed beneath grounded ice[19,20]. Similar unidirectional bed forms have also been identified further east on the East Siberian continental margin where they were related to large and coherent ice masses (East Siberian Chukchi Ice Sheet—ESCIS)[21]. The lineations identified in this study are similar to those on the East Siberian continental margin with respect to their orthogonal orientations to the proposed centre of the former ESCIS. At the present state of knowledge, the most plausible glacial scenario is a larger than originally proposed ESCIS including an ice shelf extending into the Arctic Ocean, which formed an ice rise on the Southern Lomonosov Ridge over areas presently shallower than 1,000 m. The load and erosional behaviour of this ice rise that probably occurred during extended Quaternary glaciations, for example, during Marine Isotope Stage 6 (MIS 6), may have caused physical conditions that triggered the landslides on this part of Lomonosov Ridge. A MIS 6 age of this erosional event is in line with the proposed age of a major glaciation with extended ice sheets/shelves in Eurasia and East Siberia[19–21].

Exhumation of Miocene to lower Quaternary sediments. Sediment removal from the steep slopes of the escarpments

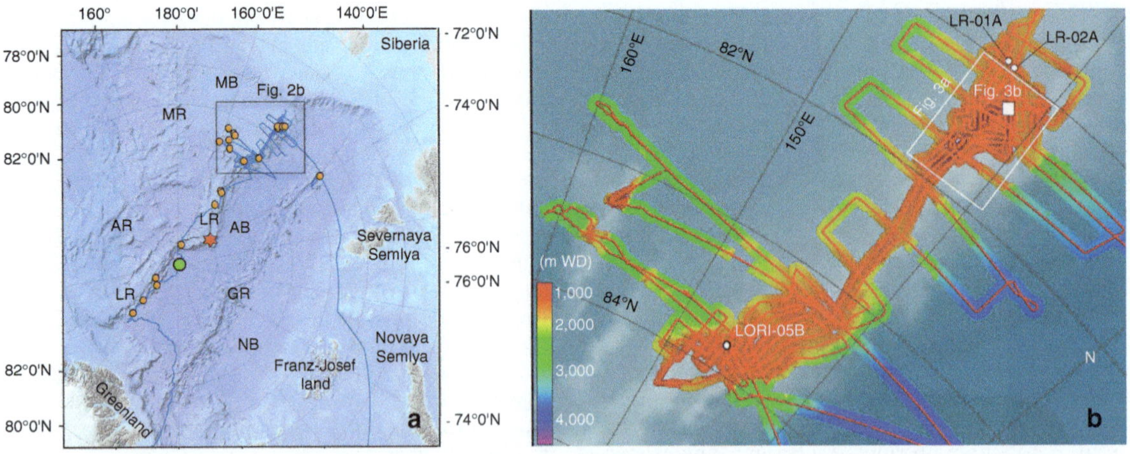

Figure 2 | Cruise track and multibeam bathymetric survey of *Polarstern* Expedition PS87. (**a**) PS87 cruise track (blue line)[17]. AB, Amundsen Basin; AR, Alpha Ridge; GR, Gakkel Ridge; LR, Lomonosov Ridge; MB, Makarov Basin; MR, Mendeleev Ridge; NB, Nansen Basin. Orange circles indicate coring stations, the red asterisk indicates the location of the ACEX Site and the green circle indicates the North Pole (*Polarstern* reached on 26 August 2014 at 10:23 UTC). (**b**) Track lines of multibeam bathymetric survey. Colour bar indicates water depth in metres (m WD). LR-01A, LR-02A and LORI-05B indicate locations of proposed IODP drill sites[51] (*cf.*, IODP Proposal 708; http://www.iodp.org/expeditions). Areas of Fig. 3a,b are indicated.

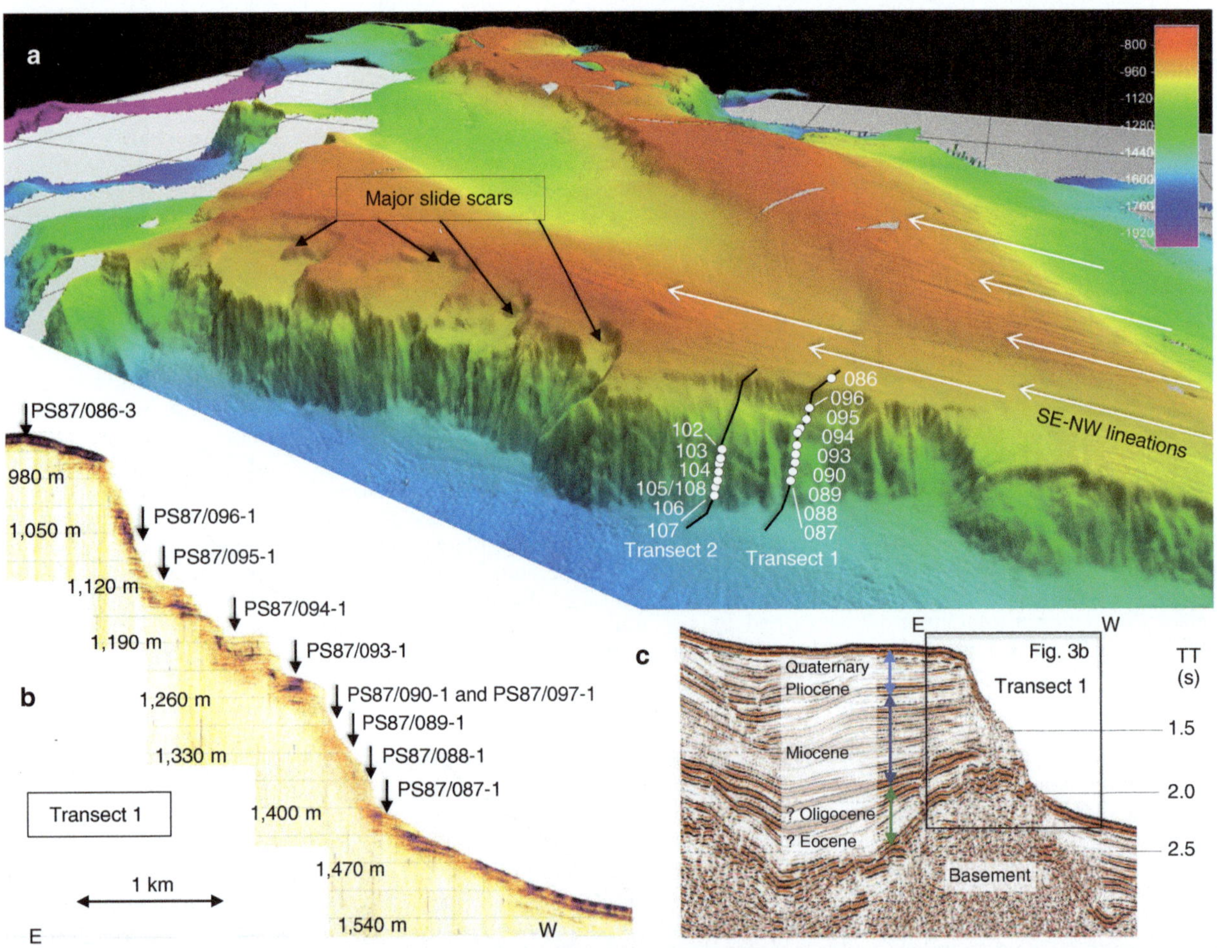

Figure 3 | Bathymetric and acoustic/seismic profiling records from southern Lomonosov Ridge. (**a**) Three-dimensional (3D) image of the swath bathymetry of southern Lomonosov Ridge, showing major slide scars and escarpments, streamlined SE-NW oriented glacial lineations formed beneath grounded ice sheets/streams and transects 1 and 2 with locations of sediment cores (*cf.*, Supplementary Table 1). (**b**) PARASOUND profile across Transect 1 with locations of coring stations. (**c**) Processed multi-channel seismic profile AWI-20140311 across Transect 1, showing prominent reflectors and age assigments based on correlation between regional seismic lines and well data[22,23].

coring from *Polarstern*, rather than expensive drilling. The PARASOUND and MCS profiles suggest that these sedimentary sections are composed of Eocene, Oligocene, Miocene, Pliocene and Quaternary strata (Fig. 3b,c). The age control for the stratigraphic units is based on links of seismic lines to drill site data from the Chukchi Shelf, the ACEX drill site on central Lomonosov Ridge and onshore geology from the New Siberian Islands[22,23]. After evaluation of the multibeam bathymetry and PARASOUND data, we selected two transects across the steep western slope of Lomonosov Ridge for an extensive sediment coring programme (Fig. 3a, Transect 1 and Transect 2). In total, 16 sediment cores were recovered from water depths between 900 m (top of Lomonosov Ridge) and 1,500 m (foot of Lomonosov Ridge)[17] (Fig. 3a,b and Supplementary Table 1).

Whereas most of the sediment cores of Transect 1 are composed of predominantly upper Quaternary (post slide) deposits, some contain prominent unconformities with lower/pre-Quaternary sediments underneath (see Methods and Supplementary Figs 1 and 2). At these unconformities, a 50- to 80-m-thick overburden has been removed, as demonstrated in compaction experiments (see 'Sediment load and compaction experiments'). Unfortunately, the microfossil assemblages (that is, palynomorphs and agglutinated benthic foraminifers) do not allow a precise age determination of the sediments underlying the Quaternary near-surface deposits in most of these cores (Supplementary Table 1). The predominance of Quaternary sediments in the cores of Transect 1, however, seems to be supported by the biomarker composition determined in selected cores. Close to zero concentrations or the absence of specific biomarkers indicative for phytoplankton and sea-ice algae productivity point to surface-water conditions similar to those of the central Arctic Ocean during late Quaternary times[16], characterized by a thick perennial sea-ice cover and SSTs $\ll 0$ °C (Supplementary Table 2 and Fig. 1).

The only core providing a clear indication that old sediments are cropping out near the seafloor is Core PS87/106. In this core

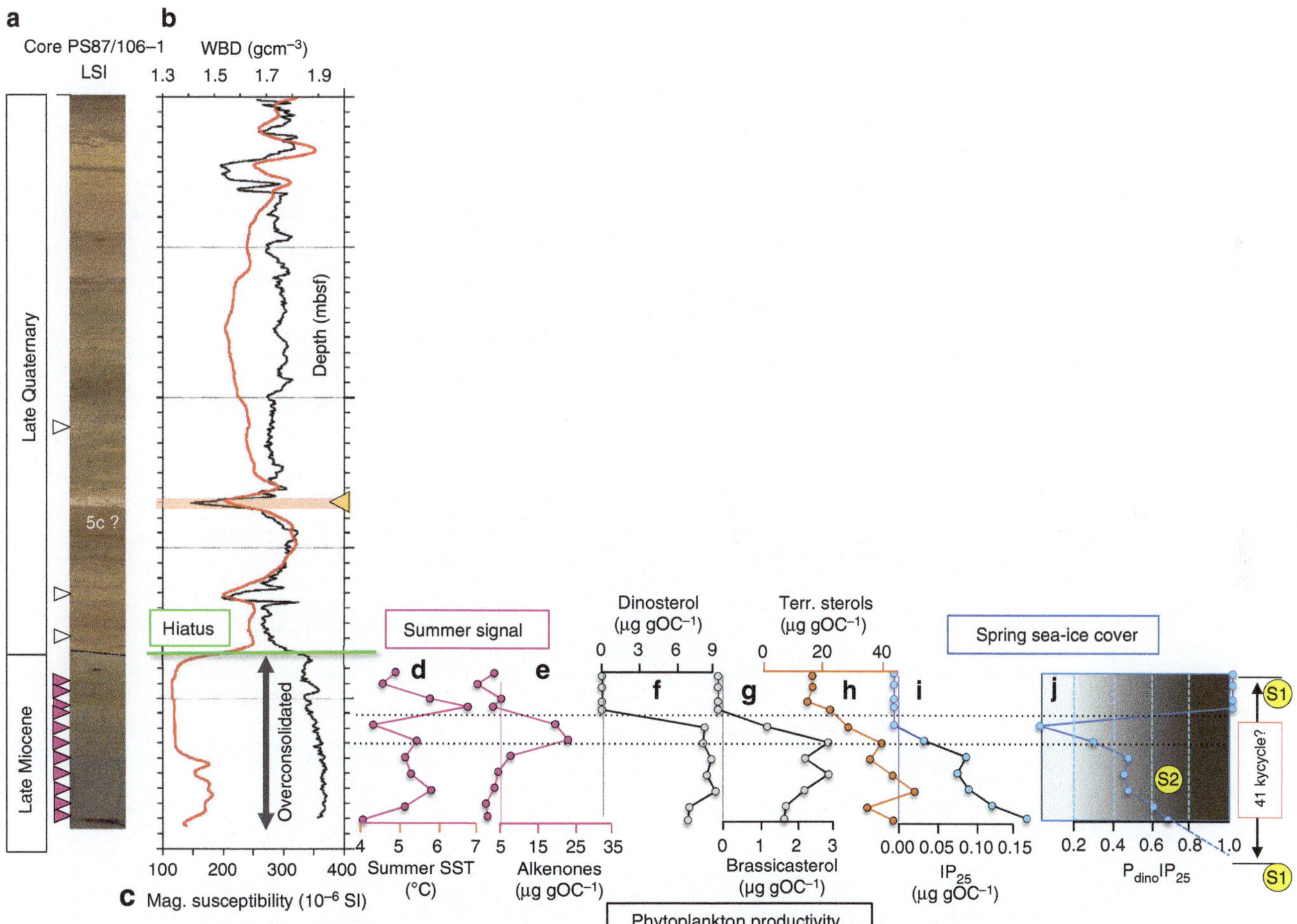

Figure 4 | Proxy evidence from Core PS87/106 for late Miocene Arctic Ocean climate conditions. (**a**) Line-scan image (LSI), (**b**) WBD, (**c**) MS, (**d**) alkenone-based SST, (**e**) concentrations of alkenones, (**f**) dinosterol and (**g**) brassicasterol as proxy for primary productivity, (**h**) concentrations of terrigenous sterols (sum of campesterol and β-sitosterol), (**i**) concentrations of sea-ice proxy IP_{25} and (**j**) sea-ice index $P_{dino}IP_{25}$ (for data see Supplementary Table 2). Purple and open triangles indicate the presence or absence of alkenones, respectively. Green line marks depth of the hiatus. Below the hiatus, sediments are overconsolidated. Zero or near-zero concentrations of IP_{25} and phytoplankton biomarkers are indicative for a closed (spring) sea-ice cover ($P_{dino}IP_{25}$ is indetermable and set to '1'; see Fig. 1 and Methods for further details). PIP_{25} values are used to identify low (0–20%), common (20–50%), ice-edge (50–70%) and extended (>70%) sea-ice concentrations (see Fig. 1). The section below the hiatus represents about one cold ('glacial')-warm ('interglacial') cycle. S1 and S2 indicate two scenarios of different sea-ice concentrations (that is, maximum and transitional/stable ice edge, respectively) within this climate cycle. The interval between the two dashed horizontal lines marks period with minimum sea-ice cover. Using mean sedimentation rates of ~3.2 cm ky^{-1} (as calculated independently from close-by gravity cores[50] and seismic data[51]), the duration of the climate cycle is ~40 ky.

recovered at the lower slope at Transect 2 (Fig. 3a), a sharp unconformity or hiatus occurs at 370 cmbsf. This unconformity is characterized by a marked change in colour, an abrupt increase in wet-bulk density (WBD; related to enhanced sediment consolidation underneath the hiatus), a significant drop in magnetic susceptibility (MS) and a major change in the biomarker composition, and it separates this interval from the overlying young and unconsolidated (upper Quaternary) sediments (Fig. 4 and Supplementary Table 2). The common occurrence of the acritarch *Decahedrella martinheadii* clearly indicates a late Miocene age for the lower part of the sedimentary sequence (Supplementary Fig. 3). The co-occurrence of the dinoflagellate cyst *Nematosphaeropsis labyrinthus* in combination with the absence of the *Batiacasphaera micropapillata* complex, however, may suggest a placement of the PS87/106 sediments within the upper part of its stratigraphic range, that is, upper Tortonian to lower Messinian, when compared with IODP/ACEX Hole M2A and ODP Hole 907A[24–27]. Reworking of these late Miocene palynomorphs is excluded due to excellent preservation of the encountered delicate palynomorph specimens (see Methods for details of the age model of Core PS87/106).

Late Miocene SST and sea-ice records. The biomarker data of Core PS87/106 suggest significantly different late Miocene paleoenvironmental conditions in comparison with those predominant during Quaternary times (Fig. 4 and Supplementary Table 2). In the upper Miocene sediments, elevated concentrations of alkenones and alkenone-derived SST between 4 and 7 °C (or even 6–9 °C in case other calibrations are used; *cf.* Supplementary Table 2) indicate relatively warm, open-water and productive paleoenvironmental conditions in the central Arctic Ocean during the summer season. This is also supported by SST values of »0 °C determined in some samples from the ACEX Site (Supplementary Tables 2 and 3).

Our results reveal for the first time the occurrence of the biomarker sea-ice proxy IP_{25} in sediments as old as late Miocene. This proxy was developed by Belt *et al.*[10] and was before our study only found in Quaternary and Pliocene sediments[12,13]. The presence of IP_{25} in the PS87/106 sediments is indicative for the presence of (spring) sea ice in the late Miocene central Arctic Ocean (Fig. 1). In comparison with IP_{25} values from the Arctic Ocean surface sediments[16] and sediment trap data[28], the absolute IP_{25} concentrations ranging between 0.05 and 0.15 µg gOC^{-1} (Fig. 4 and Supplementary Table 2) are more than one order of magnitude lower. These differences are caused by an early degradation of biomarkers that already starts in the water column and reaches its maximum in the uppermost centimetres of the sediments[28,29]. On the other hand, both IP_{25} and phytoplankton biomarker concentrations determined in Core PS87/106 are in the same range than those determined in early-mid Holocene Arctic sediments[8,30].

Using the 'PIP$_{25}$ Index' as a more semi-quantitative proxy of paleo-sea-ice cover (see Fig. 1 and Methods), our data from Core PS87/106 point to a variable spring sea-ice coverage of ~20–70% in the lower part and ~100% in the upper part of the sequence

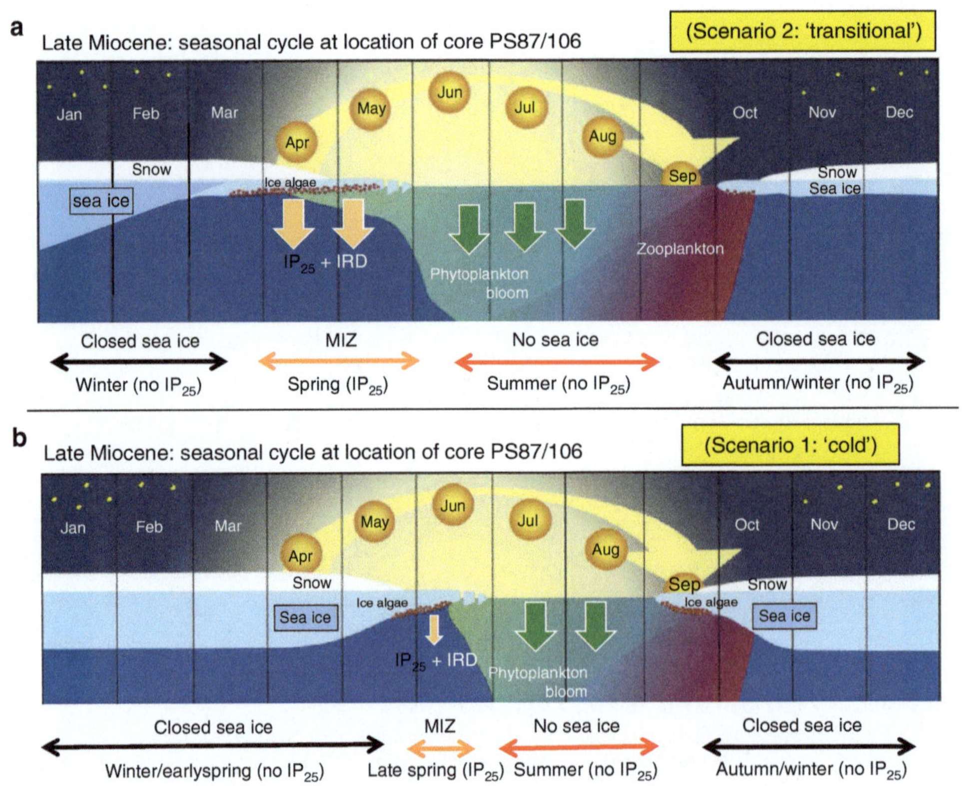

Figure 5 | Schematic illustrations of the seasonal sea-ice cycle during the late Miocene. The seasonal sea-ice cycle and related principal processes controlling productivity and carbon flux at location of central Arctic Ocean Core PS87/106 during the late Miocene are shown for two different scenarios. (**a**) Scenario 2 ('warmer/transitional situation') = extended period of spring sea-ice algae productivity and increased IP_{25} and phytoplankton biomarker fluxes. (**b**) Scenario 1 ('cold situation') = restricted period of late spring sea-ice algae productivity and very reduced fluxes (almost to zero) of IP_{25} and spring phytoplankton biomarkers; *cf.*, Fig. 1). MIZ, Marginal Ice Zone, that is, ice-edge situation. The dark period, height of the sun and changing thickness of snow and ice over the year, as well as phytoplankton, zooplankton and sea-ice productivity are shown. IP_{25} values for the different seasons are indicated (after ref. 72, supplemented).

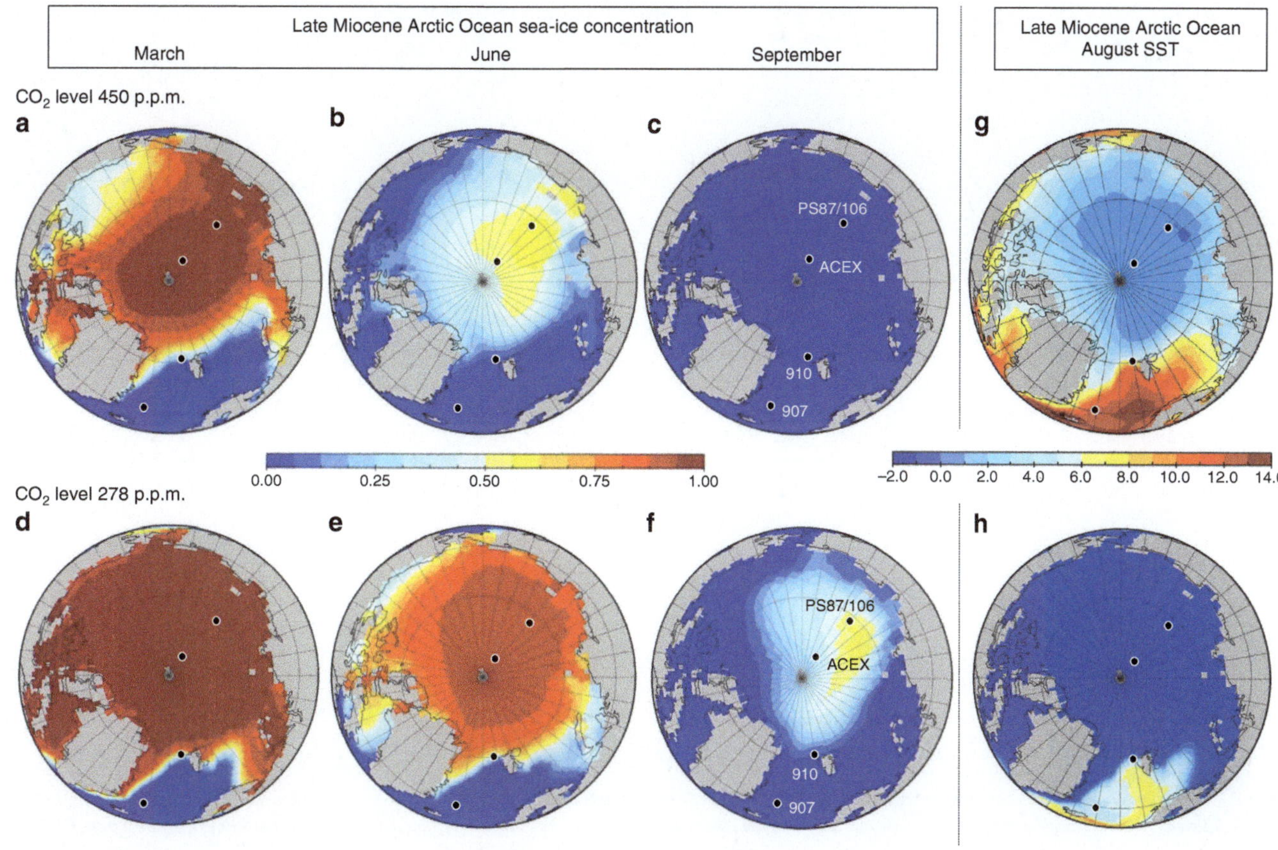

Figure 6 | Late Miocene Arctic Ocean climate simulations for high and low pCO_2 levels using a coupled atmosphere-ocean general circulation model. (**a–c**,**d–f**) Sea-ice concentrations for March, June and September at 450 and 278 p.p.m. CO_2 levels, respectively[32]. (**g**,**h**) Late Miocene August SST at 450 and 278 p.p.m. CO_2 levels, respectively.

(Fig. 4). The combination of IP_{25} and SST data indicates that the central Arctic Ocean must have been relatively warm and ice-free during summer throughout the time interval recovered in the sedimentary section of Core PS87/106 and variable sea ice must have existed during spring when daylight conditions allowed sea-ice algae production (Figs 1 and 5, and Supplementary Table 2). Furthermore, this implies the presence of an extended sea-ice cover during the dark, cold winter season. These new data clearly support that periods with only a seasonal sea-ice coverage must have occurred in the central Arctic Ocean during most of the late Miocene (Fig. 5; see further discussion below).

Simulations of late Miocene Arctic Ocean climate. Model simulations of global climate conditions reconstructing a warm late Miocene climate apply atmospheric CO_2 concentrations in the range of 280–700 p.p.m.[31–34]. Different reconstructions of late Miocene CO_2 levels have narrowed the uncertainties, with atmospheric CO_2 values likely to be below the present day concentrations[35,36]. Nevertheless, the associated uncertainties in the late Miocene CO_2 levels are at least 200 p.p.m.[34,37,38]. Therefore, we have re-analysed Miocene climate simulations[32,39] with CO_2 concentrations of 278 and 450 p.p.m. using a coupled AOGCM. The investigation focus is on seasonal changes in the central Arctic Ocean as simulated with the AOGCM[32,39] (for details and background, see Methods).

Our simulated mean August SST and mean sea-ice concentrations for March, June and September/August for high (450 p.p.m.) and low (278 p.p.m.) CO_2 levels (Fig. 6 and Supplementary Table 4) indicate the following: (1) a winter season with a closed sea-ice cover in the central Arctic Ocean decreasing towards the marginal seas for 450 p.p.m. CO_2 (Fig. 6a) and a closed sea-ice cover in the entire Arctic Ocean including the marginal seas for 278 p.p.m. CO_2 (Fig. 6d); (2) spring season sea-ice concentrations of 20–60% for 450 p.p.m. CO_2 (Fig. 6b) and even 80–90% for 278 p.p.m. CO_2 (Fig. 6e); and (3) a summer season with ice-free conditions and SSTs $>0\,°C$ for 450 p.p.m. CO_2 (Fig. 6c,g) and a reduced but still present sea-ice cover of 10–50% and SSTs $<-1\,°C$ for 278 p.p.m. CO_2 (Fig. 6f,h).

The AOGCM-based monthly mean sea-ice concentrations for 450 and 278 p.p.m. CO_2 levels for late Miocene and preindustrial conditions at the ACEX, PS87/106, ODP 910 and ODP 907 sites are shown in Fig. 7 and—together with SST values—listed in Supplementary Table 4. Based on these results, the central Arctic Ocean was more or less ice free during summer and autumn (that is, between mid-June to mid-November) for 450 p.p.m. CO_2, whereas sea ice still existed during summer times for 278 p.p.m. CO_2 simulations. At the location of ODP Site 907, ice-free conditions occurred throughout the year for both simulations. In addition to the late Miocene and preindustrial simulations, future monthly mean sea-ice concentrations for AD 2100 and AD 2200 (following the Intergovernmental Panel on Climate Change medium stabilization scenario RCP4.5) (ref. 40) are shown (Fig. 7). These simulations suggest that the summer sea-ice concentrations will become significantly reduced to 20–40% during the next two centuries.

In combination with our new SST and sea-ice proxy data, these results suggest that either late Miocene CO_2 levels have been relatively high or, alternatively, the applied model has an overly weak sensitivity especially in the northern high latitudes, as in the

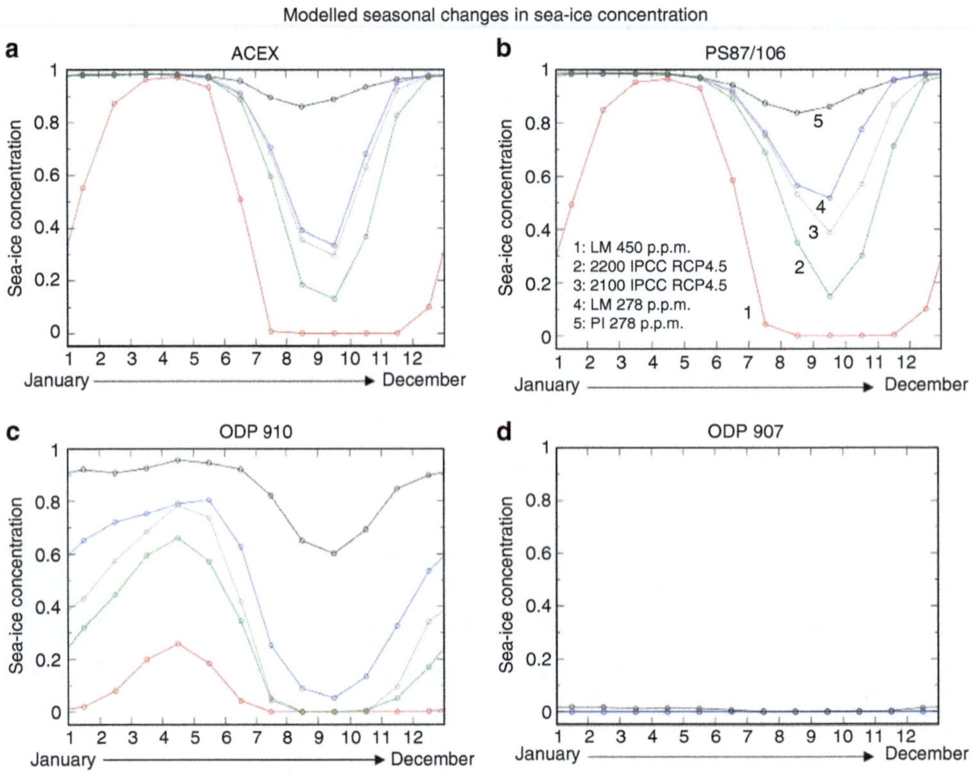

Figure 7 | Modelled monthly sea-ice concentration at the four different core locations. (**a**) ACEX Site, (**b**) Core PS87/106, (**c**) ODP Site 910 and (**d**) ODP Site 907. Climatological monthly mean sea-ice concentrations for late Miocene conditions with atmospheric CO_2 concentrations of 450 p.p.m. (red curve) and 278 p.p.m. (blue curve), modelled sea-ice concentration for preindustrial conditions (black curve) with an atmospheric CO_2 concentration of 278 p.p.m.[32], as well as future conditions (monthly mean 2,085–2,115, brown curve; monthly mean 2,185–2,215, green curve) following the Intergovernmental Panel on Climate Change (IPCC) medium stabilization scenario RCP4.5 (Gierz *et al.*[40]) are shown.

simulation with a relatively low CO_2 level of 278 p.p.m. summer SSTs are too low to explain our sedimentary SST proxy indicative of 4–7 °C. (Fig. 4 and Supplementary Tables 3 and 4). The latter reflects a characteristic challenge of current model approaches to simulate warm climates in the geological past[41,42]. Hence, the new findings from the Arctic region provide an enhanced basis for groundtruthing global climate reconstruction and modelling.

Discussion

Late Miocene climatic conditions significantly warmer than today have been reconstructed from marine and terrestrial proxy records from different localities around the globe[43–47]. However, quantitative SST proxy data from the High Arctic are exclusively restricted to a few terrestrial records[48,49]. With our study, we show for the first time that the late Miocene central Arctic Ocean was relatively warm with SSTs of ~5 °C (Figs 4 and 8) and ice-free during summer, whereas sea ice occurred during spring and autumn/winter. During the late Miocene a general cooling trend is obvious as recorded in the SST record of the sub-Arctic ODP Site 907 (Fig. 8 and Supplementary Table 5). Such a cooling trend seems to be supported by the limited number of SST values available from the ACEX site (Fig. 8). Comparing this general cooling trend with the SST values from Core PS87/106, the sedimentary section of Core PS87/106 below the hiatus probably represents a time span within the upper Tortonian to lower Messinian as supported by palynomorph stratigraphy (Fig. 8). Furthermore, the new Arctic SST data fit in very well with the grander long-term Cenozoic cooling pattern (see Supplementary Fig. 4 and references in the figure legend).

Although on a first view this short sedimentary section of Core PS87/106 only represents a short snapshot of late Miocene Arctic climate, more detailed information about the late Miocene climate on a regional to even global scale can be obtained from our record. Based on the biomarker data, the 1.3-m-thick late Miocene section of Core PS87/106 probably represents almost one cycle with extended and reduced spring sea-ice conditions (Fig. 4). The (almost) absence of phytoplankton biomarkers and IP_{25} (PIP_{25}= '1'; see Fig. 1) and low concentrations of terrigenous biomarkers may be explained by an extended to closed sea-ice cover and a very restricted spring season (Scenario 1 in Figs 4 and 5). Scenario 2, on the other hand, represents a transitional phase with a stable ice edge during an extended and productive spring season, characterized by maximum input of phytoplanktion biomarkers and IP_{25} (resulting in PIP_{25} values of 0.4–0.7), as well as maximum input of terrigenous biomarkers (and ice-rafted debris (IRD)) (Scenario 2 in Figs 4 and 5). The interval between scenarios 1 and 2 is characterized by very low to zero IP_{25} concentrations and increased concentrations of phytoplankton biomarkers, resulting in a distinct PIP_{25} minimum (Fig. 4). This interval is interpreted as a period of minimum spring sea-ice extent. Furthermore, maximum values of alkenones may reflect increased productivity of haptophyte algae during the summer season.

Using mean sedimentation rates of ~3.2 cm ky^{-1} as calculated independently from close-by gravity cores[50] and seismic data[51], the duration of this cycle is about 40 ky, that is, very similar to the 41 ky obliquity cycle (Fig. 4 and also see ref. 52 and references therein). Hence, our record may represent just one obliquity cycle with ice-free conditions during summers in both the cold ('glacial') and the warm ('interglacial') phase of this climate cycle (Figs 4 and 5). As Core PS87/106 probably is of upper Tortonian to lower Messinian age (see above),

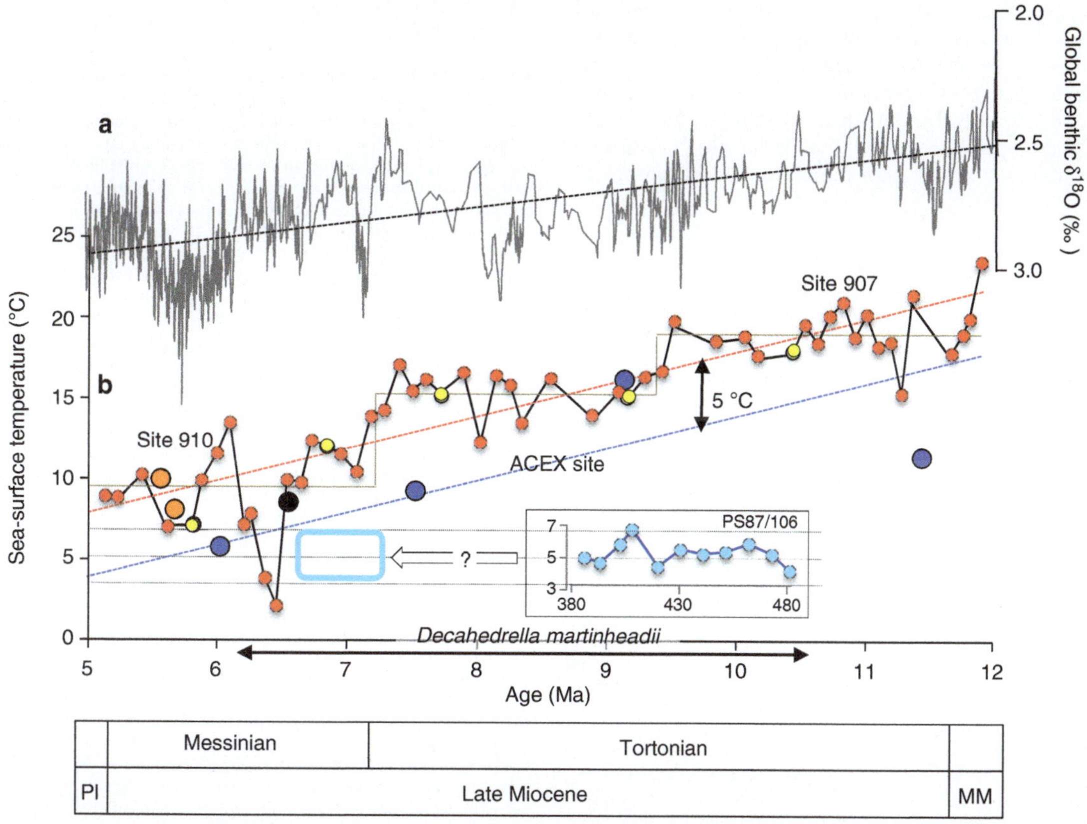

Figure 8 | Global climate (benthic $\delta^{18}O$) stack and (sub-) Arctic SST records for the Late Miocene time interval. (**a**) Global benthic $\delta^{18}O$ stack from ref. 43. (**b**) Late Miocene (Messinian/Tortonian) alkenone-based $U^{K'}_{37}$-SST of Site 907 (yellow circles)[27] and this study (red circles; Supplementary Table 5), TEX_{86}-SST values of Site 910 (orange circles)[12] and alkenone-based $U^{K'}_{37}$-SST of the ACEX Site (blue circles; this study/Supplementary Table 2). The ages of the five ACEX samples are based on ^{10}Be stratigraphy[73]. General cooling trends in the 907 SST record (stippled red line), the ACEX SST record (stippled blue line, assuming a parallel SST trend to Site 907 with an offset of about 5 °C) and the benthic $\delta^{18}O$ record (stippled black line) are indicated. The alkenone-based $U^{K'}_{37}$-SST of Core PS87/106 are plotted versus depth (centimetres below seafloor) (*cf.*, Supplementary Table 2). The stratigraphic range of the acritarch *D. martinheadii* in the sub-Arctic/Arctic realm is indicated[24,25]. As the central Arctic Ocean ACEX and the PS87/106 SST values should be similar, the PS87/106 record probably repesents a time span within the upper Tortonian to lower Messinian (light blue box). Pl, Pliocene; MM, Middle Miocene.

ice-free summer conditions should have occurred in the central Arctic Ocean during the warmer Middle Miocene to early Late Miocene time interval *a fortiori* (Fig. 8). In contrast to several previous studies, we therefore propose that a seasonal sea-ice cover was predominant in the central Arctic Ocean during (most of) the late Miocene time interval (see discussion below).

Our new semi-quantitative sea-ice and SST records are an important contribution to the ongoing and controversial debate about the reconstruction of the early (pre-Quaternary) Arctic Ocean sea-ice cover. Within this debate, the distinction between seasonal and perennial sea ice is critical, because year-round sea ice in the central Arctic implies very different climate feedback mechanisms, that is, Earth's albedo and heat exchange conditions, than an environment with ice-free conditions during summer. As outlined above, our new proxy data and modelling data clearly indicate a late Miocene seasonal sea-ice cover. Similar or even warmer climatic conditions also occurred in the Middle Miocene Arctic Ocean when looking at the alkenone-based summer SSTs of 10–13 °C reconstructed from ACEX sediments (Supplementary Fig. 4 and also see ref. 53). Abundant marine palynomorphs and foraminifers found in Miocene and early Pleistocene sediments of the ACEX section also point to at least periods with seasonally ice-free conditions[24,54]. These data are in contrast to an Arctic Ocean perennial sea-ice cover from middle Miocene onwards, as proposed by Darby[55] and Krylov *et al.*[56] based on their provenance studies of IRD in ACEX sediments. Taking the IRD with a North American or East Siberian origin found in the ACEX sediments and using modern sea-ice drift trajectories and velocities, these authors concluded that more than 1 year was needed to transport the sediments entrained in the sea ice to the ACEX location. Hence, the sea ice must have survived the summer melting season to reach the ACEX site.

Using a 'coupled sea-ice slab-ocean model including sediment transport by sea ice', Tremblay *et al.*[57] have most recently demonstrated that sea-ice drift was probably significantly faster under warmer climatic conditions with less or much thinner sea ice than today. They conclude that the presence of ice-rafted sediment of Eurasian and North American origin at the North Pole is not a definite indication of a perennial sea ice cover in the Arctic Ocean. This may demonstrate that the main assumption of Darby[55] and Krylov *et al.*[56] should be regarded critically.

Our proxy and model reconstruction of late Miocene Arctic climate is certainly a substantial step forward to improve the understanding of the pre-Quaternary Arctic Ocean sea-ice and SST history. However, to decipher the pre-Quaternary climate history of this unique and sensitive but still not well-known region on Earth in more detail, long continuous sedimentary records to be obtained only by scientific drilling are needed. These records are planned to be recovered within a new IODP drilling campaign scheduled for 2018 (IODP Proposal 708; http://

www.iodp.org/expeditions). Based on the new PS87 seismic data, ~200 m of Plio–Pleistocene, >600 m of Miocene and >300 m of Oligocene–Eocene may be recovered at the proposed drill sites[51] (see Fig. 2 for locations). The outcome of such a new programme will certainly help to improve our understanding of the complex ocean–atmosphere–ice system in the polar high northern latitudes and its role in the past, modern and future global climate.

Methods

Shipboard surveys and measurements. The methods used on board *Polarstern* during Expedition PS87 are shortly outlined in the following. For a more detailed information about the use and interpretation of the proxy data we refer to the different chapters of the Cruise Report[17].

The bathymetric survey was performed using the hull-mounted ATLAS Hydrographic HYDROSWEEP DS3, a deep-sea multi-beam swath sonar system with a resolution of up to 320 receive beams per ping, a swath width of 4–5 times the water depth and a vertical resolution of ~0.5% of the water depth. It was operated in the chirp mode with a frequency of 14–16 kHz. The mean sound velocity of the water column was calculated from conductivity-temperature-density (CTD), expandable CTD (XCTD) and Valeport Sound Velocity Profiler data.

Sub-bottom profiling data were acquired using the parametric hull-mounted system ATLAS Hydrographic PARASOUND DS III-P70. Primary operating frequencies were 18.75 and 22.95 kHz with a secondary sediment-penetrating frequency of 4.2 kHz, a beam angle of 4° and a pulse length of 2. The vertical resolution is ~0.2 m. As a result of the narrow beam angle, reflections from strata dipping by >4° cannot be received by the vessel. This explains why the thin veneer of post-slide sediments covering the slump scar is not resolved as it is above and below the headwall (Fig. 3b). The headwall has an inclination >4°. In contrast, the older (pre-Quaternary) sediments exhibit nearly horizontal bedding (Fig. 3), which are thus acoustically resolved along the headwall to their near-seafloor location. PARASOUND data visualization and processing was performed using ATLAS PARASTORE-3 software. The vertical scale on profiles has been converted from travel time to metres using a constant sound velocity of 1.5 km s^{-1}, which explains minor differences in water depth between PARASOUND and swath-sonar data.

For the MCS data acquisition, a 3,000-m-long streamer (240 active channels, group interval of 12.5 m) and an air gun array of four G-Guns (total volume of 32 l, fired with 200 bar every 15 s) were used. Processing included sorting, that is, common depth point sorting with 25 m spacing, frequency filtering (20–180 Hz), velocity analysis, multiple suppression and stacking.

Whole-core measurements included non-destructive, continuous determinations of core geometry (diameter), WBD, P-wave velocity (Vp) and loop-sensor MS at 10 mm intervals, using a standard Multi-Sensor Core Logger (GEOTEK Ltd., UK). The principle of logging cores is described in more detail in the GEOTEK manual 'Multi-Sensor Core Logging', which can be downloaded from the web (http://www.geotek.co.uk).

Line-scan images (Supplementary Figs 1 and 2) were acquired with a Jai CV L107 camera with RGB (red-green-blue) channels at 630, 535 and 450 nm, respectively, mounted to an Avaatech XRF core scanner. The camera contains three charge-coupled device sensors and a beam splitter to separate the RGB signal. Images were acquired with a down-core resolution of ~70 µm.

Stratigraphic framework and marine palynology process. The general lithostratigraphic framework and age model of the upper Quaternary sedimentary sections recovered during Expedition PS87 are robust and based on lithostratigraphy, colour imaging, WBD and MS records characterized by very prominent minima and maxima, and correlation with other dated sediment cores from Lomonosov Ridge[17,50]. Based on this concept, MIS 6 to 1 were identified in most of the cores (Supplementary Fig. 1). Some cores from very steep sections of the slope along Transect 1 (Fig. 3b), on the other hand, contain multiple unconformities and a correlation to the reference Core PS87/086 was only partly possible (Supplementary Fig. 2). WBD typically increases sharply below the unconformities due to higher degrees of compaction, supporting a hiatus and 'older' sediments below (Supplementary Fig. 2). For this study dealing with the late Miocene climate history, only the identification and dating of these 'older' (Neogene) sedimentary sections are relevant. For assessing the age of these sediments, assemblages of agglutinated benthic foraminifers and palynomorphs were used and compared with biostratigraphic records obtained from the ACEX and North Atlantic sites[17,24–27] (Supplementary Table 1).

For sample preparation and processing of the agglutinated benthic foraminifers we refer to the PS87 Cruise Report[17] and further references therein. The procedure for studying the palynomorphs are as follows: dinoflagellate cysts and acritarchs were investigated in 25 samples (mainly core catcher) from 15 cores recovered along transects 1 and 2 (Fig. 3a and Supplement Table 1), with special emphasis on Core PS87/106. Sediment was freeze dried, weighed and processed using standard palynological maceration techniques including repeated treatment with cold HCl (10%) and cold HF (38–40%), no oxidation and sieving over a 10-µm-nylon mesh. The residue has been mounted with glycerine jelly on microscope slides, which were then scanned for dinoflagellate cysts and acritarchs using a light microscope at ×400 original magnification.

Age model of the late Miocene section of core PS87/106. Based on microfossil data (that is, palynomorphs and agglutinated benthic foraminifers), the core catcher samples from the sediment cores were barren or give a Pleistocene age (Supplementary Table 1). The only core providing any clear indication that old sediments are cropping out in the shallow sub-seafloor is Core PS87/106. Well-preserved specimens of organic-walled palynomorphs (dinoflagellate cysts and acritarchs) have been recorded in successive samples from the base of the core (including core catcher) up to 420 cmbsf (that is, 50 cm below the hiatus; Supplementary Table 1). Of the encountered species, the acritarch *D. martinheadii* provides evidence that the lower part of the core is composed of sediments of late Miocene age. This species is endemic to the high northern latitudes and its stratigraphic range has been discussed previously based on comprehensive reviews of its occurrence at several DSDP, ODP and IODP sites from the Central Arctic Ocean, Norwegian-Greenland Sea, Labrador Sea, Baffin Bay and Irminger Sea[24–26]. It is restricted to the late Miocene in the Arctic and subarctic realm, and based on the pristine paleomagnetic record of Iceland Sea ODP Site 907, its stratigraphic range is independently calibrated against the astronomically tuned Neogene Timescale, thus providing absolute age control (Supplementary Fig. 3). A near-synchronous highest occurrence at *ca.* 6.3–6.2 Ma has been defined from several northern high latitude sites, suggesting this species to be an excellent marker across the subpolar/polar North Atlantic and Arctic Ocean[25]. In addition, a highest common occurrence is recognized at 6.5 Ma in Iceland Sea ODP Hole 907A and more generally at *ca.* 6.7–6.3 Ma across the Norwegian-Greenland Sea. The lowest occurrence of *D. martinheadii* has been calibrated to 10.5 Ma in ODP Hole 907. According to Schreck *et al.*[25], however, the lowest occurrence is not very well constrained at other sites but certainly younger than 11 Ma across the northernmost North Atlantic and Arctic Ocean.

In Core PS87/106, the consistent occurrence of *D. martinheadii* is accompanied by low numbers of the dinoflagellate cyst *N. labyrinthus*. Such co-occurrence has also been observed within the upper part of its stratigraphic range in IODP Hole M2A[24] and ODP Hole 907 (refs 25,27). Furthermore, the *B. micropapillata* complex has not been recorded in PS87/106 samples. This dinoflagellate cyst dominates assemblages in the late Serravallian of IODP/ACEX Hole M2A, decreases significantly across the Tortonian and disappears close to the Tortonian/Messinian boundary[26]. In ODP Hole 907A, *B. micropapillata* complex dominates the assemblage until *ca.* 8.2 Ma after which it only occurrs sporadically until its highest common occurrence at *ca.* 4.5 Ma[26,27]. Therefore, the co-occurrence of *D. martinheadii* and *N. labyrinthus* in combination with the absence of *B. micropapillata* complex may allow to place the analysed interval of Core PS87/106 into the upper Tortonian to lower Messinian. However, we note that the highest occurrence derived from ODP Site 907 may represent a minimum age for this bioevent in the Central Arctic Ocean, as successive Neogene cooling may led to an earlier disappearance of species in the higher latitudes.

D. martinheadii and *N. labyrinthus* are both very delicate species that bear processes and trabeculae (ribbon-like bars), which tend to crumple easily. All specimens encountered during palynological analyses, however, are well preserved indicating *in situ* deposition.

Biomarker analyses. Extraction of 5–10 g of freeze-dried sediments was carried out using an accelerated solvent extractor (DIONEX, ASE200; 100 °C, 5 min, 1,000 psi) with dichloromethane:methanol (2:1, v/v) as the solvent. For quantification internal standards, 7-hexylnonadecane (7-HND, 0.076 µg per sample for IP_{25} quantification), squalane (2.4 µg per sample) and cholesterol-d_6 (cholest-5-en-3β-ol-D_6, 10 µg per sample for sterol quantification) were added before analytical treatment. Separation of the hydrocarbon and sterol fractions was carried out via open column chromatography (hydrocarbon fraction with 5 ml *n*-hexane, the sterol fraction with 6 ml *n*-hexane:ethylacetate (5:1, v/v)). The latter fraction was silylated with 500 ml BSTFA (bis-trimethylsilyl-trifluoroacet-amide) (60 °C, 2 h). IP_{25} and sterols were analysed by gas chromatography (GC)/mass spectrometry. Component assignment was based on comparison of GC retention times with those of reference compounds and published mass spectra (Supplementary Figs 5 and 6). The Kovats Index calculated for IP_{25} is 2,086. For the monounsaturated HBI alkene (HBI monoene) most recently found in ancient Arctic sediments and characterized by very similar chromatographic and mass spectral properties[58], the Kovats Index has been calculated as 2,090. As this new HBI monoene is absent in the investigated cores of this study, we have calculated the index from lower Pliocene sediment samples from ODP Site 911 to show that both compounds can clearly be separated by our analytical approach (see Supplementary Fig. 5a,b). The detection limit for quantification of IP_{25} (Agilent 7890B GC, Agilent 5977A Extractor MSD with Performance Turbo Pump) is 0.005 ng μl^{-1} in SIM (selected ion monitoring) mode. To obtain mass spectra in TIC (total ion current) the limit is 0.05 ng μl^{-1}. The retention indices for brassicasterol (as 24-methylcholesta-5, 22E-dien-3β-O-Si(CH3)3), campesterol (as 24-methylcholest-5-en-3β-O-Si(CH3)3) and β-sitosterol (as 24-ethylcholest-5-en-3β-O-Si(CH3)3) were calculated to be 1.018, 1.042 and 1.077 (normalized to cholest-5-en-3β-ol-D_6 set to be 1.000), respectively.

For the quantification of IP_{25}, its molecular ion (m/z 350) in relation to the abundant fragment ion m/z 266 of the internal standard (7-HND) was used

(SIM mode). The different responses of these ions were balanced by an external calibration (Supplementary Fig. 5c and also see ref. 28). Brassicasterol (24-methylcholesta-5,22E-dien-3β-O-Si(CH_3)$_3$), campesterol (24-methylcholest-5-en-3β-O-Si(CH_3)$_3$) and β-sitosterol (24-ethylcholest-5-en-3β-O-Si(CH_3)$_3$) were quantified as trimethylsilyl ethers using the molecular ions *m/z* 470, *m/z* 472 and *m/z* 486, respectively, in relation to the molecular ion *m/z* 464 of cholesterol-D_6.

More details about the identification and quantification of IP_{25} and the sterols are described elsewhere[8,28,29,58–61].

For more semi-quantitative estimates of the present and past sea-ice coverage, Müller *et al.*[14] combined the sea-ice proxy IP_{25} and phytoplankton biomarkers in a phytoplankton-IP_{25} index, the so-called 'PIP_{25} index' (Fig. 1):

$$PIP_{25} = [IP_{25}]/([IP_{25}] + ([\text{phytoplanktonmarker}] \times c)) \quad (1)$$

with c = mean IP_{25} concentration/mean phytoplankton biomarker concentration for a specific data set or core. As phytoplankton biomarkers brassicasterol and dinosterol were used, resulting in $P_{bras}IP_{25}$ and $P_{dino}IP_{25}$ values, respectively (see refs 8,16,29,59 for discussion of advantages and limitations of the PIP_{25} approach). Most recently, Smik *et al.*[62] introduced a HBI–III alkene as phytoplankton biomarker replacing the sterols in the PIP_{25} calculation. This modified PIP_{25} approach is far less dependent on the balance factor c and based on biomarkers from the same group of compounds (that is, HBIs) with more similar diagenetic sensitivity, certainly an important improvement for paleo-sea-ice reconstructions and comparison of records from different Arctic areas.

Our reconstruction of SST is based on long-chain C_{37} alkenones synthesized by haptophyte algae[63]. The $C_{37:3}$- and $C_{37:2}$-alkenones were present in all samples, whereas the $C_{37:4}$-alkenone was not found. For alkenone ($C_{37:2}$ and $C_{37:3}$) analysis, extraction of additional 6 g of freeze-dried sediment was carried out using the ASE method under same conditions as decribed above but with dichloromethane as the solvent. The separation of compounds was carried out by open column chromatography using 5 ml *n*-hexane, followed by 5 ml *n*-hexane:dichloromethane (1:1, v/v) and 5 ml dichloromethane for eluation of the alkenones. As internal standard *n*-$C_{36:0}$ (10 μg per sample) was added before any analytical treatment. The alkenones were analysed by GC. Individual alkenone ($C_{37:3}$ and $C_{37:2}$) identification is based on retention time and the comparison with an external standard (Supplementary Fig. 7). To exclude a possible coeluation of the alkenones with other compounds, the extracts were measured first as total extract, second after additional column cleaning with dichloromethane and third after saponification[64]. The instrument stability has been continuously controlled by re-runs of an external alkenone standard (extracted from cultures of *Emiliania huxleyi* with known growth temperature) during the analytical sequences. The range of the total analytical error calculated by replicate analyses is $< 0.4\,°C$.

For calculation of SST, we used the simplified $U_{37}^{K'}$ Index[63]:

$$Uk_{37}^{'} = C_{37.2}/(C_{37.2} + C_{37.3}) \quad (2)$$

$U_{37}^{K'}$ was converted to SST according to the World Ocean core top versus annual temperature calibration ($U_{37}^{K'} = 0.033T + 0.044$) (ref. 64), the calibration most often used in the literature. Resulting SSTs vary between 4.2 and 6.7 °C (SST-1; Fig. 4 and Supplementary Table 2). For the central Arctic Ocean, these SSTs certainly have to be interpreted as summer SSTs (instead of annual mean) due to the darkness during late autumn to winter (*cf.*, Fig. 5). In addition, we also have used the Müller *et al.*[64] calibration versus summer SST (SST-2), the Sikes *et al.*[65] calibration versus summer SST obtained from the polar Southern Ocean (SST-3) and the Prahl and Wakeham[63] calibration obtained from cultural experiments (SST-4). Whereas the SST-4 values are more or less the same as the SST-1 values, the SST-2 and SST-3 values (calculated as 'summer SST') are higher and vary between 6 and 9 °C (Supplementary Table 2). Based on these results, we interpret our late Miocene summer SSTs of ~4–7 °C (mean of 5.3 °C) more as minimum values. For the Müller *et al.*[64] calibration, the standard error is reported as ± 0.050 $U_{37}^{K'}$ units or ± 1.5 °C for the entire temperature range from 0 to 27 °C. In the lower temperature range $< 10\,°C$, however, the scatter of the $U_{37}^{K'}$ values is significantly higher than the mean. Thus, one should not overinterpret the SST variability between 4 and 7 °C. In any case and independently of the calibration approach, summer SSTs were significantly higher than zero, preventing sea-ice formation during summer.

The results of the biomarker analyses (that is, alkenones, selected sterols, SSTs, IP_{25} and PIP_{25}) carried out on samples from various PS87 sediment cores, are listed in Supplementary Table 2. All biomarker data (expressed in μg gOC^{-1} and μg gSediment^{-1}) are available online at http://dx.doi.org/10.1594/PANGAEA.855509.

Model simulation. For the investigation of the late Miocene Arctic climatic conditions (that is, climatological sea-ice cover and SST) we have re-analysed Miocene climate simulations[32,39]. The simulations have been performed with a coupled AOGCM. The atmosphere model component ECHAM5 (ref. 66) was used at T31 resolution (~3.75°) with 19 vertical levels. The ocean component MPI-OM[67], including the dynamics of sea ice formulated using viscous-plastic rheology[68], has an average horizontal resolution of 3° × 1.8° with 40 uneven vertical layers. This modelling approach has been used and evaluated for investigations of the Miocene climate[32,39]. For the re-analyses, we have used data from two model runs with the same late Miocene set-up[32], except different atmospheric CO_2 concentrations. One simulation is based on a CO_2 concentration of 278 p.p.m.[32] and one uses a CO_2 concentration of 450 p.p.m.[39]. For further details of the AOGCM model configuration and the boundary conditions, we refer the reader to refs 32,39.

Sediment load and compaction experiments. The preconsolidation stress of a geological sample experienced in the past can be assessed by incrementally loading the specimen in a uniaxial deformation apparatus (so-called 'oedometer'), where deviations in the settling behaviour can be converted to the thickness of the missing overburden. We used this approach to estimate the thickness of sediment removed at prominent unconformities observed in some of our studied sediment cores (Fig. 4 and Supplementary Figs 2 and 8).

We conducted our experiments using a combined GIESA oedometer—direct shear apparatus in which both uniaxial compression tests and shear tests can be conducted in consecutive steps[69] (Supplementary Fig. 8b). The sample cell is a cylindrical volume within a stack of two steel plates. If desired, relative displacement of the plates enforces simple shear deformation in the sample, to measure undrained shear strength. Porous metal frits allow fluid communication with an open pore fluid reservoir (containing distilled water) and dissipation of excess pore pressure. Normal load is applied to the sample with a vertical ram and shear is induced by holding the upper plate fixed, while the lower plate is driven horizontally.

For the experiments in this study, we only used the consolidation function of the system. As is done routinely in such tests[70], we loaded each sample incrementally by starting with applied normal stresses of 10 kPa. This value was assumed to be below the *in situ* stress the samples had experienced before. After 24 h, the normal load is doubled. Tests were run over many days until the desired maximum normal load was reached (10,240 kPa for Core PS87/096 and 5,120 kPa for Core PS87/106). The initial water content of each sample was taken from an aliquot before the specimen was mounted into the GIESA oedometer. We measured the wet weight of the aliquot, subjected it to 48 h of gentle drying in an oven (60 °C) and then measured the dry weight, to calculate the void ratio e (where e = volume of voids/volume of solids).

Oedometer results are plotted as effective normal stress versus void ratio, the latter of which is derived from the settling (that is, change in sample thickness as determined using a vertical displacement transducer) after each loading increment. The graph of each experiment show a smooth function when e is plotted against normal stress, the latter on a logarithmic axis (Supplementary Fig. 8c). In an undeformed sample, the regular loading results in the so-called 'virgin consolidation curve', whereas samples that were previously subject to loading usually deviate from the smooth curvature and show a distinct change in gradient of the graph. The high-stress end of the graph is generally linear and serves to assess the maximum preconsilidation stress the sample has experienced following the procedure first established by A. Casagrande[69].

The stress measurements are then used to the thickness (h) of the overburden based on the effective load (h = preconsolidation stress/(bulk density of the sample − density of seawater) × g). The degree of overconsolidation of a given sample is then calculated as the ratio between preconsolidation stress and normal stress at that depth below the seafloor.

We have tested one pair of samples in cores PS87/096 and PS87/106, respectively, and loaded these samples to at least 5,120 kPa effective stress. Our results attest that the samples taken above the hiatuses are normally consolidated, while those underneath the discontinuity are overconsolidated. In Core PS87/096, the estimated preconsolidation stress below the hiatus is 590 kPa, which amounts to the removal of a 80-m-thick sediment package (Supplementary Fig. 8c). In Core PS87/106, the preconsolidation stress is between 320 and 480 kPa, which corresponds to 48 and 65 m of overburden, respectively (*cf.*, Fig. 4).

For Core PS87/096, the calculated thickness of removed sediments are more or less identical to those estimated from the ages of the sediments above and below the hiatus (that is, ~100 ka and < 2.5 Ma, respectively) and mean sedimentation rates of 3.2 cm ky^{-1} (refs 50,71), resulting in a maximum sediment removal of 80 m (Supplementary Fig. 8a).

References

1. ACIA Members. *Arctic Climate Impact Assessment Report* (Cambridge Univ. Press, 2005).
2. Stocker, T. *et al. Climate Change 2013, The Physical Science Basis* 1535 (Intergovernmental Panel on Climate Change (IPCC), Cambridge Univ. Press, 2013).
3. Stroeve, J., Holland, M. M., Meier, W., Scambos, T. & Serreze, M. Arctic sea ice decline: faster than forecast. *Geophys. Res. Lett.* **34,** L09501 (2007).
4. Backman, J., Moran, K., McInroy, D. B. & Mayer, L. A.the Expedition 302 Scientists. *Proceedings IODP 302* (Integrated Ocean Drilling Program Management International, College Station, 2006).
5. Moran, K. *et al.* The Cenozoic palaeoenvironment of the Arctic Ocean. *Nature* **441,** 601–605 (2006).
6. Stein, R. *et al.* Cenozoic Arctic Ocean climate history: some highlights from the IODP Arctic Coring Expedition (ACEX). *Dev. Marine Geol.* **7,** 259–293 (2014).
7. Thomas, D. N. & Diekmann, G. S. *Sea Ice* (Blackwell Publishing, 2010).
8. Stein, R., Fahl, K. & Müller, J. Proxy reconstruction of Arctic Ocean sea ice history: 'From IRD to IP_{25}'. *Polarforschung* **82,** 37–71 (2012).

9. De Vernal, A., Gersonde, R., Goosse, H., Seidenkrantz, M.-S. & Wolff, E. W. Sea ice in the paleoclimate system: the challenge of reconstructing sea ice from proxies—an introduction. *Quat. Sci. Rev.* **79,** 1–8 (2013).
10. Belt, S. T. *et al.* A novel chemical fossil of palaeo sea ice: IP25. *Org. Geochem.* **38,** 16–27 (2007).
11. Brown, T. A., Belt, S. T., Tatarek, A. & Mundy, C. J. 2014. Source identification of the Arctic sea ice proxy IP_{25}. *Nat. Commun.* **5,** 4197 (2014).
12. Knies, J. *et al.* The emergence of modern sea ice cover in the Arctic Ocean. *Nat. Commun.* **5,** 1–5 (2014).
13. Stein, R. & Fahl, K. Biomarker proxy IP_{25} shows potential for studying entire Quaternary Arctic sea-ice history. *Org. Geochem.* **55,** 98–102 (2013).
14. Müller, J. *et al.* Towards quantitative sea ice reconstructions in the northern North Atlantic: a combined biomarker and numerical modeling approach. *Earth Planet. Sci. Lett.* **306,** 137–148 (2011).
15. Müller, J., Massé, G., Stein, R. & Belt, S. T. Variability of sea-ice conditions in the Fram Strait over the past 30,000 years. *Nat. Geosci.* **2,** 772–776 (2009).
16. Xiao, X., Fahl, K., Müller, J. & Stein, R. Sea-ice distribution in the modern Arctic Ocean: biomarker records from Trans-Arctic Ocean surface sediments. *Geochim. Cosmochim. Acta* **155,** 16–29 (2015).
17. Stein, R. (ed.). *The Expedition PS87 of the Research Vessel Polarstern to the Arctic Ocean in 2014, Reports on Polar and Marine Research* Vol. 688 (Alfred Wegener Institute for Polar and Marine Research. http://epic.awi.de/37728/1/BzPM_0688_2015.pdf, 2015).
18. Kristoffersen, Y., Coakley, B. J., Hall, J. K. & Edwards, M. Mass wasting on the submarine Lomonosov Ridge, central Arctic Ocean. *Mar. Geol.* **243,** 132–142 (2007).
19. Jakobsson, M. *et al.* Arctic Ocean glacial history. *Quat. Sci. Rev.* **92,** 40–67 (2014).
20. Jakobsson, M. *et al.* An Arctic Ocean ice shelf during MIS 6 constrained by new geophysical and geological data. *Quat. Sci. Rev.* **29,** 3505–3517 (2010).
21. Niessen, F. *et al.* Repeated Pleistocene glaciation of the East Siberian Continental Margin. *Nat. Geosci.* **6,** 842–846 (2013).
22. Hegewald, A. & Jokat, W. Tectonic and sedimentary structures in the northern Chukchi region, Arctic Ocean. *J. Geophys. Res.* **118,** 3285–3296 (2013).
23. Weigelt, E., Jokat, W. & Franke, D. Seismostratigraphy of the Siberian Sector of the Arctic Ocean and adjacent Laptev Sea Shelf. *J. Geophys. Res. Solid Earth* **119,** 5275–5289 (2014).
24. Matthiessen, J., Brinkhuis, H., Poulsen, N. & Smelror, M. *Decahedrella martinheadii* Manum 1997—a stratigraphicall and paleoenvironmentally useful Miocene acritarch of the high northern latitudes. *Micropaleontology* **55,** 171–186 (2009).
25. Schreck, M., Matthiessen, J. & Head, M. J. A magnetostratigraphic calibration of Middle Miocene through Pliocene dinoflagellate cyst and acritarch events in the Iceland Sea (Ocean Drilling Program Hole 907A). *Rev. Palaeobot. Palynol.* **187,** 66–94 (2012).
26. Schreck, M. & Matthiessen, J. in *Biological and Geological Perspectives of Dinoflagellates.* (eds Lewis, J., Marret, F. & Bradley, L.) 301–314 (The Micropalaeontological Society, Special Publications: Geological Society, 2013).
27. Schreck, M., Méheust, M., Stein, R. & Matthiessen, J. Response of marine palynomorphs to Neogene climate cooling in the Iceland Sea (ODP Hole 907A). *Mar. Micropaleontol.* **101,** 49–67 (2013).
28. Fahl, K. & Stein, R. Modern seasonal variability and deglacial/Holocene change of central Arctic Ocean sea-ice cover: new insights from biomarker proxy records. *Earth Planet. Sci. Lett.* **351-352C,** 123–133 (2012).
29. Belt, S. T. & Müller, J. The Arctic sea ice biomarker IP_{25}: a review of current understanding, recommendations for future research and applications in palaeo sea ice reconstructions. *Quat. Sci. Rev.* **79,** 9–25 (2013).
30. Müller, J. *et al.* Holocene cooling culminates in sea ice oscillations in Fram Strait. *Quat. Sci. Rev.* **47,** 1–14 (2012).
31. Micheels, A. *et al.* Analysis of heat transport mechanisms from a Late Miocene model experiment with a fully-coupled atmosphere-ocen general circulation model. *Palaeogeogr. Palaeoclimatol. Palaeoecol.* **304,** 337–350 (2011).
32. Knorr, G., Butzin, M., Micheels, A. & Lohmann, G. A warm Miocene climate at low atmospheric CO_2 levels. *Geophys. Res. Lett.* **38,** L20701 (2011).
33. Steppuhn, A. *et al.* The sensitivity of ECHAM4/ML to a double CO_2 scenario for the Late Miocene and the comparison to terrestrial proxy data. *Glob. Planet. Change* **57,** 189–212 (2007).
34. Bradshaw, C. D. *et al.* The relative roles of CO_2 and palaeogeography in determining late Miocene climate: results from a terrestrial model-data comparison. *Clim. Past* **8,** 1257–1285 (2012).
35. Beerling, D. J. & Royer, D. L. Convergent Cenozoic CO_2 history. *Nat. Geosci.* **4,** 418–420 (2011).
36. Van de Wal, R. S. W., de Boer, B., Lourens, L. J., Köhler, P. & Bintanja, R. Reconstruction of a continuous high-resolution CO_2 record over the past 20 million years. *Clim. Past* **7,** 1459–1469 (2011).
37. Breecker, D. O. & Retallack, G. J. Refining the pedogenic carbonate atmospheric CO_2 proxy and application to Miocene CO_2. *Palaeogeogr. Palaeoclimatol. Palaeoecol.* **406,** 1–8 (2014).
38. Zhang, Y. G., Pagani, M., Liu, Z., Bohaty, S. M. & DeConto, R. A 40-million-year history of atmospheric CO_2. *Phil. Trans. R. Soc. A* **371,** 20130096 (2013).
39. Knorr, G. & Lohmann, G. Climate warming during Antarctic ice sheet expension at the Middle Miocene transition. *Nat. Geosci.* **7,** 376–381 (2014).
40. Gierz, P., Lohmann, G. & Wei, W. Response of Atlantic overturning to future warming in a coupled atmosphere-ocean-ice sheet model. *Geophys. Res. Lett.* **42,** 6811–6818 (2015).
41. PALAEOSENS Project-Members. Making sense of palaeoclimate sensitivity. *Nature* **491,** 683–691 (2012).
42. Huber, M. A sensitivity to history. *Nat. Geosci.* **6,** 15–16 (2013).
43. Zachos, J. C., Dickens, G. R. & Zeebe, R. E. An early Cenozoic perspective on greenhouse warming and carbon-cycle dynamics. *Nature* **451,** 281–283 (2008).
44. LaRiviere, J. P. *et al.* Late Miocene decoupling of oceanic warmth and atmospheric carbon dioxide forcing. *Nature* **486,** 97–100 (2012).
45. Zhang, Y. G., Pagani, M. & Liu, Z. A 12-million-year temperature history of the tropical Pacific Ocean. *Science* **344,** 84–87 (2014).
46. Pound, M. J. *et al.* A Tortonian (Late Miocene, 11.61-7.25 Ma) global vegetation reconstruction. *Palaeogeogr. Palaeoclimatol. Palaeoecol.* **300,** 29–45 (2011).
47. Lunt, D. J. *et al.* A methodology for targeting palaeo proxy data aquisition: a case study for the terrestrial late Miocene. *Earth Planet. Sci. Lett.* **271,** 53–62 (2008).
48. Wolfe, J. A. An analysis of Neogene climates in Beringia. *Palaeogeogr. Palaeoclimatol. Palaeoecol.* **108,** 207–216 (1994).
49. Utescher, T., Bondarenko, O. V. & Mosbrugger, V. The Cenozoic Cooling—continental signals from the Atlantic and Pacific side of Eurasia. *Earth Planet. Sci. Lett.* **415,** 121–133 (2015).
50. Stein, R. *et al.* Accumulation of particulate organic carbon at the Eurasian continental margin during late Quaternary times: controlling mechanisms and paleoenvironmental significance. *Glob. Planet. Change* **31,** 87–102 (2001).
51. Stein, R., Jokat, W., Niessen, F. & Weigelt, E. Exploring the long-term Cenozoic Arctic Ocean Climate History—A challenge within the International Ocean Discovery Program (IODP). *Arktos.* doi:10.1007/s41063-015-0012-x (2015).
52. Tian, J., Yang, M., Lyle, M. W., Wilkens, R. & Shackford, J. K. Obliquity and long eccentricity pacing of the Middle Miocene climate transition. *Geochem. Geophys. Geosys* **15,** 1740–1755 (2013).
53. Weller, P. & Stein, R. Paleogene biomarker records from the central Arctic Ocean (IODP Expedition 302): Organic-carbon sources, anoxia, and sea-surface temperature. *Paleoceanography* **23,** PA1S17, doi:10.1029/2007PA001472 (2008).
54. Cronin, T. M., Smith, S. A., Eynaud, F., O'Regan, M. & King, J. Quaternary paleoceanography of the central Arctic based on Integrated Ocean Drilling Program Arctic Coring Expedition 302 foraminiferal assemblages. *Paleoceanography* **23,** PA1S18 (2008).
55. Darby, D. A. Arctic perennial ice cover over the last 14 million years. *Paleoceanography* **23,** PA1S07, doi:10.1029/2007PA001479 (2008).
56. Krylov, A. A. *et al.* A shift in heavy and clay mineral provenance indicates a middle Miocene onset of a perennial sea-ice cover in the Arctic Ocean. *Paleoceanography* **23,** PA1S06, doi:10.1029/2007PA001497 (2008).
57. Tremblay, L.-B., Schmidt, G. A., Pfirman, S., Newton, R. & Derepentigny, P. Is ice-rafted sediment in a North Pole marine record evidence of perennial sea-ice cover? *Phil. Trans. R. Soc. A* **A373,** 20140168 (2015).
58. Cabedo-Sanz, P. & Belt, S. T. Identification and characterisation of a novel mono-saturated highly branched isoprenoid (HBI) alkene in ancient Arctic sediments. *Org. Geochem.* **81,** 34–39 (2015).
59. Belt, S. T. *et al.* Identification of paleo Arctic winter sea ice limits and the marginal ice zone: optimised biomarker-based reconstructions of late Quaternary Arctic sea ice. *Earth Planet. Sci. Lett.* **431,** 127–139 (2015).
60. Belt, S. T. *et al.* A reproducible method for the extraction, identification and quantification of the Arctic sea ice proxy IP_{25} from marine sediments. *Analyt. Methods* **4,** 705–713 (2012).
61. Belt, S. T. *et al.* An inter-laboratory investigation of the Arctic sea ice biomarker proxy IP_{25} in marine sediments: key outcomes and recommendations. *Clim. Past Discuss.* **10,** 155–166 (2014).
62. Smik, L., Cabedo-Sanz & Belt, S. T. Semi-quantitative estimates of paleo Arctic sea ice concentration based on source-specific highly branched isoprenoid alkenes: a further development of the PIP25 index. *Org. Geochem.* **92,** 63–69 (2016).
63. Prahl, F. G. & Wakeham, S. G. Calibration of unsaturation patterns in long-chain ketone compositions for palaeotemperature assessment. *Nature* **330,** 367–369 (1987).
64. Müller, P. J., Kirst, G., Ruhland, G., von Storch, I. & Rosell-Melé, A. Calibration of the alkenone paleotemperature index U^{k}_{37}-based on core-tops from the eastern South Atlantic and the global ocean (60 degrees N -60 degrees S). *Geochim. Cosmochim. Acta* **62,** 1757–1772 (1998).
65. Sikes, E. L., Volkman, J. K., Robertson, L. G. & Pichon, J.-J. Alkenones and alkenes in surface waters and sediments of the Southern Ocean: implications for paleotemperature estimation in polar regions. *Geochim. Cosmochim. Acta* **61,** 1495–1505 (1997).

66. Roeckner, E. *et al.* The atmospheric general circulation model ECHAM5. Part 1. Model description. *Max Planck Inst. Meteorol. Rep.* **349,** ISSN 0937-1060 (2003).
67. Marsland, S. J., Haak, H., Jungclaus, J. H., Latif, M. & Röske, F. The Max-Planck-Institute global ocean/sea ice model with orthogonal curvilinear coordinates. *Ocean Model* **5,** 91–127 (2003).
68. Hibler, W. III. A dynamic thermodynamic sea ice model. *J. Phys. Oceanogr.* **9,** 815–846 (1979).
69. Ikari, M. J. & Kopf, A. J. Cohesive strength of clay-rich sediment. *Geophys. Res. Lett.* **38,** L16309, 5 (2011).
70. Craig, R. F. *Craig's Soil Mechanics* 446 (Spon Press, 1974).
71. Stein, R. *Arctic Ocean Sediments: Processes, Proxies, and Palaeoenvironment* 2, 587 (Elsevier, 2008).
72. Wassmann, P. Arctic marine ecosystems in an era of rapid climate change. Progr. *Oceanography* **90,** 1–17 (2011).
73. Frank, M. *et al.* Beryllium isotopes in central Arctic Ocean sediments over the past 12.3 million years: stratigraphic and paleoclimatic implications. *Paleoceanography* **23,** PA1S02, doi:10.1029/2007PA001478 (2008).

Acknowledgements

We gratefully thank Captain Schwarze and his crew of RV *Polarstern* for the excellent support and cooperation during the entire cruise. We thank the PS87 Geoscience Party for support in getting geophysical and geological shipboard data and sediments during the expedition (see 'Supplementary Note 1'). We thank three anonymous reviewers for numerous constructive suggestions for improvement of the manuscript. Thanks to Simon Belt (Biogeochemistry Research Centre, University of Plymouth/UK) for providing the 7-HND standard for IP_{25} quantification. This publication is a contribution to the Research Programme PACES II, Topic 3 (The earth system from a polar perspective: data, modelling and synthesis) of the Alfred Wegener Institute Helmholtz Centre for Polar und Marine Research (AWI). The study used samples and data provided by AWI (grant number AWI-PS87_01) and the IODP Program. This research is also supported by KOPRI's Basic Research Project (PE15062) and a National Research Foundation of Korea Grant funded by the Korean Government (2015M1A5A1037243).

Author contributions

R.S. and W.J. proposed the main research goals of the PS87 Expedition. R.S. was chief scientist of the PS87 Expedition and leader of the geology group, developed the lithostratigraphic framework of the PS87 sediment cores, and wrote the first draft of the manuscript. K.F. conducted all biomarker analyses, evaluation and quality control. M.S. and M.K. identified the assemblages of palynomorphs and benthic agglutinated foraminifers, respectively, used for biostratigraphy and age control. G.K. and G.L. developed the model and experimental design. G.K. carried out the experiments. F.N. compiled the physical property and PARASOUND data. L.J. aquired and processed multibeam bathymetry data. M.F. was responsible for XRF scanning and line-scan imaging of the sediment cores. J.M. led the offshore geological activities on deck and in the laboratories. W.J. led the offshore geophysical activities and data acquisition. C.G. processed the multi-channel seismic data. A.K. carried out the sediment load and compaction experiments. All authors contributed to the data interpretation and writing of the final version of the manuscript.

Additional information

Competing financial interests: The authors declare no competing financial interests.

9

Destructive tsunami-like wave generated by surf beat over a coral reef during Typhoon Haiyan

Volker Roeber[1] & Jeremy D. Bricker[1]

Storm surges cause coastal inundation due to setup of the water surface resulting from atmospheric pressure, surface winds and breaking waves. Here we show that during Typhoon Haiyan, the setup generated by breaking waves near the fringing-reef-protected town of Hernani, the Philippines, oscillated with the incidence of large and small wave groups, and steepened into a tsunami-like wave that caused extensive damage and casualties. Though fringing reefs usually protect coastal communities from moderate storms, they can exacerbate flooding during strong events with energetic waves. Typical for reef-type bathymetries, a very short wave-breaking zone over the steep reef face facilitates the freeing of infragravity-period fluctuations (surf beat) with little energy loss. Since coastal flood planning relies on phase-averaged wave modelling, infragravity surges are not being accounted for. This highlights the necessity for a policy change and the adoption of phase-resolving wave models for hazard assessment in regions with fringing reefs.

[1] International Research Institute of Disaster Science, Tohoku University 468-1-E304 AzaAoba, Aramaki, Aoba-ku, Sendai 980-0845, Japan. Correspondence and requests for materials should be addressed to V.R. (email: roeber@irides.tohoku.ac.jp) or to J.D.B. (email: bricker@irides.tohoku.ac.jp).

The town of Hernani is a fishing and agricultural community exposed to the open Pacific Ocean in Eastern Samar, in the Philippines (Fig. 1). The tsunami-like wave that struck the town during Typhoon Haiyan on 8 November 2013 (Fig. 2)[1] astonished both villagers and disaster managers, as the coast near Hernani is sheltered by a broad fringing coral reef, which was expected to serve as a reliable wave defence[2]. During the authors' interviews of town residents as well as the survey conducted by Nobuoka *et al.*[3], witnesses reported that three to four destructive bores struck the town, and that the duration of the flooding caused by each bore was on the order of minutes. Since storm-induced ocean waves generally have periods of <20 s, the nature of this event has remained a mystery.

Media have reported this event as a meteotsunami[4] or as storm surge[5], albeit storm surge refers to the elevated still water level due to barometric-, wind- and breaking-wave-induced setup, but not to wave-induced oscillations[6]. A meteotsunami can form when the forward propagation speed of a storm is close to the free long wave speed of the water body over which it travels (Proudman resonance)[7,8]. The translation speed of Typhoon Haiyan was at most $14\,\mathrm{m\,s^{-1}}$ (refs 9,10) but the propagation speed of a free long wave is $c=(gh)^{1/2}$, where g is the acceleration due to gravity and h is the local water depth. This would have required a water depth on the order of 20 m or less for the storm to have been able to keep up with the free long wave generated. However, the water depth offshore of Hernani quickly drops to kilometres (Fig. 1), precluding the possibility of a meteotsunami at this location.

On the basis of the authors' own interviews on site, local residents believed that the typhoon triggered a landslide, which in turn caused a near-field tsunami. However, no tsunamigenic earthquake or landslide activity was reported. Furthermore, a similar tsunami-like flood was reported to have occurred during the 12 October 1897 typhoon[11], underscoring the likelihood that the bore was a direct consequence of the typhoon, and not a tsunami that happened to occur simultaneously.

The grave significance of the damage in Hernani is that the mechanism that caused the destruction is unaccounted for in coastal hazard assessment, evacuation planning or structural building codes. In the deep water offshore of Hernani, wind-driven setup only grew to a few decimetres, similar in size to the setup caused by the barometric pressure drop[12]. This is in contrast to the situation of the area around Tacloban city (Fig. 1), which is located at the end of a shallow bay with a maximum water depth of 50 m. The strong winds from Typhoon Haiyan, together with a local basin and shelf seiche, pushed enough water onshore to cause up to 6 m of storm surge to flood the city[13]. In addition, the local wind-generated waves on top of this surge also destroyed buildings along the shoreline[12]. Storm surge and damage of the type seen in Tacloban was predicted with great accuracy[14], as similar wind-driven surges have been studied for decades, with some outstanding examples being Hurricane Katrina in the United States, the Bhola cyclone in Bangladesh and the Ise-Bay typhoon in Japan. Each of these events affected low-lying land at the head of a shallow bay, similar to the setting of Tacloban. Wind-driven storm surge is well understood because

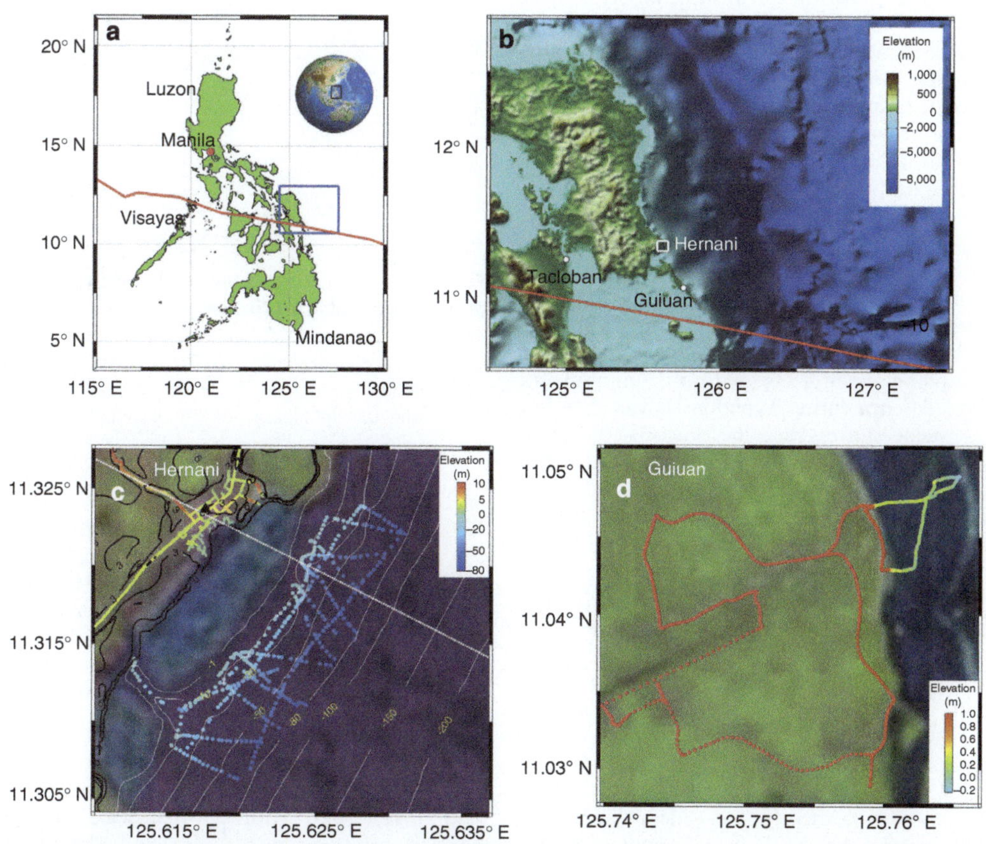

Figure 1 | Topography and bathymetry of Hernani. (**a**) World locator map[49] (inset) and locator map within the Philippines. (**b**) Ocean depth[34] near Leyte and Samar Islands. The red line is the storm track of Typhoon Haiyan[9]. (**c**) Map of measured topography and bathymetry in Hernani (Methods). The white dotted line indicates the transect used in the one-dimensional model. The black triangle is the location of the house seen in Fig. 2. (**c**) Map of reef topography survey in Guiuan. Data available from the U.S. Geological Survey.

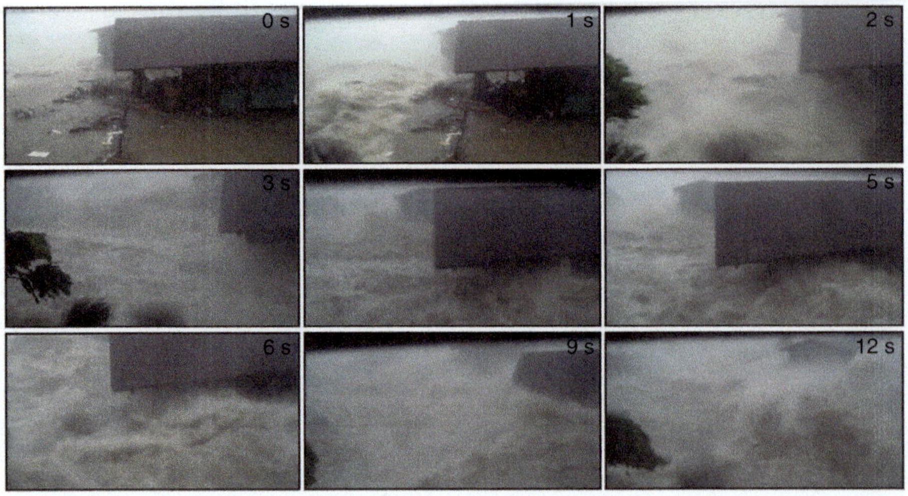

Figure 2 | Frames from the video of Gensis[1] showing a house being swept away by an infragravity-period wave. Onshore-directed flow lasted at least 20 s (the duration of the film). The base of the house is located 3 m above MSL. Images used with permission of Plan International.

of its frequent occurrence and large geographic extent. Furthermore, the physics of wind-driven storm surge is a simple balance between pressure gradient and parameterized wind stress, allowing it to be calculated with depth-averaged hydrostatic approximations of the Navier–Stokes equations. The physics responsible for damage in Hernani, however, are different.

Here results show that a numerical storm surge model coupled with a phase-averaged wave model cannot reproduce the destructive long-period wave that struck the town. A phase-resolving wave model, however, successfully captures this phenomenon because it resolves individual waves, the incidence of wave groups and the steepening of the resulting surf beat into a bore. The study illustrates that the abrupt wave-breaking process over the steep reef face that freed the initially bound underlying group wave energy was the driving factor of the destructive tsunami-like bore. Resonant amplification over the reef flat, which is a common process associated with tsunami waves[15], did not contribute significantly to the described wave phenomenon. Consequently, the wave would have been amplified if the incident wave group period had been closer to the reef's resonant period. Though Hernani's fringing reef protects the town from moderate storms, it exacerbated damage during Typhoon Haiyan therefore demonstrating that major storms bare the potential of locally generating tsunami-type flooding. This process is not being accounted for in disaster management plans or building codes. Phase-resolving wave models are suitable tools to account for this phenomenon in hazard planning, especially along reef-protected coasts around the world.

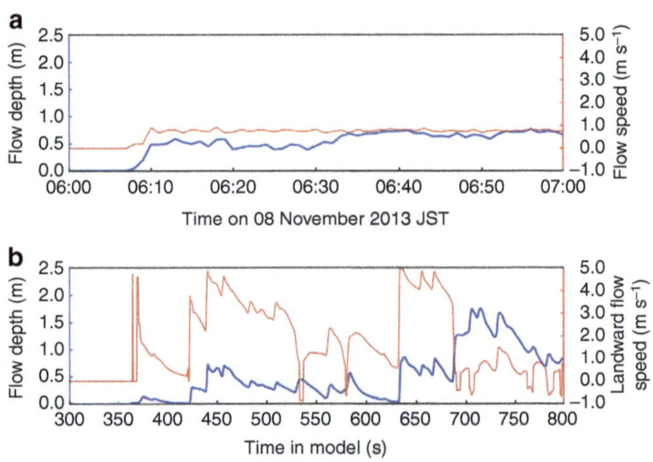

Figure 3 | Computed wave time series and spectra illustrating the physics responsible for the wave of Fig. 2a. (**a**) Time series of water surface elevation and flow speed from the Delft-3D/SWAN model at the location of the house in Fig. 2b. (**b**) Time series of water surface elevation and flow velocity (positive for landward flow) from the two-dimensional BOSZ model at the location of the house in Fig. 2.

Results

Storm surge and phase-averaged wave model. With deep water offshore of Hernani, a parametric hurricane model (Methods) together with the Delft-3D/SWAN[16,17] flow and phase-averaged spectral wave model calculates a combined wind- and pressure-driven surge of about 1.5 m above mean sea level (MSL)[12]. This modelling approach is the type used by the Federal Emergency Management Administration and the National Flood Insurance Program to determine the coastal hazard zone for flood insurance rate map assessment in the United States. The breaking-wave-induced radiation stress gradient induced an additional setup component of 2 m, resulting in a flow depth of 0.5 m and a flow speed up to 0.8 m s^{-1} (Fig. 3a) at the location of the house in Fig. 2, where the ground is 3 m above MSL. In addition, the model calculated short waves of 11 s peak period and 0.5 m significant wave height at this location. However, the computed surge was neither deep enough nor energetic enough to result in the flooding seen in Fig. 2, in which the flow depth and speed were around 2 m and 5 m s^{-1}, respectively, and landward-directed flow duration at the house was a minute or longer[3].

Background on tsunami-like surf beat. Previously, Nakaza and Hino[18] observed a similar wave phenomenon in Okinawa, which they investigated with a small-scale laboratory experiment. Nwogu and Demirbilek[19] utilized a phase-resolving Boussinesq wave model to analyse a similar set of laboratory experiments. The fundamental dynamics involved in this phenomenon are multiple. The presence of alternating large and small incident wave groups (also called sets) causes the wave setup on the beach to vary in time, resulting in surf beat with an infragravity period on the order of minutes[20,21]. The minimum wave envelope (group) return period[22] is a function of the shape of the offshore wave spectrum (Methods, equations (1) and (2)). In the frequency domain, surf beat energy may overlap with the natural quarter-wave oscillator frequency of the reef (Methods, equation (3)), and thus can amplify due to resonance. As the surf beat shoals, it can steepen and form a bore[18,23] (Methods). Individual incoming ocean waves also propagate on top of the wave setup, interact with it and run-up on land as shorter bores.

Phase-resolving wave model. To recreate the tsunami-like wave that struck Hernani, the offshore SWAN wave spectrum was input to the two-dimensional (plan view) Boussinesq-type phase-resolving wave model BOSZ[24,25] (Methods). Figure 3b shows water level and flow speed time series in close agreement with the flow at the house in Fig. 2. Figure 4b shows the spatial extent of the 1-m high bore that strikes the house at $t = 625$ s in Fig. 3b. In Fig. 3b, infragravity-period surges with steep bore-type fronts strike the house at model time 420 and 625 s. The interval between these bores is close to the theoretical return period of incident wave groups of 230 s evaluated from the offshore SWAN wave spectrum (Methods). The reef's resonant period, however, is ~480 s (Methods) and therefore does not significantly amplify the peak energy of the surf beat. In addition to infragravity-period surf beat, individual gravity waves propagate atop the water surface.

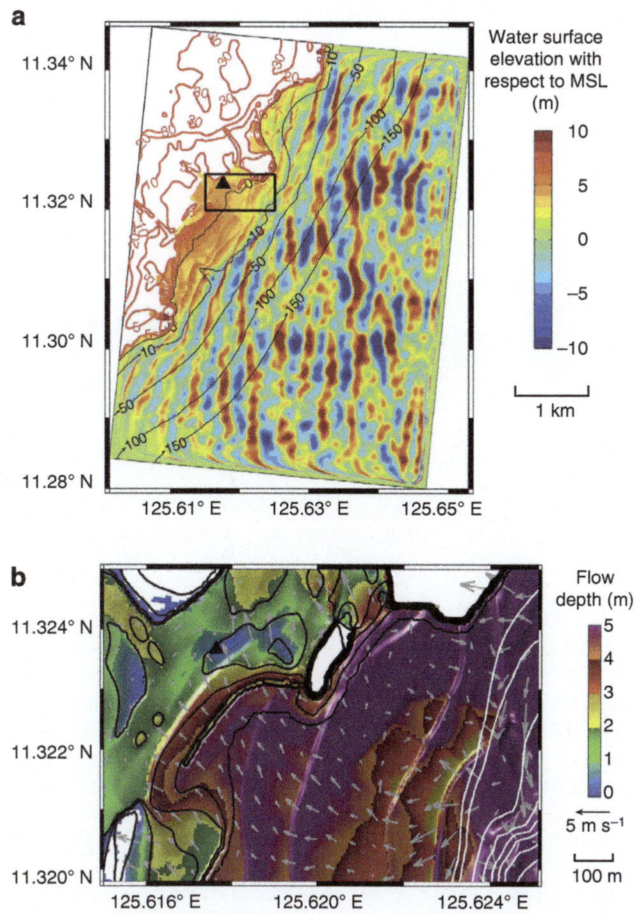

Figure 4 | Flow depth throughout Hernani when a bore strikes the house of Fig. 2. (**a**) Snapshot at $t = 625$ s of the two-dimensional BOSZ simulation. The black triangle indicates the location of the house in Fig. 2. Black and red lines are bathymetric and topographic contours, respectively, in metres above MSL. Wavemakers are located 290 m inside the domain from the South, East and North offshore boundaries. Sponge layers to absorb short-period oscillations are placed between the wavemakers and the boundary walls. The computed domain is 5 km long in the East-West direction and 6.9 m long in the North-South direction. An animation of this figure is available as Supplementary Movie 1. (**b**) Close-up on the reef and town. White lines are sea bottom elevation contours beginning at MSL and continuing downward at 5-m intervals. Black lines are land elevation contours beginning at 1 m above MSL and continuing upward at 1-m intervals to 10 m above MSL (higher elevations omitted). Arrows are flow velocity vectors. An animation of this figure is available as Supplementary Movie 2.

Wave group energy versus reef resonant amplification. To determine the relative contribution of offshore wave groups and reef resonance to amplification of the surf beat, BOSZ was evaluated in one-dimension along the transect shown in Fig. 5. For the actual bathymetry ('reef') case, most of the water level variance on the beach is at $f = 0.0035$ Hz (frequency f is the inverse of period T) as illustrated in Fig. 6. This is close to the theoretical wave group return frequency of $f = 0.004$ Hz. The hypothetical case of 'no reef' in Fig. 6 represents the bathymetry of the channel surveyed in Fig. 1, where freshwater river outflow prevented the reef from growing. For this 'no-reef' case, a peak in variance exists near the wave group return frequency, but it is much smaller than in the real reef case. For the hypothetical case of the town's seawall being infinitely high ('reef high wall' case in Fig. 6), another peak is present at $f = 0.002$ Hz, which corresponds to the reef's resonant frequency. Resonance has a much larger effect here than in the real bathymetry case because the high seawall limits the domain to the exact size of the quarter-wave oscillator, whereas in the real reef case the domain is lengthened by the run-up of the surf beat on land. The case of a steep 'drop-off' immediately offshore of the beach has a peak at the wave group frequency of 0.004 Hz of slightly larger magnitude than that of the 'reef'. Since the reef face bathymetry is identical for both 'reef' and

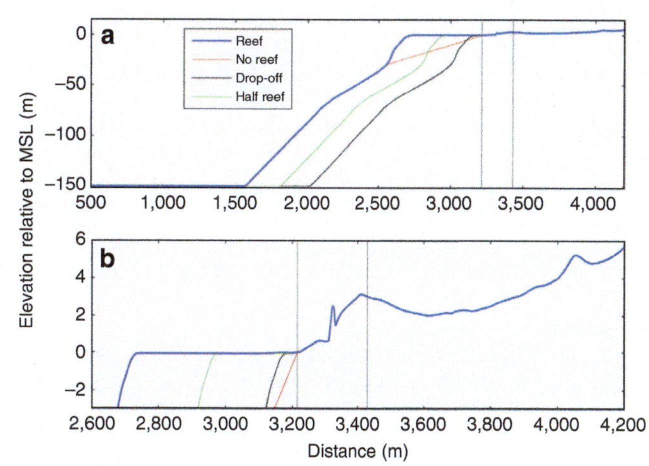

Figure 5 | Bathymetry and topography of transect used in one-dimensional BOSZ computations. The abscissa indicates the distance x along the transect of Fig. 1c from offshore (Southeast) to onshore (Northwest). (**a**) Full domain. (**b**) Zoomed in topography of the reef and onshore region. The right-hand thin black line indicates the location of the house swept away by the tsunami as seen in Fig. 2 and the left-hand thin black line is the location of the beach. The kink between the two thin black lines is the town's seawall. The red, green and medium black lines are hypothetical bathymetries, while the thick blue line is actual bathymetry.

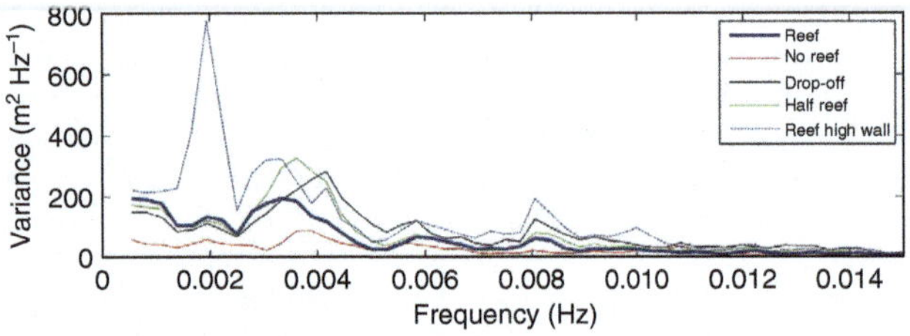

Figure 6 | Spectral analysis showing the physics responsible for the wave in Hernani. Variance spectral density of water surface elevation from the one-dimensional BOSZ model at the location of the beach.

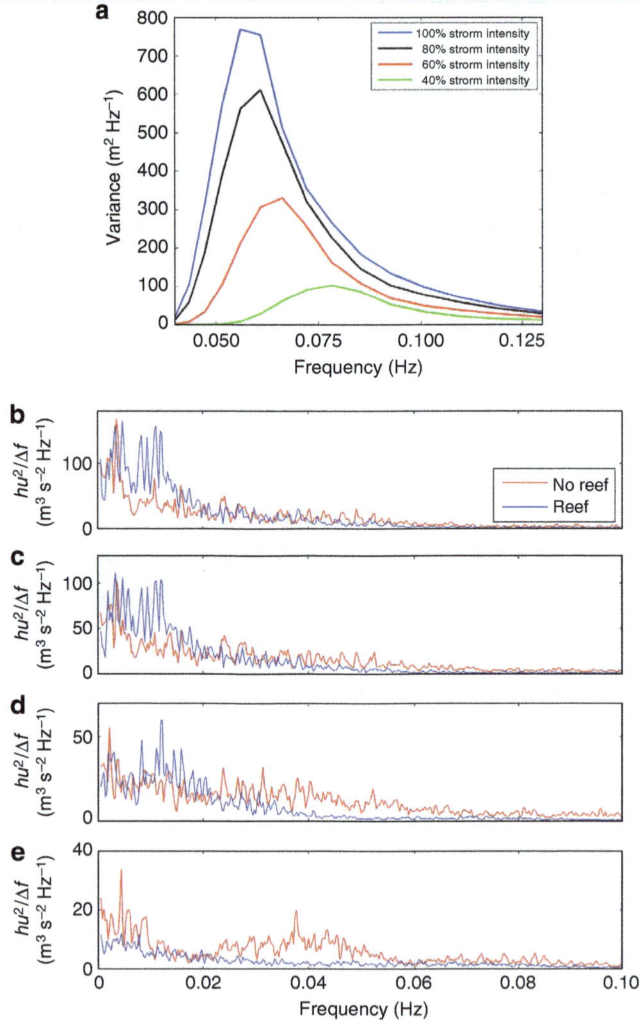

Figure 7 | Effect of the reef on storm damage. (**a**) Wave variance spectra for various typhoon intensities output from SWAN in 150 m water depth offshore of Hernani. These spectra are used to drive the wavemaker in the BOSZ and OpenFOAM domains of Fig. 5. The 100% intensity storm is the SWAN result using measured storm track data[9], while the others use storm track data with pressure drop and maximum wind speed reduced to the percent amount indicated. The resulting significant wave heights for decreasing storm intensity are 19.7, 17.1, 13.8 and 7.9 m, respectively. (**b–e**) Spectral density of hu^2, which is proportional to hydrodynamic force of equation (6), from the one-dimensional BOSZ model at the location of the beach for the 100, 80, 60 and 40% intensity storms, respectively.

'drop-off' cases, the similarity of the wave spectra between these two scenarios indicates that the mechanism responsible for the tsunami-like wave of Fig. 2 was mostly a result of the surf beat energy released through the abrupt wave-breaking process over the steep reef face. Resonant amplification of the surf beat over the reef flat played a secondary role. In contrast to the previous scenarios, the 'no-reef' case has a gradual wave-breaking zone, resulting in less energy available for the transfer of energy from gravity to infragravity frequencies (Methods).

Interestingly, the wave that struck Hernani was not a worst-case scenario: if a similar storm to Typhoon Haiyan was to strike a site fronted by a narrower reef with an equally steep reef face, resonance over the reef would become significant, with the potential of forming an even larger surf beat and subsequently more intense flooding. This is illustrated in Fig. 6 by the hypothetical case of a reef with only half the cross-shore length of the Hernani reef. The resonant frequency of this 'half reef' is 0.0042 Hz, which is close to the wave group return frequency and results in stronger amplification of the group wave energy compared with the 'reef' case.

Damage from large storms exacerbated by the reef. The effect of the presence of the reef on damage onshore was further investigated using one-dimensional BOSZ simulations. Storms of various intensities were modelled by scaling down the central pressure and maximum wind speed of Typhoon Haiyan (Fig. 7a). For each hypothetical storm, a drag law[26] (Methods) approximates the hydrodynamic force F on buildings near the beach. Figure 7b–e shows the spectra of F (proportional to hu^2, where h is water depth and u is flow speed) on the beach for the 'reef' and 'no-reef' cases. For storms of up to about 50% the strength of Typhoon Haiyan, the reef protects the shore from incident waves. During larger storms, however, it elevates the potential for damage due to an increasing hydrodynamic force in the infragravity frequency band.

Discussion

The tsunami-like wave that struck Hernani was the result of abrupt breaking of waves over the steep reef face generating an energetic surf beat. Had the surf beat been in resonance with the reef flat[15,27], the wave would have been even larger. Results show that at this site, the reef protects the shoreline during moderate storms, but can exacerbate damage during strong storms like Haiyan. This phenomenon was successfully reproduced by a phase-resolving wave model, indicating the necessity for adoption of such models in coastal hazard assessment for areas around the world that face a fringing reef-type bathymetry. Since the outlined tsunami-type flooding originated mainly from the energetic storm waves, the potential of such a destructive wave is not limited to tropical and subtropical areas with coral reefs. Alternately, sites in the path of extra-tropical storms can be exposed to similar scenarios if the local bathymetry favours the abrupt release of surf beat, which in turn can freely propagate onshore.

Methods

Field surveys. Land surface and coral reef elevations were measured from 25 to 27 May 2014 via walking and driving surveys with two Ashtech ProMark 100 differential global positioning system (DGPS) units, in base/rover differential configuration, and post-processed for vertical accuracy with GNSS Solutions software. The water depth seaward of the coral reef was measured from a hired fishing canoe, using a Hondex PS-7 handheld digital sounder, which has an operational range of up to 80 m. Sounder readings were photographed with a Canon PowerShot D20 global positioning system (GPS) camera to record the geographic location of each measurement. All elevation and depth measurements were referenced to MSL using the TPXO global tide model[28].

Since the elevation of the coral reef flat in Hernani was not measured during the 25–26 May field campaign in Hernani, the reef flat elevation of a similar reef stretch in Guiuan (35 km Southeast of Hernani; Fig. 1d) was measured during low tide on 27 May, with the same DGPS base/rover configuration as described above, and corrected to local MSL using TPXO[28]. The measured reef flat elevation varied from 10 cm above MSL to about 10 cm below MSL, though many of the deeply ponded areas on the reef flat were avoided during the survey. Therefore, for the storm surge modelling to follow, the reef flat in Hernani was assumed to have an elevation of 5 cm below MSL in the area surrounded by the topography and bathymetry measurements show in Fig. 1.

Parametric hurricane model. The behaviour of Typhoon Haiyan was hindcast using typhoon track data from the Japan Meteorological Agency[9], which was input into Holland's parametric hurricane model[29] for air-pressure field estimation, followed by the moving-typhoon model of Fujii and Mitsuda[30] as described in Veltcheva and Kawaii[31] for estimation of the wind field. Since the typhoon track data lacked information on radius to maximum wind, the relation of Quiring[32] was applied[12]. The hindcast pressure and wind fields were input into Delft-3D[16] and SWAN[17], a combined hydrodynamic and phase-averaged wave model package to hindcast the still water level and significant wave heights induced by the typhoon. An important adjustment to the default SWAN setup was the use of an air–sea drag coefficient limiter of 0.003 to prevent unphysical wave growth[33].

Delft-3D/SWAN storm surge and phase-averaged wave model. The Delft-3D/SWAN model was run using domain decomposition. The large domain (125° < East longitude < 130°, 10° < North latitude < 15°) had a resolution of 0.01°(about 1 km), and the small domain (125.59° < East longitude < 125.66°, 11.29° < North latitude < 11.36°) had a resolution of 0.0004°(about 40 m). Coarse bathymetry data was taken from GEBCO[34] and topography from SRTM[35]. Within the region of the field survey of Fig. 1c, the bathymetry and topography were interpolated from surveyed data. The Global Tide database TPXO[28] provided the tidal boundary condition to the hydrodynamic model. Bed roughness was parameterized with a uniform Manning's n of 0.035 s m$^{-1/3}$ (ref. 36). The Delft-3D model was run with a uniform time step of 6 s, while SWAN was run in stationary mode at 15-min intervals (a test was run in non-stationary mode, but results did not differ noticeably from stationary mode).

SWAN output was produced for various storm intensities to quantify the effect of the overall wave energy on the formation of the bore that struck Hernani. In Fig. 7a, the SWAN model result from Typhoon Haiyan[9] is labelled the 100% intensity storm. The hypothetical 80% storm was determined by using a storm track with 80% of the pressure drop in the eye and 80% of the maximum wind speed of the Typhoon Haiyan track data, then re-evaluating the radius to maximum winds, the parametric hurricane model and the Delft-3D/SWAN model. The procedure was repeated for the 60 and 40% storms.

BOSZ Boussinesq-type wave model. To generate the input wave time series for the phase-resolving BOSZ wave model[24,25], SWAN directional wave spectra output were extracted in 150 m water depth offshore of Hernani. The SWAN spectra are composed of 25 logarithmically spaced frequency bins between 0.04 and 0.3 Hz and 72 directional increments of 5° each. Since SWAN does not provide the phase angles between the spectral components, the common approach[37] is to treat each spectral component as a linear wave with a random initial phase. However, the discreteness of wave components leads to recycling of the superposed input time series and therefore to artificial wave groups with a return period of $1/df$. To avoid any artificial wave group generation, the SWAN spectra were resampled to a logarithmic spacing with a smallest interval of $df = 1/3{,}600$ s, which excludes repetitions of the input signal over the 1-h model run. Furthermore, to ensure quality of the BOSZ results, the high frequency tail of the SWAN spectra were truncated at 0.13 Hz and proportionally redistributed over the remaining frequency bins according to the constraints of the dispersion approximations in the model's governing equations, as described in Li *et al.*[38].

The wavemaker source function used in BOSZ[24,25] is built from those of Wei *et al.*[39] and Schäffer and Sørensen[40], and described in Li *et al.*[38]. To ensure uniform wave energy distribution over the portion of the BOSZ two-dimensional computational domain of uniform 150 m water depth, all three open ocean-facing domain boundaries contain wave-generating source functions. Wave shadowing due to the directionality of the spectral components is only visible in very limited regions along the Northern and Southern boundary (Fig. 4a). Therefore, the overall results of this study are not affected. For the one-dimensional analyses, the components from the two-dimensional wave spectrum are collapsed into a one-directional distribution of the wave energy over all frequency bins.

The resulting time series were used to evaluate the effects of surf beat and resonance over the reef in two dimensions (plan view) as well as along a one-dimensional transect passing through the location of the house seen washed away in Gensis' film[1]. The two-dimensional BOSZ model run was based on a grid with a uniform 5 m × 5 m resolution over a domain extending 5 km in the East–West direction, 6.9 km North–South, and rotated 5° clockwise to allow wave input centred around the peak direction as computed by SWAN (Fig. 4a). Owing to constraints in computational resources, only one two-dimensional computation was run for the results shown in Fig. 3b. Other cases were calculated with one-dimensional bathymetries based on a 5-m resolution (Fig. 5). For each bathymetry, 10 different one-dimensional BOSZ runs were evaluated, each with a unique random phase seed. Ensemble averages over these 10 runs for each bathymetry then lead to robust frequency spectra. Manning's n of 0.035 s m$^{-1/3}$ (ref. 36) was used uniformly and results of a sensitivity run with no friction were not significantly different.

Validation of BOSZ Boussinesq-type wave model. As no measurements of actual wave heights are available from the site, the phase-resolving depth-resolving Volume-of-Fluid model OpenFOAM[41], version 2.3.0, facilitates a redundant check of the validity of the conclusions reached with BOSZ. OpenFOAM was applied along the same one-dimensional transect with an identical input wave time series. The OpenFOAM model setup is described in detail in Bricker and Roeber[42]. Figure 8 shows a comparison between the results from BOSZ and OpenFOAM on both sides of the seawall shown in Fig. 5. The reasonable agreement between these two different types of phase-resolving wave models gives credence to the BOSZ results, in lieu of field measurements to compare with. In addition, BOSZ has been intensively validated with laboratory experiments conducted at the O.H. Hinsdale Wave Research Laboratory at Oregon State University that were specifically designed for wave propagation and bore formation over fringing reef-type

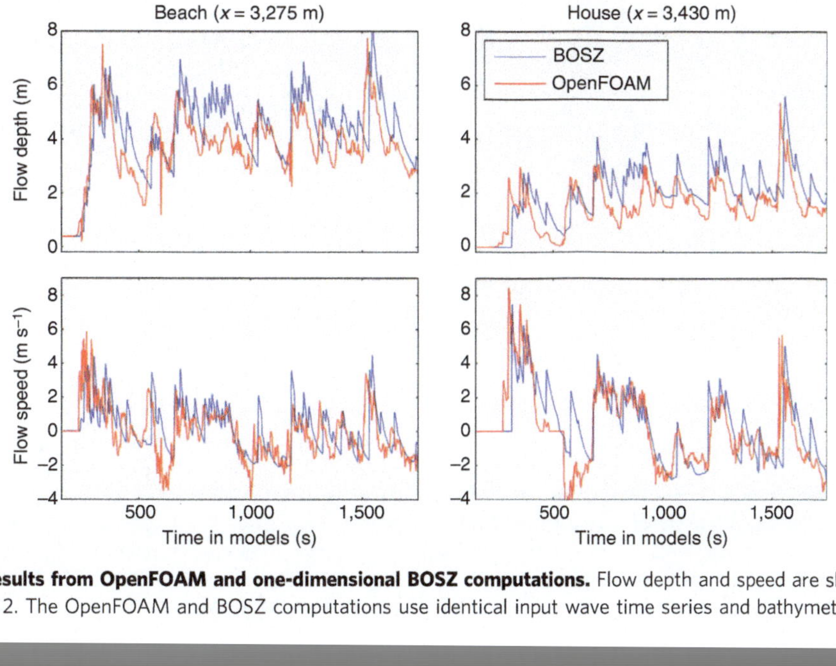

Figure 8 | Comparison of results from OpenFOAM and one-dimensional BOSZ computations. Flow depth and speed are shown at the locations of the beach and the house of Fig. 2. The OpenFOAM and BOSZ computations use identical input wave time series and bathymetry (Fig. 5).

bathymetries[43]. BOSZ and the data used to drive it are available via the contact on the authors' website at http://hydraulic.lab.irides.tohoku.ac.jp/.

Detailed results from BOSZ wave model. The model results can be understood through comparison with analytical oscillation periods of the water surface. In the incident wave time series, the minimum wave group return period is given by equation (1) (ref. 22)

$$T_{\text{envelope, min}} = \sqrt{2\pi e \frac{\mu_0}{\mu_2}} \tag{1}$$

where $\mu_0 = m_0$ and $\mu_2 = m_2 - m_1^2/m_0$. The spectral moments m_r are given by equation (2)

$$m_r = \int_0^\infty f^r E(f) df \tag{2}$$

where f is the frequency (Hz), E is the spectral energy density ($m^2\,Hz^{-1}$) and r is an integer. The actual wave group return period can be longer than $T_{\text{envelope,min}}$. For Typhoon Haiyan $T_{\text{envelope,min}}$ is 230 s, as represented by the 100% storm of Fig. 7a with a high frequency tail cutoff at 0.13 Hz. It should be noted that the cutoff position can slightly influence the group return period based on a shift in the spectral moments.

The resonant period of a quarter-wave oscillator such as a reef flat is given by equation (3)

$$T_{\text{resonance}} = \frac{4L}{(2n-1)\sqrt{gh}} \tag{3}$$

where $T_{\text{resonance}}$ is the resonant period, L is the cross-shore length of the reef, g is the acceleration due to gravity, h is the mean water depth over the reef and n is a positive integer defining the amount of nodes. This type of oscillation consists of a node at the reef face and an antinode at the beach. Assuming $L = 750$ m (the distance between the seawall and the 30-m bathymetric contour), $h = 4$ m (the mean water depth over the reef) and $n = 1$ (the fundamental period of oscillation), the reef resonant period is 480 s (corresponding to a frequency $f = 0.0021$ Hz).

The period of a basin seiche, a numerical artifact of the closed model domain, is given by equation (4)

$$T_{\text{basin}} = \frac{2L}{n\sqrt{gh}} \tag{4}$$

with the same variable definitions as in equation (3). This type of oscillation consists of an antinode at each side of the domain. For the one-dimensional transect of Fig. 5, the fundamental ($n = 1$) seiche period is the time it takes a long wave to propagate from the left wall of the domain, reflect from the reef face (at $x = 2{,}700$ m, which is the shoreward-most extent of this energy band in Fig. 9) and then propagate back to the left boundary. Numerically solving for the long wave speed in discrete cells across the variable-depth basin, this fundamental period is found to be 165 s ($f = 0.006$ Hz).

Figure 9 shows that for the real reef bathymetry, incident waves from offshore break on the reef crest at a distance of $x = 2{,}700$ m from the start of the transect. Landward of this location, energy in the gravity wave range is low, but a non-negligible amount of energy exists even beyond the seawall at $x = 3{,}314$ m, indicating ocean waves propagate atop the surf beat onto land. At infragravity frequencies, the largest energy is near $f = 0.0035$ Hz, which is the incident wave group period. This is strongest over the reef, but also energetic landward of the seawall. Energy is also substantial over the reef at its resonant frequency at $f = 0.0021$ Hz, though this band is not as strong in the area between the beach and the house location. Energy at low frequencies $f < 0.001$ Hz indicate the static setup induced by waves breaking over the reef. Energy in deep water at $f = 0.0021$ Hz and $f = 0.0035$ Hz is a result of reflection of the surf beat from the shore back into deep water as a free wave[21]. The fundamental basin seiche mode shows up as energy near $f = 0.006$ Hz and is an artifact of the closed model domain. Consequently, energy at $f = 0.012$ Hz is the first superharmonic of the basin seiche. The basin seiche shoals between the foot of the shelf slope at $x = 1{,}500$ m and the reef crest; shoaling transfers energy of a wave to its superharmonics[44]. This results in the energy bands in Fig. 9a that connect the antinodes of the basin seiche and its superharmonics to the reef crest and gravity wave frequencies.

For the 'no-reef' case, gravity wave energy dissipates gradually between the reef crest and the beach at $x = 3{,}200$ m. Almost no energy in the gravity wave range exists landward of the seawall at $x = 3{,}314$ m. At infragravity frequencies, the same basin seiches as in the 'reef' case are present, though they contain no energy landward of the seawall. Significant energy near the wave group frequency $f = 0.0035$ Hz does exist landward of the seawall; however, the amount of energy is smaller than in the real reef case. The only other infragravity energy present is at a very low frequency, representing the static setup.

Figure 9 shows that between $x = 2{,}700$ m (offshore extent of the surf zone) and $x = 3{,}200$ m (beach), the 'no-reef' case displays more energy than the 'reef' scenario. However, landward of the beach, much more infragravity energy exists with the

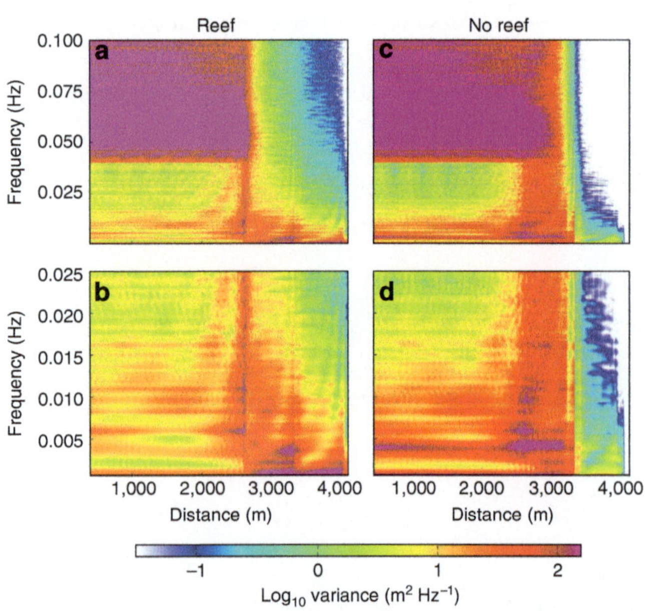

Figure 9 | Ensemble-averaged power spectral density of water surface elevation at all points along the one-dimensional transect from BOSZ. (**a,b**) Plots from the real reef case and (**c,d**) plots from no-reef case. Panels **a** and **c** show the full frequency range, while panels **b** and **d** are zoomed in on infragravity frequencies.

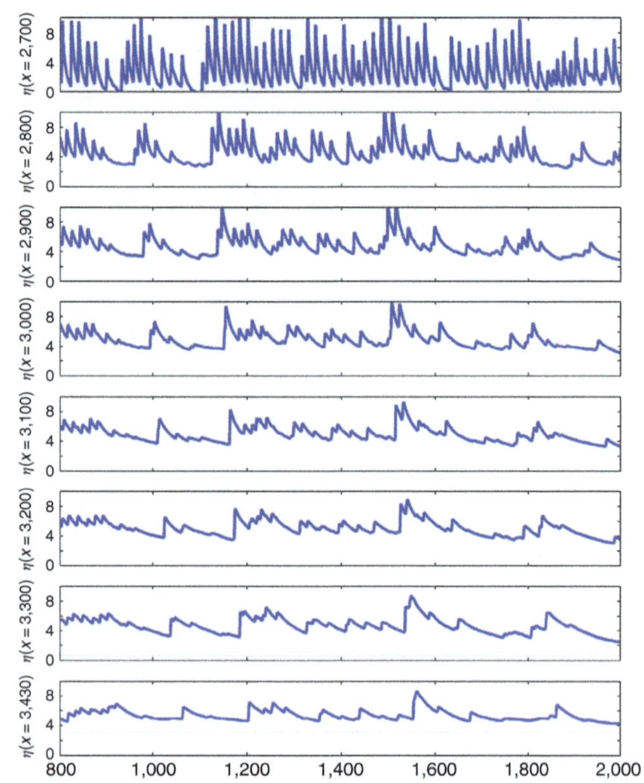

Figure 10 | Bore formation and propagation. Time series of water level (metres above MSL) at increments from the breaking location on the reef face ($x = 2{,}700$ m), atop the reef ($x = 2{,}800$-$3{,}200$ m), up the beach in front of the seawall ($x = 3{,}000$ m) and at the location of the house in Fig. 2 ($x = 3{,}430$ m). All x values refer to distance along the transect of Fig. 5. At the locations on shore ($x = 3{,}300$ and 3,430 m), the grey line indicates the elevation of the ground.

'reef' bathymetry. Schäffer[45] proved that two effects control the amount of infragravity energy present at a beach: the incident bound wave groups and the time-varying migration of the gravity wave break point. Since these processes oppose each other to some extent, short surf zones experience less dissipation of infragravity energy than broad ones. Interestingly, the variance spectral density of 200 $m^2 Hz^{-1}$ present over a frequency range of 0.002 Hz in Fig. 6 is similar in magnitude to the computed variance spectral densities of past seismic tsunamis events over reefs in Samoa[15] and along the shoreline of Chile[46].

Bore formation. van Dongeren *et al.*[23] used laboratory experiments and an infragravity wave model to show that surf beat caused by wave groups can break and form a bore over gently sloping beaches. Battjes *et al.*[47] formulated a dimensionless normalized bed slope parameter for infragravity waves as shown in equation (5)

$$\beta = \frac{h_x}{\omega}\sqrt{\frac{g}{h_b}} \quad (5)$$

where h_x is the bed slope, ω is the radial frequency of the infragravity wave, g is the acceleration due to gravity and h_b is the depth of wave breaking. Their Fig. 2 shows that for the mild slope case of $\beta < 0.5$, shoaling amplifies the surf beat and it can form a bore. For the steep slope case of $\beta > 0.5$, most of the infragravity energy is reflected, so a bore does not form. In a subsequent study, van Dongeren *et al.*[48] simulated the behaviour of surf beat over a reef and found that over a rough reef a bore does not form because bottom friction alone is sufficient to dissipate the energy of the surf beat, but over a frictionless reef a small bore forms. However, this study only considered small incident waves, on which frictional dissipation has a large influence.

In the case of Hernani, the slope of the reef face is $h_x = 0.2$ and the slope of the beach is $h_x = 0.017$ (Fig. 5). The offshore significant wave height is 19.7 m. Assuming a wave-breaking parameter of 0.75 (ref. 43), the water depth where waves break would be approximately $h_b = 15$ m. For the wave spectrum of Typhoon Haiyan (the 100% storm of Fig. 7a), the wave group period is $T_{envelope,min} = 230$ s, resulting in $\omega = 0.027$ radians per s. This results in $\beta = 6.5$ for the reef face and $\beta = 0.5$ for the beach. Thus, bore formation is not expected over the steep reef face between the break point and the reef crest, but is possible over the beach shoreward of the reef. BOSZ computations (Fig. 10) show that at some times the surf beat rises gradually over the reef and then steepens into a bore on the beach. This is the case where the bore strikes the location of the house at $t = 1870$ s. At other times, such as at $t = 1,200$ s, a bore forms at the reef crest and maintains its shape across the entire reef and then up the beach. Over the reef itself, short-period gravity waves are dissipated, but the infragravity signal is not significantly affected. In the steep 'drop-off' case (not shown), infragravity bore formation is similar, but short-period gravity waves are less damped than in the case with the reef.

Though Fig. 10 shows multiple waves striking the house over 1,200 s of BOSZ computation, it is important to note that this result was obtained with the assumption of a stationary input sea state based on the SWAN and Delft-3D outputs at 06:00 JST. At this time the sea state was still near its maximum and the still water setup had also reached its highest level (Fig. 3a). However, after 06:00 JST, SWAN computes that offshore wave heights dropped rapidly (not shown), even though the pressure- and wind-induced setup persisted. Since the pressure- and wind-driven setup lagged the offshore sea state, only a short time window existed for the surf beat to reach its most destructive form. It might be due to this short time window that residents reported only three or four infragravity waves striking the town, while Fig. 10 shows a greater number of waves.

Drag law for hydrodynamic force. The hydrodynamic force on a building[26] can be estimated by equation (6)

$$F = \frac{1}{2}\rho C_D B h u^2 \quad (6)$$

where ρ is the water density, C_D is an empirical drag coefficient, B is the width of an exposed building in the cross-flow direction, h is the flow depth and u is the flow speed.

References

1. Gensis, N. Plan International Video: Typhoon Washes House Away http://plan-international.org/about-plan/resources/videos/video-typhoon-washes-house-away/ (2013).
2. Ferrario, F. *et al.* The effectiveness of coral reefs for coastal hazard risk reduction and adaptation. *Nat. Commun.* **5,** 3794 (2014).
3. Nobuoka, H. *et al.* Coastal flood disaster due to historic Typhoon Haiyan in Hernani, Eastern Samar Province, Philippines. *J. Jpn Soc. Civil Eng. Ser. B2* **70,** I_1426–I_1430 in Japanese (2014).
4. Storm surge. *Wikipedia* http://en.wikipedia.org/wiki/Storm_surge (2014).
5. BBC News Asia. Typhoon Haiyan Storm Surge Destroys House in Seconds http://www.bbc.com/news/world-asia-24985592 (2013).
6. US Army Corps of Engineers, Coastal and Hydraulics Laboratory, Glossary of Coastal Engineering Terms http://chl.erdc.usace.army.mil/glossary (2015).
7. Monserrat, S., Vilibic, I. & Rabinovich, A. B. Meteotsunamis: atmospherically induced destructive ocean waves in the tsunami frequency band. *Nat. Hazards Earth Syst. Sci.* **6,** 1035–1051 (2006).
8. Hibiya, T. & Kajiura, K. Origin of the Abiki phenomenon (a kind of seiche) in Nagasaki Bay. *J. Oceanogr. Soc. Jpn* **38,** 172–182 (1982).
9. Japan Meteorological Agency. Regional Specialized Meteorological Center Tokyo – Typhoon Center, Best Track Data http://www.data.jma.go.jp/fcd/yoho/data/typhoon/T1330.pdf (2013).
10. Takagi, H. *et al.* Track analysis, simulation, and field survey of the 2013 Typhoon Haiyan and Storm Surge. *J. Flood Risk Manag.* doi:10.1111/jfr3.12136 (2015).
11. Bankoff, G. in *A World of Water: Rain, Rivers, and Seas in Southeast Asian Histories.* (ed. Boomgaard, P.) (KITLV Press, 2007).
12. Bricker, J. *et al.* Spatial variation of damage due to storm surge and waves during Typhoon Haiyan in the Philippines. *J. Jpn Soc. Civil Eng. Ser. B2* **70,** I_231-I_235 (2014).
13. Mori, N. *et al.* Local amplification of storm surge by Super Typhoon Haiyan in Leyte Gulf. *Geophys. Res. Lett.* **41,** 5106–5113 (2014).
14. Sun Star. Storm Surge: Biggest Threat from Tropical Cyclone http://www.sunstar.com.ph/breaking-news/2013/11/11/storm-surge-biggest-threat-tropical-cyclone-313195 (2014).
15. Roeber, V., Yamazaki, Y. & Cheung, K. F. Resonance and impact of the 2009 Samoa tsunami around Tutuila, American Samoa. *Geophys. Res. Lett.* **37,** L21604 (2010).
16. Deltares. *Delft3D-FLOW User's Manual.* Version 3.15 (2011).
17. Booij, N., Ris, R. & Holthuijsen, L. A third-generation wave model for coastal regions, part I, model description and validation. *J. Geophys. Res.* **104,** 7649–7666 (1999).
18. Nakaza, E. & Hino, M. Bore-like surf beat in a reef zone caused by wave groups of incident short period waves. *Fluid Dyn. Res.* **7,** 89–100 (1991).
19. Nwogu, O. & Demirbilek, Z. Infragravity wave motions and runup over shallow fringing reefs. *J. Waterway Port Coastal Ocean Eng.* **136,** 295–305 (2010).
20. Longuet-Higgins, M. S. & Stewart, R. W. Radiation stress and mass transport in gravity waves, with application to 'surf beats'. *J. Fluid Mech.* **13,** 481–504 (1962).
21. Masselink, G. Group bound long waves as a source of infragravity energy in the surf zone. *Cont. Shelf Res.* **15,** 1525–1547 (1995).
22. Longuet-Higgins, M. S. Statistical properties of wave groups in a random sea state. *Phil. Transact. R. Soc. Lond. A* **312,** 219–250 (1984).
23. van Dongeren, A. *et al.* Shoaling and shoreline dissipation of low-frequency waves. *J. Geophys. Res.* **112,** C02011 (2007).
24. Roeber, V., Cheung, K. F. & Kobayashi, M. H. Shock-capturing Boussinesq-type model for nearshore wave processes. *Coastal Eng.* **57,** 407–423 (2010).
25. Roeber, V. & Cheung, K. F. Boussinesq-type model for energetic breaking waves in fringing reef environments. *Coastal Eng.* **70,** 1–20 (2012).
26. Applied Technology Council. *Guidelines for Design of Structures for Vertical Evacuation from Tsunamis.* Report No. FEMA P646(Federal Emergency Management Administration (FEMA), Washington, DC, USA, 2008).
27. Pequignet, A. C. N., Becker, J. M., Merrifield, M. A. & Aucan, J. Forcing of resonant modes on a fringing reef during tropical storm Man-Yi. *Geophys. Res. Lett.* **36,** L03607 (2009).
28. TPXO. The OSU TOPEX/POSEIDON Global Inverse Solution http://volkov.oce.orst.edu/tides/global.html (2014).
29. Holland, G. J. An analytic model of the wind and pressure profiles in hurricanes. *Mon. Weather Rev.* **108,** 1212–1218 (1980).
30. Fujii, T. & Mitsuda, Y. *Synthesis of a Stochastic Typhoon Model and Simulation of Typhoon Winds.* Annual Report No. 29B-1 229–239 (Disaster Prevention Research Institute, Kyoto University, Kyoto, Japan, 1986).
31. Veltcheva, A. & Kawai, H. *Investigation of the Typhoon Pressure and Wind Field with Application for Storm Surge Estimation.* Report No. 41 23–44 (Port and Airport Research Institute, Yokosuka, Japan, 2002).
32. Quiring, S., Schumacher, A., Labosier, C. & Zhu, L. Variations in mean annual tropical cyclone size in the Atlantic. *J. Geophys. Res.* **116,** D09114 (2011).
33. Powell, M. D., Vickery, P. J. & Reinhold, T. A. Reduced drag coefficient for high wind speeds in tropical cyclones. *Nature* **422,** 279–283 (2003).
34. GEBCO. General Bathymetric Chart of the Oceans. British Oceanographic Data Center http://www.gebco.net (2014).
35. NASA. Shuttle RADAR Topography Mission (SRTM) http://www2.jpl.nasa.gov/srtm/ (2014).
36. Bretschneider, C. L., Krock, H. J., Nakazaki, E. & Casciano, F. M. *Roughness of Typical Hawaiian Terrain for Tsunami Run-up Calculations: A User's Manual.* J.K.K. Look Laboratory ReportUniversity of Hawaii (1986).
37. Dean, R. G. & Dalrymple, R. A. *Water Wave Mechanics for Engineers and Scientists* 353 ppWorld Scientific, 1991).
38. Li, N. *et al.* Integration of coastal inundation modeling from storm tides to individual waves. *Ocean Modell.* **83,** 26–42 (2014).

39. Wei, G., Kirby, J. T. & Sinha, A. Generation of waves in Boussinesq models using a source function method. *Coastal Eng.* **36,** 271–299 (1999).
40. Schäffer, H. A. & Sørensen, O. R. On the internal wave generation in Boussinesq and mile-slope equations. *Coastal Eng.* **53,** 319–323 (2006).
41. Weller, H. G., Tabor, G., Jasak, H. & Fureby, C. A tensorial approach to computational continuum mechanics using object-oriented techniques. *Comput. Phys.* **12,** 620–631 (1998).
42. Bricker, J. D. & Roeber, V. Mechanisms of damage during Typhoon Haiyan: storm surge, waves, and 'tsunami-like' surf beat. in *Proceedings of the 36th IAHR World Congress*, 28 June–3 July (The Hague, The Netherlands) (accepted for publication) (2015).
43. Roeber, V. & Cheung, K. F. in *Proc. and Results of the 2011 NTHMP Model Benchmarking Workshop* (http://nws.weather.gov/nthmp/documents/nthmpWorkshopProcMerged.pdf 361–406 (NOAA, Washington, DC, USA, 2012).
44. Holthuijsen, L. H. *Waves in Oceanic and Coastal Waters* 387 (Cambridge University Press, 2007).
45. Schäffer, H. A. Infragravity waves induced by short-wave groups. *J. Fluid Mech.* **247,** 551–588 (1993).
46. Yamazaki, Y. & Cheung, K. F. Shelf resonance and impact of near-field tsunami generated by the 2010 Chile earthquake. *Geophys. Res. Lett.* **38,** L12605 (2011).
47. Battjes, J. A., Bakkenes, H. J., Janssen, T. T. & van Dongeren, A. R. Shoaling of subharmonic gravity waves. *J. Geophys. Res.* **109,** C02009 (2004).
48. van Dongeren, A. *et al.* Numerical modeling of low-frequency wave dynamics over a fringing corsal reef. *Coastal Eng.* **73,** 178–190 (2013).
49. Bezdek, A. & Sebera, J. MATLAB script for 3D visualizing geodata on a rotating globe. *Comput. Geosci.* **56,** 127–130 (2013).

Acknowledgements

Prof. Kwok Fai Cheung provided resources on a computer cluster at the University of Hawaii for the BOSZ computation. Shuichi Kure facilitated the IRIDeS field surveys in the Philippines. Troy Heitmann, Seth Guikema, Celso Ferreira, Hiroshi Takagi and Maarten van Ormondt contributed useful discussions. Akane Bessho provided graphic design. This work was funded by an IRIDeS Tokutei Project grant, and by the JSPS–NSF Cooperative Program for Interdisciplinary Joint Research Projects in Hazards and Disasters, project entitled 'Evolution of Urban Regions in Response to Recurring Disasters'.

Author contributions

V.R. developed and applied the BOSZ Boussinesq wave model; J.D.B. applied the parametric hurricane, Delft-3D/SWAN and OpenFOAM models.

Additional information

Competing financial interests: The authors declare no competing financial interests.

10

Morphodynamics of submarine channel inception revealed by new experimental approach

Jan de Leeuw[1], Joris T. Eggenhuisen[1] & Matthieu J.B. Cartigny[2]

Submarine channels are ubiquitous on the seafloor and their inception and evolution is a result of dynamic interaction between turbidity currents and the evolving seafloor. However, the morphodynamic links between channel inception and flow dynamics have not yet been monitored in experiments and only in one instance on the modern seafloor. Previous experimental flows did not show channel inception, because flow conditions were not appropriately scaled to sustain suspended sediment transport. Here we introduce and apply new scaling constraints for similarity between natural and experimental turbidity currents. The scaled currents initiate a leveed channel from an initially featureless slope. Channelization commences with deposition of levees in some slope segments and erosion of a conduit in other segments. Channel relief and flow confinement increase progressively during subsequent flows. This morphodynamic evolution determines the architecture of submarine channel deposits in the stratigraphic record and efficiency of sediment bypass to the basin floor.

[1] Faculty of Geosciences, Utrecht University, PO Box 80021, 3508TA Utrecht, The Netherlands. [2] National Oceanography Centre, European Way, Southampton, Hampshire SO14 3ZH, UK. Correspondence and requests for materials should be addressed to J.d.L. (email: J.deleeuw@uu.nl).

Extensive channelized seascapes have been revealed by seafloor surveys[1–4]. The channels are characterized by a continuous thalweg along which sediment-laden turbidity flows dominantly bypass sediment[5]. Submarine channels can be up to several kilometres wide and hundreds of kilometres long, and provide the transport pathways for large quantities of sediment, nutrients and carbon into the deeps of the world's ocean[1,6], where the material is collected in basin-floor fans that form the largest sediment accumulations on the planet.

Seafloor and outcrop evidence demonstrates that channels are associated with erosion into underlying deposits[7] and aggradation of deposits in levees, channel fills and splays[8]. Fundamentally, different causalities have been suggested in the spatial and temporal relations between erosive and depositional changes to the submarine landscape. Some studies[9,10] envision an evolution where 'only after an initial erosional phase and channel establishment are turbidity currents able to construct aggrading levees'[9]. This contrasts with suggestions that genetically linked precursor lobe morphologies may form an initial depositional template for subsequent channel incision[8,11–13], and that channels may be entirely depositional both outside and inside the confining conduit[7,14].

Subsurface and outcrop observations on channel morphology and channel deposits are static. Similarly, the presently available direct observations of active submarine channels[4,15–18] do not span enough time to study morphodynamics of channel inception and evolution. In recent times, the extension of a submarine channel has been monitored in a quickly evolving system[19]. However, the data set did not provide direct information about the flow conditions during this morphological evolution. Therefore, modelling studies remain important in the investigation of the morphodynamic interplay between channel form and turbidity currents.

A limited number of experiments successfully produced subaqueous channels using a saline flow over a mobile substrate[20–22]. However, these flows could not produce depositional morphologies, as there was no suspended sediment load, which is vital for levee formation[22]. Therefore, these experiments provide limited insight into contributions of deposition and erosion during channel inception. Rowland *et al.*[23] reviewed the full range of published numerical and physical experiments that have tried to achieve self-channelization[24–26] by sediment-laden flows and concluded that channelization was not achieved in any of the cases.

We present experiments that for the first time capture self-channelization by turbidity currents. This was achieved by scaling sediment suspension in the experimental turbidity currents to the real world systems. This new scaling approach is called Shields scaling and focuses on two scaling parameters that regulate sediment suspension: (1) the Shields parameter and (2) a Reynolds scale of the sediment grains. The observed morphodynamic channel evolution establishes that channel inception can either commence with deposition of confining morphology by turbidity currents or erosion of a channel conduit. Thus, channel inception is not exclusively possible following erosion.

Results

Scaling approach. Classical turbidity current experiments[23,24,27–35] have focused on two non-dimensional scaling characterizations of the fluid flow: the Froude number (*Fr*), which is the ratio between momentum and gravitational forces of the flow, and the Reynolds number (*Re*), characterizing the ratio between the momentum and the viscous forces that determine the turbulent state of the flow. As it is not possible to keep both *Fr* and *Re* equal to the natural analogues while scaling down flow size, it is common to keep the *Fr* similar to natural values and to only require a *Re* above the laminar-turbulent threshold[36,37]. This Froude scaling approach has proven to be valuable in understanding the flow dynamics of turbidity currents but it does not guarantee that flows are able to transport sediment in suspension.

Many Froude-scaled experiments displayed rapid sediment depletion and were therefore limited in clarifying patterns of deposition and erosion. Depletive flows rapidly lose their complete sediment load, because they do not have enough turbulent mixing to compensate for settling of sediment from suspension. To predict whether currents are able to entrain and transport sediment in suspension, it is important to consider the force ratios acting on the sediment grains. This leads to two additional constraints: the Shields parameter, being the ratio between the turbulent shear, as expressed by the shear velocity, and the gravity-induced settling[38]; and the particle *Re*, which is the ratio of grain size to the boundary layer thickness[39]. The former is more commonly quoted in turbidity current studies[29] as the ratio between the shear velocity (u^*) and the settling velocity (u_s), but is here expressed as the Shields parameter. The latter is a Reynolds scale with significance for particle suspension near the bed. It describes the roughness of the sediment surface, which determines whether flow at the boundary is smooth and dominated by viscous forces, or rough and dominated by turbulent forces and shedding of turbulent eddies from particles at the bed surface[40]. If the boundary is smooth, a thin layer of laminar flow protects the bed and grains that settle into this near-bed boundary layer will no longer interact with suspending turbulent structures and are likely to remain deposited. In the transitionally rough regime, there is interaction of turbulent eddies with the bed but viscous forces also have a significant role. As experiments on channel inception are dependent on realistic turbulence–sediment interactions, both in the boundary layer and in suspension, it follows that such Shields scaling constrains must be satisfied.

The Shields scaling approach mirrors Froude scaling of the flow dynamics in the sense that one scale, namely the ratio of turbulent forces and gravity forces acting on the particle (the Shields parameter), is kept equal to real world values, whereas the other scale (the Reynolds particle scale) is relaxed, as long as rough to transitionally rough boundary layer conditions are maintained, to keep a realistic turbulent near-bed regime and aid sediment pick-up into suspension. These two scales form the axes of the classic Shields mobility diagram (Fig. 1), which enables a comparison between the present experiments, natural turbidity current conditions and previous experimental studies.

Comparison with natural currents and previous experiments. *In situ* measurements of turbidity currents in the Monterey Canyon[15,16] are used to estimate the position of a representative natural turbidity current on the Shields diagram (Supplementary Table 1 and Methods). The flows had a transitionally rough boundary and the Shields parameter plots above the suspension threshold (Fig. 1). Similarly, the boundary layer was transitionally rough in the present experiments and shear stresses were sufficiently high to support sustained suspension transport. As a result, these currents were sediment-bypass dominated along a significant part of the experimental domain. The experiments presented in this study were performed under Shields scaling conditions that are representative for the natural environment. Some previous confined slope experiments[24,41] also plot in the natural turbidity current regime. The experiments presented here are, however, the first to satisfy both Froude and Shields scaling (Fig. 1) on an unconfined and erodible slope, making them suitable to study flow–substrate interactions during channel inception.

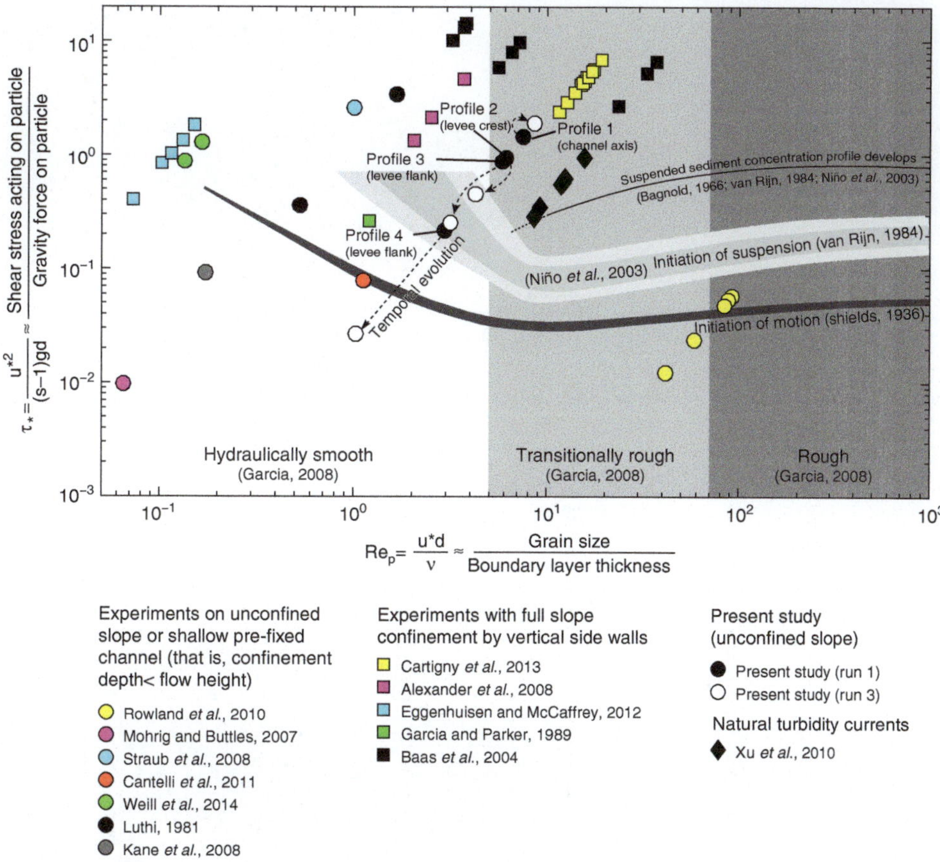

Figure 1 | Regime diagram for sediment transport. Shields mobility diagram giving an overview of experimental conditions in previous studies[22-24,28-32,34,35,41,48], natural flows[15] and the experiments presented in this study. Regime boundaries based on refs 38,40,49-51.

It emerges that many previous studies violated the proposed Shields scaling requirements, because the flows had smooth boundary layers and/or had shear stresses that were below the threshold for initiation of suspension. In the cases where flows had a smooth boundary layer and low Shields parameter[28,30,32,34], flows were always depositional. Flows in other experiments[29,31,33] had a higher Shields parameter but still a smooth boundary layer. Finally, the experiments of Rowland *et al.*[23] fulfilled the roughness requirement; however, there the Shields parameter was only approximately equal to the critical value for initiation of bedload motion (Fig. 1). None of these experiments led to channelization morphodynamics.

Morphological evolution. Three turbidity currents with the same characteristics were released successively on a constant and initially featureless sand slope (boundary conditions in Table 1 and set-up shown in Supplementary Fig. 1). The first turbidity current deposited two sub-parallel ridges, while the flow largely bypassed in between the two ridges (Fig. 2). This pattern of deposition resulted in a morphology that confined the subsequent flows. The relief of this confinement was increased during the subsequent second and third run due to continued deposition on the ridges. A circular scour with a diameter of 70 cm and a final depth of 8 cm is created in between the levees on the upper slope domain throughout the three runs. This contributes to the channel relief in that reach. On the lower slope, erosion in between the ridges was only initiated during the second run. The ridges are built of layers deposited by successive turbidity currents that are thinning away from the channel axis and therefore the ridges can be qualified as levees[42].

Table 1 | Boundary conditions of the experiments.

Input sediment concentration	17% vol
Suspension discharge	30 $m^3 h^{-1}$
Median grain size	141 μm
Bed slope	11°

The cross-sectional geometry of the experimental channel compares well with submarine channels on the modern seafloor that have remained unfilled. The depth:width ratio of the Lucia Chica channel (Fig. 3) is 1:12, whereas the aspect ratio formed in the experiment varies between 1:9 and 1:23.

The amount of sediment bypass on the slope increases in each run as is indicated by an increasing fraction of the sediment reaching the base of slope. The fraction of sediment that reaches the base of slope increases from 66% in run 1 to 80% in run 3 (Fig. 4).

Evolution of the flow field. The effect of the evolving topography on the flow field is shown by velocity profiles along a slope-perpendicular section (Fig. 5a and Supplementary Figs 2–4). At the beginning of run 1, when the slope was not yet modified, there was little across-flow variation in the downstream velocity profile. The confining morphology established by the end of run 1 resulted in an increase in flow velocity inside the confinement (U_{max} increases from 0.83 to 1.00 m s^{-1} at velocity profile 1; height of U_{max} was 1.2 cm) and a decrease in flow velocity outside of the confinement (U_{max} decreases from 0.64 to 0.38 m s^{-1} at velocity profile 3). It is noteworthy that the

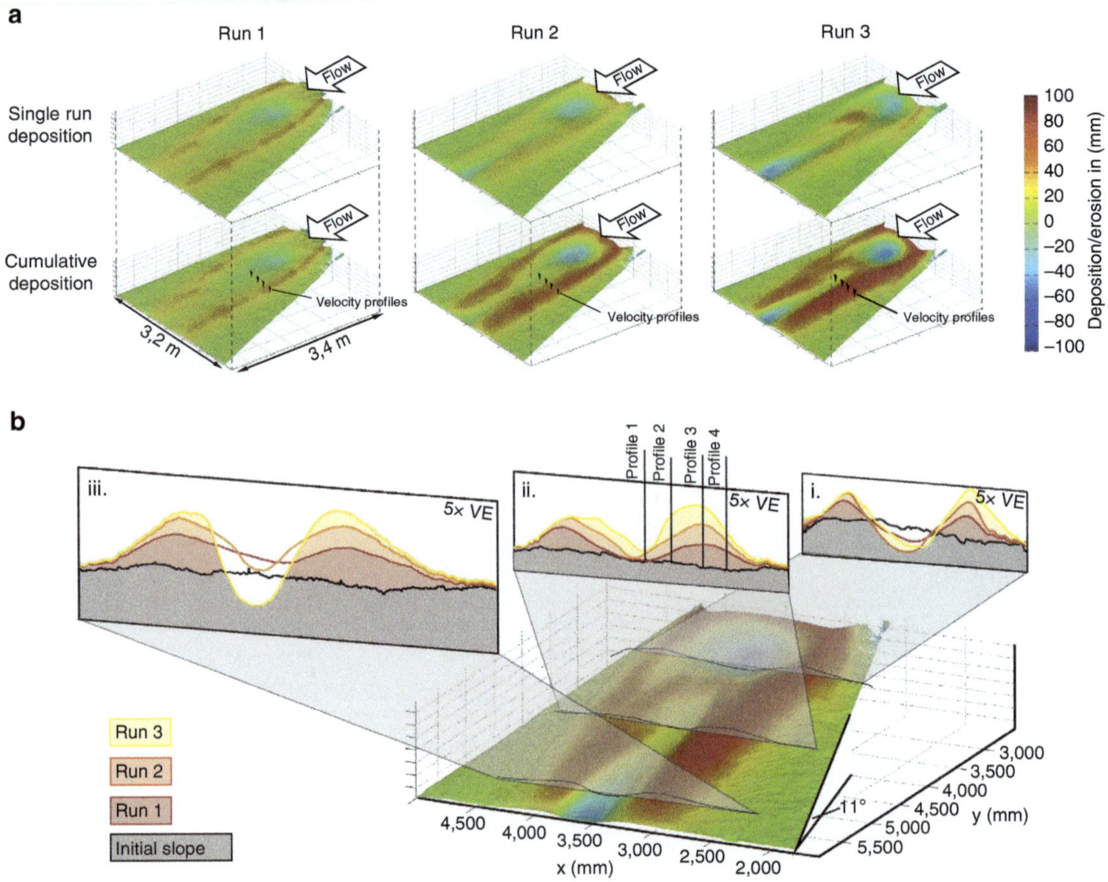

Figure 2 | Maps of deposition and erosion. (**a**) Digital elevation models of the deposits formed by sandy turbidity currents. Colours indicate the thickness of the deposits/depth of erosion. Both the cumulative erosion/deposition and the erosion/deposition after each single run are shown. (**b**) DEM of the final deposit with cross-sections at the lower, middle and upper slope. Cross-sections have five times vertical exaggeration. Positions of the probes that measured downstream velocity profiles are indicated on cross-section ii.

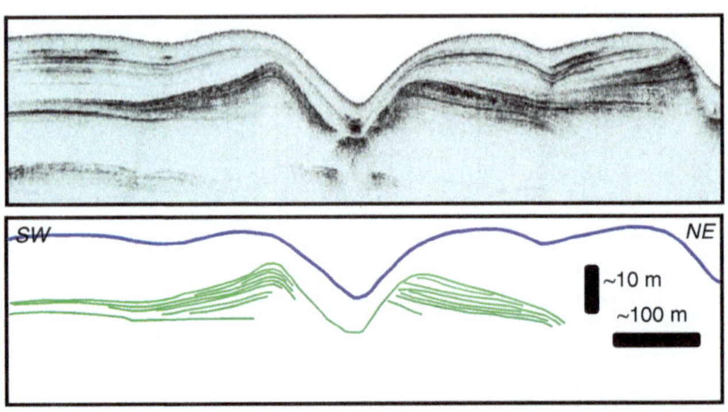

Figure 3 | Example of a submarine channel on the modern seafloor. Chirp profile through a submarine channel that is part of the Lucia Chica channel system offshore central California[10]. The green lines indicate the turbidity current deposits. Compare this figure with the cross-sections of the experimental deposits in Fig. 2b. The blue line indicates the top of the hemipelagic drape (figure reprinted with permission from the publisher).

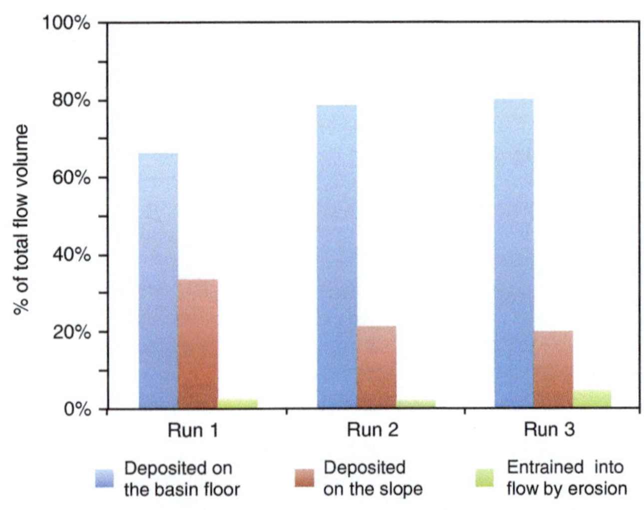

Figure 4 | Volume of deposits on the slope. Fractions of the total sediment load of the flow deposited and eroded on the slope and on the basin floor. The volume of sediment supplied to the experiments is equal for each of the runs.

change in the flow field during run 1 was caused by a channel with a depth ($h_{Umax} = 2.6$ cm) that was only a fraction of the flow height ($h = 7.3$ cm). The increase in channel depth during runs 2 and 3 does not result in a systematic change in flow velocity at any of the profiling locations. These results confirm previous pre-fixed channel experiments[28], which showed that a flow is already effectively confined within a conduit once the channel depth is greater than the height of the velocity maximum.

The spatial and temporal variations in flow velocity affected the ability of the flows to transport their sediment and these changes

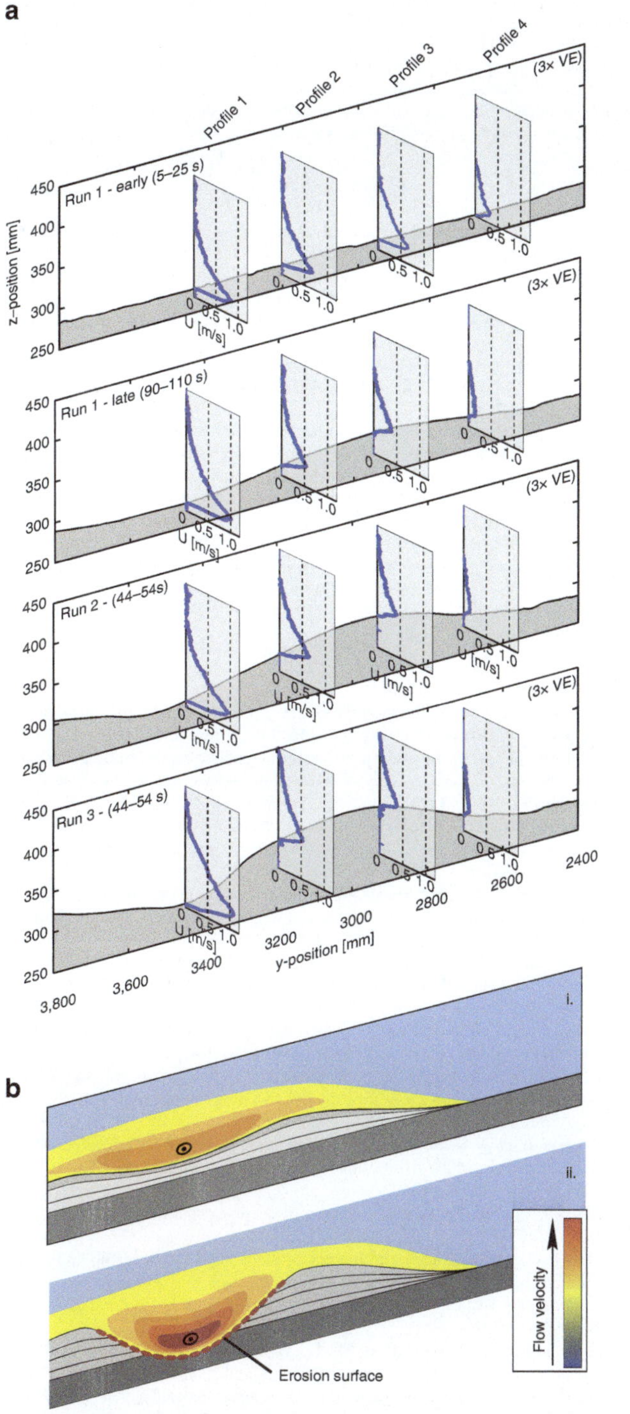

Figure 5 | Changes in flow velocity as a result of increasing confinement. (**a**) Topographic profiles with time-averaged velocity profiles measured along the same transect at different time intervals (early run 1 (5–25 s), late run 1 (90–110 s), middle run 2 (44–54 s) and middle run 3 (44–54 s)). The full velocity time series are supplied in Supplementary Note 3. It is noteworthy that there is three times vertical exaggeration in the topographic profiles. (**b**) Model for the co-evolution of the flow field and the topography derived from the experiments: (i) broad and weakly confined flows build a subtle depositional confinement, because deposition rates are slightly lower in the axis. (ii) A threshold at which incision start is reached, causing a rapid increase in the confinement relief. It is noteworthy that the erosion of the channel floor was only observed downstream of the location of velocity profile 1.

can be tracked within the Shields diagram (Fig. 1). The flow along the centreline of the slope (Fig. 5a, velocity profile 1) has a transitionally rough boundary layer and is able to bypass/erode sediment until the base of the slope. Flow at the off-axis locations of velocity profile 2 and 3 plot near the boundary between smooth and transitionally rough flow during run 1. The position of these points shifts towards the hydraulically smooth regime and below the suspension initiation threshold during run 3. Thus, the conditions at the locations of velocity profiles 2 and 3 are at or below the conditions for sediment bypass, and thus there is continuous deposition at these localities. The flow furthest away from the axis (Fig. 5a, velocity profile 4) plots within the field where flows have a smooth boundary layer and are below the boundary for suspended sediment transport during all runs. This indicates that flows that carry suspended load are highly depletive in these realms.

The temporal increase in axial flow velocity, which is caused by the progressively increasing confinement, causes an increase in the Shields parameter (Fig. 1, profile 1). The resulting small shift on the Shields diagram of the position of the flow at profile 1 appears to have little effect on the ability of the flow to transport sediment at this locality, because little deposition or erosion is observed here throughout the three runs. Although axial flow velocity was not monitored in the lower channel section, a larger shift on the Shields diagram can be inferred there because of the observed transition from deposition on the channel floor during run 1 to erosion on the channel floor during run 2 and 3 (Fig. 2b, cross-section iii.). The axial erosion is a further contributor to flow confinement in a dynamic feedback and consequently increases the rates of erosion.

In contrast, the flow conditions at the off-axis locations (velocity profiles 2, 3 and 4) are shifting in the Shields mobility diagram towards positions below the suspension initiation threshold and indeed there is continuous deposition at these localities.

In summary, spatial and temporal variation in the ability to transport sediment is predicted from the velocity measurements. The relative positions and temporal evolution on the Shields diagram predict the deposition of levees alongside a fairway dominated by sediment bypass and reflects progressive confinement increase during channel inception (Fig. 5b).

Discussion

Significant debate has surrounded the nature of the relief that is created during the initial phase of channel formation. It has been argued that initial relief that turbidity currents create at a site of repeated activity is likely erosional[9,10], which implies that levees commonly form from overspill after formation of an entrenched channel confinement. Trains of erosional scours are widely observed on the floors of channels on the modern ocean floor[3,4,9,42–44] and are indeed a probable initial feature of channelization in many cases. Similarly, the scour in front of the outlet in the present experiments contributes to the initial confinement on the upper slope. However, the initial confinement along the middle and lower slope is created purely by depositional patterns arising from low deposition rates below the flow axis compared with the flow margins. Thus, the incipient levees formed by lateral variations in sediment transport processes and not by overspill from an already established channel. This morphodynamic development confirms the role that depositional templates may play in initial confinement[8,11,12], and the experimental deposit cross-sections are strongly reminiscent of classic observational suggestions of depositional channel architecture[7,45].

Channel axis erosion caused by initial depositional confinement represents transition from depositional channelization to erosional channelization (Fig. 5b). This is a confirmation of a 'channelization threshold' at which a subtle confinement created by small depositional gradients causes incision, followed by a channelization feedback[11]. It emerges that channelization of turbidity currents can arise from both depositional and erosive sculpting of the seafloor, and may transition from depositional to erosive confinement. These flows thus have various intrinsic tendencies for channelization, which explains the ubiquitous presence of channels on submarine slopes.

The channel inception debate is just one of the many aspects of seafloor morphodynamics that can now be subjected to thorough testing in the laboratory environment. Previously, this was not possible, because experimental turbidity currents did not show realistic patterns of deposition and erosion as a result of inadequate scaling of the suspended sediment transport.

We conclude that confinement can progressively evolve from a depositional or erosional template, promoting gradual enhancement of sediment bypass on the slope. The increase in sediment bypass during the early phase of channel evolution will result in autogenic progradation of the system and deposits with according stacking patterns. Coarse-grained deposits can be expected at the base of finer-grained levees; such coupled stratigraphic bodies can be explained with a single genetic sequence of progressive channel inception, without the need to invoke changes in external mechanisms.

Methods

Description of the set-up. The experiments were conducted in the Eurotank Flume Laboratory at Utrecht University. The Eurotank measures 6 × 11 m in planform and was filled with water up to a level of 1.2 m above the horizontal floor (Supplementary Fig. 1). The bathymetry at the bottom of the tank consisted of a slope of 11° and a horizontal basin floor at the base of this slope. The slope was covered with a 10-cm-thick layer of loose sand that had the same grain-size distribution as the turbidity currents. A wooden duct was present at the top of the slope, to resemble a non-erodible canyon setting at the top of the slope. At the other end of the slope, a 10-cm-high ridge was placed, to provide down-dip accommodation for the sediment that reached the base of the slope.

Sediment suspensions. Before each experiment, the sediment mixture was prepared outside the tank in a 1.1-m^3 mixing tank, with two propellers designed to homogenize sediment–water mixtures up to 30% volumetric sediment concentrations. The quartz sand used to make the suspensions had a median grain size (D_{50}) of 141 μm, a D_{10} of 44 μm and a D_{90} of 199 μm (Supplementary Fig. 5), and had a specific density of 2,650 kg m^{-3}. The grain size was analysed using a Malvern Mastersizer particle sizer.

Data collection. During the experiments, a slurry pump was used to supply the suspension to the set-up. A discharge meter (Krohne Optiflux 2300) was mounted in the supply pipe. The discharge was regulated by a Labview control system that adjusted the pump speed whenever the measured discharge deviated from a set reference value. The discharge during each of the experiments presented here was 30 $m^3 h^{-1}$. The experiments lasted ~100 s before the mixing tank was drained. Four Ultrasonic Doppler Velocity Profiler probes (UVP Duo MX, 1 MHz) were installed on an aluminium frame, to monitor the flow field during the experiments. These probes were set up at around 0.15 m above the erodible basin floor, with their beam pointing diagonally down into the flow at an angle of 60° relative to the initial local slope of the flume floor. The planview location of the probes is indicated in Fig. 2. Each of these probes measured a full profile of bed-parallel flow in the direction of the probe orientation. The profiles had a spatial resolution of 0.64 mm and the measurement frequency was 1.81 Hz. Individual velocity profiles have a spiky appearance due to the turbulent nature of the flows. Therefore, time-averaging was applied to create smoother profiles as presented in Fig. 5a. After each experimental run, the basin was drained to expose the deposit. Next, a digital elevation model (DEM) with a horizontal resolution of 2 × 2 mm was created using a laser scanner. By subtracting the DEM of the experimental deposits and a DEM of the sediment bed before the experiments, a map of deposition and erosion was created for each experiment.

Determination of flow conditions. The following flow parameters were required to determine the position of each reviewed experiment on the Shields diagram (Fig. 1):

Grain size (d): Here, the median grain size of the initial sediment mixture was used.

Kinematic viscosity (ν): Here, the viscosity of clear water at 20° was used (1×10^{-6}).

Shear velocity (U^*): When estimates are supplied in experimental studies, then they are followed. Elsewise, the shear velocities were determined as[46] $U^* = \sqrt{g'hS}$.

For the present experiments, the shear velocity was determined using[39]:

$$U^* = U_{max}\kappa\left[\ln\left(\frac{h_{max}}{0.1D_{90}}\right)\right]^{-1}$$

where h_{max} is the height of the velocity maximum and U_{max} is the maximum velocity.

Reduced gravity (g'): $g(\Delta\rho/\rho)$, where ρ is the density of the suspension and $\Delta\rho$ is the excess density of the sediment submerged in the ambient fluid. To calculate the density of the suspension, it is assumed that the density of the sediment concentration is equal to the concentration of the initial mix.

Flow height up to the velocity maximum (h): If only the total flow thickness was given, it was assumed that the height of the velocity maximum is at one-fourth of the total flow thickness.

Bed slope (S): sin(bed slope in degrees).

Flow conditions for turbidity currents in the Monterey canyon were determined using information reported in ref. 15. A representative median grain size for the turbidity currents was estimated from sediment cores of the Monterey canyon floor. Core data in ref. 47 shows that a broad range of grain sizes (ranging from silt to boulders) were deposited on the canyon floor. Middle sand (diameter of 350 μm) was chosen as a representative grain size, because it was the most common grain size in the cores. The D_{90} was estimated at 500 μm. The shear velocity was determined using the formula that is also used for this purpose for the present experiments.

References

1. Talling, P., Wynn, R., Masson, D. & Frenz, M. Onset of submarine debris flow deposition far from original giant landslide. *Nature* **450,** 541–544 (2007).
2. Damuth, J. E. *et al.* Meandering channels meandering and bifurcation patterns on the Amazon deep sea fan as revealed by pong-range side-scan sonar. *Geology* **11,** 94–98 (1983).
3. Maier, K. L. *et al.* The elusive character of discontinuous deep-water channels: new insights from Lucia Chica channel system, offshore California. *Geology* **39,** 327–330 (2011).
4. Hay, A. E. Turbidity currents and submarine channel formation in Rupert Inlet, British Columbia: 2. The roles of continuous and surge-type flow. *J. Geophys. Res. Ocean* **92,** 2883–2900 (1987).
5. Stevenson, C. J., Jackson, C. A., Hodgson, D. M., Hubbard, S. M. & Eggenhuisen, J. T. Deep-water sediment bypass. *J. Sediment. Res.* **85,** 1058–1081 (2015).
6. Galy, V. *et al.* Efficient organic carbon burial in the Bengal fan sustained by the Himalayan erosional system. *Nature* **450,** 407–410 (2007).
7. Clark, J. D. & Pickering, K. T. Architectural elements and growth patterns of submarine channels: application to hydrocarbon exploration. *Am. Assoc. Pet. Geol. Bull.* **80,** 194–221 (1996).
8. Morris, E. *et al.* Sedimentology, stratigraphic architecture, and depositional context of submarine frontal-lobe complexes. *J. Sediment. Res.* **84,** 763–780 (2014).
9. Fildani, A. *et al.* Erosion at inception of deep-sea channels. *Mar. Pet. Geol.* **41,** 48–61 (2013).
10. Maier, K. L. *et al.* Deep-sea channel evolution and stratigraphic architecture from inception to abandonment from high-resolution Autonomous Underwater Vehicle surveys offshore central California. *Sedimentology* **60,** 935–960 (2013).
11. Eggenhuisen, J. T., McCaffrey, W. D., Haughton, P. D. W. & Butler, R. W. H. Shallow erosion beneath turbidity currents and its impact on the architectural development of turbidite sheet systems. *Sedimentology* **58,** 936–959 (2011).
12. Ortiz-Karpf, A., Hodgson, D. M. & McCaffrey, W. D. The role of mass-transport complexes in controlling channel avulsion and the subsequent sediment dispersal patterns on an active margin: the Magdalena Fan, offshore Colombia. *Mar. Pet. Geol.* **64,** 58–75 (2015).
13. Sun, T., Ghayour, K., Hall, B. & Miller, J. in *Proceedings of the 30 th GCSSEPM Foundation Bob F Perkins Research Conference* Houston, Texas, USA (2011).
14. Stevenson, C. J. *et al.* The flows that left no trace: very large-volume turbidity currents that bypassed sediment through submarine channels without eroding the sea floor. *Mar. Pet. Geol.* **41,** 186–205 (2013).
15. Xu, J. P. Normalized velocity profiles of field-measured turbidity currents. *Geology* **38,** 563–566 (2010).
16. Xu, J. P., Noble, M. A. & Rosenfeld, L. K. *In-situ* measurements of velocity structure within turbidity currents. *Geophys. Res. Lett.* **31,** L09311 (2004).
17. Cooper, C., Wood, J. & Andrieux, O. in *Offshore Technology Conference 2012 Proceedings* Houston, Texas, USA (2013).
18. Hughes Clarke, J. E., Brucker, S., Muggah, J., Church, I. & Cartwright, D. The Squamish Delta Repetitive Survey Program: a simultaneous investigation of

prodeltaic sedimentation and integrated system accuracy. *U.S. Hydrographic Conference* 1–16 (2011).
19. Hughes Clarke, J. E. *et al.* Monitoring the evolution of submarine channels on Fjord Prodeltas and associated depositional basins. *AAPG Search Discov.* 41678 (2015).
20. Metivier, F., Lajeunesse, E. & Marie-Christine, C. Submarine Canyons in the bathtub. *J. Sediment. Res.* **75,** 6–11 (2005).
21. Hoyal, D., Sheets, B. A., Edwards, C. M. & Bloch, R. B. Controls from channel formation in deep water distributive systems. *AAPG Search Discov.* 50090 (2008).
22. Weill, P., Lajeunesse, E. & Devauchelle, O. Experimental investigation on self-channelized erosive gravity currents. *J. Sediment. Res.* **84,** 487–498 (2014).
23. Rowland, J. C., Hilley, G. E. & Fildani, A. A test of initiation of submarine leveed channels by deposition alone. *J. Sediment. Res.* **80,** 710–727 (2010).
24. Baas, J., Kesteren, W., Van & Postma, G. Deposits of depletive high-density turbidity currents: a flume analogue of bed geometry, structure and texture. *Sedimentology* **51,** 1053–1088 (2004).
25. Imran, J., Parker, G. & Katopodes, N. A numerical model of channel inception on submarine fans. *J. Geophys. Res.* **103,** 1219 (1998).
26. Yu, B. *et al.* Experiments on self-channelized subaqueous fans emplaced by turbidity currents and dilute mudflows. *J. Sediment. Res.* **76,** 889–902 (2006).
27. Middleton, G. small-scale models of turbidity currents and the criterion for auto-suspension. *J. Sediment. Petrol* **36,** 202–208 (1966).
28. Mohrig, D. & Buttles, J. Deep turbidity currents in shallow channels. *Geology* **35,** 155–158 (2007).
29. Straub, K. M., Mohrig, D., McElroy, B., Buttles, J. & Pirmez, C. Interactions between turbidity currents and topography in aggrading sinuous submarine channels: a laboratory study. *Geol. Soc. Am. Bull.* **120,** 368–385 (2008).
30. Cantelli, A., Pirmez, C., Johnson, S. & Parker, G. Morphodynamic and stratigraphic evolution of self-channelized subaqueous fans emplaced by turbidity currents. *J. Sediment. Res.* **81,** 233–247 (2011).
31. Luthi, S. Experiments on non-channelized turbidity currents and their deposits. *Mar. Geol.* **40,** 59–68 (1981).
32. Kane, I. A., McCaffrey, W. D. & Peakall, J. Controls on sinuosity evolution within submarine channels. *Geology* **36,** 287–290 (2008).
33. Alexander, J. *et al.* Laboratory sustained turbidity currents form elongate ridges at channel mouths. *Sedimentology* **55,** 845–868 (2008).
34. Eggenhuisen, J. T. & McCaffrey, W. D. The vertical turbulence structure of experimental turbidity currents encountering basal obstructions: implications for vertical suspended sediment distribution in non-equilibrium currents. *Sedimentology* **59,** 1101–1120 (2012).
35. Garcia, M. & Parker, G. Experiments on hydraulic jumps in turbidity currents near a canyon-fan transition. *Science* **245,** 393–396 (1989).
36. Peakall, J., Ashworth, P. & Best, J. in *The Scientific Nature of Geomorphology.* (eds Rhoads, B. L. & Thorn, C. E.) 221–253 (John Wiley and Sons, 1996).
37. Parsons, J. & Garcıa, M. Similarity of gravity current fronts. *Phys. Fluids* **10,** 1–5 (1998).
38. Shields, A. *Anwendung der Aehnlichkeitsmechanik und der Turbulenzforschung auf die Geschiebebewegung* (Preussischen Versuchsanstalt für Wasserbau, 1936).
39. Rijn, L. C. Van. *Principles of Sediment Transport in Rivers , Estuaries and Coastal Seas* (Aqua Publications, 1993).
40. Garcia, M. *Sedimentation Engineering: Processes, Measurements, Modeling and Practise* (American Society of Civil Engineers, 2008).
41. Cartigny, M. J. B., Eggenhuisen, J. T., Hansen, E. W. M. & Postma, G. Concentration-dependent flow stratification in experimental high-density turbidity currents and their relevance to turbidite facies models. *J. Sediment. Res.* **83,** 1046–1064 (2014).
42. Kane, I. A. & Hodgson, D. M. Sedimentological criteria to differentiate submarine channel levee subenvironments: exhumed examples from the Rosario Fm. (Upper Cretaceous) of Baja California, Mexico, and the Fort Brown Fm. (Permian), Karoo Basin, S. Africa. *Mar. Petrol. Geol.* **28,** 807–823 (2011).
43. Heiniö, P. & Davies, R. J. Trails of depressions and sediment waves along submarine channels on the continental margin of Espirito Santo Basin, Brazil. *Geol. Soc. Am. Bull.* **121,** 698–711 (2009).
44. Covault, J. A., Kostic, S., Paull, C. K., Ryan, H. F. & Fildani, A. Submarine channel initiation, filling and maintenance from sea-floor geomorphology and morphodynamic modelling of cyclic steps. *Sedimentology* **61,** 1031–1054 (2014).
45. Mutti, E. & Normark, W. R. in *Deep Water Clastic Deposits: Models and Case Histories.* (eds Legget, J. K. & Zuffa, G. G.) 1–38 (Graham and Trotman, 1987).
46. Kneller, B. & Mccaffrey, W. D. The interpretation of vertical sequences in turbidite beds: the influence of longitudinal flow. *J. Sediment. Res.* **73,** 706–713 (2003).
47. Paull, C. K. *et al.* Origins of large crescent-shaped bedforms within the axial channel of Monterey Canyon, offshore California. *Geosphere* **6,** 755–774 (2010).
48. Alexander, J. *et al.* Laboratory sustained turbidity currents form elongate ridges at channel mouths. *Sedimentology* **55,** 845–868 (2007).
49. Bagnold, R. An approach to the sediment transport problem from general physics. *U.S. Geol. Surv. Prof. Paper* **422,** I1–I37 (1966).
50. van Rijn, L. Sediment transport, part 2: Suspended load transport. *J. Hydraul. Eng* **110,** 1613–1641 (1984).
51. Niño, Y., Lopez, F. & Garcia, M. Threshold for particle entrainment into suspension. *Sedimentology* **50,** 247–263 (2003).

Acknowledgements

This is a contribution of the Eurotank Studies of Experimental Deepwater Sedimentology (EuroSEDS), financially supported by the Dutch Organisation of Scientific Research (NWO), ExxonMobil, Shell and Statoil. Patrick Michielsen and Martijn ter Braack have assisted during the laboratory experiments.

Author contributions

M.J.B.C. and J.T.E. developed the scaling approach. J.d.L. conducted the experiments, analysed the results and wrote the initial manuscript. All authors edited the manuscript

Additional information

Competing financial interests: The authors declare no competing financial interests.

11

Increased reservoir ages and poorly ventilated deep waters inferred in the glacial Eastern Equatorial Pacific

Maria de la Fuente[1], Luke Skinner[2], Eva Calvo[1], Carles Pelejero[1,3] & Isabel Cacho[4]

Consistent evidence for a poorly ventilated deep Pacific Ocean that could have released its radiocarbon-depleted carbon stock to the atmosphere during the last deglaciation has long been sought. Such evidence remains lacking, in part due to a paucity of surface reservoir age reconstructions required for accurate deep-ocean ventilation age estimates. Here we combine new radiocarbon data from the Eastern Equatorial Pacific (EEP) with chronostratigraphic calendar age constraints to estimate shallow sub-surface reservoir age variability, and thus provide estimates of deep-ocean ventilation ages. Both shallow- and deep-water ventilation ages drop across the last deglaciation, consistent with similar reconstructions from the South Pacific and Southern Ocean. The observed regional fingerprint linking the Southern Ocean and the EEP is consistent with a dominant southern source for EEP thermocline waters and suggests relatively invariant ocean interior transport pathways but significantly reduced air–sea gas exchange in the glacial southern high latitudes.

[1] Departament de Biologia Marina i Oceanografia, Institut de Ciències del Mar, CSIC, Passeig Marítim de la Barceloneta 37–49, Barcelona 08003, Spain. [2] Godwin Laboratory for Palaeoclimate Research, Department of Earth Sciences, University of Cambridge, Downing Street, Cambridge CB2 3EQ, UK. [3] Institució Catalana de Recerca i Estudis Avançats, Barcelona, Spain. [4] Grup de Recerca de Geociències Marines, Departament d'Estratigrafia, Paleontologia i Geociències Marines, Universitat de Barcelona, C/Martí i Franquès, Barcelona 08028, Spain. Correspondence and requests for materials should be addressed to M.dlF. (email: mfuente@icm.csic.es).

The Pacific accounts for ~50% of the total ocean volume, such that changes in its average chemistry may exert a significant influence on that of the global average ocean and therefore that of the atmosphere and the Earth's climate. Indeed, it has long been proposed that a reduction in the ocean's average rate of overturn[1] and/or ocean-atmosphere CO_2 exchange[2] may have been the single most important contributor to reduced atmospheric CO_2 during past glacial periods, and that this would be confirmed by the identification of a significant increase in the average extent of radiocarbon depletion (that is, the ocean-atmosphere radiocarbon disequilibrium or radiocarbon 'ventilation age') in the Pacific Ocean during the last glacial period[3]. Despite a surge of marine radiocarbon studies focusing on this issue in recent years, and despite circumstantial support for this hypothesis from atmospheric radiocarbon concentration and production reconstructions[4–6], the glacial radiocarbon inventory of the Pacific Ocean and its evolution across the last deglaciation remain poorly constrained.

Thus, radiocarbon ventilation reconstructions from throughout the mid-depth (~500–1,500 m) and deep (>1,500 m) Pacific are in apparent conflict, with some indicating very little or no change in radiocarbon ventilation between the Last Glacial Maximum (LGM) and the pre-industrial period[3,7,8–13], some indicating transient anomalies of varying magnitude and sense[14–17], and with others showing quite significant changes, expressed as a net change between the LGM and the pre-industrial[18–21] or as a positive anomaly in the deglaciation[22–25] (Table 1). Although much of the apparent conflict between the various Pacific radiocarbon ventilation records may reflect small-scale aspects of the glacial circulation as compared with the present (resulting in highly localized signals or errors in water-mass attribution), it could also reflect to a large extent the different methods and assumptions that are adopted in the various studies.

Perhaps the most important methodological consideration is whether or not independent calendar age control, and therefore surface reservoir age reconstructions are available. It is notable that most of the existing radiocarbon ventilation records from the Pacific are based on the commonly used method of calculating benthic-planktonic radiocarbon age offsets (B–P) (Table 1), which ignores the possible impact of variable surface reservoir ages on the inferred radiocarbon disequilibrium of deep waters relative to the atmosphere (that is, the benthic-atmosphere radiocarbon age offset (B-Atm)). Furthermore, it is also notable that studies adopting this method or the ^{14}C projection-age method[26], and assuming invariant surface reservoir ages, tend not to find evidence of poorly ventilated glacial mid/deep Pacific ocean waters[3,7–13,15–17] (Table 1). However, numerical model experiments have suggested that surface reservoir age should in principle have varied since the last glacial period[27], and this has been confirmed via a range of techniques applied, not only in the Pacific Ocean[17,20,21,24], but also in the Southern Atlantic Ocean[28] and other basins[29,30]. Notably, Siani *et al.*[24] suggested that data from a previous study at a nearby location (core SO161-SL22 (ref. 12)) would indeed show significant changes in ventilation ages over the glacial period if their surface reservoir ages were taken into account. These observations serve to underline the importance of assessing the variability of the surface reservoir ages, especially when using the B–P offset approach to calculate radiocarbon ventilation ages.

Today, the Southern Ocean is the main region where water from the ocean interior makes its first contact with the surface, and where surface water first enters the ocean interior[31]. It is also the main source of deep and abyssal water exported to the rest of the global ocean[32], and it therefore plays a key role in the water-mass budget and the ventilation (that is, the mixing of surface water properties into the ocean interior) of the Pacific Ocean[33]. It is for this reason that the Southern Ocean has been proposed as the main area for the release of 'old' carbon-rich deep-water from the ocean interior at the close of the last glacial, possibly due to increased upwelling and/or air–sea gas exchange as a result of altered winds[34] and/or sea-ice/shallow sub-surface density stratification[28,35]. A connection between the high and low latitude Pacific is provided through upwelled shallow sub-surface/intermediate waters from the Southern Ocean (See Methods for further details) in the Eastern Equatorial Pacific (EEP) via the Equatorial Undercurrent (EUC), identified for both the modern[36–38] and the glacial circulation[34,39–42]. This far-field dynamical connection would imply that changes occurring in the Southern Ocean should exert an influence on the hydrography of the thermocline in low latitude areas like the EEP. The EEP thus provides an excellent study location for assessing past changes in Pacific radiocarbon ventilation, in particular where these are believed to originate in the Southern Ocean.

Here we present deep and shallow sub-surface (that is, deep thermocline) radiocarbon ventilation age reconstructions for the EEP, from sediment core ODP Site 1240 located in the Panama Basin (0°01.31′ N, 86°27.76′ W; 2,921 m water depth). This core site is currently bathed by Upper Circumpolar Deep Water (UCDW) and/or Pacific Deep Water (PDW), with southern sourced shallow sub-surface/intermediate waters contributing to the EEP thermocline via the EUC. Deep-water radiocarbon ventilation ages are estimated from B–P age offsets (B: mixed benthic foraminifera; P: *Neogloboquadrina dutertrei*, a deep thermocline-dwelling planktonic foraminifer, 100–150 m water depth[43]), plus a variable shallow sub-surface reservoir age (yielding B-Atm age offsets), which is derived on the basis of chronostratigraphic alignment to Greenland ice-core records. Notably, the revised chronology that we derive for this core implies a significant increase in both deep-water and shallow sub-surface radiocarbon reservoir/ventilation ages during the last glacial period in the EEP, consistent with a significant change in the radiocarbon concentration of waters exported from the Southern Ocean to the rest of the global ocean during the last glacial period.

Results

Radiocarbon dates and B–P offsets in core ODP Site 1240. Seventeen paired planktonic and benthic radiocarbon measurements and the offset between them (B–P) are shown in Fig. 1a,b, respectively, and in Supplementary Table 1. Of the data shown in Fig. 1a,b, three B–P pairs (depth intervals 26, 223 and 289; see Fig. 1b in grey) were flagged up as potential outliers in our present study, on the basis that these three pairs yield B–P offsets that fall anomalously far away from the three-point running mean of the data set and that the benthic ages converge on their planktonic counterparts far more rapidly than could be expected due to real changes in deep-ocean circulation/ventilation. However, we show these tentative flyers in Fig. 1a,b, and note that it is possible (if not presently demonstrable) that other studies in the region might in future provide support for their veracity.

The 14 retained radiocarbon B–P pairs (Fig. 1b) indicate a change of ~870 year between the last part of the last glacial period (average B–P ~1,805 year) and the Holocene (average B–P ~935 year, including the modern B–P offset of ~1,250 year; see Methods for further detail). Across the deglaciation, decreasing B–P offsets indicate enhanced deep-ocean ventilation that peaks in the early Holocene and then drops again to modern values across this period. These results thus provide clear evidence for a drop in deep equatorial Pacific radiocarbon ventilation ages since the LGM.

Reservoir age assessment and age model. An assessment of shallow sub-surface reservoir age variability at the location of

Table 1 | Summary of radiocarbon ventilation records from the Pacific Ocean over the last deglaciation grouped by water depth (>1,500 m/<1,500 m) and area.

Area	Coordinates	Core name	Water depth (m)	Study	Method to calculate deep ventilation age	Reservoir age variability assessment?	Type of deep ventilation reconstructions*
Sediment cores <1,500 m water depth							
EEP	0.5° S 82.1° W	VM19-27	1,370	8	B-P	No (constant R.age)	0
	1°13′ S 89° 41′ W	MV21-30	617	16	B-P	No (constant R.age)	0+
SEP	36°13′ S 73° 40′ W	SO161-SL22	1,000	12	B-P	No (constant R.age)	0
NEP	23.5° N 111.6° W	MV99-MC19/GC31/ PC08	705	15	Proj.age	No	0+
NWP	41°07.1′ N 142° 24.2′ E	MR01-K03 PC4/PC5	1,366	14	Proj.age	No	0+
	52° N-56° N 144° E-169° E	Several cores from Bearing Sea and Okhotsk Sea	630-1,765	25	B-P	No (constant R.age)	1+
	34° N 137° E	Several cores from off the Tokai area	1,076-1,356	23	B-P	Yes (TEPHRAS)	1+
Sediment cores >1,500 m water depth							
EEP	3°37.2′ S 83° 58′ W	TR163-31B	3,210	18	B-P	No (constant R.age)	1
	1.5° S 85.8° W	RC11-238	2,570	8	B-P	No (constant R.age)	0
	5.5° S 106.8° W	VM21-40	3,180	8	B-P	No (constant R.age)	0
	0° 01.31′ N 86° 27.76′ W	ODP1240	2,921	This study	B-P	Yes (chronostratigraphy)	1+
SEP	46° S 75° W	MD07-3088	1,536	24	B-P	Yes (TEPHRAS, partially)	1+
SWP	40° 2.935′ S 177° 59.68′ E	MD97-2121	2,314	21	B-P	Yes (TEPHRAS)	1
	36° S-45° S 177° E-179° E	Several cores from Bay of plenty and Chatman Rise	1,300-2,700	19	B-P	Yes (TEPHRAS)	1
NEP	54.37° N 148.45° W	ODP 887	3,648	22	B-P	No	1+
	42.1° N 125.8° W	W8709A-13PC	2,710	17	Proj.age	Partially, only after 15 kyr	0+
NWP	18° 54′ N 115° 46′ E	Sonne 50-37KL	2,695	3	B-P	No (constant R.age)	0
	36° 04′ N 141° 49′ E	MD01-2420	2,101	13	Proj.age	No	0
	51° 27′ N 167° 73′ E	MD01-2416	2,317	20	B-P	Yes (^{14}C Plateau tuning)	1
	20° 07′ N 117° 23′ E	GIK 17940	1,727	20	B-P	Yes (^{14}C Plateau tuning)	1
	7° 12′ N 112° 5′ E	V35-5	1,953	7	B-P	No (constant R.age)	0
WEP	6° N 126° E	MD98-2181	2,100	9	B-P	No (constant R.age)	0
	1° S 146° E	MD98-2138	1,900	9	B-P	No (constant R.age)	0
	1.1° N 130° E	MD01-2386	2,820	10,11	B-P	No (constant R.age)	0
CEP	10° 17′ S 111° 20′ W	TT154-10	3,225	7	B-P	No (constant R.age)	0

B-P, benthic-planktonic radiocarbon offset method; CEP, Central Equatorial Pacific; EEP, Eastern Equatorial Pacific; LGM, Last Glacial Maximum; NEP, North East Pacific; NWP, North West Pacific; Proj.age, projection-age method from ref. 26; SEP, South East Pacific; SWP, South West Pacific; WEP, Western Equatorial Pacific.
*0 indicates little or no change in radiocarbon ventilation between the LGM and the pre-industrial period; 0+—transient anomalies of varying magnitude and sense; 1—quite significant changes, expressed as a net change between the LGM and the pre-industrial period; 1+—quite significant changes, expressed as a net change between the LGM and the pre-industrial period but with a positive anomaly in the deglaciation.

ODP1240 (Fig. 2) has been performed by chronostratigraphic alignment between the $\delta^{18}O_{ice}$ and the $[Ca^{2+}]$ records from three Greenland ice cores (NGRIP, GRIP and GISP2 (ref. 44)) placed on the recent GICC05 time scale[45] and two independent records from Site 1240: (a) Sea Surface Temperature (SST) based on the $U^{K'}_{37}$ index (Calvo *et al.*, unpublished) and (b) $\delta^{18}O_{sw}$ obtained from the planktonic foraminifer *Globigerinoides ruber*[40]. Our proposed alignments are based on a host of previous studies, including previous palaeoclimate reconstructions/interpretations and numerical model studies all of which support the underlying premise of our stratigraphic alignments[46–48]. The outcome of this previous work is that anomalies in sea surface temperatures recorded by alkenones and surface salinities reflected in surface water $\delta^{18}O$ reconstructions at our location should coincide in time with North Atlantic/Greenland temperature anomalies. There is a set of well tested physical mechanisms underlying this correspondence that relate to changes in cross-equatorial temperature gradients and their impact on easterly winds, equatorial upwelling and moisture transport across the Isthmus of Panama.

For this alignment, four stratigraphic tie-points from the ice-core records were selected coinciding with major climatic changes at the Heinrich Stadial 2 (HS2) end, the HS1 onset, the HS1/Bølling-Allerød transition and the Younger Dryas end (Fig. 2; Supplementary Table 2). The HS1 onset was designated only through the change in the $[Ca^{2+}]$ record at 17.6 kyr since precipitation $\delta^{18}O$ records from the various Greenland ice cores present differences between 19–16 kyr due to different site locations (Fig. 2a,b). An additional age constraint for the ODP1240 core-top was derived based on a single calibrated radiocarbon date (from the study by Pena *et al.*[40]), corrected for the modern shallow sub-surface reservoir age in the area (this is ~735 year; see Methods for further detail).

On the basis of the above chronologies, two sets of reservoir age estimates were generated (that is, one from each ODP1240 record used for stratigraphic alignment) (Fig. 2e; Supplementary Table 2). These differ due to slight mismatches between the patterns of SST and $\delta^{18}O_{sw}$ variability before 17 kyr (Fig. 2c,d). Reservoir ages were calculated as the difference between the interpolated ODP1240 planktonic ^{14}C for each tie-point[28] and the atmospheric ^{14}C-age (IntCal09 curve[49]) at the calendar age of each tie-point (note that we have not used the more recent IntCal13 (ref. 50) radiocarbon calibration curve to avoid possible

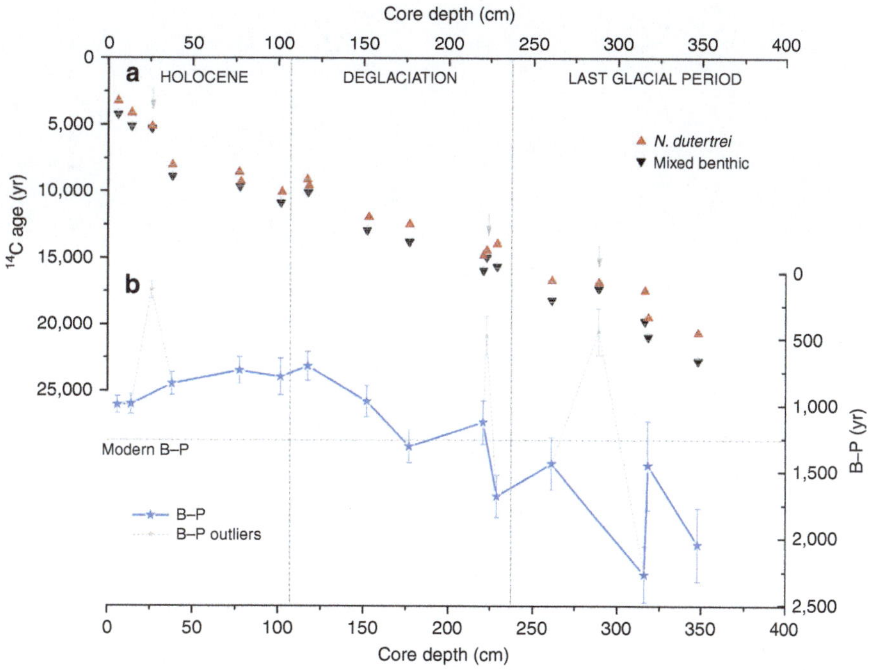

Figure 1 | Summary of planktonic and benthic radiocarbon dates and their offsets from core ODP1240. (**a**) Radiocarbon dates from *N. dutertrei* (red triangles) and mixed benthic foraminifera (black triangles) (all dates were $\delta^{13}C$-normalized, without reservoir age correction and uncalibrated to calendar years), with associated 1σ-uncertainties. The discarded B-P offsets are indicated by a grey vertical arrow. Note that planktonic dates from intervals 77/78 and 117/118 were averaged for B-P calculation to be able to compare with the coeval benthic radiocarbon dates, which were combined due the low carbonate abundance of each interval (see Methods and Supplementary Table 1). (**b**) Benthic-Planktonic radiocarbon offset (B-P) (thick light blue line and stars) plus the discarded B-P offsets (stapled light grey line) (error bars represent combined 1σ error in ^{14}C dates). Modern B-P offset is shown by a dashed horizontal line and relevant climatic periods are separated by dashed vertical lines.

problems of inconsistency when comparing with previously published radiocarbon studies that are referenced to IntCal09; only minor differences would arise if IntCal13 were used instead). This approach provides reservoir age estimates only at the four proposed calendar age tie-points. An alternative approach of interpolating calendar ages (and on this basis ^{14}C atm) for all the planktonic radiocarbon dates in ODP1240 would yield spurious reservoir ages where no direct calendar age constraints exist (as would occur in the Holocene, for example, where no tie-points were designated).

Four age models were thus derived using the Bayesian age-depth modelling programme Bchron[51] and the IntCal09 radiocarbon calibration curve[49] (Fig. 3; Supplementary Table 3): (a) two of them by calibrating 17 *N. dutertrei* radiocarbon dates corrected for the two sets of reservoir age estimates; and (b) two more using only the stratigraphic tie-points (Fig. 3). The purpose of this was to check the consistency of the age models when ignoring the radiocarbon age constraints between stratigraphic tie-points, as well as the radiocarbon age reversals that are automatically rejected by Bchron as outliers in the first two age models. Notably, all of these four chronologies are in good agreement, indicating a good degree of consistency between the various chronostratigraphic constraints that we have adopted, within their respective uncertainty ranges. Only across the Holocene, where no stratigraphic tie-points were selected, do large differences between the radiocarbon and stratigraphic age models emerge, reflecting significant changes in sedimentation rate that cannot be captured by our stratigraphic alignment over this interval. Finally, circumstantial support for the age models is also provided by the fact that deglacial minima in the *N. dutertrei* $\delta^{13}C$ record from ODP1240 (ref. 40) correspond with similar $\delta^{13}C$ minima and peak upwelling intensities observed in the Southern Ocean[35], which also broadly coincide with two prominent minima in the atmospheric $\delta^{13}CO_2$ record (Supplementary Fig. 1).

The two sets of reservoir age estimates show a broadly similar pattern of variability across the last deglaciation, indicating large changes in the shallow sub-surface ocean ventilation over this time interval (Fig. 2e). Quantitatively, small differences are found between both SST and $\delta^{18}O_{sw}$ alignments, yielding glacial shallow sub-surface reservoir ages of ~970 years and ~1,105 years, respectively, that is ~405 and ~540 years older than during the Holocene (when reservoir ages are ~565 years on average, including the modern value of ~735 years). As shown in Fig. 2e, shallow sub-surface reservoir ages begin to decline coincident with the beginning of the deglaciation (HS1 onset), continuing to drop across HS1 until minimum reservoir ages (that is, maximum ventilation) are reached around the HS1/B-A transition.

Ventilation ages in the EEP. Using the observed radiocarbon offsets between B-P (Figs 1b and 4a) and between planktonic-atmosphere (that is, reservoir ages; Figs 2e and 4b), the offset between B-Atm (that is, deep-ocean ventilation; calculated as B-Atm = B-P + R.age; Fig. 4c) can be estimated. Thus, Fig. 4 displays the ventilation history at the location of ODP1240, taking into account all four possible age models and their associated reservoir age implications.

As observed in the shallow sub-surface reservoir age reconstructions, the deep-ocean ventilation (Fig. 4c) also exhibits different amplitude and slightly different trends since the LGM depending on the various alignments (SST or $\delta^{18}O_{sw}$), but still broadly similar patterns of variability. Quantitatively, the averages from both SST and $\delta^{18}O_{sw}$ alignments indicate B-Atm values of

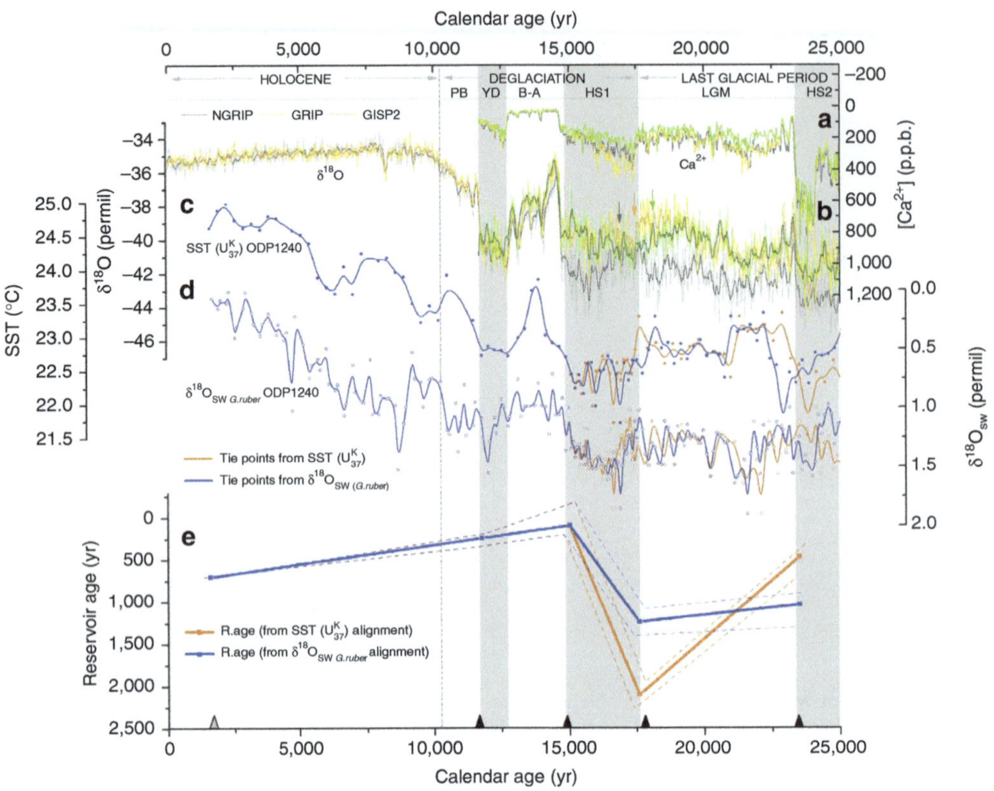

Figure 2 | Surface reservoir age assessment for ODP1240 sediment core through the alignment between two independent ODP1240 records and three Greenland ice cores over the last 25 kyr. (**a**) [Ca^{2+}] and (**b**) $\delta^{18}O$ records from Greenland ice cores[44]: NGRIP (grey), GRIP (yellow) and GISP2 (green). Small vertical arrows above Greenland $\delta^{18}O$ records indicate the apparent HS1 onset for each ice core. (**c**) SSTs from the $U^{K'}_{37}$ index and (**d**) $\delta^{18}O_{sw\ (G.\ ruber)}$[40] records from ODP1240 core. In blue, the alignment based on the SST tie-points, and in orange based on the $\delta^{18}O_{sw\ (G.\ ruber)}$ tie-points. (**e**) Estimated reservoir age scenarios for ODP1240 core using the two alignments: Thick central lines represent the best reservoir age estimate and stapled lines their upper and lower reservoir age limits (by taking an alignment uncertainty of ±200 year) (in orange, based on the SST correlation, and in blue based on the $\delta^{18}O_{sw\ (G.\ ruber)}$ correlation). Selected tie-points are indicated at the bottom by black triangles. An additional core-top date (from the study by Pena *et al.*[40]), corrected and calibrated is also shown (grey triangle). Light grey bands highlight the coldest periods of the last 25 kyr and delimit relevant climatic periods.

~2,775 years and ~2,910 years, respectively, during the last part of the glacial period, which would correspond to a ~1,275 and ~1,410 years older deep glacial ocean than during the Holocene (on average ~1,500 years including the modern B-Atm value of ~1,985 year at 3,000 m water depth; see Methods for further detail). Enhanced ventilation of the deep eastern Pacific is thus observed from the start of HS1, with peak ventilation (minimum B-Atm) being reached at the end of the deglaciation, in the early Holocene (Fig. 4c).

Discussion

The ventilation reconstructions from Site ODP1240 indicate a glacial-interglacial B–P change of ~870 years that, taking into account the shallow sub-surface (deep thermocline) reservoir age change of ~405/540 years (from both SST and $\delta^{18}O_{sw}$ alignments, respectively), imply a glacial-interglacial B-Atm change of ~1,275/1,410 years (Fig. 4; Supplementary Table 3). Taken together, these results indicate that both deep and shallow sub-surface reservoir ages (Fig. 4c,b, respectively) were significantly increased at the LGM and that both evolved in a similar manner over the last 25 kyr, *albeit* with a larger change in the deep ventilation ages (as indicated by the change in B–P across the deglaciation; Fig. 4a). Therefore, although the B–P data themselves indicate for the first time a clear deglacial change in deep equatorial Pacific ventilation (Figs 1b and 4a), the B-Atm estimates further suggest that the B–P offsets in fact underestimate this change in deep-ocean ventilation (Fig. 4a,c).

A poorly ventilated glacial shallow sub-surface and deep EEP is thus demonstrated, both by the B–P offsets and B-Atm offsets, with implications for the glacial ocean's total radiocarbon inventory and its role in glacial-interglacial CO_2 change[6]. Indeed, the radiocarbon ventilation rate of both the thermocline and the deep EEP are observed to increase broadly in time with the onset of deglaciation, the rise in atmospheric CO_2[52] and the drop in atmospheric ^{14}C[4,50] (Fig. 5a,b). These results would thus support the hypothesized role of reduced ocean-atmosphere CO_2 exchange rates in sequestering carbon in the glacial ocean interior, and the release of this carbon to the atmosphere across the last termination. The demonstration of reduced radiocarbon ventilation rates in both the deep and shallow sub-surface Pacific Ocean is crucial in this respect, as this basin accounts for ~50% of the ocean's total volume. Thus, small ventilation or carbon sequestration effects in the Pacific can have a particularly large impact on global ocean chemistry, as well as that of the atmosphere.

The changes in B–P offsets between the LGM and modern observed in ODP1240 clearly reflect a discernable trend in deep equatorial Pacific ventilation across the last deglaciation that is consistent with other B–P records from the northern and

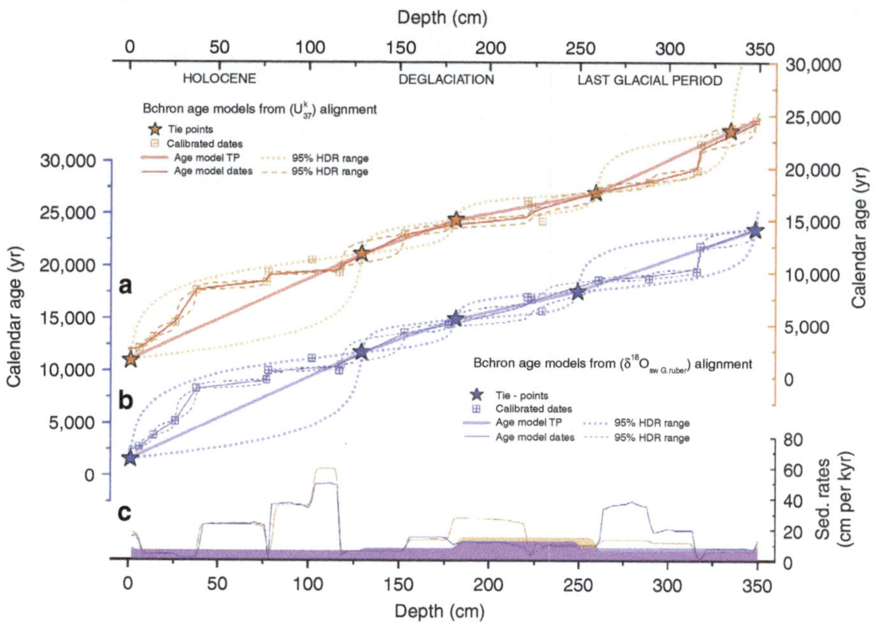

Figure 3 | Age models for core ODP1240. Four possible Bchron age models and their uncertainties (95% HDR range) for core ODP1240 derived from; (**a**) two from the $\delta^{18}O_{(Greenland\ ice\ cores)}$—$SST_{(ODP1240)}$ records alignment (orange curves) and (**b**) two from the $\delta^{18}O_{(Greenland\ ice\ cores)}$—$\delta^{18}O_{sw\ (G.\ ruber)\ (ODP1240)}$ records alignment (blue curve). Thick solid lines and stars represent the age models derived only from the tie-points (the error is 200 year). Thin solid lines and squares denote age models using the planktonic radiocarbon measurements corrected for the correspondent interpolated reservoir age obtained from the alignments. (**c**) Sedimentation rates (SR) obtained from the four age models; Shaded areas correspond to SR using only the tie-points and solid lines using the radiocarbon dates. As for **a** and **b**, orange colours refer to the SST alignment, and blue colours refer to the $\delta^{18}O_{sw\ (G.\ ruber)}$ alignment.

southern high latitude Pacific Ocean (southwest Pacific (SWP)[19,21]; southeast Pacific (SEP)[24]; northeast Pacific (NEP)[22]). All of these studies indicate slightly higher B–P age offsets during the last glacial period (~200–600 years) and, although they are not extremely large, they should not be downplayed. Indeed, it has been argued that the full 190‰ drop in atmospheric $\Delta^{14}C$ since the LGM (ignoring changes in radiocarbon production) could in principle be explained by rather small changes in the mean ventilation age of the whole upper ocean (for example, ~100 years over 60% of the ocean) if these were combined with a slightly larger change in the mean ventilation of the deepest ocean (for example, ~3,000 years on average, over 35% of the ocean)[28].

Although B–P offsets are clearly observed to decrease across the last deglaciation (Figs 1b and 4a), a larger drop in B-Atm offsets is inferred on the basis of our shallow sub-surface reservoir age reconstructions. In a first instance, we explore the regional coherence of our shallow sub-surface reservoir age reconstructions via a comparison of planktonic stable isotope records from our core and core TR163-31B (ref. 18; located in the same area and also with *N. dutertrei* radiocarbon dates), performed by placing each core on its own conventional ^{14}C-age scale (Supplementary Fig. 2). For this purpose we use the radiocarbon data from ref. 18. The fact that the isotope stratigraphies and radiocarbon data from the two cores are consistent (Supplementary Fig. 2) indicates that both locations have experienced broadly similar reservoir age changes. On this basis, a 'regional radiocarbon calibration' can be used to transfer calendar ages (and therefore reservoir ages) from ODP1240 to TR163-31B. This in turn allows B-Atm offsets to be derived for core TR163-31B, using the few available benthic radiocarbon dates from ref. 18, as described below.

While the comparison described above demonstrates that similar reservoir age changes were experienced at the locations of ODP1240 and TR163-31B, it does not provide an independent confirmation of the reservoir ages we derive for ODP1240. However, direct support for our observation of significantly enhanced glacial shallow sub-surface reservoir ages is provided by the few studies in the South Pacific Ocean that have also sought to constrain reservoir age variability using paleoceanographic observations (Fig. 5c). Thus, in the SWP a pioneering study in the Bay of Plenty, using ^{14}C-dated tephra, found a considerable increase in shallow sub-surface reservoir ages during the last glacial period[19] that, with the recently updated age for the Kawakawa tephra[53], increases the inferred glacial reservoir ages to ~3,200 years. The latter is an extreme value for a shallow sub-surface radiocarbon reservoir age in the modern context and, although it may yet be proven correct, we tentatively report the Bay of Plenty data in Fig. 5 excluding the glacial point. Indeed, a recent study in the same area (SWP) in core MD97-2121, also using the revised age of the Kawakawa tephra and radiocarbon dates on the same deep dwelling planktonic species (*Globigerina inflata*), found a ~337 years increase in shallow sub-surface reservoir ages, from 700 to ~1,037 years, during the last glacial period[21]. The latter values compare well with other reservoir age estimates from core MD07-3088 in the SEP, also based on tephrostratigraphy and marine radiocarbon dates in *G. bulloides*, which indicate ~500 years older reservoir ages over the latest part of the deglaciation, as compared with modern[54,55]. These reservoir age estimates have been extended beyond the range of available ^{14}C-dated tephra by assuming an inverse relationship between B–P offsets and planktonic versus benthic foraminiferal $\Delta\delta^{13}C$ from the same core, suggesting that increased surface reservoir ages also prevailed during the earlier part of the deglaciation[24].

The coherence of these various surface reservoir age estimates[19,21,24] (including those from this study) from different regions and water column depths may be explained by the

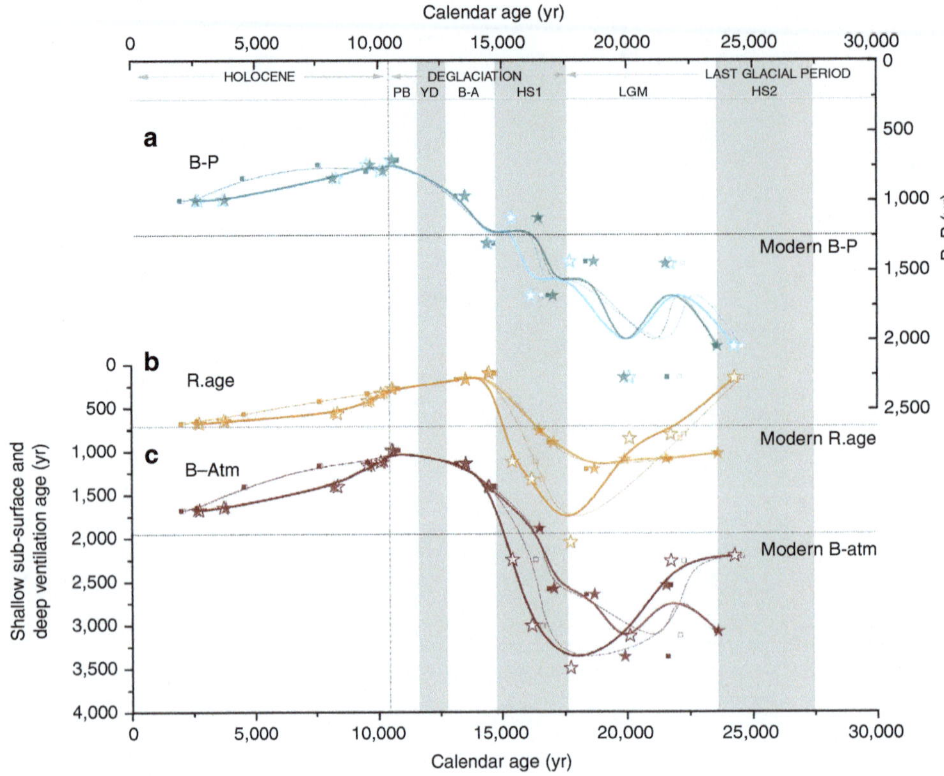

Figure 4 | Ventilation scenarios for core ODP1240 over the past 25 kyr using the four estimated age models from Fig. 3 in each case. Age models are b-spline smoothed. (**a**) B-P offset; (**b**) Shallow sub-surface reservoir age; (**c**) Ventilation age (B-Atm offset). Thick solid lines with stars denote age models using the planktonic radiocarbon measurements corrected for the correspondent interpolated reservoir age obtained from the alignments, and thin stapled lines with squares represent the age model derived only from the tie-points (hollow/solid stars and squares corresponds to $U^{K'}_{37}/\delta^{18}O_{sw(G.ruber)}$ alignments, respectively, in all cases). Modern offsets are shown by dashed horizontal lines. Light grey bands highlight the coldest periods of the last 25 kyr and delimit relevant climatic periods. Note the different *y*-axis scale in **a–c**.

transport pathways that link them together in the modern ocean circulation, as noted by a large number of studies[36,38,56]. Under the premise that the glacial circulation pathways were not wildly different from the modern (that is, a broadly similar pattern of buoyancy and wind forcing), a numerical model simulation[21] can be used to show a direct dynamical connection between all these areas and depths (Fig. 6). This illustrates that an increase in the shallow sub-surface reservoir age across the Southern Ocean (for example, as proposed by Skinner *et al.*[28]) should also be conveyed to the shallow sub-surface of the EEP and SWP via the dominant transport pathways in the ocean interior (Fig. 6), as is also indicated for the surface SEP, where *G. bulloides* (dated by Siani *et al.*[24]) currently thrives.

In agreement with this proposition, ODP1240 shallow sub-surface reservoir ages also generally exhibit trends and values that are consistent with reconstructions from the Southern Ocean, including the records based on chronostratigraphically dated planktonic foraminifera from core MD07-3076 from the sub-Antarctic Atlantic[28] and Drake Passage corals from water depths of 300–800 m and 700–1,800 m (hereafter referred to as 'shallow corals' and 'deep corals', respectively)[57] (Fig. 5c). The planktonic foraminifera from core MD07-3076 would have lived north of the Sub-Antarctic Front where some intermediate waters are subducted and the 'shallow corals' from the Drake Passage have been interpreted to provide a record of Antarctic Intermediate Water (AAIW) ventilation ages. Interestingly, the 'deep corals', although initially proposed to reflect the radiocarbon ventilation age of UCDW, are rather similar to the rest of the shallow sub-surface/intermediate water candidate reconstructions (Fig. 5c). This could suggest that, at least over the glacial/deglacial period, the deep Drake Passage corals might have been substantially influenced by SO-intermediate waters, instead of UCDW as previously proposed[57]. Such an interpretation would also help to reconcile the large differences between glacial ventilation ages observed in the deep Drake Passage corals[57] versus the deep sub-Antarctic Atlantic[28], which would reflect a large radiocarbon ventilation age gradient between AAIW and LCDW, rather than between UCDW and LCDW.

This consistency between shallow sub-surface reservoir ages in the EEP and the Atlantic and Pacific sectors of the Southern Ocean would support the idea of the proposed link between these two regions through shallow sub-surface/intermediate waters transported by the EUC to the base of the EEP thermocline; an idea that has been advanced previously based on other records[34,39–41]. We therefore propose here that the variability we see in *N. dutertrei* from ODP1240 is primarily related to the changing Southern Ocean end-member signature that is being 'upwelled' in (or at least transported to) the base of the EEP thermocline (where *N. dutertrei* lives), rather than, shifts in *N. dutertrei's* habitat towards the Southern Ocean end-member, for example due to 'more upwelling'. Nevertheless, we underline that this interpretation is not inconsistent with evidence for variable upwelling in the EEP, or indeed for changes in the contribution of southern sourced water in the EUC[42]. Such changes would only exacerbate the impacts that we infer by making *N. dutertrei* experience sub-surface waters that are more similar to an 'older/

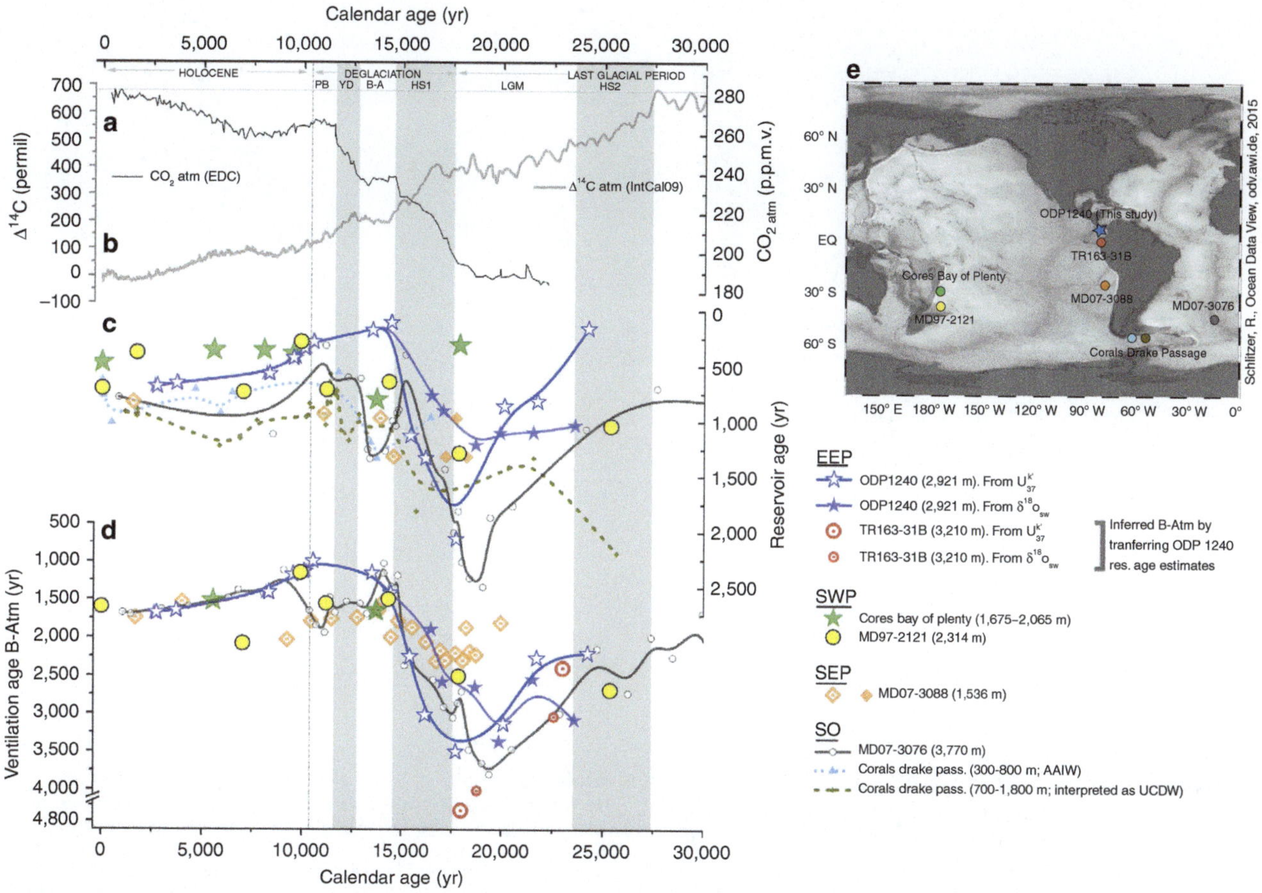

Figure 5 | Comparison of reservoir and ventilation age reconstructions from the South Pacific and Southern Ocean over the last deglaciation grouped according to their likely water-mass affiliations. (**a**) Atmospheric CO_2 concentrations from European Project for Ice Coring in Antarctica (EPICA) Dome C (EDC) ice core (for the deglacial period see ref. 52; for the Holocene period see ref. 68) placed on the age scale of ref. 69. (**b**) Atmospheric radiocarbon activity ($\Delta^{14}C$) changes (IntCal09 calibration curve[49]). (**c**) Surface/shallow sub-surface reservoir age records and (**d**) deep-water ventilation reconstructions from several locations in the Pacific and Southern Oceans over the last 25 kyr (B-spline smoothed): ODP1240 (2,921 m; *N. dutertrei*; dark blue hollow/solid stars from SST/$\delta^{18}O_{sw}$ alignments, respectively; this study); TR163-31B (3,210 m; *N. dutertrei*; red hollow big/small circles from SST/$\delta^{18}O_{sw}$ alignments in ODP1240, respectively[18]); Bay of Plenty (1,675–2,065 m; *Globorotalia inflata*; green stars[19]); MD97-2121 (2,314 m; *G. inflata*; yellow dots[21]); MD07-3088 (1,536 m; *Globigerina bulloides*; light orange diamonds from[54,55]; dark orange diamonds from[24]); MD07-3076 (3,770 m; *G. bulloides* and *G. inflata*; dark grey dots[28]); Corals from the Drake Passage (300–800 m 'shallow corals'; small light blue triangles and 700–1,800 m 'deep corals' small dark green dots[57]). Note that the age models used for ODP1240 are those obtained using the calibrated radiocarbon dates and not the tie-points. Note the *y*-axis breaks in **d**. (**e**) Locations of marine records shown in **c**,**d**.

younger' Southern Ocean origin, even if the latter effect was the largest, as we propose on the basis of the regional comparisons and associations shown in Fig. 5 and the numerical model 'thought experiment' illustrated in Fig. 6.

Regardless of their dynamical associations, or indeed their causes, increased glacial surface/shallow sub-surface reservoir ages have been reconstructed across the South Pacific and Southern Ocean using a range of different approaches, including U-series dating of corals[57], tephrochronology[19,21,24] and chronostratigraphic alignments[28] as well as radiocarbon 'plateau tuning' records in the North Pacific[20]. Along with the results of this study, these observations challenge the common assumption of invariant reservoir ages, in particular at high latitudes but also in tropical upwelling areas, as we emphasize here.

In addition to this general agreement of surface/shallow sub-surface reservoir age reconstructions, a broad similarity between deep ventilation age variability across the Equatorial and South Pacific as well as across the Southern Ocean is also apparent (Fig. 5d). Thus, B-Atm estimates from core ODP1240 indicate a significant glacial/interglacial change in radiocarbon ventilation, as also observed in the nearby core TR163-31B when the 'regional radiocarbon calibration' implied by our chronology for ODP1240 is applied to the few available benthic radiocarbon dates from this core[18] (Fig. 5d). Similarly, deep-water ventilation changes observed in the SWP[19,21] and the SEP[24] all show a broadly similar deglacial trend compared with the EEP, as well as comparable B-Atm values. This coherence is highly significant, as it would confirm that a large volume of the ocean interior was indeed very poorly ventilated during the last glacial period; a proposition that has so far been supported by a paucity of data.

The modern deep South and Equatorial Pacific is currently influenced by UCDW and/or PDW, which occupy similar density and depth ranges and are characterized by low oxygen, high nutrients and high ^{14}C-age[33]. A hydrographic/dynamical connection between the various ventilation age reconstructions compared in Fig. 5d is therefore entirely plausible; however, it could be seen to conflict with the previous attribution of an AAIW association for core MD07-3088 from the SEP[24]. Alternatively, the radiocarbon ventilation record from 1,536 m in the SEP[24] may in fact reflect the influence of deeper and older

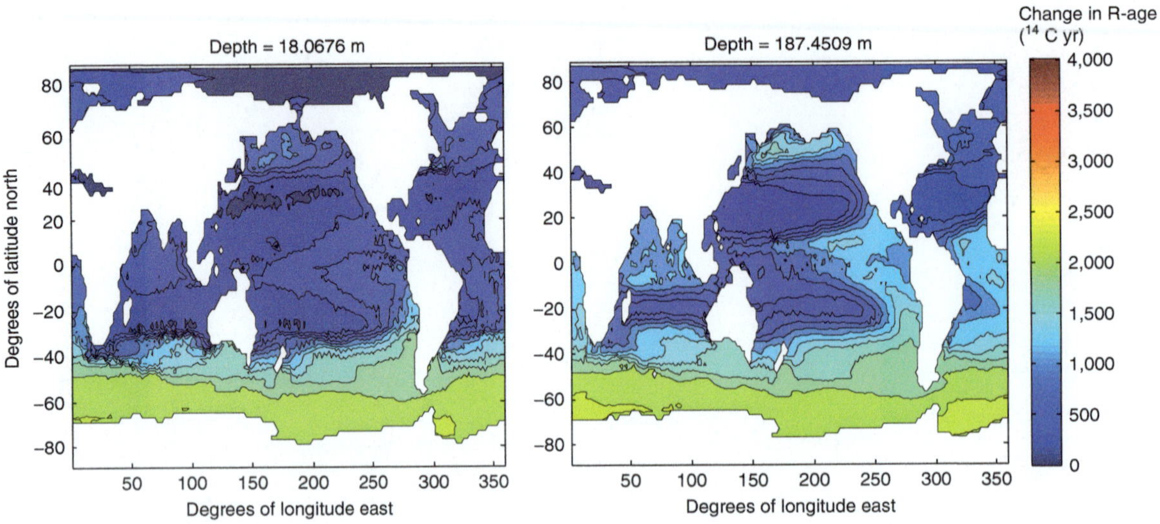

Figure 6 | Numerical model simulation. Marine radiocarbon ages (relative to the atmosphere), as reported by Skinner *et al.*[21], showing ages for 18 and 187 m water depth. The simulation adopts the model framework of DeVries *et al.*[70] and assumes the modern circulation, *albeit* with >1,000 year surface reservoir ages arbitrarily imposed in the Southern Ocean. The simulation thus demonstrates how an increase in Southern Ocean surface reservoir ages would be transmitted to other regions of the global ocean, including the EEP in particular, simply due to the dominant transport pathways of the ocean circulation. Coherent surface reservoir age changes in the Southern Pacific, Southern Atlantic and EEP could conceivably be explained in this way.

water masses such as PDW/UCDW, instead of AAIW as previously proposed[24]. This speculation is supported by the fact that the SEP core MD07-3088 is currently located just at the boundary between AAIW and PDW. In any event, the deep-water radiocarbon ventilation ages from ODP1240 (this study), and other cores from the deep Eastern Equatorial and South Pacific are all in good agreement with deep radiocarbon ventilation ages from core MD07-3076 from the sub-Antarctic Atlantic[28] (Fig. 5d), which is consistent with a Southern Ocean source of deep waters to the deep Pacific.

In this study, we have argued that even relatively small changes in B–P radiocarbon ventilation ages are relevant to the question of deglacial radiocarbon and carbon cycling, especially where these are observed over a large ocean volume like the Pacific. Furthermore, due to the possibility (indeed the likelihood) of significant changes in shallow sub-surface reservoir ages since the last glacial period, relatively small deglacial changes in B–P offsets may not necessarily rule out the occurrence of much larger changes in ocean interior radiocarbon ventilation, even in the low latitudes, as we demonstrate here for the EEP. Indeed, our results underline the crucial importance of constraining surface reservoir ages (or radiocarbon-independent calendar ages). Furthermore, these findings are found to be consistent with other surface- and deep-water ventilation age reconstructions from the southern Pacific and Atlantic. This may reflect glacial circulation patterns that were not very different from modern, *albeit* with altered surface boundary conditions (air–sea exchange efficiency) in the Southern Ocean in particular, which we identify as the main source region for poorly ventilated deep glacial waters in the shallow sub-surface and deep EEP. These results have implications for the ocean's role in glacial-interglacial carbon cycling, demonstrating an increase in the Pacific Ocean's mean carbon residence time, which very likely would have contributed to lower atmospheric CO_2 during the last glacial period. Furthermore, it would appear that this change in the ventilation state of the glacial Pacific may have stemmed in large part from less effective ocean-atmosphere CO_2 exchange in the Southern Ocean in particular.

Methods

Area of study. Sediment core ODP Site 1240 (Ocean Drilling Programme, Leg 202) was collected from the northern flank of the Carnegie Ridge (0°01.31′ N, 86°27.76′ W; 2,921 m water depth) in the southwestern margin of the Panama Basin[58]. This core location is characterized by high-accumulation-rates (> 10–15 cm per kyr[58,40]), which minimizes possible biases in benthic versus planktonic radiocarbon dating introduced by bioturbation[59]. The Panama Basin presents two main inflow passages for the deepest waters; one along the Ecuador Trench and another one across the broad central saddle of the Carnegie Ridge (sill depths ~2,900 m and ~2,300 m, respectively)[60,61]. Thus, sediment core ODP1240 must be currently bathed by UCDW and/or PDW (the latter also called Common Water[62]) as both occupy approximately the same density and depth range in the Pacific[33]. Both UCDW and PDW are characterized by low oxygen (especially the UCDW), high nutrients and large ^{14}C-age although the origin is different. PDW is formed internally in the Pacific from bottom waters upwelling and diffusion, while UCDW is originated by the mixture of PDW and Indian Deep Water that upwells just below the surface in the Antarctic Circumpolar Current[33]. The shallow sub-surface circulation over our core location is governed by the EUC between 50 and 300 m water depth flowing eastward along the tropics[36–38]. The EUC is partially fed by intermediate water masses (AAIW and Sub-Antarctic Mode Water) originated from upwelled CDW in the Southern Ocean[56], that are subsequently subducted by winter convection across the Antarctic Polar Front and Sub-Antarctic Front, respectively, spreading its signature into the Pacific interior. The EUC is also partially fed by Sub-Antarctic Water, which is a central water mass originated in the Subantartic zone that subducts at the Subtropical Front (STF; ~30° S) with its core flowing at 150–200 m water depth within the Subtropical gyre[63,64].

Sample treatment and analysis. About 19 radiocarbon dates on *N. dutertrei* (a thermocline-dwelling planktonic foraminifer species, thriving at 100–150 m depth[43]) and 17 samples of mixed benthic foraminifera were performed from core ODP1240 (that is, 17 paired B–P radiocarbon dates). *N. dutertrei* was picked from the >212 μm fraction (4–6 mg carbonate per sample), while mixed benthic foraminifera were from the >150 μm fraction (1–6 mg carbonate per sample; adjacent sediment samples were combined when the amount of carbonate was <1 mg).

Sample treatment was performed at the Godwin Laboratory for Palaeoclimate Research of the University of Cambridge (UK). To avoid sample loss, a minimal cleaning approach was carried out: removal of particles, fibres or other detritus or small carbonate clasts with a fine hair brush followed by repeated rinses in deionized water. Cleaned foraminifera were placed into clean glass vials, rinsed with methanol, dried and evacuated prior to graphitization. The calcite was hydrolysed in 0.5 ml of phosphoric acid and then reduced to graphite using a standard hydrogen and iron catalyst method[65]. Primary and secondary standards were graphitized in parallel with samples. NIST OXII was used as a primary reference standard (for normalization), and IAEA-C7 and -C8 were used as secondary standards (for quality control). When samples were <3 mg carbonate, size-matched aliquots of radiocarbon-dead Iceland Spar calcite were graphitized for small sample background (modern contamination) correction[66]. The radiocarbon analyses were carried out at The ^{14}CHRONo Centre of the Queen's University of Belfast (UK) using a National Electrostatic Corporation (NEC) compact model 0.5 MV accelerator mass spectrometer, with online $\delta^{13}C$ measurement.

Chronology. The ODP1240 core chronology for the last 25 kyr was rebuilt in this study. The previous age model for this core[40] was based on calibrated radiocarbon dates corrected, as is customarily done, for a constant surface reservoir age of ~472 years. However, in this study we assessed the variability of shallow sub-surface reservoir ages using independent chronostratigraphic constraints. We found that shallow sub-surface reservoir ages varied significantly over the last 25 kyr and produced new age models that take account of this variability, using the Bayesian depth-age modelling programme Bchron[51] and the IntCal09 radiocarbon calibration curve[49].

Radiocarbon calculations in modern water column. The modern radiocarbon values for ODP1240 core site were obtained from section P19 of the World Ocean Circulation Experiment[63], very close to our core site. Using the '$\Delta^{14}C_{natural}$' and applying the equation of Stuiver and Polach[67] ($t = -8{,}033\ln(1 + (\Delta^{14}C/1{,}000))$), we calculated the current shallow sub-surface and deep ventilation ages in this site. Thus, the ventilation ages would summarize as: modern shallow sub-surface ventilation age (reservoir age) at ~150 m water depth is ~735 radiocarbon years; modern deep ventilation age (B-Atm) at ~3,000 m water depth is ~1,985 radiocarbon years; modern B–P radiocarbon offset between 150 and 3,000 m water depth is ~1,250 radiocarbon years.

References

1. Toggweiler, J. R. Variation of atmospheric CO_2 by ventilation of the ocean's deepest water. *Paleoceanography* **14,** 571–588 (1999).
2. Stephens, B. B. & Keeling, R. F. The infuence of Antarctic sea ice on glacial—interglacial CO_2 variations. *Nature* **404,** 171–174 (2000).
3. Broecker, W. S., Peng, T.-H., Trumbore, S., Bonani, G. & Wolflp, W. The distribution of radiocarbon in the glacial ocean. *Glob. Biogeochem. Cycles* **4,** 103–117 (1990).
4. Hughen, K. A. *et al.* ^{14}C Activity and Global Carbon cycle changes over the past 50,000 years. *Science* **303,** 202–207 (2004).
5. Muscheler, R. *et al.* Changes in the carbon cycle during the last deglaciation as indicated by the comparison of ^{10}Be and ^{14}C records. *Earth Planet. Sci. Lett.* **219,** 325–340 (2004).
6. Broecker, W. S. & Barker, S. A 190‰ drop in atmosphere's $\Delta^{14}C$ during the 'Mystery Interval' (17.5 to 14.5 kyr). *Earth Planet. Sci. Lett.* **256,** 90–99 (2007).
7. Broecker, W. S. *et al.* Preliminary estimates for the radiocarbon age of deep water in the glacial ocean. *Paleoceanography* **3,** 659–669 (1988).
8. Broecker, W. S., Clark, E., Hajdas, I. & Bonani, G. Glacial ventilation rates for the deep Pacific Ocean. *Paleoceanography* **19,** 1–12 (2004).
9. Broecker, W. S. *et al.* Ventilation of the glacial deep Pacific Ocean. *Science* **306,** 1169–1172 (2004).
10. Broecker, W. S. *et al.* Radiocarbon age of late glacial deep water from the equatorial Pacific. *Paleoceanography* **22,** 1–6 (2007).
11. Broecker, W. S., Clark, E. & Barker, S. Near constancy of the Pacific Ocean surface to mid-depth radiocarbon-age difference over the last 20 kyr. *Earth Planet. Sci. Lett.* **274,** 322–326 (2008).
12. De Pol-Holz, R., Keigwin, L., Southon, J., Hebbeln, D. & Mohtadi, M. No signature of abyssal carbon in intermediate waters off Chile during deglaciation. *Nat. Geosci.* **3,** 192–195 (2010).
13. Okazaki, Y., Sagawa, T., Asahi, H., Horikawa, K. & Onodera, J. Ventilation changes in the western North Pacific since the last glacial period. *Clim. Past* **8,** 17–24 (2012).
14. Ahagon, N., Ohkushi, K., Uchida, M. & Mishima, T. Mid-depth circulation in the northwest Pacific during the last deglaciation: Evidence from foraminiferal radiocarbon ages. *Geophys. Res. Lett.* **30,** 2097 (2003).
15. Marchitto, T. M., Lehman, S. J., Ortiz, J. D., Flückiger, J. & van Geen, A. Marine radiocarbon evidence for the mechanism of deglacial atmospheric CO_2 rise. *Science* **316,** 1456–1459 (2007).
16. Stott, L., Southon, J., Timmermann, A. & Koutavas, A. Radiocarbon age anomaly at intermediate water depth in the Pacific Ocean during the last deglaciation. *Paleoceanography* **24,** PA2223 (2009).
17. Lund, D. C., Mix, A. C. & Southon, J. Increased ventilation age of the deep northeast Pacific Ocean during the last deglaciation. *Nat. Geosci.* **4,** 771–774 (2011).
18. Shackleton, N. *et al.* Radiocarbon age of last glacial Pacific deep water. *Nature* **335,** 708–711 (1988).
19. Sikes, E. L., Samson, C. R., Guilderson, T. P. & Howard, W. R. Old radiocarbon ages in the southwest Pacific Ocean during the last glacial period and deglaciation. *Nature* **303,** 555–559 (2000).
20. Sarnthein, M., Schneider, B. & Grootes, P. M. Peak glacial ^{14}C ventilation ages suggest major draw-down of carbon into the abyssal ocean. *Clim. Past* **9,** 2595–2614 (2013).
21. Skinner, L. *et al.* Reduced ventilation and enhanced magnitude of the deep Pacific carbon pool during the last glacial period. *Earth Planet. Sci. Lett.* **411,** 45–52 (2015).
22. Galbraith, E. *et al.* Carbon dioxide release from the North Pacific abyss during the last deglaciation. *Nature* **449,** 890–893 (2007).
23. Ikehara, K. *et al.* Paleoceanographic control on a large marine reservoir effect offshore of Tokai, south of Japan, NW Pacific, during the last glacial maximum-deglaciation. *Quat. Int.* **246,** 213–221 (2011).
24. Siani, G. *et al.* Carbon isotope records reveal precise timing of enhanced Southern Ocean upwelling during the last deglaciation. *Nat. Commun.* **4,** 2758 (2013).
25. Max, L. *et al.* Pulses of enhanced North Pacific Intermediate Water ventilation from the Okhotsk Sea and Bering Sea during the last deglaciation. *Clim. Past* **10,** 591–605 (2014).
26. Adkins, J. F. & Boyle, E. Changing atmospheric $\Delta^{14}C$ and the record of deep water paleoventilation ages. *Paleoceanography* **12,** 337–344 (1997).
27. Butzin, M., Prange, M. & Lohmann, G. Readjustment of glacial radiocarbon chronologies by self-consistent three-dimensional ocean circulation modeling. *Earth Planet. Sci. Lett.* **317-318,** 177–184 (2012).
28. Skinner, L., Fallon, S., Waelbroeck, C., Michel, E. & Barker, S. Ventilation of the deep Southern Ocean and deglacial CO_2 rise. *Science* **328,** 1147–1151 (2010).
29. Bard, E. *et al.* The North Atlantic atmosphere-sea surface ^{14}C gradient during the Younger Dryas climatic event. *Earth Planet. Sci. Lett.* **126,** 275–287 (1994).
30. Siani, G. *et al.* Mediterranean Sea surface radiocarbon reservoir age changes since the last glacial maximum. *Science* **294,** 1917–1920 (2001).
31. Primeau, F. Characterizing transport between the surface mixed layer and the ocean interior with a forward and adjoint global ocean transport model. *J. Phys. Oceanogr.* **35,** 545–564 (2005).
32. Orsi, A. H., Johnson, G. C. & Bullister, J. L. Circulation, mixing, and production of Antarctic Bottom Water. *Prog. Oceanogr.* **43,** 55–109 (1999).
33. Talley, L. D., Pickard, G. L., Emery, W. J. & Swift, J. H. *Descriptive Physical Oceanography: An introduction* (Elsevier, 2011).
34. Anderson, R. *et al.* Wind-driven upwelling in the Southern Ocean and the deglacial rise in atmospheric CO_2. *Science* **323,** 1443–1448 (2009).
35. Skinner, L., Waelbroeck, C., Scrivner, A. E. & Fallon, S. J. Radiocarbon evidence for alternating northern and southern sources of ventilation of the deep Atlantic carbon pool during the last deglaciation. *Proc. Natl Acad. Sci USA* **411,** 1–5 (2014).
36. Toggweiler, J. R., Dixon, K. & Broecker, W. S. The Peru Upwelling and the ventilation of the South Pacific Thermocline. *J. Geophys. Res.* **96,** 20467–20497 (1991).
37. Lukas, R. The termination of the Equatorial Undercurrent in the eastern Pacific. *Prog. Oceanogr.* **16,** 63–90 (1986).
38. Tsuchiya, M., Lukas, R. & Fine, R. A. Source waters of the Pacific equatorial undercurrent. *Prog. Oceanogr.* **23,** 101–147 (1989).
39. Spero, H. J. & Lea, D. The cause of carbon isotope minimum events on glacial terminations. *Science* **296,** 522–525 (2002).
40. Pena, L., Cacho, I., Ferretti, P. & Hall, M. A. El Niño–Southern Oscillation–like variability during glacial terminations and interlatitudinal teleconnections. *Paleoceanography* **23,** 1–8 (2008).
41. Calvo, E., Pelejero, C., Pena, L., Cacho, I. & Logan, G.A. Eastern equatorial pacific productivity and related-CO_2 changes since the last glacial period. *Proc. Natl Acad. Sci. USA* **108,** 5537–5541 (2011).
42. Pena, L. *et al.* Rapid changes in meridional advection of Southern Ocean intermediate waters to the tropical Pacific during the last 30 kyr. *Earth Planet. Sci. Lett.* **368,** 20–32 (2013).
43. Fairbanks, R., Sverdlove, M. S., Free, R., Wiebe, P. & Be, A. Vertical distribution and isotopic fractionation of living planktonic foraminifera from the Panama Basin. *Nature* **298,** 841–844 (1982).
44. Rasmussen, S. O. *et al.* Synchronization of the NGRIP, GRIP, and GISP2 ice cores across MIS 2 and palaeoclimatic implications. *Quat. Sci. Rev.* **27,** 18–28 (2008).
45. Svensson, A. *et al.* The Greenland Ice Core Chronology 2005, 15–42 ka. Part 2: comparison to other records. *Quat. Sci. Rev.* **25,** 3258–3267 (2006).
46. Kienast, M. *et al.* Eastern Pacific cooling and Atlantic overturning circulation during the last deglaciation. *Nature* **443,** 846–849 (2006).
47. Leduc, G. *et al.* Moisture transport across Central America as a positive feedback on abrupt climatic changes. *Nature* **445,** 908–911 (2007).
48. Dubois, N., Kienast, M., Kienast, S. & Timmermann, A. Millennial-scale Atlantic/East Pacific sea surface temperature linkages during the last 100,000 years. *Earth Planet. Sci. Lett.* **396,** 134–142 (2014).
49. Reimer, P. J. *et al.* IntCal09 and marine09 radiocarbon age calibration curves, 0-50,000 years cal BP. *Radiocarbon* **51,** 1111–1150 (2009).

50. Reimer, P. J. *et al.* Intcal13 and marine13 radiocarbon age calibration curves 0 – 50,000 years cal BP. *Radiocarbon* **55,** 1869–1887 (2013).
51. Parnell, A. C., Haslett, J., Allen, J. R. M., Buck, C. E. & Huntley, B. A flexible approach to assessing synchroneity of past events using Bayesian reconstructions of sedimentation history. *Quat. Sci. Rev.* **27,** 1872–1885 (2008).
52. Monnin, E. *et al.* Atmospheric CO_2 concentrations over the last glacial termination. *Science* **291,** 112–114 (2001).
53. Vandergoes, M. J. *et al.* A revised age for the Kawakawa/Oruanui tephra, a key marker for the Last Glacial Maximum in New Zealand. *Quat. Sci. Rev.* **74,** 195–201 (2013).
54. Haberle, S. G. & Lumley, S. H. Age and origin of tephras recorded in postglacial lake sediments to the west of the southern Andes, 44°S to 47°S. *J. Volcanol. Geotherm. Res.* **84,** 239–256 (1998).
55. Carel, M., Siani, G. & Delpech, G. Tephrostratigraphy of a deep-sea sediment sequence off the south Chilean margin: New insight into the Hudson volcanic activity since the last glacial period. *J. Volcanol. Geotherm. Res.* **208,** 99–111 (2011).
56. Kessler, W. The circulation of the eastern tropical Pacific: a review. *Prog. Oceanogr.* **69,** 181–217 (2006).
57. Burke, A. & Robinson, L. F. The Southern Ocean's role in carbon exchange during the last deglaciation. *Science* **335,** 557–561 (2012).
58. Mix, A. C., Tiedemann, R. & Blum, P. Leg 202 summary. *Proc. Ocean Drill. Prog. Initial Rep.* **202 (2003).**
59. Barker, S., Broecker, W. S., Clark, E. & Hajdas, I. Radiocarbon age offsets of foraminifera resulting from differential dissolution and fragmentation within the sedimentary bioturbated zone. *Paleoceanography* **22,** 1–11 (2007).
60. Laird, N. P. Panama Basin deep water- properties and circulation. *J. Mar. Res.* **29,** 226–234 (1971).
61. Lonsdale, P. & Malfait, B. Abyssal dunes of foraminiferal sand on the Carnegie Ridge. *Geol. Soc. Am. Bull.* **85,** 1697–1712 (1974).
62. Montgomery, R. B. Water characteristics of Atlantic Ocean and of world ocean. *Deep Sea Res.* **5,** 134–148 (1958).
63. Tsuchiya, M. & Talley, L. D. A Pacific hydrography section at 88°W: water-property distribution. *J. Geophys. Res.* **103,** 899–12,918 (1998).
64. Llanillo, P. J., Karstensen, J., Pelegrí, J. L. & Stramma, L. Physical and biogeochemical forcing of oxygen and nitrate changes during El Niño/El Viejo and La Niña/La Vieja upper-ocean phases in the tropical eastern South Pacific along 86° W. *Biogeosciences* **10,** 6339–6355 (2013).
65. Vogel, J. S., Southon, J. R., Nelson, D. E. & Brown, T. A. Performance of catalytically condensed carbon for use in accelerator mass spectrometry. *Nucl. Instrum. Methods Phys. Res.* **B5,** 289–293 (1984).
66. Santos, G. M., Southon, J., Griffin, S., Beaupre, S. R. & Druffel, E. R. M. Ultra small-mass AMS ^{14}C sample preparation and analyses at KCCAMS/UCI facility. *Nucl. Instrum. Methods Phys. Res. B* **259,** 293–302 (2007).
67. Stuiver, M. & Polach, H. A. Radiocarbon, discussion reporting of ^{14}C data. *Radiocarbon* **19,** 355–363 (1977).
68. Flückiger, J., Monnin, E., Stauffer, B., Schwander, J. & Stocker, T. F. High-resolution Holocene N_2O ice core record and its relationship with CH_4 and CO_2. *Glob. Biogeochem. Cycles* **16,** 1–8 (2002).
69. Lemieux-Dudon, B. *et al.* Consistent dating for Antarctic and Greenland ice cores. *Quat. Sci. Rev.* **29,** 8–20 (2010).
70. Devries, T. The oceanic anthropogenic CO_2 sink: Storage, air-sea fluxes and transports over the industrial era. *Glob. Biogeochem. Cycles* 1–17 (2014).

Acknowledgements

We would like to thank J. Peña-Izquierdo and P.J. Llanillo for insightful discussions on modern water-mass pathways in the Pacific Ocean. P. Povea and J. Torner are also thanked for assistance with foraminifera identification and picking. We are also grateful for the assistance of A.E. Scrivner with graphitization and R. Reimer with AMS analysis. We acknowledge funding by the Spanish Ministry of Economy and Competitiveness through grants CTM2009-08849 (ACDC Project) and CTM2012-32017 (MANIFEST Project). M.dlF. was funded by a FPI studentship (BES-2010-039700) from the Spanish Government, and L.S. acknowledges support from the NERC grant NE/L006421/1. I.C. thanks the ICREA Academia program from Generalitat de Catalunya and C.P. the Generalitat de Catalunya through grant 2014SGR1029.

Author contributions

M.dlF. led the writing of the paper with contributions from all other authors. Radiocarbon dating was performed by M.dlF. with technical support from L.C.S. Alkenone SST data were generated by E.C. and C.P., and L.C.S. performed the model simulation.

Additional information

Competing financial interests: The authors declare no competing financial interests.

12

Modelled drift patterns of fish larvae link coastal morphology to seabird colony distribution

Hanno Sandvik[1], Robert T. Barrett[2], Kjell Einar Erikstad[1,3], Mari S. Myksvoll[4], Frode Vikebø[4], Nigel G. Yoccoz[3,5], Tycho Anker-Nilssen[6], Svein-Håkon Lorentsen[6], Tone K. Reiertsen[3], Jofrid Skarðhamar[4], Mette Skern-Mauritzen[4] & Geir Helge Systad[3]

Colonial breeding is an evolutionary puzzle, as the benefits of breeding in high densities are still not fully explained. Although the dynamics of existing colonies are increasingly understood, few studies have addressed the initial formation of colonies, and empirical tests are rare. Using a high-resolution larval drift model, we here document that the distribution of seabird colonies along the Norwegian coast can be explained by variations in the availability and predictability of fish larvae. The modelled variability in concentration of fish larvae is, in turn, predicted by the topography of the continental shelf and coastline. The advection of fish larvae along the coast translates small-scale topographic characteristics into a macroecological pattern, *viz.* the spatial distribution of top-predator breeding sites. Our findings provide empirical corroboration of the hypothesis that seabird colonies are founded in locations that minimize travel distances between breeding and foraging locations, thereby enabling optimal foraging by central-place foragers.

[1] Centre for Biodiversity Dynamics, Department of Biology, Norwegian University of Science and Technology, 7491 Trondheim, Norway. [2] Department of Natural Sciences, Tromsø University Museum, PO Box 6050 Langnes, 9037 Tromsø, Norway. [3] Norwegian Institute for Nature Research, FRAM—High North Research Centre for Climate and the Environment, 9296 Tromsø, Norway. [4] Institute of Marine Research and Hjort Centre for Marine Ecosystem Dynamics, PO Box 1870 Nordnes, 5817 Bergen, Norway. [5] Department of Arctic and Marine Biology, University of Tromsø, PO Box 6050 Langnes, 9037 Tromsø, Norway. [6] Norwegian Institute for Nature Research, PO Box 5685 Sluppen, 7485 Trondheim, Norway. Correspondence and requests for materials should be … (email: hanno@evol.no)

Advantages and disadvantages of colonial breeding are associated with social interactions, predator avoidance and food acquisition (exchange of information), for which selection pressures vary across species[1–3]. Whereas 13% of all birds are colonial breeders, ~98% of the some 330 seabird species breed in colonies comprised of often very dense aggregations of breeding territories[4,5]. Possible explanations include the minimization of travel distances between the nest and foraging locations (geometrical model[1,6]), enhanced food finding efficiency through information transfer (information centre hypothesis[7]), limited nest-site availability[1], predator avoidance[8] and mate-choice mechanisms[2].

The distance between neighbouring colonies is often negatively related to their size[9], indicating a regulating role of intraspecific competition and emphasising the importance of food availability[10–13]. The mechanisms behind dispersal of individuals from and their immigration into already existing colonies are increasingly understood[14,15]. However, the question remains what determined the placement of seabird colonies along a coast in the first place. Whereas recent colonization of empty spaces may have been determined by human-induced habitat changes[16], the initial colonization of an area can be expected to be linked to the predictability of habitat quality[17]. Habitat quality may be rather straightforward to quantify in terrestrial ecosystems[18–20], but this is not as easily accomplished for species feeding on oceanic prey[21,22]. Of the main factors contributing to nesting habitat suitability, features such as substrate and the absence of terrestrial predators are predictably determined by a site's physical properties, whereas food availability in the local marine environment is much less predictable in time and space. The formation and maintenance of large seabird colonies in relation to semi-stable physical oceanic phenomena, such as upwellings, fronts, gyres and polynyas, nonetheless indicate that there may, under some conditions, be a certain degree of predictability of food supply within the foraging range of seabirds[1,23,24]. Until now, however, no studies have documented more than theoretical associations between food availability and the distribution of seabird colonies, and little data exist to evaluate the predictions empirically[21,22,25,26].

We carried out a quantitative test of the association between modelled food availability/predictability and colony locations, using empirical data from the coast of Northern Norway (66–71°N). More than 90% of the two million pairs of Norwegian cliff-nesting seabirds (mainly Atlantic puffin *Fratercula arctica*, black-legged kittiwake *Rissa tridactyla* and common guillemot *Uria aalge*) occur north of 66°N (Fig. 1) (ref. 27). This distribution has been attributed to favourable physical oceanographic conditions linked to the northward-flowing Norwegian Coastal Current (NCC) and to bathymetry, because the exceptional marine productivity at the continental shelf break is close to the coastline (Fig. 1) (refs 27,28). Such productivity is essential to sustain the large numbers of breeding seabirds, whose foraging range is constrained by flight costs and the energetic demands of raising offspring[29].

Along the long stretch of Northern Norway's coast—roughly 1,200 km along its shortest line—there are 20 large seabird colonies (here defined as 10,000 breeding pairs or more). As suitable habitat (high cliffs, turf-covered islands and promontories that are all but inaccessible to terrestrial predators) is available along almost all the coast, and human disturbance is minimal for most of its parts, nest-site availability is not a factor limiting the distribution of seabird colonies in Northern Norway. We hypothesize that large seabird colonies are located in those areas where large numbers of planktonic organisms, for example, small crustaceans (especially *Calanus finmarchicus*), fish and their spawning products (eggs and larvae), occur predictably within the foraging range of breeding birds during their breeding period. Such organisms are important food items for both seabirds and some of their fish prey (for example, capelin *Mallotus villosus*, 0- and I-group herring *Clupea harengus*, sandeels *Ammodytes* spp.), and an independent quantification of their abundance is thus a potential proxy of food availability for seabirds.

Coupled ocean circulation and biophysical models allow the quantification of seasonally predictable biological production[30,31]. Here, we use an ocean circulation model hindcast to quantify the drift of particles, such as eggs and larvae, from fish spawning grounds along the coast and their interaction with the physical environment[32], providing a proxy of the abundance of ichthyoplankton throughout the study area with a spatial resolution of 4 km^2. The output of this model has been validated against empirical observations in the Barents Sea[32,33], and the proxy has already proven to be useful by giving novel insights into several aspects of seabird ecology, such as the day-to-day and interannual variation in stress hormones[34], fledging weight[35] and population dynamics[36] of common guillemots.

Using this approach, we relate the positions of seabird colonies directly to the modelled small-scale temporal and spatial variation of plankton along a coastline. Our findings provide an empirical corroboration of the hypothesis that the initial formation of colonies minimizes travel distances between breeding and foraging locations in a marine habitat.

Results

Physical correlates of high larval density. Despite the fact that the NCC transports spawning products along the entire

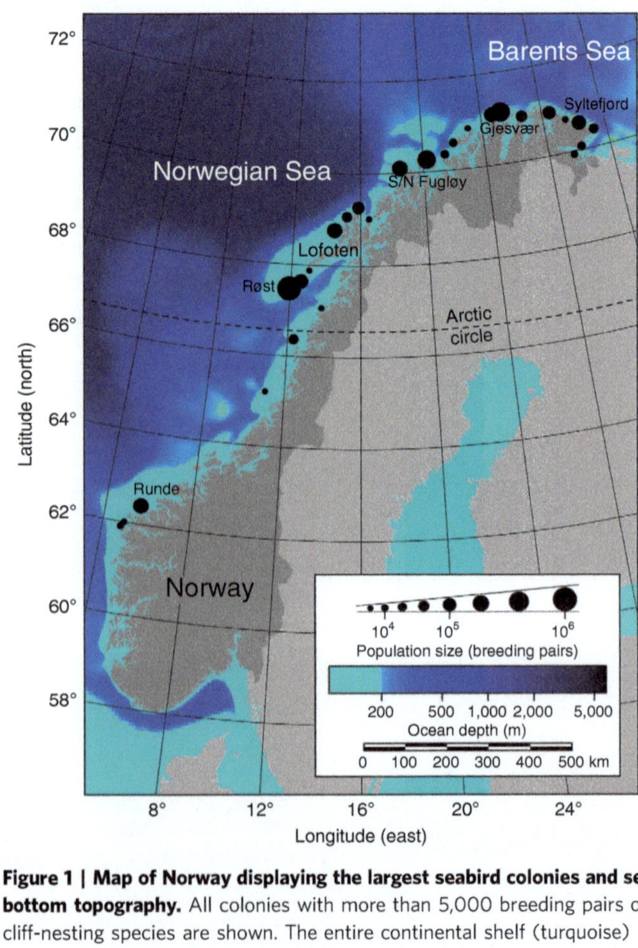

Figure 1 | Map of Norway displaying the largest seabird colonies and sea bottom topography. All colonies with more than 5,000 breeding pairs of cliff-nesting species are shown. The entire continental shelf (turquoise) north of 61°N was used as initiation area of the generic larval drift model.

Norwegian coast into the Barents Sea (Supplementary Fig. 1a), the larval drift models show clearly that these particles are not uniformly distributed along the coast. Instead, their distribution is rather patchy with widely varying concentrations (Fig. 2a). Physical features explain 39% of the variation in particle density (Supplementary Table 1). This is mainly due to a gradient on a large spatial scale, *viz* a decrease in particle concentration towards the east. Since the particle trajectories spread out as they approach the Barents Sea, the maximum is reached at 68° 15′ N, where the currents are funnelled round the Lofoten Islands by the narrowing continental shelf.

The smaller-scale patchiness is caused by the interaction between the NCC and coastal morphology, two aspects of which have been quantified: in areas where the continental shelf is narrow, the NCC is constricted near the coast, so that bathymetry explains 7% of the variation in particle density. A further 10% are explained by the structure of the coastline. This is likely because the NCC interacts with bank structures, islets and promontories

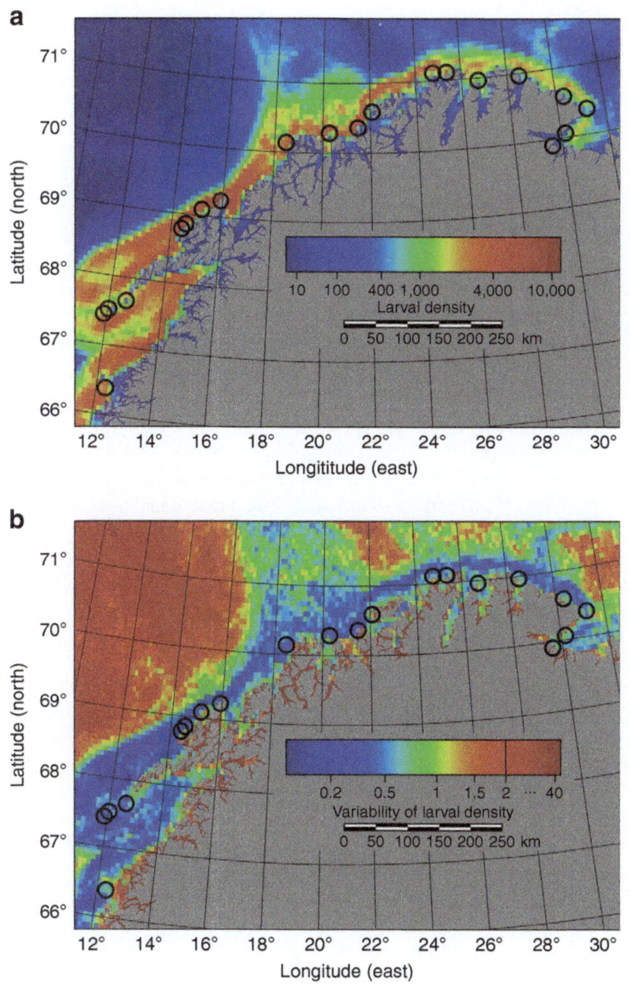

Figure 2 | Association between abundance of fish larvae and the location of seabird colonies along the coast of Northern Norway. Circles indicate foraging ranges with radii of 10 km around the 20 largest seabird colonies (at least 10,000 breeding pairs). (**a**) Map of seabird colonies and sites of high modelled larval abundance. There are more high-abundance grid cells within the circles than expected by chance ($P = 0.014$). (**b**) Map of seabird colonies and sites of high predictability (low variability) of modelled larval density. Variability is measured as quartile coefficients of dispersion. There are more low-variability grid cells within the circles than expected by

along the coast, generating stationary eddies that increase the residence time of passive particles such as crustaceans, fish eggs and larvae.

Relation of seabird colonies to larval density. The existing seabird colonies are located in or close to areas where particle concentrations are higher than the average for the respective coastal segments (Fig. 2a), with the association between seabird colonies and grid cells with high particle abundance being stronger than expected by chance ($P = 0.014$). The association between seabird colonies and grid cells with low interannual variability in prey abundance is even stronger ($P < 0.001$; Figs 2b and 3a). This means that seabird colonies along the North Norwegian coast are systematically located to give a much better and more stable, that is, predictable, access to suitable food than randomly chosen locations. The importance of low temporal variability in prey abundance is further highlighted by the fact that the colony locations are strongly associated with grid cells that have high minimum food availability across years ($P = 0.013$), but less so with grid cells that have high maximum food availability ($P = 0.059$; Supplementary Table 2).

Simulations based on fish species rather than generic particles, that is, simulations that take into account the spawning sites, growth patterns and vertical migration behaviour of particular species, showed divergent results. Herring larvae alone did not yield a significant association with colony sites ($P = 0.36$), whereas simulations based on cod (*Gadus morhua*) larvae ($P = 0.003$) or combinations of both species did ($P < 0.025$; Supplementary Table 2). Cod larvae thus produced a model with a much stronger association between larval abundance and colony placement than the generic particle model. A *post-hoc* analysis of the different spawning grounds of cod revealed that this very significant pattern was due mainly to one spawning ground, the Lofoten area ($P = 0.005$; Supplementary Table 3). This indicates that any perturbations to cod spawning in this area may have serious consequences for seabirds breeding in and to the north of Lofoten, an area that holds 19 out of the 21 large seabird colonies in Norway. The poorer performance of the herring simulation is explained by the spawning grounds of herring being further south, resulting in considerably lower concentrations of herring than cod larvae along the northern and eastern parts of the coast.

Significance of foraging distance. The strength of the association between seabird colonies and grid cells with high particle abundance varies depending on the foraging range of the seabirds (Fig. 3b,c). In the initial model, a foraging radius of 10 km is assumed. However, significant associations are obtained for foraging radii of up to 25 km, and associations are strongest for radii between 6 and 8 km ($P < 0.01$, Fig. 3b,c). This can be taken to indicate that seabird colonies are usually situated within 5–25 km of productive food patches.

Although recent tracking studies have revealed that several seabirds are capable of foraging at much larger distances from their breeding colony (>400 km)[37], average trip length under normal conditions is much shorter (<50 km)[38,39]. Long trips are costly and are most likely forced upon birds only under unusual food conditions[37]. Our results indicate that seabirds are rather successful in choosing breeding sites that minimize travel distances between food patches and the nest.

Robustness of results. The association between seabird colonies and particle abundance, which is estimated using randomization tests, holds for a broad range of model parameterizations and is further corroborated by *post-hoc* tests (Fig. 3 and Supplementary

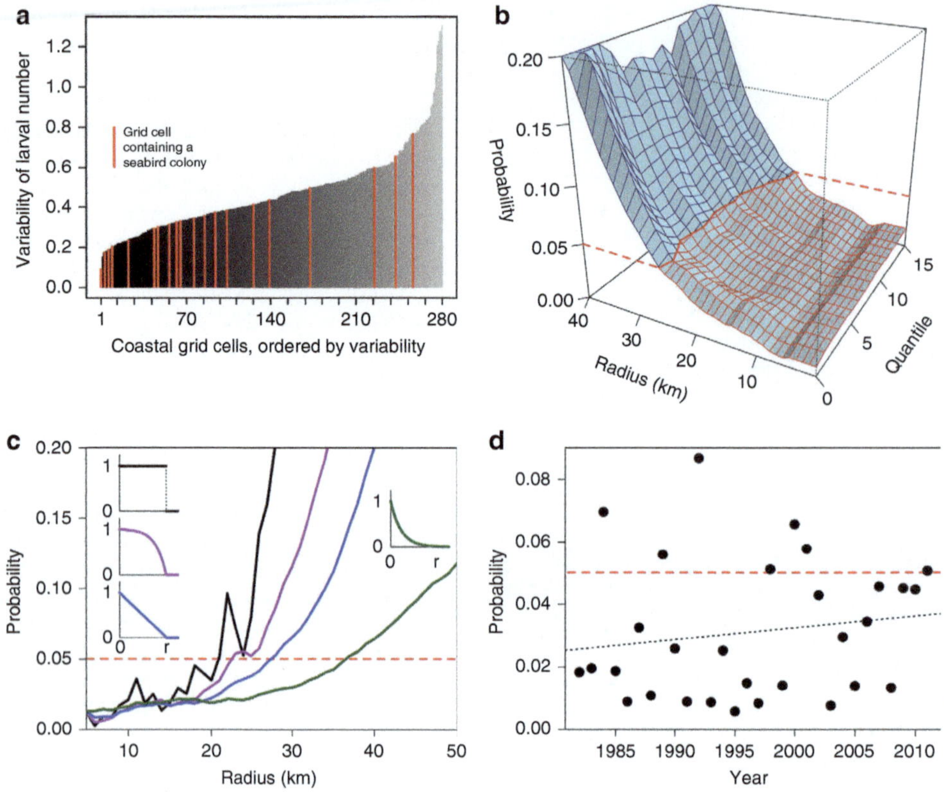

Figure 3 | Results of simulations and relaxation of the assumptions on which the initial model is based. (**a**) Histogram of the 280 grid cells along the coast, sorted by increasing variability of larval abundance (as measured by the quartile coefficient of dispersion). Grid cells containing the 20 largest seabird colonies, highlighted as red bars, are concentrated in the left part of the histogram, that is, they have significantly lower variability in larval abundance than randomly chosen grid cells. (If uniformly distributed, the 20 red bars would lie exactly between the 21 ticks of the *x*-axis.) (**b–d**) The panels report the probability that a sample of 20 randomized colony positions obtains a better availability of fish larvae than the actual positions of the 20 largest seabird colonies in Northern Norway. (Each point in the parameter space is based upon simulations with 100,000 replicates. The portion of the parameter space below the red line is statistically significant at the 5% level). (**b**) Probabilities given different radii of feeding ranges around seabird colonies (relaxation of assumption (vii)) and given different 30-quantiles of particle abundance for each grid cell (relaxation of assumption (iv)). The 0th 30-quantile corresponds to the minimum particle density within the 30-year period, the 15th 30-quantile to the median. The initial model uses a radius of 10 km and the 3rd 30-quantile. (**c**) Probabilities given different feeding radii (*r*) and different weighting schemes of particles within the feeding radius specified (relaxation of assumption (viii)): none-or-all threshold (black), convex weighting function (purple), linear weighting function (blue), concave weighting function (green). Small panels visualize the weighting functions. The initial model uses a linear weighting scheme. (**d**) Probabilities for single years. There is no significant trend in the probabilities (dotted line; slope 0.00038 ± 0.00047, $R^2 = 0.02$, $P = 0.42$).

Tables 2 and 3), indicating that it is very robust to changes in the background assumptions that had to be made. By modifying some of the initial assumptions, model fit could be substantially improved (all details in Supplementary Table 2), for example, by considering conditions in July alone rather than the average of the breeding season ($P = 0.002$), by using a wider definition of suitable coastline ($P < 0.001$ for the widest definition) or by using smaller or larger size thresholds for the inclusion of colonies than 10,000 breeding pairs ($P = 0.006$ for colonies with at least 5,000 breeding pairs). On the other hand, results were rather insensitive to transformation of larval abundance data (Supplementary Table 2), to the quantile used to represent larval counts (Fig. 3b) or to the weighting function of different foraging distances within the maximum foraging radius (Fig. 3c).

The *post-hoc* tests showed that the association was not merely due to one influential colony location; and that data for a single year of an even more highly resolved larval model (800 m) are compatible with the 4 km model (Supplementary Table 3). When decreasing the grid resolution of the simulations, the association becomes gradually weaker as sample size decreases and the grid cell size exceeds foraging radius; however, even for a fourfold decrease in resolution, the association remains significant for cod ($P = 0.025$) and marginally so for the generic particles ($P = 0.051$). When considering probabilities for single years, the association between particles and seabird colonies is significant at the 5% level in 23 out of 30 years in the period from 1982 to 2011, the non-significant years do not cluster, and there is no temporal trend in the probabilities (Fig. 3d).

Discussion

During their breeding season, seabirds are central-place foragers depending on spatially patchy and temporally variable food. According to the geometrical model and optimal foraging theory, such conditions favour the evolution of colonial breeding[6]. Our findings accord with the hypothesis that seabird colonies are established in locations that minimize distances between breeding and foraging locations: colonies are considerably closer to the nearest high-density prey patch than suggested by the potential flight range of seabirds and than expected by chance. Although documented for colonial breeders in terrestrial ecosystems[18–20], this has been difficult to show in marine ecosystems, presumably because the small-scale distribution of prey items is harder to quantify (but see refs 21,22).

This analysis has been made possible by a coupled ocean circulation and biophysical model of ichthyoplankton abundance. At the same time, the use of a modelling approach introduces some additional uncertainty, because the empirical validation[32] is necessarily restricted to a fraction of the area and time period covered by the models. It is also obvious that the findings are open to alternative interpretations, and that the model does not identify all factors relevant for the establishment and maintenance of seabird colonies. There is, for instance, a suspicious absence of seabird colonies in two regions where Fig. 2a suggests high levels of larvae (*viz*, 67.7°–68.7°N and 69.3°–70.1°N). However, the interannual variability of larval abundance is somewhat higher in these regions (0.39 ± 0.12 (quartile coefficient of dispersion (QCD) ± s.d.; see the Methods: Simulation: iv), $N = 52$ grid cells) than in the surrounding colony locations (0.27 ± 0.02, $N = 4$; cf. Fig. 2b). Of course, a number of other causes can be involved in the absence of colonies, including habitat quality, predation pressure, human disturbance and stochastic factors.

One major cost of coloniality is increased competition among foragers at food patches[10–13]. To sustain coloniality, the food patches must thus be sufficiently large that sharing by foragers is not prohibitively costly, yet they must also be sufficiently scarce and ephemeral that information exchange about their location will result in a higher food intake relative to the alternative strategy of breeding alone[24]. In our case, the areas of temporary food retention are continually resupplied with fish larvae during their drift northwards in the NCC, thus reducing prey depletion and enhancing the food source for seabirds. Areas of particle accumulation may also exhibit high primary production caused by turbulence, which might synergistically improve the foraging conditions for seabird prey.

Although the data on larval abundance used here spanned a 30-year period, the spatial larvae aggregation patterns we have identified are almost certainly more permanent in nature, especially to the degree they are linked to coastal topography and bathymetry. According to historical[40], archaeological[41,42] and sedimentological evidence[43–45], seabird colonies are known to have been occupied for hundreds, and even thousands, of years. No such datings are available for Norway. However, the long-term importance of the Lofoten area as a cod spawning area is documented by the fact that this area has exported stockfish at least since the Viking age (that is, for 900 years)[46]. It is therefore possible that the distribution of large colonies along the Norwegian coast constitutes a macroecological pattern that has been stable for a long time, potentially since the end of the last glaciation, *c.* 9,000 years ago.

Methods

Oceanography of the study system. Variability in oceanographic features of the Northeastern North Atlantic causes pronounced variations in the distribution and abundance of larvae and juvenile fish in the marine pelagic ecosystem[32]. The physical conditions in the model domain are dominated by the influx of relatively warm Atlantic water through the Faeroe–Shetland Channel[47], continuing as a two-branch flow, with the outer branch known as the Norwegian Atlantic Slope Current (NASC) along the shelf break (Supplementary Fig. 1a)[48]. On the shelf, the less saline NCC, originating from south of Norway but fed by rivers all along the coast, is trapped between the NASC and the coast[49]. Interannual variations in its physical features affect the dispersal of ichthyoplankton[50]. The frontal zone area near the northern Norwegian shelf break separates the northward-flowing NASC and NCC, and is associated with meso-scale meanders, filaments and eddies[51]. Short-term variability of the flow is forced by variable winds and tide, whereas long-term features are controlled by topography such as banks and troughs.

Larval drift model. By using a daily hydrodynamic model archive with 4 km × 4 km horizontal resolution (vertical resolution into 32 sigma-levels; baroclinic time step 160 s; barotropic time step 5 s), we simulate the horizontal dispersion of passive particles along the Norwegian coast. The ocean circulation model provides a 30-year time series of hydrography from 1982 to 2011 stored on a daily basis[33]. This spatial resolution is sufficient in order to resolve the eddy field adequately[52]. An individual-based larval drift model was used to advect passive particles from potential spawning areas along the entire Norwegian coast. A full description of the features of this model has been provided by Myksvoll *et al.*[35] The larval drift model, as well as the ocean circulation model underlying it, have been validated empirically using observations from the Norwegian and Barents Sea[32,33].

In a generic model, the entire continental shelf (that is, all grid points with bottom depth less than 200 m; Fig. 1) from 61°N and northwards were used as initial release locations for passive particles (6,156 grid points in total). One particle was released per grid point per day from 1 March to 30 April, adding up to 369,360 passive particles in total. After release, the particles drifted freely with the ocean currents for 3 weeks, representing the fish egg stage, followed by a phase with vertical migration depending on swimming capability and light availability, representing the fish larval stage[53]. Although designed to represent the physiology and behaviour of fish larvae, the model also captures the drift of other, more or less passively drifting organisms, such as crustaceans or fish at the small fry stage.

Particle abundance was interpolated to a gridded data set with a latitudinal resolution of 0.075° (*c.* 8.3 km) and a longitudinal resolution of 0.150° (*c.* 5.4 to 6.8 km). The same grid was used as the basis for aggregation of seabird numbers and for simulation (*cf.* Supplementary Fig. 1b).

In addition to the generic model, we run two models for specific target species, *viz* Northeast Arctic cod and Norwegian spring-spawning herring. The cod and herring spawning areas overlap with the original release areas in the generic model, but they consist of several separated areas of limited geographical extent (Supplementary Fig. 1c)[54,55]. The cod larvae model includes a three weeks long free drifting stage before the eggs hatch into larvae with vertical migration[53]. The herring larvae model only includes the larval stage, as herring have adhesive benthic eggs. The models released a total of 94,500 cod eggs and 230,400 herring eggs from 1 March to 30 April.

Coastal morphology. The near-shore accumulation of particles is partly governed by physical properties of the coast. We quantified two aspects of coastal morphology for each coastal grid cell, *viz* the width of the continental shelf (bathymetry) and the degree of protrusion of the coastline (topography). Width of the continental shelf was defined as the shortest distance between the points of the coastline within the grid cell and the 200-m isobath (depth-contour), rounded to the nearest kilometre (*cf.* Fig. 1).

Protrusion was quantified by fitting a straight line through portions of coastline as follows: separately for each grid cell, other grid cells within a 40-km radius were considered, sampling from each grid cell the point protruding most towards the ocean (Supplementary Fig. 1d). The straight line was fitted to minimize squared distances for this sample of points (major axis regression), excluding the focus cell. The shortest distance (km) of the most protruding point in the focus cell to this line is referred to as the degree of protrusion of this grid cell. Headlands and outlying islands are characterized by positive protrusions, bays and fjords by negative protrusions. This algorithm could not be unambiguously applied to the southernmost grid cell of the Lofoten archipelago and the islands southwest of Lofoten (Røst, Værøy and others; in total, 9 grid cells, hereafter referred to as 'ambiguous'; *cf.* Supplementary Fig. 1d). In these cases, protrusion was defined as the distance from the nearest coastal point.

The effect of coastal morphology on particle density was investigated using linear models. Model selection was based on Akaike's Information Criterion. To account for ambiguities regarding protrusion, models were rerun excluding ambiguous grid cells. To correct for spatial autocorrelation, *P*-values were obtained using randomization tests (see below, 'Spatial autocorrelation'). The proportion of randomized linear models obtaining an equal or larger R^2 than the original linear model was used as a corrected *P*-value.

Seabird colony data. The position and size of all seabird colonies north of 66°N that at any one time comprised 1,000 or more pairs were extracted from the Norwegian seabird database (http://www.seapop.no; *cf.* Supplementary Table 4). Entries in this database date back to the mid-twentieth century. Species included in the counts were cliff-nesting seabirds known to feed at least partly on fish larvae or the forage fish predators of larvae, that is, razorbill (*Alca torda*), Atlantic puffin (*Fratercula arctica*), northern fulmar (*Fulmarus glacialis*), black-legged kittiwake (*Rissa tridactyla*), common guillemot (*Uria aalge*) and Brünnich's guillemot (*Uria lomvia*). Apart from the fulmar that may forage far out at sea[56] (but whose breeding numbers are small in Norway), these are species with, under normal conditions, foraging ranges of *c.* 20–50 km from the nest[38,39], that is, considerably larger than the dimension of grid cells.

A grid cell of the larval drift model was considered to contain a seabird colony if the maximum sum of breeding pairs of all species within the grid cell at any point in time was at least 10,000 (but other population thresholds were also considered;

see 'Simulation'). For each grid cell, we considered the maximum overall population count available across years, rather than the count from a specific year, because we were interested in the position of colonies in a long-term perspective, independent of short-term variation in the size of breeding populations. This procedure yielded a sample of 20 seabird colonies of at least 10,000 breeding pairs.

Spatial autocorrelation. The spatial autocorrelation of observations introduces a statistical dependence that precludes the use of standard tests. Probabilities were therefore estimated using randomization tests, that is, a Monte Carlo method involving the repeated random reordering of the data. Based on at least 100,000 replications, probabilities were defined as the proportion of reordered data sets that obtained a test statistic as extreme as, or more extreme than, the one observed.

As the former method may not remove all covariation between variables, it was supplemented by the toroidal-shift method as a *post-hoc* test. This test entails shifting or rotating one point pattern against the other, while leaving both point patterns in their original order[57,58]. This was accomplished by shifting grid cells along the coastline, treating the latter as a closed ring (torus), that is, by connecting the two ends of the coastline. To avoid edge effects in models involving colony locations, both ends of the coastline were defined as the grid cell placed midway between the ultimate and penultimate colonies, reducing the number of colonies included by 2. The sample size of this test equals the number of coastal grid cells.

Simulation. Simulations were used in order to estimate the probability that randomly distributed seabird colonies would have equally good or better access to food particles than actual seabird colonies. Each simulation consisted of a randomization test that randomly defined *n* different coastal grid cells as seabird colonies (where *n* is the number of actual colonies of a certain size in Northern Norway), aggregated larval abundance within a certain feeding radius around each colony, summed the *n* values to obtain the total aggregation, and compared this value to the total aggregation around the actual colonies. This procedure was repeated 1,000,000 times for the initial model.

Before simulation, certain assumptions had to be made, involving (i) the prey species modelled; (ii) the months considered; (iii) how to quantify larval abundance; (iv) how to take the inter-annual variability of larval abundance into account; (v) the selection of grid cells deemed suitable for seabird colonies; (vi) the size threshold for actual colonies included; (vii) the radius of the foraging range and (viii) the weighting of larval numbers within the feeding range. The optimal parameterization and its importance for the results were unknown in advance. Therefore, the initial model was built upon a reasonable set of *a priori* assumptions, all of which were relaxed in separate analyses (*cf.* Supplementary Table 2), each based on simulations with 100,000 replicates:

(i) The initial model was based on generic particles rather than specific fish species. This means that the model captures features common to all planktonic prey items. It was contrasted with separate simulations based on cod larvae, herring larvae and combinations of these two. In one combined model, fish larvae abundance in each grid cell was defined as the sum of cod and herring larvae in each grid cell, scaled by the overall geometric mean for each species. In the other combinations, herring was considered up to a latitude of 68° 30′ N or 69° 30′ N, respectively, and cod north of these thresholds.

(ii) Output from larval drift models was extracted for May, June and July of each year, which constitute the main breeding season of the seabirds considered. The initial model is based on the mean larval abundance for this 3-month period for each year. Single months were also tested.

(iii) The larval abundances across grid cells were heavily skewed, with many grid cells having low abundances, and a few cells with very high abundances. A selection of grid cells containing one of the latter cells could theoretically bias the simulations, if the larval abundance in this cell overshadows the abundances of the remaining grid cells. To avoid this potential bias, the initial model was based on log-transformed larval abundances. This model was contrasted with (a) simulations using untransformed larval abundances and (b) simulations that merely considered the ordering of grid cells (that is, grid cells were ranked in the order of their larval abundances, and ranks, rather than larval abundances, were compared).

(iv) We hypothesize that optimal colony sites are those with high mean abundance and low inter-annual variability in abundance of larvae. As a measure that captures aspects of both central tendency and variability, we used the 1st decile (3rd 30-quantile) of larval abundances for each grid cell. This means that, for each grid cell and in 9 out of 10 years, larval abundances were equal to or higher than the specific value identified. This simulation was contrasted with others in which larval abundance was represented by 16 quantiles from the minimum (0th 30-quantile) to the median (15th 30-quantile) of the 30 annual estimates for each of the grid cells (*cf.* Fig. 3b). In addition, the arithmetic and geometric mean, the maximum value and three measures of variability were tested explicitly, *viz*, the variance, the coefficient of variation (CV = standard deviation/mean) and the QCD (QCD = interquartile range/median). The latter is more robust than the CV when distributions are skewed, and results reported for variability are therefore based on the QCD.

(v) The result of simulation depends on the selection of coast cells deemed suitable. We therefore produced four sets of grid cells, progressively omitting cells that did not have direct contact with oceanic cells or were placed inside fjords or sounds (Supplementary Fig. 1b). Our main results are reported for the most conservative selection of grid cells (that is, the smallest set). Grid cells outside the outermost colonies (that is, grid cells south/west of the southernmost actual colony on the Norwegian Sea coast and south/east of the southeasternmost actual colony on the Barents Sea coast) were disregarded.

(vi) In the initial model, the threshold for actual colonies included was arbitrarily set at ≥10,000 breeding pairs (20 colonies). In addition, we used other thresholds for the inclusion of actual seabird colonies, *viz*, ≥5,000 pairs, ≥20,000 pairs, ≥50,000 pairs and ≥100,000 pairs, resulting in samples of 9 to 27 colonies (Fig. 1 and Supplementary Fig. 1b).

(vii) Different seabird species have different foraging radii, and foraging ranges reported in the literature are likely to be greater than the distances preferred by the birds (assuming that seabirds are selected to minimize the energetic costs of flight). We therefore had no *a priori* estimate of what the relevant radius might be, and assumed 10 km as a reasonable first guess. In addition, radii of foraging range were varied by 1 km between 5 and 50 km (*cf.* Fig. 3b).

(viii) Different assumptions can be made about how seabirds choose between prey patches in the vicinity of their breeding colony. It is likely that prey close to the colony is preferred over distant prey, but the shape of this preference function has not been quantified. In the initial model, we assume that the degree of preference decreases linearly from 100% at a distance of 0 to 1% at the maximum allowed feeding distance, by weighting larval numbers accordingly. Simulations based on this assumption were contrasted with models with alternative weighting schemes, *viz*., a threshold shape (all prey patches within the feeding radius are weighted equally), a concave and a convex shape (both based on exponential functions reaching 1% at the predefined maximum radius; see Fig. 3c for details).

***Post-hoc* tests.** Based on the results obtained for the initial model and the set of models relaxing its assumptions, a number of further tests were devised (*cf.* Supplementary Table 3). Each of these *post-hoc* tests was based on simulations with 100,000 replicates.

Exclusion of the most influential colony. To ensure that the patterns found are not merely driven by one colony with extremely high larval numbers in its vicinity, the most influential colony was excluded. This was done for the initial model as well as the model of particle variability (QCD) and the model for cod larvae. In each case, the colony that obtained the highest (in case of variability, lowest) number of particles was identified, and the respective grid cell removed from the simulation.

Higher-resolution larval drift model. A sensitivity test was performed using an ocean circulation model with higher resolution (horizontal resolution 800 m × 800 m; vertical resolution into 35 sigma-levels; baroclinic time step 60 s; barotropic time step 1 s; output fields of model variables stored on hourly basis)[59,60]. As it was not feasible to process data for 30 seasons at this resolution, the model was run for cod in a single year (2010), the results were interpolated to the same grid as the original data set, and contrasted with the results of the 4-km model for the same fish species and year.

Lower-resolution coastal grid. To ensure that the associations found do not critically depend on the grid chosen, we varied the grid scale by collapsing neighbouring grid cells into larger cells and base the simulations on the new grid obtained. This was done for the generic model and the cod model. We collapsed pairs of grid cells (thus obtaining a latitudinal resolution of 0.075° and a longitudinal resolution of 0.300°) and quadruples of grid cells (thus obtaining a latitudinal resolution of 0.15° and a longitudinal resolution of 0.30°). As there are two different ways to collapse pairs and four different ways to collapse quadruples, all versions were simulated (each 100,000 times), and the average probability is reported. Averaging probabilities is adequate in this case, because each probability expresses a proportion (*viz*, the fraction of simulations that obtained equal or higher larval counts around randomized colonies than around actual colonies).

Toroidal-shift method. The toroidal-shift method (see above, 'Spatial autocorrelation') was used as an alternative test that takes the spatial autocorrelation pattern of the variables into account. It was applied to the initial model, the model of particle variability (QCD) and the model for cod larvae. As the sample size of this test is the number of grid cells rotated ($N = 260$), the P-value cannot be lower than 0.004.

Separate analyses of cod spawning grounds. As the distribution of cod larvae gave an even better fit to seabird colonies than the distribution of generic particles, and because the spawning grounds of cod are well known, we quantified the relative importance of the latter. Spawning grounds were aggregated in seven areas, and tested separately.

Coastal characteristics. As both fine-scale measures of coastal topography (protrusion, breadth of continental shelf; *cf.* 'Coastal topography' above) turned out to be important for the distribution of particles along the coast, we quantified their direct importance for the location of seabird colonies. Simulations were carried out excluding the grid cells with ambiguous definitions of protrusion (see the section 'Coastal morphology').

Single years. Finally, each year of the 30-year-period was tested separately. This was done in order to ensure that the results were not driven by a few anomalous years (*cf.* Fig. 3d).

References

1. Wittenberger, J. F. & Hunt, G. L., Jr in *Avian Biology* Vol VIII (eds Farner, D. S., King, J. R. & Parkes, K. C.) 1–78 (Academic, 1985).
2. Danchin, É. & Wagner, R. H. The evolution of coloniality: the emergence of new perspectives. *Trends Ecol. Evol.* **12,** 342–347 (1997).
3. Rolland, C., Danchin, É. & de Fraipont, M. The evolution of coloniality in birds in relation to food, habitat, predation, and life-history traits: a comparative analysis. *Am. Nat.* **151,** 514–529 (1998).
4. Hamer, K. C., Schreiber, E. A. & Burger, J. in *Biology of Marine Birds* (eds Schreiber, E. A. & Burger, J.) 217–261 (CRC, 2002).
5. Coulson, J. C. in *Biology of Marine Birds* (eds Schreiber, E. A. & Burger, J.) 87–113 (CRC, 2002).
6. Horn, H. S. The adaptive significance of colonial nesting in the Brewer's blackbird (*Euphages cyanocephalus*). *Ecology* **49,** 682–694 (1968).
7. Ward, P. & Zahavi, A. The importance of certain assemblages of birds as "information centres" for food-finding. *Ibis* **115,** 517–534 (1973).
8. Barrett, R. T. Recent establishments and extinctions of Northern Gannet *Morus bassanus* colonies in North Norway, 1995–2008. *Ornis Nor.* **31,** 172–182 (2008).
9. Furness, R. W. & Birkhead, T. R. Seabird colony distributions suggest competition for food supplies during the breeding season. *Nature* **311,** 655–656 (1984).
10. Ashmole, N. P. The regulation of numbers of tropical oceanic birds. *Ibis* **103b,** 458–473 (1963).
11. Cairns, D. K. The regulation of seabird colony size: a hinterland model. *Am. Nat.* **134,** 141–146 (1989).
12. Wakefield, E. D., Phillips, R. A. & Matthiopoulos, J. Habitat-mediated population limitation in a colonial central-place forager: the sky is not the limit for the black-browed albatross. *Proc. R. Soc. B* **281,** 20132883 (2014).
13. Oppel, S. *et al.* Foraging distribution of a tropical seabird supports Ashmole's hypothesis of population regulation. *Behav. Ecol. Sociobiol.* **69,** 915–926 (2015).
14. Boulinier, T., McCoy, K. D., Yoccoz, N. G., Gasparini, J. & Tveraa, T. Public information affects breeding dispersal in a colonial bird: kittiwakes cue on neighbours. *Biol. Lett.* **4,** 538–540 (2008).
15. Cristofari, R. *et al.* Spatial heterogeneity as a genetic mixing mechanism in highly philopatric colonial seabirds. *PLoS ONE* **10,** e0117981 (2015).
16. Oro, D. & Ruxton, G. D. The formation and growth of seabird colonies: Audouin's gull as a case study. *J. Anim. Ecol.* **70,** 527–535 (2001).
17. Kildaw, S. D., Irons, D. B., Nysewander, D. R. & Buck, C. L. Formation and growth of new seabird colonies: the significance of habitat quality. *Mar. Ornithol.* **33,** 49–58 (2005).
18. Gibbs, J. P., Woodward, S., Hunter, M. L. & Hutchinson, A. E. Determinants of Great Blue Heron colony distribution in coastal Maine. *Auk* **104,** 38–47 (1987).
19. Gibbs, J. P. Spatial relationships between nesting colonies and foraging areas of Great Blue Herons. *Auk* **108,** 764–770 (1991).
20. Gibbs, J. P. & Kinkel, L. K. Determinants of the size and location of Great Blue Heron colonies. *Colon. Waterbirds* **20,** 1–7 (1997).
21. Ainley, D. G., Ford, R. G., Brown, E. D., Suryan, R. M. & Irons, D. B. Prey resources, competition, and geographic structure of Kittiwake colonies in Prince William Sound. *Ecology* **84,** 709–723 (2003).
22. Ford, R. G., Ainley, D. G., Brown, E. D., Suryan, R. M. & Irons, D. B. A spatially explicit optimal foraging model of Black-legged Kittiwake behavior based on prey density, travel distance, and colony size. *Ecol. Model.* **204,** 335–348 (2007).
23. Brown, R. G. B. & Nettleship, D. N. The biological significance of polynyas to Arctic colonial seabirds. *Can. Wildl. Serv. Occas. Pap.* **45,** 59–65 (1981).
24. Buckley, N. J. Spatial-concentration effects and the importance of local enhancement in the evolution of colonial breeding in seabirds. *Am. Nat.* **149,** 1091–1112 (1997).
25. Buckley, F. G. & Buckley, P. A. in *Behavior of Marine Animals: Current Perspectives in Research. Volume 4: Marine Birds* (eds Burger, J., Olla, B. L. & Winn, H. E.) 69–112 (Plenum, 1980).
26. Brown, C. R., Brown, M. B. & Ives, A. R. Nest placement relative to food and its influence on the evolution of avian coloniality. *Am. Nat.* **139,** 205–217 (1992).
27. Barrett, R. T., Lorentsen, S.-H. & Anker-Nilssen, T. The status of breeding seabirds in mainland Norway. *Atl. Seab.* **8,** 97–126 (2006).
28. Skjoldal, H. R., Dalpadado, P. & Dommasnes, A. in *The Norwegian Sea Ecosystem* (ed. Skjoldal, H. R.) 447–506 (Tapir, 2004).
29. Ellis, H. I. & Gabrielsen, G. W. in *Biology of Marine Birds* (eds Schreiber, E. A. & Burger, J.) 359–408 (CRC, 2002).
30. Svendsen, E. *et al.* An ecosystem modeling approach to predicting cod recruitment. *Deep Sea Res. 2* **54,** 2810–2821 (2007).
31. Daewel, U. *et al.* Coupling ecosystem and individual-based models to simulate the influence of environmental variability on potential growth and survival of larval sprat (*Sprattus sprattus* L.) in the North Sea. *Fish. Oceanogr.* **17,** 333–351 (2008).
32. Vikebø, F. B. *et al.* Real-time ichthyoplankton drift in Northeast Arctic cod and Norwegian spring-spawning herring. *PLoS ONE* **6,** e27367 (2011).
33. Lien, V. S., Gusdal, Y. & Vikebø, F. B. Along-shelf hydrographic anomalies in the Nordic Seas (1960–2011): locally generated or advective signals? *Ocean Dyn.* **64,** 1047–1059 (2014).
34. Barrett, R. T. *et al.* The stress hormone corticosterone in a marine top-predator reflects short-term changes in food availability. *Ecol. Evol.* **5,** 1306–1317 (2015).
35. Myksvoll, M. S., Erikstad, K. E., Barrett, R. T., Sandvik, H. & Vikebø, F. Climate-driven ichthyoplankton drift model predicts growth of top predator young. *PLoS ONE* **8,** e79225 (2013).
36. Erikstad, K. E., Reiertsen, T. K., Barrett, R. T., Vikebø, F. & Sandvik, H. Seabird–fish interactions: the fall and rise of a common guillemot *Uria aalge* population. *Mar. Ecol. Prog. Ser.* **475,** 267–276; **481,** 305 (2013).
37. Ponchon, A. *et al.* When things go wrong: intra-season dynamics of breeding failure in a seabird. *Ecosphere* **5,** art4 (2014).
38. Coulson, J. C. *The Kittiwake* (Poyser, 2011).
39. Harris, M. P. & Wanless, S. *The Puffin* 2nd ed (Poyser, 2011).
40. Petersen, Æ. Rita í Breiðafjarðareyjum: varpdreifing, stofnbreytingar, landnám og talningaraðferðir. *Náttúrufræðingurinn* **79,** 45–56 (2010).
41. Falk, K. *et al.* Seabirds utilizing the Northeast Water polynya. *J. Mar. Syst.* **10,** 47–65 (1997).
42. Simeone, A. & Navarro, X. Human exploitation of seabirds in coastal southern Chile during the mid-Holocene. *Rev. Chil. Hist. Nat.* **75,** 423–431 (2002).
43. Hiller, A., Wand, U., Kämpf, H. & Stackebrandt, W. Occupation of the Antarctic continent by petrels during the past 35000 years: inferences from a ^{14}C study of stomach oil deposits. *Polar Biol.* **9,** 69–77 (1988).
44. Gaston, A. J. & Donaldson, G. Peat deposits and thick-billed murre colonies in Hudson Strait and Northern Hudson Bay: clues to post-glacial colonization of the area by seabirds. *Arctic* **48,** 354–358 (1995).
45. Sun,, L., Xie, Z. & Zhao, J. A 3,000-year record of penguin populations. *Nature* **407,** 858 (2000).
46. Barrett, J. *et al.* Detecting the medieval cod trade: a new method and first results. *J. Archaeol. Sci.* **35,** 850–861 (2008).
47. Eldevik, T. *et al.* Observed sources and variability of Nordic Seas overflow. *Nat. Geosci.* **2,** 406–410 (2009).
48. Orvik, K. A. & Skagseth, Ø. Heat flux variations in the eastern Norwegian Atlantic Current toward the Arctic from moored instruments, 1995–2005. *Geophys. Res. Lett.* **32,** L14610 (2005).
49. Skagseth, Ø., Drinkwater, K. & Terrile, E. Wind and buoyancy induced transport of the Norwegian Coastal Current in the Barents Sea. *J. Geophys. Res.* **116,** C08007 (2011).
50. Opdal, A. F., Vikebø, F. B. & Fiksen, Ø. Parental migration, climate, and thermal exposure of larvae: spawning in southern regions gives Northeast Arctic cod a warm start. *Mar. Ecol. Prog. Ser.* **439,** 255–262 (2011).
51. Orvik, K. A. A case-study of Doppler-shifted inertial oscillations in the Norwegian Coastal Current. *Cont. Shelf Res.* **15,** 1369–1379 (1995).
52. Nurser, A. J. G. & Bacon, S. The Rossby radius in the Arctic Ocean. *Ocean Sci.* **10,** 967–975 (2014).
53. Vikebø, F., Jørgensen, C., Kristiansen, T. & Fiksen, Ø. Drift, growth, and survival of larval Northeast Arctic cod with simple rules of behaviour. *Mar. Ecol. Prog. Ser.* **347,** 207–219 (2007).
54. Sætre, R., Toresen, R., Søiland, H. & Fossum, P. The Norwegian spring-spawning herring – spawning, larval drift and larval retention. *Sarsia* **87,** 167–178 (2002).
55. Sundby, S. & Nakken, O. Spatial shifts in spawning habitats for Arcto-Norwegian cod related to multidecadal climate oscillations and climate change. *ICES J. Mar. Sci.* **65,** 953–962 (2008).
56. Weimerskirch, H., Chastel, O., Cherel, Y., Henden, J.-A. & Tveraa, T. Nest attendance and foraging movements of northern fulmars rearing chicks at Bjørnøya Barents Sea. *Polar Biol.* **24,** 83–88 (2001).
57. Lotwick, H. W. & Silverman, B. W. Methods for analysing spatial processes of several types of points. *J. R. Stat. Soc. B* **44,** 406–413 (1982).
58. Fortin, M.-J., Drapeau, P. & Jacquez, G. M. Quantification of the spatial co-occurrences of ecological boundaries. *Oikos* **77,** 51–60 (1996).
59. Albretsen, J. *et al.* NorKyst-800 report no. 1: user manual and technical descriptions. *Fisken Havet* **2,** 1–46 (2011).
60. Skarðhamar, J., Skagseth, Ø. & Albretsen, J. Diurnal tides on the Barents Sea continental slope. *Deep Sea Res. 1* **197,** 40–51 (2015).

Acknowledgements

The study was funded by grants from the Norwegian Research Council (project number 216547 to K.E.E. and Centres of Excellence funding scheme, project number 223257, to H.S.) and the FRAM Centre (Flagship 'Fjord and Coast' to K.E.E.). We thank the Norwegian Environment Agency, the SEAPOP programme (http://www.seapop.no)

and all who have collected, collated and organized seabird distribution data on mainland Norway.

Author contributions

R.T.B., K.E.E. and H.S. designed the study; M.S.M., J.S. and F.V. carried out the larval drift models; H.S. performed simulations and statistical tests; H.S., R.T.B., M.S.M. and F.V. wrote the paper. All authors discussed the results and commented on the manuscript.

Additional Information

Competing financial interests: The authors declare no competing financial interests.

13

Stratified coastal ocean interactions with tropical cyclones

S.M. Glenn[1], T.N. Miles[1], G.N. Seroka[1], Y. Xu[2], R.K. Forney[1], F. Yu[3], H. Roarty[1], O. Schofield[1] & J. Kohut[1]

Hurricane-intensity forecast improvements currently lag the progress achieved for hurricane tracks. Integrated ocean observations and simulations during hurricane Irene (2011) reveal that the wind-forced two-layer circulation of the stratified coastal ocean, and resultant shear-induced mixing, led to significant and rapid ahead-of-eye-centre cooling (at least 6 °C and up to 11 °C) over a wide swath of the continental shelf. Atmospheric simulations establish this cooling as the missing contribution required to reproduce Irene's accelerated intensity reduction. Historical buoys from 1985 to 2015 show that ahead-of-eye-centre cooling occurred beneath all 11 tropical cyclones that traversed the Mid-Atlantic Bight continental shelf during stratified summer conditions. A Yellow Sea buoy similarly revealed significant and rapid ahead-of-eye-centre cooling during Typhoon Muifa (2011). These findings establish that including realistic coastal baroclinic processes in forecasts of storm intensity and impacts will be increasingly critical to mid-latitude population centres as sea levels rise and tropical cyclone maximum intensities migrate poleward.

[1] Center for Ocean Observing Leadership, Department of Marine and Coastal Sciences, School of Environmental and Biological Sciences, Rutgers University, 71 Dudley Road, New Brunswick, New Jersey 08901, USA. [2] State Key Laboratory of Estuarine and Coastal Research, East China Normal University, 3663 Zhongshan Road North, Shanghai 200062, China. [3] Institute of Oceanology, Chinese Academy of Sciences, 7 Nanhai Road, Qingdao 266071, China. Correspondence and requests for materials should be addressed to S.G. (email: glenn@marine.rutgers.edu).

Tropical cyclones are among the most destructive weather phenomena on Earth[1]. Declines in hurricane related mortalities[2] reflect improvements in global atmospheric and ensemble modelling approaches[3] that have reduced hurricane track forecast errors by factors of 2–3 (ref. 4). Despite two decades of progress in hurricane track prediction, improvements in hurricane-intensity forecast skill have lagged significantly[4]. The predictions, public response and unexpected devastation patterns related to Hurricane Irene exemplify this dichotomy. Accurate track forecasts days in advance provided time for preparations and coastal evacuations, but Irene's official forecast maximum wind speeds along the Mid-Atlantic coast were consistently $\sim 5\,\mathrm{m\,s^{-1}}$ too high[5]. Irene instead caused catastrophic inland flooding because of heavy rainfall[5], making it the eighth costliest cyclone to hit the United States since 1900 (ref. 6), with damages of ~$16 billion (ref. 5). These intensity forecast uncertainties have significant negative consequences, ranging from unnecessary preparation costs to future public skepticism[7].

Improved tropical cyclone intensity predictions include dependencies on the rapid space–time evolution of the atmosphere–ocean responses and feedbacks[8]. Coupled atmosphere–ocean models demonstrate that small shifts in sea surface temperature (SST) and stratification, even on small (100 km) horizontal scales, can have significant impacts on storm intensity[9–11]. Several studies have noted[12–16] the relationship between warm and cold mesoscale features in the deep ocean and rapid changes in intensity, but the coastal ocean has received much less attention.

Here, utilizing an ocean observing network to inform ocean and atmospheric model simulations, the role of baroclinic processes on a stratified coastal ocean and their impact on the intensity of Hurricane Irene was quantified. The high percentage of ahead-of-eye-centre[14,17,18] cooling (76–98%) observed in Irene is not reproduced by standard open ocean models that exclude these coastal baroclinic processes. Atmospheric model sensitivity studies indicate that intense in-storm sea surface cooling over a strongly stratified coastal regime is the missing contribution required to reproduce the rapid decay of Hurricane Irene's intensity. The 30-year historical buoy record shows an average of 73% of the in-storm cooling occurs ahead-of-eye-centre on the Mid-Atlantic Bight (MAB) in the stratified season. A Yellow Sea buoy observed up to 85% of in-storm cooling ahead-of-eye-centre during Super Typhoon Muifa (2011). The results demonstrate the importance of rapid ahead-of-eye-centre vertical shear-induced mixing processes and the ensuing ocean–atmosphere feedbacks for generating more accurate simulations of storm intensity.

Results

Synoptic conditions. Hurricane Irene formed east of the Caribbean's Windward Islands on 22 August 2011 and made initial United States landfall in North Carolina as a Category 1 hurricane on 27 August. It re-emerged over the ocean in the MAB before a second landfall in New Jersey as a tropical storm on 28 August (ref. 5), closely following the historical northeastward tracks of hurricanes along the northeast United States[19]. Irene accelerated and lost intensity as it crossed the MAB, moving parallel to the coast with the eye over inner-continental shelf waters (Fig. 1a). Propagation was rapid at $30–40\,\mathrm{km\,h^{-1}}$, requiring only ~9.5 h to cross from North Carolina to New Jersey landfall. Cloud bands extended over 600 km from the eye centre, obscuring the ocean from satellite infrared SST sensors during passage. Differencing 3-day composites of cloud-free satellite imagery before (24–26 August) from after (29–31 August) Irene reveals the regional pattern of MAB sea surface cooling (Fig. 1a and Supplementary Fig. 1A,B). The largest

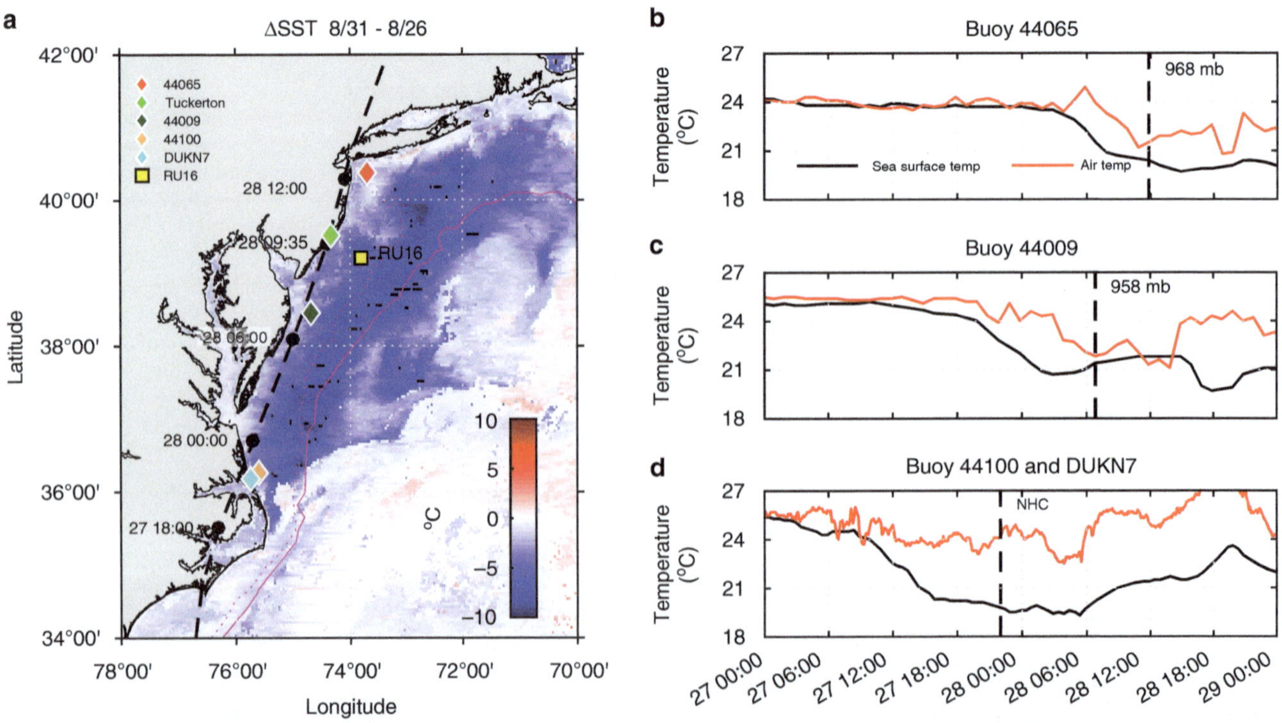

Figure 1 | Map of the study domain with satellite and buoy data. (**a**) SST difference map post-Irene (8/31) minus pre-Irene (8/26) with NHC best track (black dots connected by dashed line labelled with August date and UTC time), weather buoys/stations (coloured diamonds), underwater glider RU16 location during storm (yellow square) and bathymetry at 50 m (dotted magenta) and 200 m (solid magenta). (**b-d**) Buoy/station observed SST (blue) and air temperature (red) with vertical black dashed line/label indicating the time/value of minimum air pressure (**b**,**c**), and time of eye passage according to NHC best track data (**d**). The individual SST three-day composite maps for 24-26 August and 29-31 August are provided in Supplementary Fig. 1A,B.

cooling (5–11 °C) was observed to the right of the eye centre over the MAB's middle to outer shelf. Inner shelf cooling was slightly less, with averages of 3–5 °C of cooling within the 25-km radius eye wall (Supplementary Fig. 1C). Cooling was much less significant on the shelf seas to the south of the MAB, in the deep ocean to the east and, as previously noted in other hurricanes[20], along the very shallow unstratified coast, bays and sounds.

Observations. National Data Buoy Center (NDBC) buoys 44009 and 44065 recorded peak wind speeds (Supplementary Fig. 2) near $20\,\mathrm{m\,s^{-1}}$ from offshore as Irene approached. At these NDBC buoys and at 44100, water temperatures dropped rapidly by 3.8–6.3 °C ahead of eye centre passage (Fig. 1b–d), representing 82–98% of the in-storm cooling at these locations (Supplementary Fig. 3). At Irene's fast propagation speed, the eye was still 150–200 km to the south after the most rapid cooling was complete. As the ocean surface cooled, observed air temperatures were greater than SSTs, indicating air–sea-sensible heat fluxes were from the atmosphere into the ocean.

Atmospheric conditions (Fig. 2a) were recorded just inshore of a Slocum autonomous underwater glider[21,22] measuring subsurface ocean conditions[23] during Irene at the location shown in Fig. 1a (see Supplementary Fig. 4 for a plot of the complete glider track well before, during and after the storm). Winds initially from offshore (90°), with speeds near $20\,\mathrm{m\,s^{-1}}$ ahead of the eye, rotated rapidly to blow from onshore (270°) after the eye passed. Glider-observed subsurface temperatures (Fig. 2b) indicate that initially, typical MAB summer stratification[24] was present, with a seasonally warmed surface layer ($\sim$24 °C) above the MAB Cold Pool[25] ($<$10 °C) separated by a sharp ($<$8 m thick) thermocline. Significant cooling of the surface layer (5.1 °C) and deepening of the thermocline ($>$15 m) was observed under the leading edge of the storm. Little change in thermocline depth and much less cooling (1.6 °C) of the upper layer was observed after eye passage. Thus, ahead-of-eye-centre cooling represents 76% of in-storm cooling observed at the glider (Fig. 2b). Both the glider and buoy data suggest that much of the satellite observed SST cooling (over $\sim 100{,}000\,\mathrm{km^2}$ of continental shelf) occurred ahead-of-eye-centre.

Ocean surface currents measured by a CODAR high-frequency (HF) radar[26] network[27] illustrated the rapid response of the thin surface layer (Supplementary Fig. 5) to the changing wind direction (Fig. 2a). Time-series of the cross-shelf components of the currents (Fig. 2c) at the glider location, with positive values towards land, indicate that the onshore surface currents began building before the eye entered the MAB, increasing to a peak value $>50\,\mathrm{cm\,s^{-1}}$ towards the coast before the eye passage. Along-shelf currents throughout the water column were weak (Fig. 2d). After the eye, the winds changed direction and within a few hours, the cross-shelf surface currents switched to offshore. Despite the strong observed surface currents, the depth-averaged current (Fig. 2c) reported by the glider remained small during the storm's duration, with peaks barely exceeding $5\,\mathrm{cm\,s^{-1}}$. As in deep water, the current response is baroclinic[28,29], but the low depth-averaged current implies a strong offshore flow in the bottom layer. These bottom layer currents were estimated based on the relative layer thicknesses and the requirement that the combined surface and bottom layer-averaged currents matched the glider-observed dead-reckoned depth-averaged current. The estimated bottom layer currents accelerated in the offshore direction as the eye approached, causing significant shear between the two layers at the same time the surface layer was deepening and cooling.

... three-dimensional (3D)

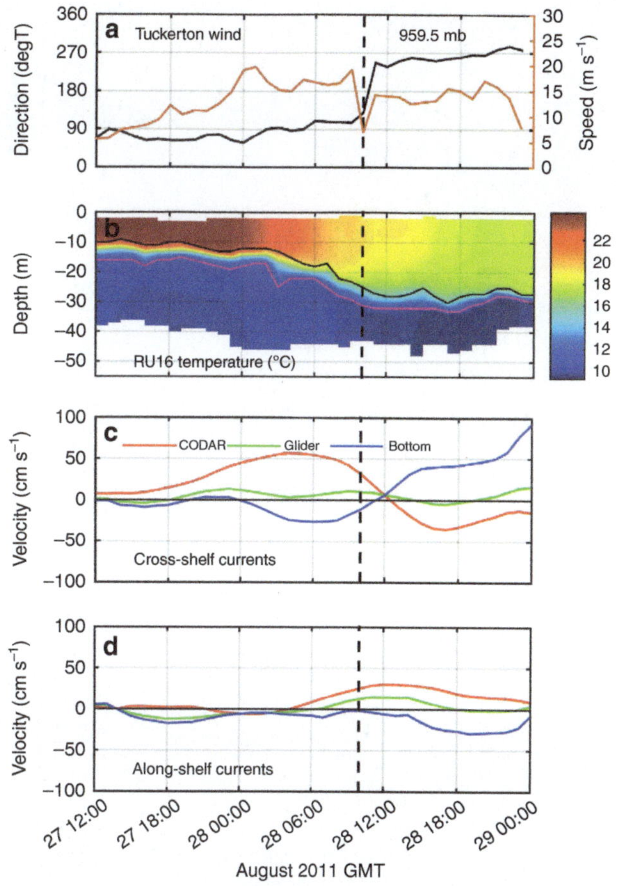

Figure 2 | Data from a local meteorological station, glider and HF radar. (**a**) Tuckerton WeatherFlow, Inc. station 10 m wind speed (orange) and direction from (black) with vertical black dashed line/label indicating the time/value of the minimum air pressure corresponding to landfall time on 28 August at 935 GMT. (**b**) Temporal evolution and vertical structure of the glider temperature during storm conditions with lines indicating top (black) and bottom (magenta) of thermocline. (**c**) Cross-shelf currents (positive onshore, negative offshore) for the surface layer (red) from CODAR HF Radar, depth-averaged (green) from the glider and bottom layer (blue) calculated from the depth-weighted average of the HF radar and glider velocities. (**d**) Same as **c** but for along-shelf currents (positive up-shelf northeastward and negative down-shelf southwestward).

System (ROMS) in the MAB[30,31] successfully reproduced the thermocline deepening and surface layer cooling (Fig. 3a) similar to the glider observations (Fig. 2b). The modelled cross-shelf velocity component (Fig. 3b) also has similarities to the combined glider and HF radar data (Fig. 2c). The surface layer flow accelerated shoreward for 12 h until eye passage, while the bottom layer responded more slowly with an offshore counter-flow. A few hours after eye passage, the cross-shelf flows reversed, also consistent with observations. The dominant terms in the cross-shelf momentum balance (Fig. 3g) indicate that the surface wind stress increased as the eye approached and decreased as it receded. Before the eye centre arrival, the presence of a coastline produced an offshore-directed pressure gradient that nearly balanced the wind stress and accelerated the offshore jet in the bottom layer. After the storm passage, the cross-shelf surface current switched to offshore; the cross-shelf pressure gradient also switched sign and was redirected towards the coast. At this point in the storm, the dominant cross-shelf momentum balance was nearly geostrophic (Fig. 3g) with a northward along-shelf surface ... (Fig. 3d).

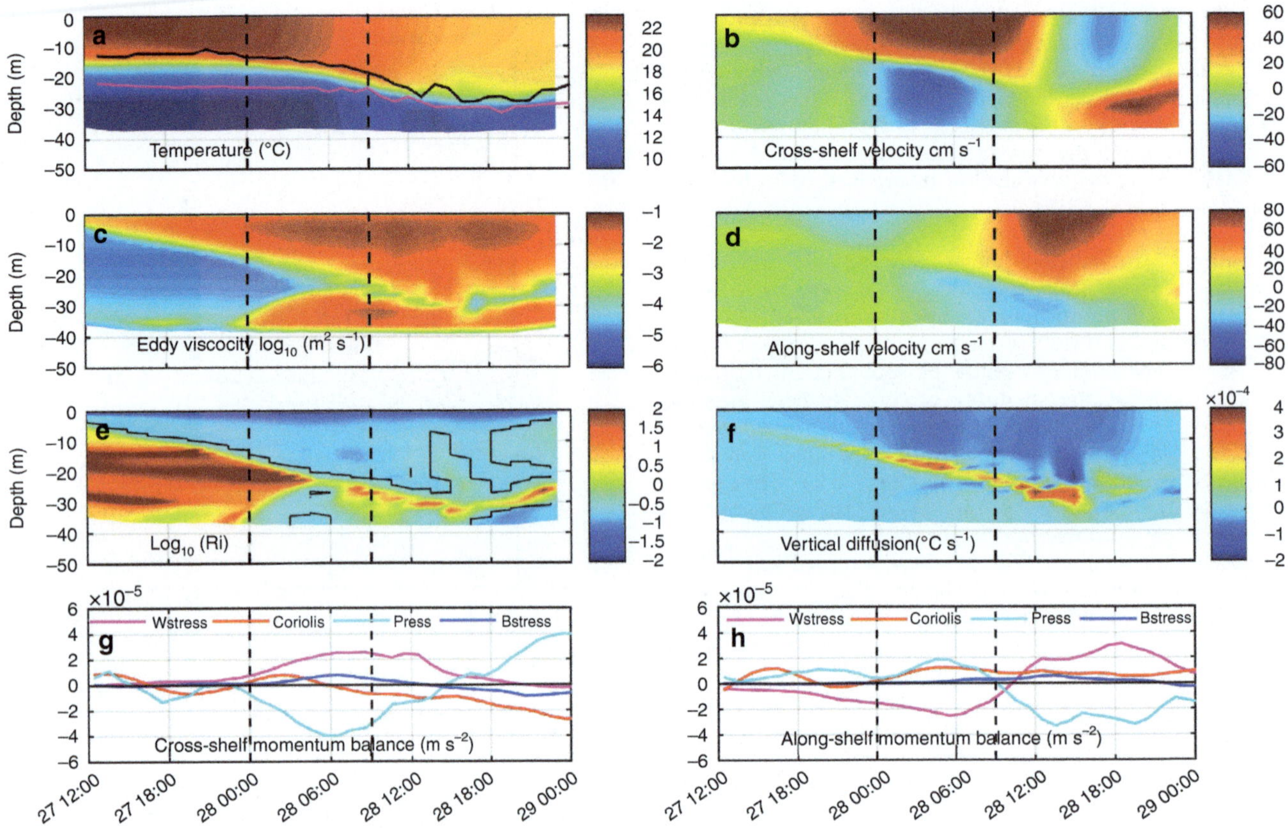

Figure 3 | ROMS ocean simulation results at the glider location. ROMS ocean simulation results at the glider location during the storm period, with first vertical black dashed line indicating initiation of the coastal baroclinic response and second vertical black dashed line indicating eye passage. (**a**) Temperature with top (black) and bottom (magenta) of thermocline as in Fig. 2b. (**b**) Cross-shelf velocity (red/yellow onshore; blue offshore). (**c**) Eddy viscosity. (**d**) Along-shelf velocity (red/yellow northward; blue southward). (**e**) Log_{10}(Richardson number) with black contour indicating Richardson number of 0.25. (**f**) Vertical diffusion temperature diagnostic equation term, showing warming (positive, red/yellow) and cooling (negative, dark blue). (**g**) Dominant depth-averaged cross-shelf momentum balance terms (positive onshore and negative offshore) from wind stress (wstress, magenta), Coriolis force (coriolis, red), pressure gradient (press, cyan) and bottom stress (bstress, blue). (**h**) Same as **g** but for along-shelf momentum balance terms (positive northward, negative southward).

The subsurface cross-shelf circulation within the two-layer coastal ocean had a significant influence on vertical mixing as illustrated by the Richardson number (Fig. 3e) and the vertical eddy viscosity (Fig. 3c). The Richardson number and the eddy viscosity show that the surface layer deepened to meet the stratification at the top of the thermocline as the surface layer accelerated with the approaching storm. As the offshore counter current accelerated in the bottom boundary layer, the lower layer Richardson number also decreased and eddy viscosity increased until the two layers interacted. The most rapid ahead-of-eye-centre cooling and deepening of the surface layer occurred when the small Richardson numbers and large vertical eddy viscosities from the surface and bottom boundary layers overlapped. The model's temperature diagnostic equation indicates that vertical diffusion (Fig. 3f) was the dominant term (Supplementary Fig. 6) acting to deepen the thermocline and cool the surface layer during the event.

Atmospheric model simulations. Atmospheric model simulations of Irene used the Weather Research and Forecasting (WRF)[32] model as applied to the US East Coast for tropical cyclone forecasting[33]. Typical surface boundary approaches in uncoupled atmospheric models use satellite SSTs over water that remain fixed when new data is not available because of cloud cover. A matrix of over 130 simulations revealed ahead-of-eye-centre cooling of the ocean's surface layer has a significant impact on intensity as reflected in the hurricane pressure (Fig. 4) and wind fields (Supplementary Fig. 7). Examining the ensemble of simulations with track errors less than one eye-wall radius, the largest wind and pressure intensity sensitivities were generated using fixed warm pre-storm and cold post-storm SST boundary conditions (Supplementary Figs 8,9). The sea level pressure (SLP) fields at landfall indicate the warm (Fig. 4a) versus the cold (Fig. 4b) SST changed the centre SLP by 7–8 hPa, with the maximum wind speed reduced by $>5\,m\,s^{-1}$ due to the cooler SST (Supplementary Fig. 7). The minimum SLP time history (Fig. 4c) of selected model runs can be compared with the National Hurricane Center (NHC) best track parameters. The best track central pressure remains constant near 952 hPa until the eye enters the MAB (28 August at about 00 h), followed by a steady increase in the central pressure to 965 hPa 13 h later as the eye leaves the MAB. Once Irene's eye entered the MAB, the cold SST air–sea flux parameterization sensitivities all produce a reduction in intensity that cluster with the best track analysis, while the warm SST air–sea flux parameterization sensitivities maintain a lower minimum SLP with little change nearly until landfall.

The top three model sensitivities are quantified by the envelope width for the minimum SLP (Fig. 4d). For both warm and cold SSTs, sensitivities to the three standard WRF air–sea flux formulations range from 0 to 2 hPa for the 13 h after the eye entered the MAB. The sensitivity to warm and cold SST

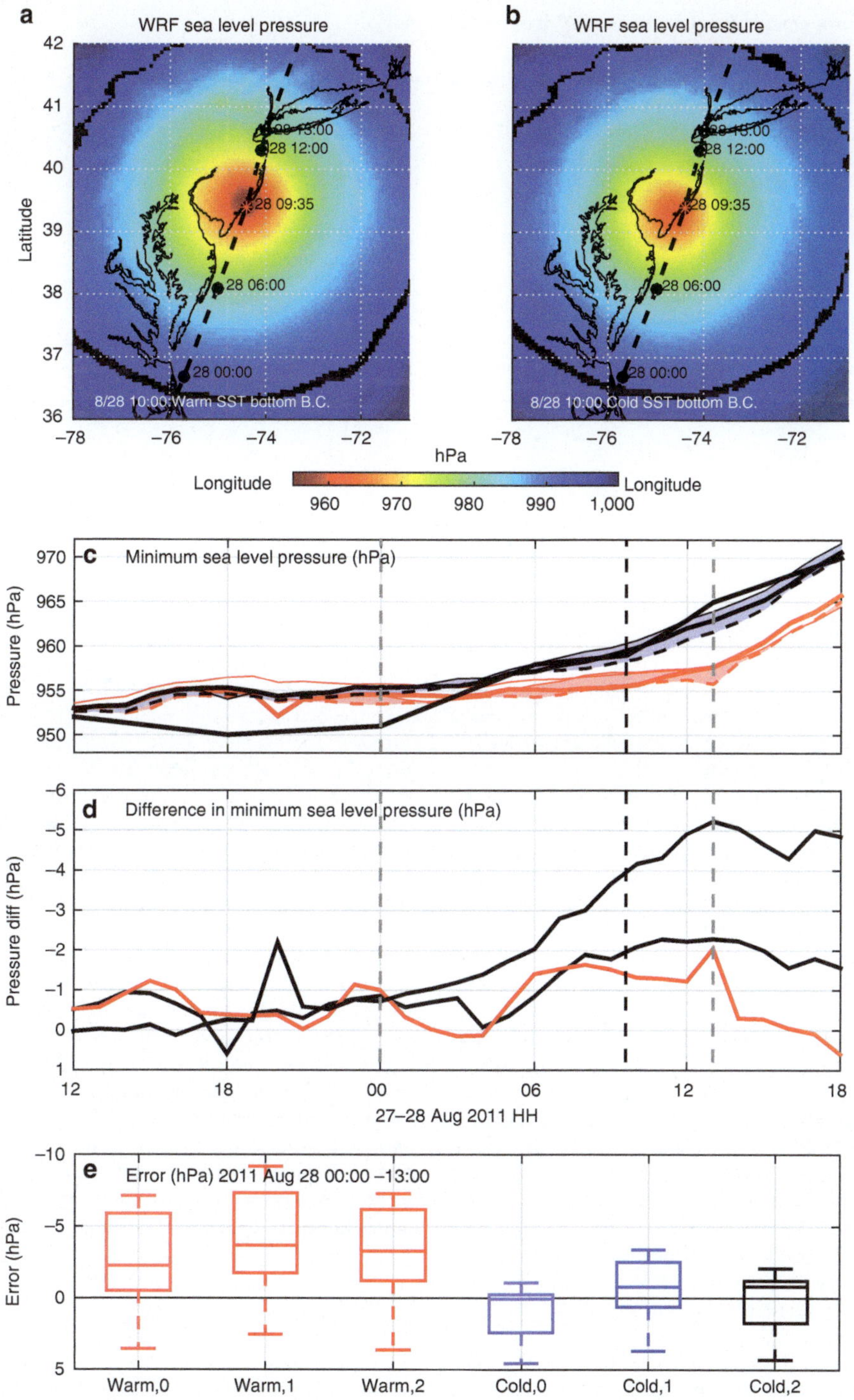

Figure 4 | WRF atmospheric model simulation results. (**a**) WRF model SLP (with surface flux option 2) at landfall (red star) for the warm SST boundary condition with NHC best track drawn as in Fig. 1a. (**b**) Same as **a** but for the cold SST. (**c**) Minimum SLP for NHC best track (black), and WRF's three air–sea flux parameterization options isftcflx = 0 (thin line); 1 (dotted line); and 2 (thick line) for the warm (red) and cold (blue) SST. Vertical grey and black dashed lines indicate eye enters MAB, makes landfall and leaves MAB. (**d**) Model SLP sensitivity to SST (black, warm minus cold SST for isftcflx = 2), and to flux parameterizations (isftcflx = 1 minus isftcflx = 0) for warm (red) and cold (blue) SST. (**e**) Box and whisker plots of SLP deviations from NHC best track when eye is over MAB for warm (red) and cold (blue) SST.

begins growing as the storm nears the MAB, climbing steadily to 5 hPa as it leaves the MAB. Statistical comparisons of each model run to the NHC best track over the MAB are quantified by the box and whisker plots (Fig. 4e) showing the median, inter-quartile range and outliers. The three warm SST air–sea flux sensitivities consistently over-predict the intensity with minimum SLPs that are too low, while the three cold SST air–sea flux sensitivities more accurately reflect the intensity reduction for all of the air–sea flux options.

Table 1 | Sea surface temperature cooling in coastal tropical cyclones.

Storm name	Buoy	Water depth (m)	Ahead-of-eye-centre cooling (°C)	In-storm cooling (°C)	% Ahead-of-eye-centre
Arthur (2014)	44014	48	1.4	2.4	58%
Irene (2011)	44009	26	4.5	5.5	82%
Barry (2007)	ALSN6	29	5.1	5.1	100%
Hermine (2004)	44009	31	0.9	1.1	82%
Allison (2001)	CHLV2	14	2.3	2.6	88%
Bonnie (1998)	CHLV2	14	4.2	4.2	100%
Danny (1997)	44009	31	2.1	3.6	58%
Arthur (1996)	44009	31	2.3	3.5	66%
Emily (1993)	44014	48	2.3	2.8	82%
Bob (1991)	44025	41	2.1	4.6	46%
Charley (1986)	44009	31	2.7	5.4	50%
Average		**31**	**2.7**	**3.7**	**73%**
Standard deviation		**11**	**1.3**	**1.4**	**19%**
Irene (2011)	44065	25	3.8	4.2	90%
Irene (2011)	RU16	37-46	5.1	6.7	76%
Irene (2011)	44100	26	6.3	6.4	98%
Muifa (2011)	37.045 N 122.66 E	31	4.1	4.8	85%

Ahead-of-eye-centre cooling (°C), in-storm cooling (°C) and % ahead-of-eye-centre observed at nearshore MAB buoys for 11 tropical cyclones that traversed the MAB continental shelf during summer stratified conditions since 1985, additional data from Hurricane Irene and Super Typhoon Muifa.

Discussion

Using Hurricane Irene as a diagnostic case study, a new feedback mechanism on storm intensity in the coastal ocean has been identified. The strong onshore winds occurring ahead-of-eye-centre in tropical cyclones and the coastal wall set up a down-welling circulation that limits the storm surge and results in significant shear across the thermocline. This shear leads to turbulent entrainment of abundant cold bottom water and mixing with warmer surface water. The resulting ocean cooling reduces surface heat fluxes to the atmosphere, weakening the storm.

Rapid tropical cyclone intensity changes over the deep ocean have been correlated with storm passage over warm and cold core eddies[12–16,34]. Also in the deep ocean, SST changes of as little as 1 °C are noted to significantly impact storm intensity[9,35]. During Hurricane Irene, ahead-of-eye-centre cooling of 3.8–6.3 °C was observed with nearshore buoys (Supplementary Fig. 3) and 5.1 °C was observed with a mid-shelf glider (Fig. 2). Storm-induced cooling in deep water is often equally distributed between the front and back half of the storm[36]. Deep ocean simulations of Irene with both a 1D ocean mixed layer model and the 3D Price–Weller–Pinkel[37] model produced 32 and 56% of the in-storm cooling ahead-of-eye-centre, respectively. In Hurricane Irene, 76% (glider) to 98% (buoy 44100) of the in-storm cooling occurred ahead-of-eye-centre, indicating that coastal baroclinic processes are enhancing the percentage of ahead-of-eye-centre cooling in Irene.

To verify that enhanced ahead-of-eye-centre coastal ocean cooling is not unique to Irene, 30 years of historical nearshore buoy data throughout the MAB were investigated. During that time period, ahead-of-eye-centre cooling was observed in all 11 tropical cyclones that tracked northeastward over the MAB continental shelf during the highly stratified summer months (June–August)[24,38] (Table 1 and Supplementary Figs 10–12). The maximum continental shelf buoy observed ahead-of-eye-centre cooling for these 11 storms averages 2.7 ± 1.3 °C, representing an average of 73% of the in-storm cooling.

An 11-year global satellite climatology[39] reveals that the shallow mid-latitude Yellow Sea and northern East China Sea also experience a large 20 °C seasonal SST cycle, similar to the MAB but over three times larger in area. A 1986 Yellow Sea shipboard conductivity temperature and depth survey reports surface to bottom temperature differences approaching 15 °C (ref. 40), also similar to the stratified summer MAB. Maps of western Pacific typhoon tracks (coast.noaa.gov/hurricanes) indicate 26 typhoons have tracked across the northern East China Sea and Yellow Sea during June–August since 1985. Like Irene, the landfalling intensity of Super Typhoon Muifa (2011) was over-predicted by standard models[41]. Satellite SST maps indicate Muifa caused significant in-storm cooling (up to 7 °C) across ~300,000 km^2 of the continental shelf[41]. Nearshore buoy observations show cooling of 4.1 °C (85% of the in-storm cooling observed at that location) was ahead-of-eye-centre (Table 1, Supplementary Fig. 13).

Globally, over the past 30 years, tropical cyclone maximum intensities have migrated poleward[42]. In the North Atlantic, hurricane intensities have increased since the early 1980s and are projected to continue to increase as the climate warms[43–46]. Combined with rapid sea level rise[47], mid-latitude population centres will experience heightened vulnerability to storm surge and inundation from increasingly powerful storms. To mitigate these risks, improved forecasting of tropical cyclone intensity over mid-latitude stratified coastal seas is vital, and will require realistic 3D ocean models to forecast enhanced ahead-of-eye-centre cooling.

Methods

Data source. The Mid-Atlantic Regional Association Coastal Ocean Observing System (MARACOOS) is a sustained regional component of the US Integrated Ocean Observing System (IOOS)[48]. Its integrated observation network of satellites, buoys, coastal meteorological stations, HF radar and autonomous underwater gliders provided the data used in this study[49].

Satellite remote sensing. National Oceanographic and Atmospheric Administration (NOAA) Advanced Very High-Resolution Radiometer (AVHRR) satellite data (Supplementary Fig. 1) were acquired through a SeaSpace TeraScan L-Band satellite ground station at Rutgers University. AVHRR data are converted to SST using the multi-channel SST algorithm[50]. To specifically map areas of rapid cooling, a 'coldest-dark-pixel' composite technique is used to identify and remove bright cloud covered pixels while retaining the darker ocean pixels. This is accomplished through the following series of tests performed on AVHRR channels 4 and 2 scans. Pixels are considered contaminated by clouds and removed if (1) AVHRR channel 4 (10.3–11.3 μm) temperatures are <5 °C (3.5 °C) in summer (winter); or (2) near infrared albedo in daytime AVHRR Channel 2 (0.725–1 μm) exceeds 2.3% (an empirically derived threshold specific to the MAB). Further tests are performed on 3 × 3 km grid boxes to account for large changes in temperature over short distances typical of cloud edges. Centre pixels are flagged as potential cloud edges and removed if (1) temperature changes in AVHRR channel 4 scans are $>1\,°C$ across the centre point of each 3×3 grid data; or (2) the change in infrared albedo across the centre of each 3×3 grid box is $>0.15\%$. After declouding is performed, the resulting 3 days of scans between 12:00 to 17:00 GMT are composited with the NASA (National Aeronautics and Space Administration) short-term Prediction Research and Transition centre (SPoRT) 2 km blended 7-day SST product. At each pixel the coldest value is retained between all daytime AVHRR scans for the past 3 days and the SPoRT SST product for that day to ensure retention of coastal upwelling zones and regions that underwent rapid mixing. Consistent with real-time processing protocols, the date assigned to each composite corresponds to the final day of the data window.

Meteorological observations. Meteorological observations were obtained from NOAA NDBC buoys, coastal towers and pier stations, and a WeatherFlow Inc. meteorological tower located in Tuckerton, New Jersey (Fig. 1a). Buoys 44009 (38.461° North and 74.703° West) and 44065 (40.369° North and 73.703° West) included wind speed and direction measured at a height of 5 m, air temperature at a height of 4 m and ocean temperatures at 0.6 m depth. Buoy 44100 (36.255° North and 75.591° West) is a Waverider buoy managed by Scripps Institution of Oceanography that measured ocean temperatures at 0.46 m depth. Station DUKN7 (36.184° North and 75.746° West) is a coastal station that measures air temperature at 15.68 m above mean sea level. The Tuckerton WeatherFlow Inc. meteorological tower (39.52° North and 74.32° West) measured wind speed and direction at 12 m. Meteorological data is plotted at the standard frequencies and averaging intervals reported by these stations.

High frequency radar. A network of over 40 CODAR Ocean Sensors SeaSonde HF Radar stations[26] are deployed along the MAB coast by a consortium of institutions coordinated through MARACOOS[27]. The stations transmit HF radio waves that are scattered off the ocean surface waves and then received back on shore. The Doppler shift in the Bragg peaks of the received signal are used to map the radial components of the total surface velocity field in front of each station[51]. Radial components from multiple stations are combined using an optimal interpolation scheme[52] to produce 1 h centre-averaged hourly surface current maps[53] with a nominal 6 km spatial resolution (Supplementary Fig. 5).

Autonomous underwater gliders. Teledyne Webb Research Slocum gliders are buoyancy-driven underwater vehicles that act as mobile sensor platforms[22]. These instrument platforms adjust small amounts of buoyancy in order to glide through the water column at 20–30 $\mathrm{cm\,s^{-1}}$ in a sawtooth pattern. At pre-programmed intervals the gliders come to the surface and transfer data back to Rutgers University in near real-time. The glider used in this study, RU16, was equipped with an un-pumped Seabird conductivity temperature and depth sensor that logged data every 4 s on downcasts and upcasts. Depth- and time-averaged velocity calculations were performed using a dead-reckoning technique typical for such platforms[22,54,55]. The measured pitch angle, fall velocity and a model of glider flight to estimate angle of attack are used to calculate an underwater horizontal displacement during each dive segment. The difference between the calculated horizontal displacement from the final pre-dive location and the actual surfacing location divided by the time underwater provides an estimate of depth- and time-averaged velocity.

A combination of dead-reckoned depth-averaged glider currents and HF radar surface currents are used to estimate bottom currents along the glider track (Fig. 2c). The following algorithm assumes that the HF radar surface currents are representative of the surface layer above the thermocline (defined as the maximum vertical temperature gradient along each profile) and requires that the depth-weighted average surface and bottom layer currents must equal the total depth-averaged glider current:

$$U_b = \frac{U_g(H_s + H_b)}{H_b} - \frac{U_s H_s}{H_b} \tag{1}$$

$$V_b = \frac{V_g(H_s + H_b)}{H_b} - \frac{V_s H_s}{H_b} \tag{2}$$

where H_s and H_b are the layer thicknesses above and below the thermocline, respectively, U_g and V_g are along- and cross-shelf depth-averaged currents, respectively, from glider dead-reckoning, U_s and V_s are surface layer-averaged currents from HF radar, and U_b and V_b are the calculated bottom layer-averaged currents (Fig. 2).

ROMS model setup. The numerical simulations were conducted using the ROMS[31], a free-surface, sigma coordinate, primitive equation ocean model (code available at http://www.myroms.org) that has been widely used in a diverse range of coastal applications. The ESPreSSO (Experimental System for Predicting Shelf and Slope Optics) model[56] covers the MAB from the centre of Cape Cod southward to Cape Hatteras, from the coast to beyond the shelf break and shelf/slope front. Gridded bathymetric data is used to construct a model grid with a horizontal resolution of 5 km (Supplementary Fig. 4) and 36 vertical levels in a terrain-following s-coordinate system. The initial conditions were developed from the same domain ROMS run with strong constrained four-dimensional variational (4D-Var) data assimilation[57]. The meteorological forcing is from the North American Mesoscale (NAM) model 12 km 3-hourly forecast data. Reanalyses of surface air temperature, pressure, relative humidity, 10 m vector winds, precipitation, downward longwave radiation and net shortwave radiation were used to specify the surface fluxes of momentum and buoyancy based on the COARE bulk formulae[58]. Boundary conditions are daily two-dimensional surface elevation, as well as three-dimensional velocity, temperature, and salinity fields from the Hybrid Coordinate Ocean Model Navy Coupled Ocean Data Assimilation forecast system. Inflows for the seven largest rivers are from daily average United States Geological Survey discharge data. Tidal boundary conditions are from the The ADvanced CIRCulation tidal model. The general length scale method k-kl type vertical mixing scheme[59,60] is used to compute vertical turbulence diffusivity.

ROMS momentum balance analysis. We extracted depth-averaged momentum balance terms from ROMS (Fig. 3g–h) at the glider sampling location in order to diagnose the dominant forces during the storm, where the acceleration terms are balanced by a combination of horizontal advection, pressure gradient, surface and bottom stresses and the Coriolis force (horizontal diffusion was small and neglected in this case):

$$\underbrace{\frac{\partial u}{\partial t}}_{\text{Acceleration}} = \underbrace{-\frac{\partial(uu)}{\partial x} - \frac{\partial(vu)}{\partial y}}_{\text{Horizonal advection}} \underbrace{- \frac{1}{\rho_0}\frac{\partial P}{\partial x}}_{\text{Pressure gradient}} + \underbrace{\left(\underbrace{\frac{\tau_s^x}{h\rho_0}}_{\text{Surface}} - \underbrace{\frac{\tau_b^x}{h\rho_0}}_{\text{Bottom}}\right)}_{\text{Stress}} + \underbrace{fv}_{\text{Coriolis}} \tag{3}$$

$$\underbrace{\frac{\partial v}{\partial t}}_{\text{Acceleration}} = \underbrace{-\frac{\partial(uv)}{\partial x} - \frac{\partial(vv)}{\partial y}}_{\text{Horizonal advection}} \underbrace{- \frac{1}{\rho_0}\frac{\partial P}{\partial y}}_{\text{Pressure gradient}} + \underbrace{\left(\underbrace{\frac{\tau_s^y}{h\rho_0}}_{\text{Surface}} - \underbrace{\frac{\tau_b^y}{h\rho_0}}_{\text{Bottom}}\right)}_{\text{Stress}} - \underbrace{fu}_{\text{Coriolis}} \tag{4}$$

where u and v are the along-shelf and cross-shelf components of velocity respectively, t is time, P is pressure, ρ_o is a reference density, τ_s and τ_b are surface and bottom stresses, h is water column depth and f is the latitude-dependent Coriolis frequency.

ROMS heat balance analysis. *Heat balance analysis.* The general conservation expression for the temperature budget in ROMS is given by

$$\frac{\partial T}{\partial t} = -\frac{\partial(uT)}{\partial x} - \frac{\partial(vT)}{\partial y} - \frac{\partial(wT)}{\partial z} + \frac{\partial A_{kt}\frac{\partial T}{\partial z}}{\partial z} + \mathcal{D}_T + \mathcal{F}_T \tag{5}$$

with the following surface and bottom boundary conditions:

$$\left(A_{kt}\frac{\partial T}{\partial z}\right)_{z=0} = \frac{Q_{net}}{\rho_0 C_p} \tag{6}$$

$$\left(A_{kt}\frac{\partial T}{\partial z}\right)_{z=-h} = 0 \tag{7}$$

Here, T is the temperature, t is time, u, v and w are the along-shelf, cross-shelf and vertical components of velocity. A_{kt} is the vertical diffusivity coefficient, $\mathcal{D}_T$ is the horizontal diffusion term and $\mathcal{F}_T$ is friction. Q_{net} is the surface net heat flux, $\rho_0 = 1025\ \mathrm{kg\,m^{-3}}$ is a reference density, $C_p = 3985\ \mathrm{J\,(kg\,°C)^{-1}}$ is the specific heat capacity of seawater and h is the water depth.

The ROMS conservation of heat equation was used to diagnose the relative contributions of the different terms responsible for the modelled temperature change. Time-series of the vertical temperature diagnostic terms were investigated along the glider track with emphasis on the temperature evolution between the top of the thermocline depth (the shallowest location where the vertical temperature gradient exceeded $0.4\,°\mathrm{C\,m^{-1}}$, black contour in Fig. 3a and Supplementary Fig. 6) and the transition layer depth (the deepest location where the vertical temperature gradient exceeded $0.7\,°\mathrm{C\,m^{-1}}$, magenta contour in Fig. 3a and Supplementary Fig. 6). Term-by-term analysis of equation 5 offered additional insights on the temperature source and sink terms. Supplementary Fig. 6A shows the temperature rate of change, which is the sum of the vertical diffusion term (Supplementary Fig. 6B) and advection term (Supplementary Fig. 6C), in which the advection term is separated into along-shelf advection (Supplementary Fig. 6D), cross-shelf advection (Supplementary Fig. 6E) and vertical advection (Supplementary Fig. 6F). The horizontal diffusion term's order of magnitude is much smaller than other terms and is not plotted. The dominant term influencing the surface mixed layer temperature change was the vertical diffusion, which is plotted in Fig. 3f.

WRF-ARW model setup. The Weather Research and Forecasting Advanced Research (WRF-ARW) dynamical core (code available at http://www.wrf-model.org)[32], Version 3.4 was used for the atmospheric simulations in this study. WRF-ARW is a fully compressible, non-hydrostatic, terrain-following coordinate, primitive equation atmospheric model. Our WRF-ARW domain extends from South Florida to Nova Scotia (Supplementary Fig. 14), with grid resolution of 6 km in the horizontal and 35 vertical levels. Lateral boundary conditions used are from the Global Forecast System (GFS) 0.5° initialized at 06 UTC on 27 August 2011.

Our simulations begin at 06 UTC on 27 August 2011 when Hurricane Irene was south of North Carolina (NC) over the South-Atlantic Bight (SAB) and end at 18

UTC on 28 August 2011 as the storm moved into New England. Simulation results shown (Fig. 4c,d and Supplementary Fig. 7C,D) begin at 12 UTC on 27 August 2011, at NC landfall time, after the model has 6 h to adjust to vortex initialization. WRF's digital filter initialization (DFI) was run to determine the sensitivities to different realizations of the GFS initializations. DFI deepened the initial vortex central pressure by over 10–960 hPa, which matches GFS initial central pressure (Supplementary Fig. 15). However, downstream sensitivity to DFI beyond 2 h was minimal.

For our control run, the following are used: longwave and shortwave radiation physics were both computed by the Rapid Radiative Transfer Model-Global scheme; the Monin–Obukhov atmospheric layer model and the Noah Land Surface Model were used with the Yonsei University planetary boundary layer scheme; and the WRF Double-Moment 6-class moisture microphysics scheme was used for grid-scale precipitation processes.

WRF sensitivity to SST. The model was run over 130 times to compare the sensitivity of certain parameter tuning. All sensitivities were compared to the control run (described above), which for surface boundary conditions over the ocean, that is, SST, used the Real-Time Global High-Resolution (RTG HR) SST analysis from 00 UTC on 27 August 2011 fixed throughout the simulation. This is the warm pre-storm SST, and has temperatures across the model domain similar to the AVHRR coldest-dark-pixel composite a day earlier (Supplementary Fig. 1A). By having the control run use Real-Time Global High-Resolution SST fixed throughout the simulation, we are consistent with what the operational NAM 12 km model used for bottom boundary conditions over the ocean.

To show the maximum impact of the ahead-of-eye-centre SST cooling on storm intensity, we compared our control run with a simulation using observed cold post-storm SST. For this, we used our AVHRR coldest-dark-pixel composite, which includes data from 29 to 31 August 2011 (Supplementary Fig. 1B). According to underwater glider and NDBC buoy observations along Irene's entire MAB track, almost all of the SST cooling occurred ahead of Irene's eye centre (Fig. 1b–d). NDBC buoy observations near Irene's track in the SAB (41013, 41036, 41037) also show ahead-of-eye-centre SST cooling, but values are on the order of 1 °C or less (Fig. 1a). Because our model simulations include only 6 h of storm presence over the SAB before NC landfall, and SST cooling in the SAB was significantly less than observed in the MAB (Fig. 1), we can conclude that the main result from our SST sensitivity is due to the ahead-of-eye-centre cooling in the MAB.

WRF sensitivity to air-sea flux parameterizations. The equations for the momentum (τ), sensible (H) and latent heat fluxes (E) are as follows:

$$\tau = -\rho C_D U^2 \tag{8}$$

$$H = -\left(\rho c_p\right) C_H U(\theta_{2m} - \theta_{sfc}) \tag{9}$$

$$E = -(\rho L_v) C_Q U(q_{2m} - q_{sfc}) \tag{10}$$

where ρ is density of air, C_D is drag coefficient, U is 10 m wind speed, c_p is specific heat capacity of air, C_H is sensible heat coefficient, θ_{2m} is potential temperature at 2 m and θ_{sfc} is potential temperature at the surface, L_v is enthalpy of vaporization, C_Q is latent heat coefficient, q_{2m} is specific humidity at 2 m and q_{sfc} is interfacial specific humidity at the surface.

Three options exist in WRF-ARW Version 3.0 and later for air–sea flux parameterizations (WRF namelist option isftcflx = 0, 1, and 2; see (ref. 61) for more details). These parameterization options change the momentum (z_0), sensible heat (z_T) and latent heat roughness lengths (z_Q) in the following equations for drag (C_D), sensible heat (C_H) and latent heat (C_Q) coefficients:

$$C_D = K^2/[\ln(z_{ref}/z_0)]^2 \tag{11}$$

$$C_H = \left(C_D^{1/2}\right)[K/\ln(z_{ref}/z_T)] \tag{12}$$

$$C_Q = \left(C_D^{1/2}\right)[K/\ln(z_{ref}/z_Q)] \tag{13}$$

where K is the von Kármán constant and z_{ref} is a reference height (usually 10 m).

Therefore, our SST sensitivity effectively changes the variables θ_{sfc} and q_{sfc} in equations 8–10 above, while our air–sea flux parameterization sensitivities change the equations for the momentum, sensible heat and latent heat coefficients (equations 11–13) going into the respective flux equations 8–10.

For our air–sea flux parameterization sensitivities in this study, we ran isftcflx = 0, 1, and 2 with both the warm (control) and cold SST boundary conditions.

Additional WRF sensitivities. We have discussed SST and air–sea flux parameterizations. WRF-ARW was run over 130 times in total, with various model configuration and physics options turned on and off.

We examined the ensemble of simulations with space/time track errors <25 km (one eye-wall radius) from available NHC best track positional data. Only preserving those simulations with accurate tracks is important because Hurricane Irene tracked close to and parallel to the Mid-Atlantic coast. The remaining sensitivities are shown in central pressure (Supplementary Fig. 8) and maximum winds (Supplementary Fig. 9). These are cumulative hourly sensitivities during Irene's presence over the MAB and NY Harbor (28 August 00-13 UTC). Supplementary Table 1 shows a list of these sensitivities, with the WRF namelist option number alongside its name (control run listed last for each sensitivity).

The sensitivity titled 'latent heat flux <0 over water' requires a brief explanation. In the WRF surface layer scheme code, there is a switch that disallows any latent heat flux less than 0 W m^{-2} (similarly, there is a switch that disallows any sensible heat flux less than −250 W m^{-2}). WRF convention for negative heat flux is downward, or atmosphere to land/water. We run WRF after removing the line of code disallowing negative latent heat flux, and compare to the control run. This switch removal only changes latent heat flux and allows it to be negative over water, as the subsequent WRF land surface scheme modifies fluxes and allows for negative latent heat flux over land.

Ahead-of-eye-centre and in-storm cooling calculations. Ahead-of-eye-centre cooling (Table 1) at NDBC buoys (Supplementary Figs 10–12) and the Yellow Sea buoy (Supplementary Fig. 13) was calculated by taking the difference between the maximum water temperature as the winds increased above 5 m s^{-1} and the minimum water temperature before or at the minimum observed SLP. In-storm cooling was determined as the difference between the same maximum water temperature as the winds increased above 5 m s^{-1} and the minimum water temperature while winds remained above 5 m s^{-1} after the pressure minimum. To calculate the average and standard deviation of cooling for the 11 storms passing through the MAB since 1985, we selected the one buoy on the continental shelf that recorded wind speed, pressure and water temperature and exhibited the greatest ahead-of-eye-centre cooling. For completeness we show Irene cooling statistics (Table 1) and time-series (Supplementary Fig. 3) for buoys 44065 and 44100 used in Fig. 1.

References

1. Keim, B. D., Muller, R. A. & Stone, G. W. Spatiotemporal patterns and return periods of tropical storm and hurricane strikes from Texas to Maine. *J. Clim.* **20,** 3498–3509 (2007).
2. Walker, N. *et al.* Hurricane prediction a century of advances. *Oceanography* **19,** 24–36 (2006).
3. Krishnamurti, T. N. *et al.* Improved weather and seasonal climate forecasts from multimodel superensemble. *Science* **285,** 1548–1550 (1999).
4. Cangialosi, J. P. & Franklin, J. L. *2012 National Hurricane Center Forecast Verification Report* pp. 79 (National Hurricane Center, 2013).
5. Avila, L. A. & Cangialosi, J. P. *Tropical Cyclone Report: Hurricane Irene* (2012); pp. 45.
6. Blake, E. S., Landsea, C. W. & Gibney, E. J. *The Deadliest, Costliest, and Most Intense United States Tropical Cyclones from 1851 to 2010* (National Weather Service National Hurricane Center, 2011) pp. 47.
7. Considine, T. J., Jablonowski, C., Posner, B. & Bishop, C. H. The value of hurricane forecasts to oil and gas producers in the Gulf of Mexico. *J. Appl. Meteorol.* **43,** 1270–1282 (2004).
8. Emanuel, K., DesAutels, C., Holloway, C. & Korty, R. Environmental control of tropical cyclone intensity. *J. Atmos. Sci.* **61,** 843–858 (2004).
9. Emanuel, K. Tropical cyclones. *Annu. Rev. Earth Planet. Sci.* **31,** 75–104 (2003).
10. Schade, L. R. & Emanuel, K. A. The ocean's effect on the intensity of tropical cyclones: results from a simple coupled atmosphere–ocean model. *J. Atmos. Sci.* **56,** 642–651 (1999).
11. Jaimes, B. & Shay, L. K. Enthalpy and momentum fluxes during Hurricane Earl relative to underlying ocean features. *Mon. Weather Rev.* **143,** 111–131 (2015).
12. Ma, Z., Fei, J., Liu, L., Huang, X. & Cheng, X. Effects of the cold core eddy on tropical cyclone intensity and structure under idealized air–sea interaction conditions. *Mon. Weather Rev.* **141,** 1285–1303 (2012).
13. Emanuel, K. A. Thermodynamic control of hurricane intensity. *Nature* **401,** 665–669 (1999).
14. Jaimes, B. & Shay, L. K. Mixed layer cooling in mesoscale oceanic eddies during Hurricanes Katrina and Rita. *Mon. Weather Rev.* **137,** 4188–4207 (2009).
15. Shay, L. K., Goni, G. J. & Black, P. G. Effects of a warm oceanic feature on Hurricane Opal. *Mon. Weather Rev.* **128,** 1366–1383 (2000).
16. Walker, N. D., Leben, R. R. & Balasubramanian, S. Hurricane-forced upwelling and chlorophyll a enhancement within cold-core cyclones in the Gulf of Mexico. *Geophys. Res. Lett.* **32,** 1–5 (2005).

17. Jacob, S. D. & Shay, L. K. The role of oceanic mesoscale features on the tropical cyclone-induced mixed layer response: a case study. *J. Phys. Oceanogr.* **33,** 649–676 (2003).
18. D'Asaro, E. The ocean boundary layer below Hurricane Dennis. *J. Phys. Oceanogr.* **31,** 561–579 (2003).
19. Hall, T. & Yonekura, E. North American tropical cyclone landfall and SST: a statistical model study. *J. Clim.* **26,** 8422–8439 (2013).
20. Cornillon, P., Stramma, L. & Price, J. F. Satellite measurements of sea surface cooling during Hurricane Gloria. *Nature* **326,** 373–375 (1987).
21. Schofield, O. *et al.* A regional Slocum Glider network in the mid-Atlantic bight leverages broad community engagement. *Mar. Technol. Soc. J.* **44,** 185–195 (2010).
22. Schofield, O. *et al.* Slocum gliders: robust and ready. *J. F. Robot.* **24,** 473–485 (2007).
23. Kohut, J. T., Haldeman, C. & Kerfoot, J. *Monitoring Dissolved Oxygen in New Jersey Coastal Waters Using Autonomous Gliders.* EPA/600/R-13/180 (US Environmental Protection Agency, Washington, DC, 2014).
24. Castelao, R., Glenn, S. & Schofield, O. Temperature, salinity, and density variability in the central middle Atlantic Bight. *J. Geophys. Res.* **115,** C10005 (2010).
25. Houghton, R. W., Schlitz, R., Beardsley, R. C., Butman, B. & Chamberlin, J. L. The Middle Atlantic Bight cold pool: evolution of the temperature structure during summer 1979. *J. Phys. Oceanogr.* **12,** 1019–1029 (1982).
26. Barrick, D. E., Evans, M. W. & Weber, B. L. Ocean surface currents mapped by radar. *Science* **198,** 138–144 (1977).
27. Roarty, H. *et al.* Operation and application of a regional high-frequency radar network in the Mid-Atlantic Bight. *Mar. Technol. Soc. J.* **44,** 133–145 (2010).
28. Shay, L. K., Chang, S. W. & Elsberry, R. L. Free surface effects on the near-inertial ocean current response to a hurricane. *J. Phys. Oceanogr.* **20,** 1405–1424 (1990).
29. Teague, W. J., Jarosz, E., Wang, D. W. & Mitchell, D. A. Observed oceanic response over the upper continental slope and outer shelf during Hurricane Ivan. *J. Phys. Oceanogr.* **37,** 2181–2206 (2007).
30. Wilkin, J. L. & Hunter, E. J. An assessment of the skill of real-time models of Mid-Atlantic Bight continental shelf circulation. *J. Geophys. Res. Ocean* **118,** 2919–2933 (2013).
31. Haidvogel, D. B. *et al.* Ocean forecasting in terrain-following coordinates: formulation and skill assessment of the Regional ocean modeling system. *J. Comput. Phys.* **227,** 3595–3624 (2008).
32. Skamarock, W. *et al. A description of the advanced research WRF version 3, NCAR Technical Note,* NCAR/TN-475 + STR (NCAR, Boulder, CO, USA, 2008).
33. Seroka, G. *et al.* in *Proceedings of the OCEANS MTS/IEEE Conference,* 1–4 (Rutgers University Coastal Ocean Observation Laboratory (RU-COOL) Hampton Roads, North Carolina, USA, 2012).
34. Lin, I.-I., Wu, C.-C., Pun, I.-F. & Ko, D.-S. Upper-ocean thermal structure and the Western North Pacific Category 5 Typhoons. Part I: ocean features and the category 5 typhoons' intensification. *Mon. Weather Rev.* **136,** 3288–3306 (2008).
35. Cione, J. J. & Uhlhorn, E. W. Sea surface temperature variability in hurricanes: implications with respect to intensity change. *Mon. Weather Rev.* **131,** 1783–1796 (2003).
36. Price, J. F. Upper ocean response to a hurricane. *J. Phys. Oceanogr.* **11,** 153–175 (1981).
37. Price, J. F. & Weller, R. A. Diurnal cycling: observations and models of the upper ocean response to diurnal heating, cooling, and wind mixing. *J. Geophys. Res.* **91,** 8411–8427 (1986).
38. Gong, D., Kohut, J. & Glenn, S. Seasonal climatology of wind-driven circulation on the New Jersey Shelf. *J. Geophys. Res. Ocean* **115,** 953–968 (2010).
39. Schofield, O. *et al.* The decadal view of the mid-Atlantic Bight from the COOLroom: is our coastal system changing? *Oceanography* **21,** 108–117 (2008).
40. Chen, C., Beardsley, R. C., Limeburner, R. & Kim, K. Comparison of winter and summer hydrographic observations in the Yellow and East China Seas and adjacent Kuroshio during 1986. *Cont. Shelf Res.* **14,** 909–929 (1994).
41. Sun, M. *et al.* Simulation of typhoon Muifa using a mesoscale coupled atmosphere-ocean model. *Acta Oceanol. Sin.* **33,** 123–133 (2014).
42. Kossin, J. P., Emanuel, K. A. & Vecchi, G. A. The poleward migration of the location of tropical cyclone maximum intensity. *Nature* **509,** 349–352 (2014).
43. Walsh, J. *et al. Our Changing Climate.* Synopsis of Climate Change Impacts in the United States. The Third National Climate Assessment, Natl Clim. Team Staff (2014).
44. Knutson, T. R. Dynamical downscaling projections of twenty-first-century Atlantic Hurricane activity: CMIP3 and CMIP5 model-based scenarios. *J. Clim.* **26,** 6591–6617 (2013).
45. Knutson, T. R. *et al.* Tropical cyclones and climate change. *Nat. Geosci.* **3,** 157–163 (2010).
46. Seneviratne, S. *et al.* in *Managing the Risk of Extreme Events and Disasters to Advance Climate Change Adaptation* (eds Field, C. *et al.*) 109–230 (Cambridge University Press, 2012).
47. Miller, K. G. *et al.* Sea-level rise in New Jersey over the past 5000 years: Implications to anthropogenic changes. *Glob. Planet. Change* **66,** 10–18 (2009).
48. Briscoe, M. G., Martin, D. L. & Malone, T. C. Evolution of regional efforts in international GOOS and US IOOS. *Mar. Technol. Soc. J.* **42,** 4–9 (2008).
49. Glenn, S. & Schofield, O. Growing a distributed ocean observatory: our view from the COOL Room. *Oceanography* **22,** 128–145 (2009).
50. McClain, E. P., Pichel, W. G., Walton, C. C., Paul, E. & Pichel, G. Comparative performance of AVHRR-based multichannel sea surface temperatures. *J. Geophys. Res. Ocean* **90,** 11587–11601 (1985).
51. Kohut, J. T. & Glenn, S. M. Improving HF Radar surface current measurements with measured antenna beam patterns. *J. Atmos. Ocean. Technol.* **20,** 1303 (2003).
52. Kohut, J., Roarty, H., Randall-Goodwin, E., Glenn, S. & Lichtenwalner, C. S. Evaluation of two algorithms for a network of coastal HF radars in the Mid-Atlantic Bight. *Ocean Dyn.* **62,** 953–968 (2012).
53. Roarty, H., Baskin, C., Smith, M. & Glenn, S. in *Proceedings of the OCEANS MTS/IEEE Conference* (Rutgers University Coastal Ocean Observation Laboratory (RU-COOL) New Brunswick, New Jersey, USA, 2014).
54. Davis, R., Eriksen, C. & Jones, C. in *Technology and Applications of Autonomous Underwater Vehicles* (ed. Griffiths, G.) 37 (Taylor & Francis, 2003).
55. Sherman, J., Davis, R. E., Owens, W. B. & Valdes, J. The autonomous underwater glider 'Spray'. *IEEE J. Ocean. Eng.* **26,** 437–446 (2001).
56. Wilkin, J. & Hunter, E. An assessment of the skill of real-time models of Mid-Atlantic Bight continental shelf circulation. *J. Geophys. Res.* **118,** 2919–2933 (2013).
57. Moore, A. M. *et al.* The Regional Ocean Modeling System (ROMS) 4-dimensional variational data assimilation systems. *Prog. Oceanogr.* **91,** 34–49 (2011).
58. Fairall, C. W., Bradley, E. F., Hare, J. E., Grachev, A. A. & Edson, J. B. Bulk Parameterization of air–sea fluxes: updates and verification for the COARE algorithm. *J. Clim.* **16,** 571 (2003).
59. Warner, J. C., Sherwood, C. R., Arango, H. G. & Signell, R. P. Performance of four turbulence closure models implemented using a generic length scale method. *Ocean Model.* **8,** 81–113 (2005).
60. Umlauf, L. & Burchard, H. A generic length-scale equation for geophysical turbulence models. *J. Mar. Res.* **61,** 235–265 (2003).
61. Green, B. W. & Zhang, F. Impacts of air–sea flux parameterizations on the intensity and structure of tropical cyclones. *Mon. Weather Rev.* **141,** 2308–2324 (2013).

Acknowledgements

Support was provided by the National Oceanic and Atmospheric Administration (NOAA) led Integrated Ocean Observing System (IOOS) through the Mid-Atlantic Regional Association Coastal Ocean Observing System (MARACOOS, NA11NOS0120038), the Environmental Protection Agency (EP-11-C-000085), the New Jersey Department of Environmental Protection (WM13-019-2013) and Board of Public Utilities (2010RU-COOL) and the NOAA Cooperative Institute for the North Atlantic Region (CINAR, NA13OAR4830233), Disaster Recovery Act. The authors thank Teledyne Webb Research and Rutgers University for student support, and the NOAA National Centers for Environmental Prediction for student engagement. We further thank David Titley, Jim Price and an anonymous reviewer for their useful comments.

Author contributions

S.M.G. synthesized and analysed the multiple data sets and wrote the manuscript in collaboration with the other authors. T.N.M. assisted in the synthesis of the *in situ* oceanographic data. G.N.S. contributed the atmospheric and storm sensitivity studies. Y.X. contributed the ocean simulations and analysis. R.K.F. performed historical buoy data and storm track analysis. F.Y. provided plots of buoy data beneath Super Typhoon Muifa. H.R. provided the observational data from the HF Radars. O.S. was involved in data collections and involved in analysis and manuscript preparation. J.K. contributed the Slocum data and was involved in analysis and manuscript preparation. All authors reviewed and edited this manuscript.

Additional information

Competing financial interests: The authors declare no competing financial interests.

Pliocene cooling enhanced by flow of low-salinity Bering Sea water to the Arctic Ocean

Keiji Horikawa[1,2], Ellen E. Martin[1], Chandranath Basak[1,†], Jonaotaro Onodera[3], Osamu Seki[4], Tatsuhiko Sakamoto[3,†], Minoru Ikehara[5], Saburo Sakai[3] & Kimitaka Kawamura[4]

Warming of high northern latitudes in the Pliocene (5.33–2.58 Myr ago) has been linked to the closure of the Central American Seaway and intensification of North Atlantic Deep Water. Subsequent cooling in the late Pliocene may be related to the effects of freshwater input from the Arctic Ocean via the Bering Strait, disrupting North Atlantic Deep Water formation and enhancing sea ice formation. However, the timing of Arctic freshening has not been defined. Here we present neodymium and lead isotope records of detrital sediment from the Bering Sea for the past 4.3 million years. Isotopic data suggest the presence of Alaskan glaciers as far back as 4.2 Myr ago, while diatom and $C_{37:4}$ alkenone records show a long-term trend towards colder and fresher water in the Bering Sea beginning with the M2 glaciation (3.3 Myr ago). We argue that the introduction of low-salinity Bering Sea water to the Arctic Ocean by 3.3 Myr ago preconditioned the climate system for global cooling.

[1] Department of Geological Sciences, University of Florida, Gainesville, Florida 32611, USA. [2] Graduate School of Environmental Studies, Nagoya University, Furo-cho, Chikusa-ku, Nagoya 464-8601, Japan. [3] Japan Agency for Marine-Earth Science and Technology, 2-15 Natsushima-cho, Yokosuka 237-0061, Japan. [4] Institute of Low Temperature Science, Hokkaido University, N19W8, Kita-ku, Sapporo, Hokkaido 060-0819, Japan. [5] Center for Advanced Marine Core Research, Kochi University, B200 Monobe, Nankoku 783-8502, Japan. † Present address: Graduate School and Faculty of Bioresources, Mie University, 1577 Kurimamachiya-cho, Tsu, Mie 514-8507, Japan (T.S.); Lamont-Doherty Earth Observatory, 61 Route 9W, PO Box 1000, Palisades, New York 10964, USA (C.B.). Correspondence and requests for materials should be addressed to K.H. (email: horikawa@sci.u-toyama.ac.jp).

The period from 4.4–4.0 million years ago (Myr ago) was the warmest interval within the Pliocene epoch and was characterized by global mean temperatures ~3 °C warmer than today[1,2]. The subsequent cooling throughout the late Pliocene culminated in the late-Pliocene climate transition (LPCT, 3.6–2.4 Myr ago), which led to major Northern Hemisphere glaciation at 2.73 Myr ago[3,4]. There is evidence for restricted, local glaciers on Arctic continents during the Pliocene warmth[5]; however, the evolution of these glaciers during the LPCT is poorly constrained, as is the role of the Central American Seaway (CAS)[6]. Tectonic shoaling of the CAS may have reached a critical threshold between ~4.7 and 4.2 Myr ago[6–9] that restricted Pacific–Caribbean surface-water exchange and redirected warm, saline waters from the equatorial Atlantic to the North Atlantic. These changes in the ocean circulation system are believed to contribute to an initial intensification of the Atlantic Meridional Overturning Circulation (AMOC) and warming of the North Atlantic prior to 3.6 Myr ago[9–12]. Yet, there is no consensus regarding what processes reduced the intensification of the AMOC and led to the late-Pliocene global cooling.

Modelling results suggest increased moisture supply (that is, snowfall) related to the increased AMOC was insufficient to produce the major glaciation of Greenland[12]. Alternative explanations emphasize increased freshwater inputs to the Arctic Ocean (AO) that could have generated lower sea-surface salinities (SSSs), leading to sea ice formation, increased ice-albedo effects and reduced heat exchange between the ocean and atmosphere[6,7]. The proposed source for this freshwater is the enhanced flow of Siberian rivers, driven by increased moisture transport to Eurasia by the AMOC[7]. Another possible source is low-salinity Bering Sea water[6,13–15]. The influence of Bering Sea water on the AO during the LPCT is particularly relevant because the flow direction through the Bering Strait is believed to have shifted from southward to its present northward direction at 3.6 Myr ago in response to changes in the steric height gradient between the North Pacific and AO, generated by closure of the CAS[11,13]. The freshwater influx to the AO today is derived from river discharge (38%), inflow of low-salinity Bering Sea water (30%) and net precipitation (24%)[16,17]. Thus, the amount of freshwater transported from the Bering Sea into the AO currently influences the strength of deep-ocean convection in the North Atlantic[18–20]. An excess influx of freshwater from the Bering Sea to the AO during the Pliocene may have compounded the effect of Siberian River inputs on the stability of the AMOC, resulting in cooling during the LPCT[6].

Low salinities in the Bering Sea today are primarily derived from the glacier-fed Alaska Coastal Current (ACC)[21] (Fig. 1). Therefore, Alaskan mountain glacier build-up and seasonal melting of ice should be necessary for establishing low salinities in the Bering Sea; however, the Pliocene evolution of this glacial input has been difficult to track due to limited records of the glacial history of Alaska and past surface-water properties in the Bering Sea. Here we generate Pliocene records of sediment provenance and surface-water properties in the Bering Sea to demonstrate the timing of cryospheric evolution in the Bering/Alaskan region and clarify the relationship between flow of low-salinity Bering Sea water to the AO and the late-Pliocene global cooling.

Results

Surface-water properties and Nd and Pb isotope records. In this study, we interpret the relative abundance of tetraunsaturated C_{37} alkenones to total C_{37} alkenones (% $C_{37:4}$ alkenone) as a proxy for temperature and/or salinity, with higher percentages reflecting cooler and/or freshwater masses[22,23]. The percentage of 'sea-ice related' diatoms is the relative abundance of *Thalassiosira antarctica* resting spores, *Fragilariopsis cylindrus* and *Fr. oceanica* to the total number of diatoms[24]. Today, these diatoms are common in surface sediment from the Bering Sea shelf and slope, where sea ice expands seasonally, but they are absent or found in low abundances in surface sediments in the central Bering Sea, where sea ice is less common[25]. An additional proxy is the percent of *Actinocyclus* spp., which is a species that is abundant in surface water with meltwater-related stratification[24,26].

At IODP (Integrated Ocean Drilling Program) Site U1341B (Expedition 323) located at Bowers Ridge in the southern Bering Sea (Fig. 1; Methods; Supplementary Fig. 1; refs 27,28), all three of these surface-water proxies record gradual increases beginning around the time of MIS (marine isotope stage) M2 glaciation at 3.3 Myr ago and continuing to 2.75–2.5 Myr ago (Fig. 2d–f), indicating cooling and/or freshening and increased meltwater-related stratification. The highest % $C_{37:4}$ alkenone values are observed during the mid-Pleistocene transition (MPT), probably reflecting additional cooling and/or freshening (Fig. 2f). The low-resolution % $C_{37:4}$ alkenone record in the Bering Sea has remarkable similarities to a high-resolution % $C_{37:4}$ alkenone record from Site 882 in the subarctic western North Pacific[29], which also records the highest values during the MPT and stable, lower values prior to 3.4 Myr ago (Figs 2f and 4e). The lack of correlation between % $C_{37:4}$ alkenones and % *Actinocyclus* spp. during the MPT is attributed to higher percentages in *T. trifulta* group and *T. antarctica* resting spores[24], which prefer cold water and reduce the % *Actinocyclus* spp.

The detrital ε_{Nd} values from Site U1341B span a wide range from −6.2 to +6.9 ε_{Nd} over the past 4.3 Myr ago, with large, rapid fluctuations from 2.8 to 2.0 Myr ago (Fig. 2c). The ε_{Nd} variability was confirmed by replicate analysis of some of the samples with high and low ε_{Nd} values (Supplementary Table 1). Further, similar highly variable detrital ε_{Nd} values are documented in glacial–interglacial sediments at Bow-8A (−6.1 to +2.2 ε_{Nd}; Supplementary Fig. 2) and at Site 884 (−1.2 to +8.6 ε_{Nd}; ref. 30) (Fig. 1). These highly variable detrital ε_{Nd} values suggest regional glacial–interglacial shifts of sediment sources and/or sediment transport processes. A lower-resolution detrital lead (Pb) isotope record ($n = 29$) also shows similar high variability (Supplementary Fig. 3; Supplementary Table 1). The long-term ε_{Nd} record (400-kyr running mean) demonstrates a marked shift from more radiogenic to less radiogenic values across the LPCT starting at 3.0 Myr ago and during the MPT starting at 1.2 Myr ago, coherent with the % $C_{37:4}$ alkenone record (Fig. 2c,f). Further, the pronounced glacial MIS M2 is marked by the lower detrital ε_{Nd} value, together with the increases in % sea-ice-related diatoms and % *Actinocyclus* spp.

Sediment provenance. Detrital material is delivered to the Bering Sea from continental sources via rivers, ice and wind, and distributed by surface and deep currents. Today, the Yukon River, the largest river in the Bering Sea, provides >63% (55 Mt yr^{-1})[31] of the total sediment load to the Bering Sea in association with abundant freshwater discharge (6,400 $m^3 s^{-1}$)[32] as it drains portions of the glaciated Alaska Range, Wrangell Mts and St Elias Mts in Alaska (Fig. 1; Supplementary Note 1). Approximately half of the annual suspended sediment exported via the Yukon River occurs during spring season snowmelt and the other half occurs during summer–autumn season glacier melt[33]. The meltwater river plumes, and counterclockwise surface currents in the Bering Sea distribute these Yukon River detrital sediments over long distances and a wide area of the eastern Bering Sea shelf[34,35] (Fig. 1).

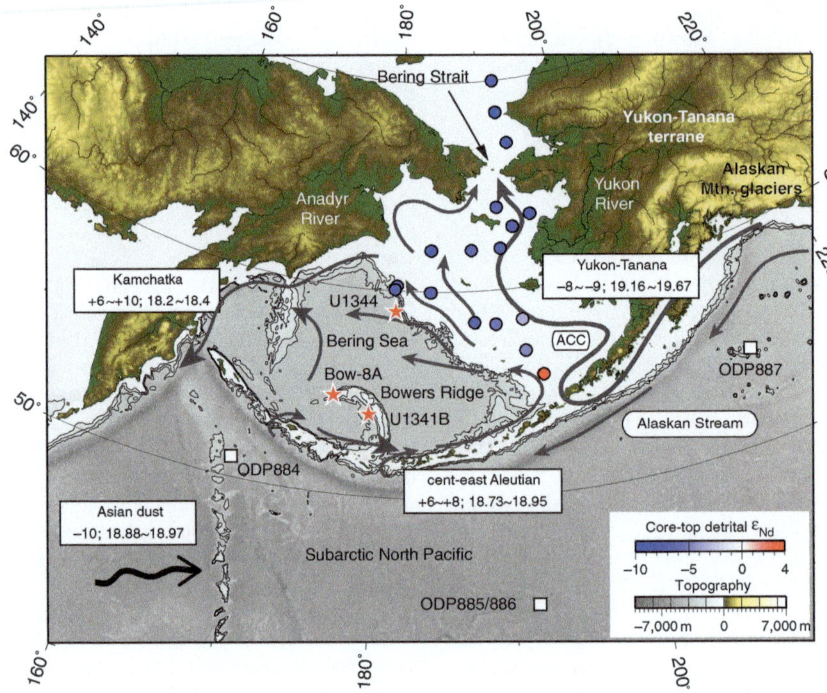

Figure 1 | Locations of studied sediment cores and major surface currents. The Bering Strait is narrow (~85 km wide) and shallow (~50 m deep) at the far northern end of the Pacific Ocean that connects to the Arctic Ocean (AO). Glacial meltwater provides a substantial portion of the total freshwater runoff into the Alaska Coastal Current (ACC), and the low-salinity ACC flows to the AO. Grey arrows indicate subsurface flow at 40 m (ref. 37). On the map, numbers in the white boxes represent the ε_{Nd} value and the $^{206}Pb/^{204}Pb$ ratio of detrital materials in the source regions. Isotopic data are compiled from the GEOROC database[38] and from the literature (see references in the figure caption of Fig. 3). Coloured circles represent sampling sites and their detrital ε_{Nd} values of surface sediments[34]. Red stars (this study) and white squares[30,40,54] are the locations of core sites cited in this study. The black arrow represents the transport of Asian dust. Bathymetry and land topography are also shown.

The Alaskan material stored on the Bering Sea shelf is transported to Bowers Ridge by the southward-flowing current. Thus, despite the distance from the Yukon outflow to Bowers Ridge, potential detrital sources at Site U1341B in the southern Bering Sea include the Yukon-Tanana terrane, Aleutian arc and aeolian dust from the Asian continent. Aeolian sources from Alaska and the Anadyr River detrital material might be less significant because the core site is not downwind from Alaska[36], and the detrital flux from the Anadyr River is a small ($2\,Mt\,yr^{-1}$)[30] and is restricted to the northern shelf margins by the southwestward-flowing Kamchatka current[37], respectively. The three primary detrital sources, Yukon-Tanana terrane, Aleutian arc and Asian dust, have very distinct Nd and Pb isotopic signatures; Yukon-Tanana terrane Alaskan material has ε_{Nd} values of -8 to -9 and $^{206}Pb/^{204}Pb$ values of 19.16–19.67 (refs 30,38,39), Aleutian arc material has ε_{Nd} values of $+6$ to $+8$ and $^{206}Pb/^{204}Pb$ values of 18.73–18.95 (refs 30,38) and Asian dust has ε_{Nd} values ca. -10 and $^{206}Pb/^{204}Pb$ values of 18.88–18.97 (ref. 40–42) (Fig. 1). Therefore, these isotope data will be sensitive monitors of provenance for the southern Bering Sea sediments.

Nd and Pb isotopic compositions of detrital sediments from U1341B create linear arrays for isotope plots of $^{206}Pb/^{204}Pb$ versus $^{207}Pb/^{204}Pb$, $^{206}Pb/^{204}Pb$ versus $^{208}Pb/^{204}Pb$ and $^{206}Pb/^{204}Pb$ versus ε_{Nd} which connect Aleutian arc material and Yukon-Tanana terrane Alaskan material (Fig. 3a–c). In contrast, data from Site 885/886 in the subarctic North Pacific[40] create a distinct linear array that connects central-eastern Aleutian Arc materials and Chinese loess (Figs 1 and 3a–c). These data show that Site 885/886 is marked by higher fluxes of the Asian dust and lower detrital fluxes from the Aleutian arcs than U1341B. However, the Asian dust fluxes estimated at Site 885/886 for the past 5 Myr ago ($8–122\,\mu g\,cm^{-2}\,yr^{-1}$; ref. 40) and the fluxes observed in the modern North Pacific ($22–100\,\mu g\,cm^{-2}\,yr^{-1}$; ref. 43) are negligible compared with the higher mass accumulation rates of detrital materials for U1341B ($8\,mg\,cm^{-2}\,yr^{-1}$, assuming $14\,cm\,kyr^{-1}$ as the average sedimentation rate, 40 wt% of detrital material content and $1.4\,g\,cm^{-3}$ of bulk sediment density[28]). Thus, we interpret the detrital sediments of U1341B as a mixture dominated by sediments from the Aleutian and Alaskan sectors throughout the past 4.3 Myr ago.

Prior to MIS M2 at 3.3 Myr ago, most of the detrital Nd isotope data plot between $+5$ and $+6$ ε_{Nd} (Fig. 2c), suggesting that sedimentation at the site was dominated by local Aleutian sources. However, there are four pulses that produced progressively less radiogenic values. These occurred at 4.2 Myr ago, 3.96 Myr ago, 3.55 Myr ago, and the MIS M2 glaciation (3.3 Myr ago). A simple binary mixing model estimates that Alaskan-sourced detrital materials accounted for 13% of the detrital sediment at 4.2Myr ago, 26% at 3.96Myr ago, 32% at 3.55Myr ago and 48% at 3.32 Myr ago (Methods), indicating that the proportion of Alaskan inputs increased in progressively younger samples.

Transport of Alaskan detrital material to Bowers Ridge. Core-top late-Holocene sediment from core Bow-8A at Bowers Ridge located near the Aleutian arc has an ε_{Nd} value of $+2.2$ (Fig. 1; Supplementary Fig. 2). Simple binary mixing (Methods) suggests the core-top sediment is predominantly composed of central-eastern Aleutian material with a minor component of Alaskan material (18%). This Yukon River-derived Alaskan material stored on the Bering Sea shelf may have been transported

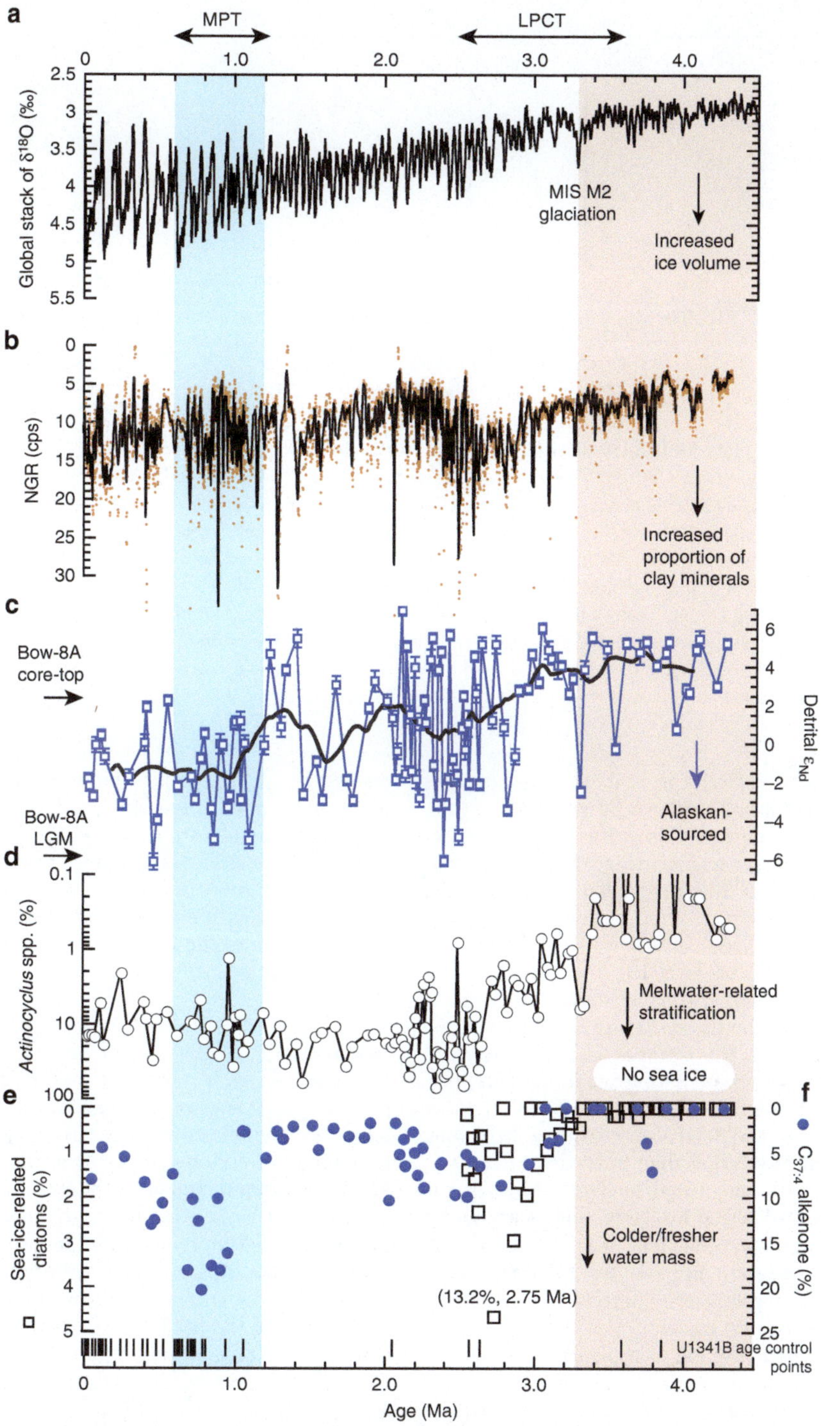

Figure 2 | Multi-proxy records from Bering Sea Site U1341B for the past 4.3 Myr ago. (**a**) Global stack of $\delta^{18}O$ record[4]. Records of (**b**) natural gamma radiation (NGR), (**c**) detrital Nd isotope values, (**d**) % *Actinocyclus* spp[24]., (**e**) % sea-ice-related diatoms[24] and (**f**) % $C_{37:4}$ alkenones from U1341B in the Bering Sea. The thick line in **c** represents the 400-kyr running mean, and the error bars correspond to $2\sigma_{\mu m}$ error (see details in Supplementary Table 1). From 4.3 to 3.3 Myr ago, there is no evidence of sea-ice-related diatoms in the southern Bering Sea. Some minor increases in % sea-ice-related diatoms before 3.3 Myr ago were not significant relative to error. These multi-proxy records from the Bering Sea highlight that the marine isotope stage (MIS) M2 glaciation at 3.3 Myr ago marks the beginning of the long-term trend towards cooling/freshening, meltwater-related stratification, expansion of sea ice and increased proportion of clay minerals in the southern Bering Sea. LPCT and MPT are Late-Pliocene climate transition and Mid-Pleistocene transition, respectively.

to Bowers Ridge by the southward-flowing current. In contrast, during the last deglaciation and the last glacial maximum (LGM), detrital sediments at Bow-8A have less radiogenic ε_{Nd} values [illegible] (Supplementary Fig. 2), which translates to a much greater contribution from Alaskan sources (60–78%). A measured decrease in diatom $\delta^{18}O$ in the northeast Bering Sea during the deglacial is believed to reflect a large flux of meltwater from retreating alpine glaciers in Alaska[44], which may have transported

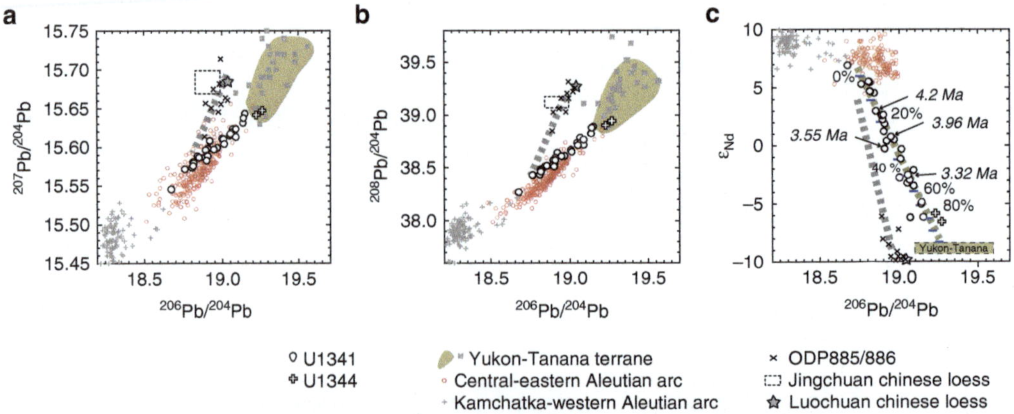

Figure 3 | Pb and Nd isotope cross-plots for U1341B detrital sediments and potential local source rocks. (**a**) $^{206}Pb/^{204}Pb$ versus $^{207}Pb/^{204}Pb$, (**b**) $^{206}Pb/^{204}Pb$ versus $^{208}Pb/^{204}Pb$ and (**c**) $^{206}Pb/^{204}Pb$ versus ε_{Nd} isotope plots. Core U1344 (open cross) was retrieved from the northern Bering slope near the Yukon River, and thus detrital sediments deposited during the LGM (last glacial maximum) and the last deglaciation are representative of Yukon River-delivered Alaskan detrital material. U1341B detrital sediments of Nd and Pb isotopes (open circle) represent a binary mixing line (coloured dashed line in **c**) showing a mixture of sediment from the Aleutian and Alaskan sectors[30,38,39], whereas detrital sediments at Site 885/886 (× -mark, ref. 40) represent a linear array (grey dashed line in **a–c**) connecting Chinese loess and central-eastern Aleutian Arc sectors[38,41,42]. The coordinates of points (bar) on the Aleutian-Yukon (Alaska) mixing line (**c**) represent the proportion of the Alaskan detrital material (Methods). Detrital ε_{Nd} record from U1341B represents four pulses that produced progressively less radiogenic ε_{Nd} values prior to 3.3 Myr ago, and these data are marked with the ages in **c**. Although detrital materials from the Yukon-Tanana terrane have a wide range of $^{206}Pb/^{204}Pb$ ratios, the average $^{206}Pb/^{204}Pb$ ratio of Alaskan-sourced material is assumed to be 19.25 for the mixing line (Methods and Supplementary Table 3).

some of the detrital sediment. Evidence from diatom assemblages suggests the presence of drift ice at northern Bowers Ridge at the LGM[45]. Given these results, it appears that meltwater plumes and sea ice rafting were responsible for transporting a large amount of Alaskan detrital materials from the prograded mouth of the Yukon River towards Bowers Ridge during deglacial and glacial periods, respectively.

During glacial periods when sea level fell, the closed Bering Strait (currently 50 m deep) would also have altered both surface and intermediate water circulation in the Bering Sea[45,46]. Yet, the tight coupling between detrital Nd isotopes and foraminiferal $\delta^{18}O$ records from Bow-8A within the Bering Sea (Supplementary Fig. 2) and Site 884 in the North Pacific south of the Aleutian Arc[30] suggests delivery of Alaskan detrital material was driven by a process beyond the Bering Sea rather than internal variations in circulation within the Bering Sea caused by the closure of the Bering Strait. Instead, changes driven by ice volume, such as the location of Alaskan river outflow, the extent of sea ice formation and the volume of meltwater, appear to provide better explanations of observed changes in the composition of detrital material during glacial–interglacial cycles.

In the early Pliocene, most of the Nd isotopic data are centred on +5 to +6 ε_{Nd}, higher than the Bow-8A core-top value (+ 2.2 ε_{Nd}) (Fig. 2c), and there is no evidence of sea-ice-related diatoms from 4.3 to 3.3 Myr ago[24] (Fig. 2e). These findings indicate that the southern Bering Sea was relatively warm and there was insufficient ice or glacial melt to transport Alaskan detrital material, consistent with the general view of Pliocene warmth[1,2,47] (Fig. 4b–d). The relatively stable and radiogenic ε_{Nd} values observed from 4.3 to 3.3 Myr ago transition to an interval of high-amplitude variations and overall decreasing ε_{Nd} values across the LPCT from 3.3 Myr ago to the MPT starting at 1.2 Myr ago. This long-term shift of detrital ε_{Nd} coincides with (1) greater variability and a general increase in global ice volume (Fig. 2a) and (2) increasing natural gamma radiation in the core, indicating an increased proportion of clay minerals particularly during glacial intervals[48] (Fig. 2b) and (3) increasing % $C_{37:4}$ alkenones, indicating cooling and/or freshening of water masses[22,23] (Fig. 2f). These proxies further support the interpretation of increasing glacial ice volume in the Bering/Alaskan region during the LPCT to MPT. The correlation between increased ice volume, increased proportions of clay minerals and lower detrital Nd isotopes argues that sea ice rafting and snow/glacier melt discharge from the Yukon River supplied a larger fraction of the Alaskan detrital material to Site U1341B for these periods.

Discussion

Previous research on Gulf of Alaska sediments shows the presence of tidewater glaciers in southeastern Alaska since the Late Miocene, with more definitive evidence for glaciers based on the presence of dropstones beginning at 4.2–4.3 Myr ago[49]. The Pliocene detrital ε_{Nd} record from U1341B shows pulses of progressively greater inputs of Alaskan detrital material beginning as early as 4.2 Myr ago (Figs 2c and 4f). The first pulse of Alaskan detrital inputs that coincides with a slight increase in % sea-ice-related diatoms and a significant increase in % *Actinocyclus* spp. occurs during the MIS M2 glaciation at 3.3 Myr ago (Fig. 2c–e). This correlation indicates the first Alaskan glaciers large enough to produce associated meltwater and sea ice appeared at 3.3 Myr ago.

In contrast, other early pulses of Alaskan detrital material inputs at 4.2, 3.96 and 3.55 Myr ago lack a response in % $C_{37:4}$ alkenones and % *Actinocyclus* spp. Although Alaskan mountain glaciers appear to have developed in the early Pliocene, the lack of coherent responses suggests that Alaskan glaciations were probably limited in size and early-glacial meltwater plumes were not large enough to produce stratification and freshening in the southern Bering Sea. However, initial deposition of Alaskan-sourced detrital material on the Bering Sea shelf may have been sourced by smaller glacial meltwater pulses and transported to Bowers Ridge by the southward-flowing current, as seen in the Holocene. The details of current processes that transported Alaskan detrital sediments to the Bowers Ridge remain uncertain, but an important point is that pulses of Alaskan detrital inputs

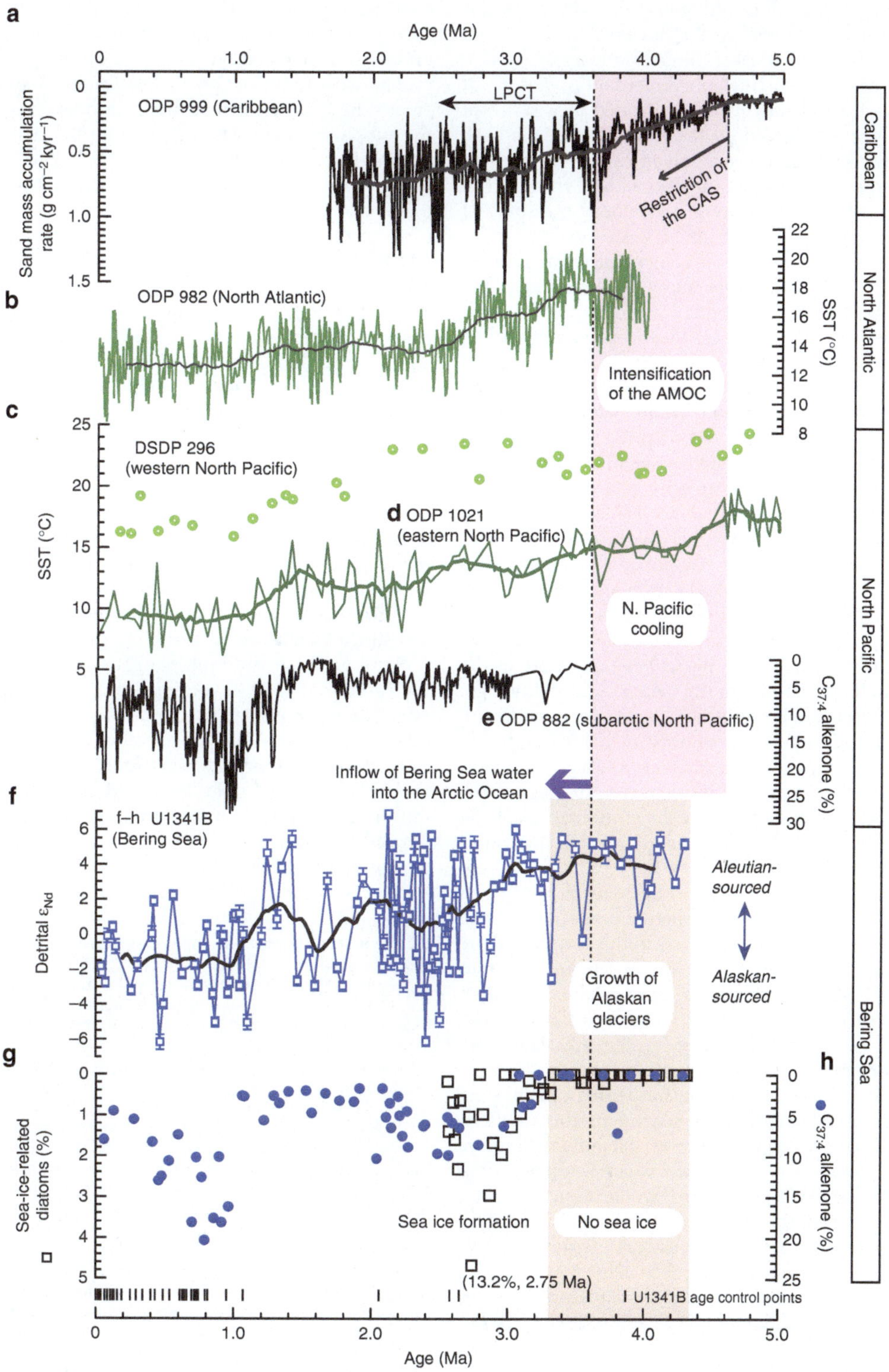

Figure 4 | Comparison of global and regional data from the Pliocene-Pleistocene. (**a**) Sand mass accumulation rate in the Caribbean Sea[9], (**b**) Alkenone-derived SST records in the North Atlantic[60], (**c**) Winter SST records based on transfer function methods of planktic foraminifera assemblages in the western North Pacific[51], (**d**) Alkenone-derived SST records in the eastern North Pacific[52], (**e**) $C_{37:4}$ alkenone abundance from the subarctic western North Pacific[29], (**f-h**) Records of detrital Nd isotopes, sea-ice-related diatom abundance[24] and $C_{37:4}$ alkenone abundance from Bering Sea Site U1341B. Restriction of the CAS (Central American Seaway) started at 4.6–4.7 Myr ago[8,9]. The presence of Alaskan mountain glaciers was as far back as 4.2 Myr ago, and the flow direction of Bering Strait throughflow has shifted from southward to its present northward direction at 3.6 Myr ago[13] (dashed line). Introduction of low-salinity Bering Sea water into the AO might have started by at least 3.3 Myr ago. Thick lines in **a**,**b**,**d** and **f** represent the 400-kyr running mean.

identified by detrital Nd isotopes in the early Pliocene support growth of Alaskan mountain glaciers as a necessary precondition [illegible] detrital material on the

Bering Sea shelf. Further, we also propose that increased volumes of melt with progressively larger glaciers from 4.2 Myr ago would have increased the depositional area of Alaskan-sourced detrital material on the Bering Sea shelf, resulting in progressively greater inputs of Alaskan detrital material to the Bowers Ridge.

Early-Pliocene evolution of Alaskan mountain glaciers ($\sim$4.2–3.3 Myr ago) capable of producing a sedimentary signature of Alaskan material at Bowers Ridge may be related to a series of ocean–climate feedbacks triggered by the restriction of the CAS starting 4.6–4.7 Myr ago (Fig. 4a). A coupled ocean–atmosphere general circulation model found that meridional heat transport in the North Pacific weakened in response to reduced transport of saline water to the equatorial Pacific as the CAS shoaled[50]. This change in heat transport facilitated cooler air temperatures around the North Pacific and more snow accumulation over Alaska and western Canada[50]. Pronounced decreases in sea-surface temperature (SST) in the western subtropical and eastern North Pacific from $\sim$4.4 to 4.6 Myr ago[51,52] (Fig. 4c,d) and persistent cooling of the Chinese Loess Plateau beginning $\sim$4 Myr ago[53] support this modelling result. We acknowledge the lack of the detrital Nd isotopic record prior to 4.3 Myr ago limits our evaluation of the relationship between restriction of the CAS and growth of Alaskan mountain glaciers. However, terrigenous mass accumulation rates from ODP 887 in the Gulf of Alaska also increased from 4.55 Myr ago, with relatively higher values after 4.24 Myr ago[54], again supporting a relationship between the sediment flux from Alaska and restriction of the CAS. On the basis of these data around Alaska, we propose that the growth of Alaskan mountain glaciers in the early Pliocene was accelerated by the reduction of northward heat transport in the North Pacific due to the restriction of the CAS at 4.6–4.7 Myr ago.

The evolution of a relatively fresh ACC is poorly constrained, but is presumably linked to the onset of Alaskan glaciation, which had started by at least 4.2 Myr ago. Smaller Alaskan mountain glaciers in the early Pliocene might have started to feed glacial meltwater to the coastal Bering Sea, but the combination of the small volume and the southward flow direction of Bering Strait throughflow limited the impact in the AO. Reversal of the Bering Strait throughflow at 3.6 Myr ago[11,13] might be related to a change in the steric sea-level gradient between the Arctic and Bering Sea basins driven by increases in SSS in the North Atlantic and decreases in SSS in the North Pacific in response to restriction of the CAS. The resulting flow might have first introduced northern Bering Sea coastal water into the AO, but freshwater inputs from the Bering Sea to the AO were not significant enough to perturb the freshwater balance in the AO.

Our evidence for long-term cooling/freshening, meltwater-related stratification and expansion of sea ice in the southern Bering Sea by 3.3 Myr ago supports the establishment of more extensive ice on Alaska by this time, which enhances the likelihood that a fresher ACC developed in the eastern Bering Sea and delivered low-salinity water to the AO. Progressively larger fluxes of meltwater from progressively larger Alaskan mountain glaciers that developed from 3.3 to 2.8–2.5 Myr ago, as documented in $C_{37:4}$ alkenone and *Actinocyclus* spp. records, would have further reduced salinity in ACC and ultimately in the AO.

Introduction of low-salinity Bering Sea water into the AO, starting by at least 3.3 Myr ago, would have perturbed the freshwater budget in the AO. Development of a low-salinity surface layer in the AO would have isolated the warmer, more saline intermediate waters from the atmosphere, and promoted expansion of sea ice in the North Atlantic that began at 3.2 Myr ago[55,56] as well as freshening of the East Greenland Current between 3.25 and 3.16 Myr ago[14]. This scenario sets up northern high-latitude cooling through ice-albedo feedbacks. Ultimately, outflow of this low-salinity Arctic water to the North Atlantic would have weakened the 'super conveyor' of North Atlantic deep water (4.0–3.0 Myr ago)[10,57], further contributing to major continental glaciation in the Northern Hemisphere. Indeed, a recent coupled ocean–atmosphere climate model conducted under boundary conditions at 112 kyr supports this sequence of events, with introduction of low-salinity water from the Bering Sea to the AO perturbing the freshwater balance in the AO and the subpolar North Atlantic, muting the AMOC and leading to cooling in the Northern Hemisphere[20]. Even if the inflow of the low-salinity Bering Sea water into the AO was not the ultimate driver of the Pliocene climate cooling, Bering Strait throughflow of freshwater transport into the AO might have worked as an effective amplifier to cool the northern high latitudes starting at 3.3 Myr ago.

In summary, our Nd isotopic data highlight the presence of Alaskan detrital material at the Bowers Ridge as far back as 4.2 Myr ago. Transport of this material by sea ice and meltwater plumes supports development of Alaskan mountain glaciers in the early Pliocene; timing that coincides with proposed North Pacific cooling in response to restriction of the CAS. Increasing fluxes of meltwater from progressively larger Alaskan mountain glaciers from 3.3 to 2.8–2.5 Myr ago are indicated by the Nd isotopic data, as well as % *Actinocyclus* spp. and % $C_{37:4}$ alkenones at Site U1341B. Introduction of low-salinity Bering Sea water into the AO starting by at least 3.3 Myr ago would have strengthened halocline development and sea-ice formation in the AO, leading to subarctic cooling/freshening from 3.2 to 3.3 Myr ago through ice-albedo feedbacks.

We argue that the multiple gateway events (restriction of the CAS and reversal of the Bering Strait throughflow) in the Pliocene established the glacier-fed, low-salinity Bering Sea water and its inflow into the AO, respectively. The effect of Bering Strait throughflow on freshwater transport into the AO during the early to late Pliocene should be tested quantitatively by numerical modelling to better understanding the outflow of low-salinity Arctic water to the North Atlantic and its impact on the stability of the AMOC in the Pliocene.

Methods

Sediment core samples. We analysed detrital Nd and Pb isotopes from sediment cores obtained at Site U1341B (54°02.00′N, 179°00.52′E, 2,140 m water depth) at Bowers Ridge in the southern Bering Sea, and at Site U1344 (59°03.02′N, 179°12.20′E, 3,173 m water depth) at the northern Bering slope during the IODP Expedition 323 (Fig. 1; ref. 27). Supplementary detrital Nd isotope data were also obtained from core Bow-8A (54°47′N, 176°55′E, 884 m water depth) near site U1341B during the R/V Hakuho-maru cruise KH99-3 (Fig. 1).

Sediments obtained from site U1341B are composed of a variable mixture of detrital, biogenic and authigenic components; there is a major contribution of biogenic opal and detrital clay/silt, and calcium carbonate content is generally low throughout the core (<1.0 wt%)[27,28]. The initial age model for Site U1341B was originally based on several paleomagnetic and siliceous microfossil datums that have been determined from an on-board study[27,28]. However, in this paper the age model has been updated using detailed onshore biostratigraphy[24], magnetostratigraphy and foraminifera $\delta^{18}O$ stratigraphy (Supplementary Fig. 1; Supplementary Table 2). The age–depth plot at Site U1341B was produced by assuming linear sedimentation rates between the age control points, and by extrapolating the sedimentation rate beyond the oldest diatom-based datum (3.87 Myr ago); this resulted in an estimated bottom sediment age of $\sim$4.3 Myr ago (Supplementary Fig. 1; Supplementary Table 2). We sampled U1341B at constant depths (4.5 and 7.5 m) in each core ($n = 115$). Variations in the sedimentation rate produced apparent variations in the sample density (time intervals range from 5 to 116 kyr) of 115 Nd isotope data points plotted against age (Fig. 2c; Supplementary Fig. 1). Furthermore, we also generated Pb isotopic compositions of detrital sediments ($n = 29$) from U1341B (Supplementary Fig. 3). Other detailed sedimentological and geochemical data relating to core U1341 can be seen in associated literature[24,48,58].

Core U1344 is characterized by higher proportions of siliciclastic components than core U1341B. The last deglaciation and the LGM sediments of core U1344 show the Nd and Pb isotopic compositions ($\varepsilon_{Nd} = -5.8$ and -6.6; $^{206}Pb/^{204}Pb = 19.23$ and 19.27) that are close to those of source rocks in the Yukon-Tanana terrane in Alaska ($\varepsilon_{Nd} = -9$ to -8; $^{206}Pb/^{204}Pb = 19.16$ to 19.67; Fig. 3a–c).

Detrital Nd and Pb isotopes. We leached bulk sediments ($\sim$1.0 g) for 24 h on an orbital shaker at room temperature in 0.02 M hydroxylamine hydrochloride–25%

Optima acetic acid, buffered to pH 4 with Optima sodium hydroxide. This reductive cleaning step also completely removed carbonate (av. 0.9 wt%, ref. 28) from the bulk sediment. Following centrifugation and triple rinses in ultrapure water, the samples were dried in an oven at 60 °C. The oxide-free bulk sediments (~40 mg) were dissolved in a 7-ml Teflon vial by digestion in a concentrated Optima HF-HNO_3 mixture (1:3 v/v) for ~48 h on a hotplate (~120 °C). Bulk rare-earth elements were separated by a primary column using cation-exchange resin (Mitsubishi diaion CK08P, 75–150 μm) and 1.7 N HCl as an eluent. Nd was then isolated by passing the REE aliquot through Eichrom Ln-spec resin (50–100 μm) with 0.25 N HCl as an eluent. Nd aliquots were dried and then redissolved in 2% HNO_3 for analysis on a Nu Plasma MC-ICP-MS with a DSN-100 nebulizer at the University of Florida[59]. The total procedural blank for Nd is negligible (14 pg) compared with the amount of Nd in the samples. Pb aliquots were redissolved in 1 N seastar HBr and passed through Dowex 1-X8 (100–200 mesh) resin. Pb was collected in 20% Optima HNO_3. The total procedural blank for Pb is 10–30 pg Pb and is negligible. Thallium (Tl)-spiked Pb concentrates dissolved in 2% HNO_3 were adjusted for appropriate dilution to obtain a 2–5-V beam on ^{208}Pb. Samples were aspirated through a DSN-100 nebulizer and Pb isotopes were measured by a Nu Plasma MC-ICP-MS. A Tl normalization technique was used for Pb isotope analyses. Long-term NBS 981 values analysed over several years at the University of Florida are $^{206}Pb/^{204}Pb = 16.937 \pm 0.004$ (2σ), $^{207}Pb/^{204}Pb = 15.489 \pm 0.003$ (2σ) and $^{208}Pb/^{204}Pb = 36.695 \pm 0.008$ (2σ).

$C_{37:4}$ alkenone abundance. Total lipids were extracted from 1–3.5 g of dried sediments with dichloromethane/methanol (95:5) using an accelerated solvent extractor (ASE 200, Dionex) at 100 °C and 1,000 p.s.i. for 10 min. The total extracts were separated into three fractions using silica gel column chromatography: aliphatic and aromatic hydrocarbons, aliphatic ketones (alkenones) and alcohol plus fatty acid fractions. The C_{37} alkenones were analysed with an HP6890 gas chromatograph (GC) equipped with an on-column injector-fused silica capillary column (CPSIL-5 CB, 50 m length × 0.32 mm inner diameter, film thickness of 0.12 mm), and a flame ionization detector. The GC oven temperature was programmed to increase from 80 to 180 °C at 10 °C min^{-1}, and then from 180 to 310 °C at 5 °C min^{-1}. The relative abundance of the tetraunsaturated C_{37} alkenones to total C_{37} alkenones (% $C_{37:4}$) was defined as % $C_{37:4} = [C_{37:4}]/([C_{37:2}] + [C_{37:3}] + [C_{37:4}]) \times 100$.

Isotope mass balance calculation. A simple isotope mass balance calculation is based on a two-end-member mixing model. If A and B are components containing different concentrations of element X, and if R^X is an isotope ratio of X, then the following equation can be obtained:

$$R_M^X = \frac{R_A^X X_A f + R_B^X X_B (1-f)}{X_A f + X_B (1-f)} \quad (1)$$

where R_M^X is the isotope ratio of X in a mixture of components A and B, X_A and X_B are concentrations of X in A and B, respectively, and f is the weight fraction of A.

In our study, based on Nd and Pb isotope analyses, we consider Aleutian arc and Alaskan materials as end-members (see Fig. 3 and details in the text). The average values of Nd and Pb ($^{206}Pb/^{204}Pb$) isotopic compositions and their concentrations in the Aleutian and Alaskan materials were estimated from the GEOROC database[38] and relevant literature[34,39] (Supplementary Table 3). However, there were no available data for Pb concentrations in the Alaskan detrital material. Therefore, we evaluated a number of binary mixing curves that had different Pb concentrations (5, 10, 15 and 20 μg g^{-1}) in Alaskan detrital material (Supplementary Fig. 4). The Nd–Pb isotopic compositions of the U1341B detrital sediments were then plotted around a mixing line with a Pb concentration of 10 μg g^{-1} (Supplementary Fig. 4). Therefore, we set a value of 10 μg g^{-1} as the Pb concentration in Alaskan-sourced detrital material, and then calculated the contribution from Alaskan-sourced detrital materials to the sediments at 4.2 (13%), 3.96 (26%), 3.55 (32%) and 3.32 Myr ago (48%).

References

1. Haywood, A. M. & Valdes, P. J. Modelling Pliocene warmth: contribution of atmosphere, oceans and cryosphere. *Earth Planet. Sci. Lett.* **218,** 363–377 (2004).
2. Fedorov, A. V. *et al.* Patterns and mechanisms of early Pliocene warmth. *Nature* **496,** 43–49 (2013).
3. Haug, G. H., Sigman, D. M., Tiedemann, R., Pedersen, T. F. & Sarnthein, M. Onset of permanent stratification in the subarctic Pacific Ocean. *Nature* **401,** 779–782 (1999).
4. Lisiecki, L. E. & Raymo, M. E. A Pliocene-Pleistocene stack of 57 globally distributed benthic $\delta^{18}O$ records. *Paleoceanography* **20,** PA1003 (2005).
5. Nielsen, T. & Kuijpers, A. Only 5 southern Greenland shelf edge glaciations since the early Pliocene. *Sci. Rep.* **3,** 1875 (2013).
6. Haug, G. H., Tiedemann, R., Zahn, R. & Ravelo, A. C. Role of Panama uplift on oceanic freshwater balance. *Geology* **29,** 207–210 (2001).
7. Driscoll, N. W. & Haug, G. H. A short circuit in thermohaline circulation: a cause for Northern Hemisphere Glaciation? *Science* **282,** 436–438 (1998).
8. Keigwin, L. Isotopic paleoceanography of the Caribbean and East Pacific: role of Panama uplift in Late Neogene time. *Science* **217,** 350–353 (1982).
9. Haug, G. H. & Tiedemann, R. Effect of the formation of the Isthmus of Panama on Atlantic Ocean thermohaline circulation. *Nature* **393,** 673–676 (1998).
10. Steph, S. *et al.* Early Pliocene increase in thermohaline overturning: a precondition for the development of the modern equatorial Pacific cold tongue. *Paleoceanography* **25,** PA2202 (2010).
11. Maier-Reimer, E., Mikolajewicz, U. & Crowley, T. Ocean general circulation model sensitivity experiment with an open Central American Isthmus. *Paleoceanography* **5,** 349–366 (1990).
12. Lunt, D. J., Valdes, P. J., Haywood, A. & Rutt, I. C. Closure of the Panama Seaway during the Pliocene: implications for climate and Northern Hemisphere glaciation. *Clim. Dyn.* **30,** 1–18 (2008).
13. Marincovich, L. Central American paleogeography controlled Pliocene Arctic Ocean molluscan migrations. *Geology* **28,** 551–554 (2000).
14. Sarnthein, M. *et al.* Mid-Pliocene shifts in ocean overturning circulation and the onset of Quaternary-style climates. *Clim. Past* **5,** 269–283 (2009).
15. Marincovich, L. & Gladenkov, A. Y. Evidence for an early opening of the Bering Strait. *Nature* **397,** 149–151 (1999).
16. Serreze, M. C. *et al.* The large-scale freshwater cycle of the Arctic. *J. Geophys. Res.* **111,** C11010 (2006).
17. Woodgate, R. A. & Aagaard, K. Revising the Bering Strait freshwater flux into the Arctic Ocean. *Geophys. Res. Lett.* **32,** L24603, doi:10.1029/2004GL021747 (2005).
18. Hu, A. *et al.* Response of thermohaline circulation to freshwater forcing under present-day and LGM conditions. *J. Clim.* **21,** 2239–2258 (2008).
19. Shaffer, G. & Bendtsen, J. Role of the Bering Strait in controlling North Atlantic ocean circulation and climate. *Nature* **367,** 354–357 (1994).
20. Hu, A. *et al.* Influence of Bering Strait flow and North Atlantic circulation on glacial sea-level changes. *Nat. Geosci.* **3,** 118–121 (2010).
21. Kipphut, G. W. Glacial meltwater input to the Alaska Coastal Current: evidence from oxygen isotope measurements. *J. Geophys. Res.* **95,** 5177–5181 (1990).
22. Rosell-Melé, A. Interhemispheric appraisal of the value of alkenone indices as temperature and salinity proxies in high-latitude locations. *Paleoceanography* **13,** 694–703 (1998).
23. Sicre, M.-A., Bard, E., Ezat, U. & Rostek, F. Alkenone distributions in the North Atlantic and Nordic sea surface waters. *Geochem. Geophys. Geosyst.* **3,** doi:10.1029/2001GC000159 (2002).
24. Onodera, J., Takahashi, K. & Nagatomo, R. Diatoms, silicoflagellates, and ebridians at Site U1341 on the western slope of Bowers Ridge, IODP Expedition 323. *Deep Sea Res. Pt II (in the press)* doi:10.1016/j.dsr2.2013.03.025.
25. Sancetta, C. Distribution of diatom species in surface sediments of the Bering and Okhotsk seas. *Micropaleontology* **28,** 221–257 (1982).
26. Sancetta, C. & Silvestri, S. Pliocene-Pleistocene evolution of the North Pacific Ocean-Atmosphere system, interpreted from fossil diatoms. *Paleoceanography* **1,** 163–180 (1986).
27. Takahashi, K., Ravelo, A. C. & Zarikian, C. A. IODP expedition 323—Pliocene and Pleistocene paleoceanographic changes in the Bering Sea. *Sci. Drill.* **11,** 4–13 (2011).
28. Takahashi, K., Ravelo, A. C. & Zarikian, C. A. *Proceedings of the Integrated Ocean Drilling Program* (Integrated Ocean Drilling Program Management International, Inc., 2011).
29. Martinez-Garcia, A., Rosell-Mele, A., McClymont, E. L., Gersonde, R. & Haug, G. H. Subpolar link to the emergence of the Modern Equatorial Pacific Cold Tongue. *Science* **328,** 1550–1553 (2010).
30. VanLaningham, S., Pisias, N. G., Duncan, R. A. & Clift, P. D. Glacial–interglacial sediment transport to the Meiji Drift, northwest Pacific Ocean: evidence for timing of Beringian outwashing. *Earth Planet. Sci. Lett.* **277,** 64–72 (2009).
31. Eberl, D. D. Quantitative mineralogy of the Yukon River system: changes with reach and season, and determining sediment provenance. *Am. Mineral.* **89,** 1784–1794 (2004).
32. Brabets, T. P. & Walvoord, M. A. Trends in streamflow in the Yukon River Basin from 1944 to 2005 and the influence of the Pacific Decadal Oscillation. *J. Hydrol.* **371,** 108–119 (2009).
33. Dornblaser, M. M. & Striegl, R. G. Suspended sediment and carbonate transport in the Yukon River Basin, Alaska: fluxes and potential future responses to climate change. *Water Resour. Res.* **45,** W06411 (2009).
34. Asahara, Y. *et al.* Provenance of terrigenous detritus of the surface sediments in the Bering and Chukchi Seas as derived from Sr and Nd isotopes: Implications for recent climate change in the Arctic regions. *Deep Sea Res. Pt II* **61–64,** 155–171 (2012).
35. Nagashima, K. *et al.* Contribution of detrital materials from the Yukon River to the continental shelf sediments of the Bering Sea based on the electron spin resonance signal intensity and crystallinity of quartz. *Deep Sea Res. Pt II* **61–64,** 145–154 (2012).

36. Wang, J., Hu, H., Mizobata, K. & Saitoh, S. Seasonal variations of sea ice and ocean circulation in the Bering Sea: a model-data fusion study. *J. Geophys. Res.* **114,** C02011 (2009).
37. Stabeno, P. J. & Reed, R. K. Circulation in the Bering Sea Basin observed by satellite-tracked drifters: 1986-1993. *J. Phys. Oceanogr.* **24,** 848–854 (1994).
38. GEOROC database. at http://georoc.mpch-mainz.gwdg.de/georoc/.
39. Aleinikoff, J. N., Lang Farmer, G., Rye, R. O. & Nokleberg, W. J. Isotopic evidence for the sources of Cretaceous and Tertiary granitic rocks, east-central Alaska: implications for the tectonic evolution of the Yukon–Tanana Terrane. *Can. J. Earth Sci.* **37,** 945–956 (2000).
40. Pettke, T., Halliday, A. N., Hall, C. M. & Rea, D. K. Dust production and deposition in Asia and the north Pacific Ocean over the past 12 Myr. *Earth Planet. Sci. Lett.* **178,** 397–413 (2000).
41. Jones, C. E., Halliday, A. N., Rea, D. K. & Owen, R. M. Eolian inputs of lead to the North Pacific. *Geochim. Cosmochim. Acta* **64,** 1405–1416 (2000).
42. Sun, J. & Zhu, X. Temporal variations in Pb isotopes and trace element concentrations within Chinese eolian deposits during the past 8 Ma: Implications for provenance change. *Earth Planet. Sci. Lett.* **290,** 438–447 (2010).
43. Uematsu, M. *et al.* Transport of mineral aerosol from Asia over the North Pacific Ocean. *J. Geophys. Res.* **88,** 5343–5352 (1983).
44. Sancetta, C., Heusser, L., Labeyrie, L., Naidu, A. S. & Robinson, S. W. Wisconsin—Holocene paleoenvironment of the Bering Sea: evidence from diatoms, pollen, oxygen isotopes and clay minerals. *Mar. Geol.* **62,** 55–68 (1984).
45. Katsuki, K. & Takahashi, K. Diatoms as paleoenvironmental proxies for seasonal productivity, sea-ice and surface circulation in the Bering Sea during the late Quaternary. *Deep Sea Res. Pt II* **52,** 2110–2130 (2005).
46. Horikawa, K., Asahara, Y., Yamamoto, K. & Okazaki, Y. Intermediate water formation in the Bering Sea during glacial periods: evidence from neodymium isotope ratios. *Geology* **38,** 435–438 (2010).
47. Salzmann, U. *et al.* Climate and environment of a Pliocene warm world. *Palaeogeogr. Palaeoclimatol. Palaeoecol.* **309,** 1–8 (2011).
48. Aiello, I. W. & Ravelo, A. C. Evolution of marine sedimentation in the Bering Sea since the Pliocene. *Geosphere* **8,** 1231–1253 (2012).
49. Lagoe, M. B., Eyles, C. H., Eyles, N. & Hale, C. Timing of late Cenozoic tidewater glaciation in the far North Pacific. *Geol. Soc. Am. Bull.* **105,** 1542–1560 (1993).
50. Motoi, T., Chan, W.-L., Minobe, S. & Sumata, H. North Pacific halocline and cold climate induced by Panamanian Gateway closure in a coupled ocean-atmosphere GCM. *Geophys. Res. Lett.* **32,** L10618 (2005).
51. Wang, L. Sea surface temperature history of the low latitude western Pacific during the last 5.3 million years. *Palaeogeogr. Palaeoclimatol. Palaeoecol.* **108,** 379–436 (1994).
52. LaRiviere, J. P. *et al.* Late Miocene decoupling of oceanic warmth and atmospheric carbon dioxide forcing. *Nature* **486,** 97–100 (2012).
53. Jiang, H., Mao, X., Xu, H., Thompson, J. & Ma, X. ~4 Ma coarsening of sediments from Baikal, Chinese Loess Plateau and South China Sea and implications for the onset of NH glaciation. *Palaeogeogr. Palaeoclimatol. Palaeoecol.* **298,** 201–209 (2010).
54. Rea, D. K. & Snoeckx, H. Sediment fluxes in the Gulf of Alaska: paleoceanographic record from Site 887 on the Patton-Murray seamount platform. *Proc. ODP Sci. Res.* **145,** 247–255 (1995).
55. Jansen, E., Fronval, T., Frack, F. & Channell, J. E. T. Pliocene-Pleistocene ice rafting history and cyclicity in the Nordic Seas during the last 3.5 Myr. *Paleoceanography* **15,** 709–721 (2000).
56. Lacasse, C. & van den Bogaard, P. Enhanced airborne dispersal of silicic tephras during the onset of Northern Hemisphere glaciations, from 6 to 0 Ma records of explosive volcanism and climate change in the subpolar North Atlantic. *Geology* **30,** 623–626 (2002).
57. Raymo, M. E., Grant, B., Horowitz, M. & Rau, G. H. Mid-Pliocene warmth: stronger greenhouse and stronger conveyor. *Mar. Micropaleontol.* **27,** 313–326 (1996).
58. März, C., Schnetger, B. & Brumsack, H. J. Nutrient leakage from the North Pacific to the Bering Sea (IODP Site U1341) following the onset of Northern Hemispheric Glaciation? *Paleoceanography* **28,** 68–78 (2013).
59. Kamenov, G., Perfit, M., Mueller, P. & Jonasson, I. Controls on magmatism in an island arc environment: study of lavas and sub-arc xenoliths from the Tabar–Lihir–Tanga–Feni island chain, Papua New Guinea. *Contrib. Mineral. Petrol.* **155,** 635–656 (2008).
60. Lawrence, K. T., Herbert, T. D., Brown, C. M., Raymo, M. E. & Haywood, A. M. High-amplitude variations in North Atlantic sea surface temperature during the early Pliocene warm period. *Paleoceanography* **24,** PA2218, doi:10.1029/2008pa001669 (2009).

Acknowledgements

This work was supported by JSPS KAKENHI Grant Numbers 08J06613, 23810009, 23241015 and 20244084. We thank JSPS Excellent Young Researchers Overseas Visit Program for their financial support, and G. Kamenov (University of Florida) and Y. Asahara (Nagoya University) for laboratory assistance. This research used samples and data provided by the Integrated Ocean Drilling Program (IODP) Expedition 323. We acknowledge the great support of the IODP Expedition 323 shipboard scientists and staff, especially the co-chief scientists K. Takahashi and A.C. Ravelo.

Author contributions

K.H. and E.E.M. conceived the study. K.H. analysed the Nd isotope. C.B. analysed the Pb isotope. O.S. and M.I. analysed the alkenone. J.O. analysed the diatom. T.S. and S.S. measured foraminiferal $\delta^{18}O$ and developed the $\delta^{18}O$ stratigraphy for core U1341B. All the authors contributed to the interpretation of the data and preparation of the manuscript.

Additional information

Competing financial interests: The authors declare no competing financial interests.

15

Radiocarbon constraints on the extent and evolution of the South Pacific glacial carbon pool

T.A. Ronge[1], R. Tiedemann[1], F. Lamy[1], P. Köhler[1], B.V. Alloway[2], R. De Pol-Holz[3], K. Pahnke[4], J. Southon[5] & L. Wacker[6]

During the last deglaciation, the opposing patterns of atmospheric CO_2 and radiocarbon activities ($\Delta^{14}C$) suggest the release of ^{14}C-depleted CO_2 from old carbon reservoirs. Although evidences point to the deep Pacific as a major reservoir of this ^{14}C-depleted carbon, its extent and evolution still need to be constrained. Here we use sediment cores retrieved along a South Pacific transect to reconstruct the spatio-temporal evolution of $\Delta^{14}C$ over the last 30,000 years. In ~2,500–3,600 m water depth, we find ^{14}C-depleted deep waters with a maximum glacial offset to atmospheric ^{14}C ($\Delta\Delta^{14}C = -1{,}000‰$). Using a box model, we test the hypothesis that these low values might have been caused by an interaction of aging and hydrothermal CO_2 influx. We observe a rejuvenation of circumpolar deep waters synchronous and potentially contributing to the initial deglacial rise in atmospheric CO_2. These findings constrain parts of the glacial carbon pool to the deep South Pacific.

[1] Alfred-Wegener-Institut Helmholtz-Zentrum für Polar- und Meeresforschung, Department for Marine Geology, PO Box 120161, Bremerhaven 27515, Germany. [2] School of Geography, Environment and Earth Sciences, Victoria University of Wellington, PO Box 600, 6012 Wellington, New Zealand. [3] GAIA-Antárctica Universidad de Magellanes, Department of Paleclimatology, Oceanography, Punta Arenas 01855, Chile. [4] Max Planck Research Group—Marine Isotope Geochemistry, Institute for Chemistry and Biology of the Marine Environment, Department of Marine Isotope Geochemistry, Carl von Ossietzky University, PO Box 2503, Oldenburg 26111, Germany. [5] School of Physical Science, Department of Earth Science, University of California, Irvine, California 92697-4675, USA. [6] Laboratory of Ion Beam Physics (HPK), Eidgenössische Technische Hochschule, Schafmattstrasse 20, Zürich 8093, Switzerland. Correspondence and requests for materials should be addressed to T.A.R. (email: Thomas.Ronge@awi.de).

The deep ocean contains the largest carbon reservoir within the global carbon cycle that might interact with the atmosphere on glacial/interglacial timescales. Therefore, the deglacial rise in atmospheric CO_2 by ~90 p.p.m.v. (ref. 1) was probably linked to significant modifications in oceanic circulation that resulted in increasing rates of CO_2 outgassing[2–4]. Thus, in order to sequester large amounts of atmospheric CO_2, the deep glacial ocean must have been effectively cut off from gas exchange with the atmosphere. Throughout a glacial period, the isolation of deep waters from the surface, and hence the atmosphere, leads to an accumulation of carbon (CO_2) and nutrients in the deep ocean, which is accompanied by a progressive depletion of radiocarbon (^{14}C). Consequently, the older a water mass gets, the more enriched in ^{14}C-depleted CO_2 it becomes.

So far, only isolated occurrences of old glacial water masses have been identified in the North and South Pacific, as well as in the South Atlantic, which suggests that the storage of CO_2 occurred in the deep glacial ocean[4–8] (Supplementary Fig. 1 and Supplementary Table 1). In particular, the overturning circulation of the Southern Ocean (SO), where nowadays ~65% of all deep waters make first contact with the atmosphere[9], controls the ventilation of the oceans interior. However, the surface residence time of upwelled waters before re-subduction is an important factor controlling the efficiency of air–sea gas exchange. Changes in the climate system of the SO, such as the intensification or weakening of stratification or westerly winds have the potential to significantly alter the oceanic uptake or release of CO_2 (refs 2, 3 and 10) and likewise the radiocarbon budget of deep waters. Therefore, the circum-Antarctic upwelling region is considered the most likely deglacial pathway of stored old carbon from the abyss to the atmosphere. In this oceanic window, carbon-rich deep waters like Pacific Deep Water (PDW) are mixed and upwelled and provide a major source for Antarctic Intermediate Water (AAIW), formed close to the Subantarctic Front (SAF)[11]. Hence, AAIW is able to propagate the circulation- and outgassing signals into the major ocean basins. In this context, numerous intermediate-water records have been analysed to track the timing and pathways of SO deep water upwelling[12–16]. The spatial and temporal dimension of the glacial reservoir itself as well as the pathway, magnitude and process of the deglacial CO_2 release remain elusive, although evidence for carbon storage in the deep glacial southwest Pacific is increasing[5,6].

Here, to better constrain the glacial carbon pool, its vertical extent and evolution, we use $\Delta^{14}C$-records from six sediment cores at the New Zealand Margin (NZM; Fig. 1a and Supplementary Fig. 2), covering the major South Pacific water masses AAIW and Upper Circumpolar Deep Water/Lower Circumpolar Deep Water between ~830 and ~4,300 m water depth (Fig. 1b). To assess the lateral extent of the glacial carbon pool, we have additionally analysed an open-ocean sediment core from the East Pacific Rise (EPR; PS75/059-2; 3,613 m) located more than 4,000 km east of the NZM (Fig. 1a). Our NZM depth transect is well suited for the analysis of SO water mass ventilation as we can record ^{14}C-depleted deep waters on their way to the upwelling region further south, as well as recently subducted intermediate waters moving towards the north (Fig. 1b).

We show that throughout the water column, a wide range of radiocarbon activities ($\Delta^{14}C$) indicates a highly stratified South Pacific during the last glacial. This stratification implicates pronounced sequestration of CO_2 in circumpolar deep waters below a water depth of ~2,000 m. Building on the hypothesis of increased glacial outgassing of volcanic CO_2 along mid-ocean ridges (MORs)[17–19], we use a simple box model to highlight that the most extreme ^{14}C-depletion between ~2,500 and ~3,600 m water depth might be explained by a combination of aged ^{14}C-depleted waters and the additional admixture of ^{14}C-dead hydrothermal CO_2. At the end of the glacial period, our deep water $\Delta^{14}C$-values increase throughout the water column in unison to rising atmospheric CO_2-values[1]. On the basis of these patterns, we conclude that the deep South Pacific was an important contributor to the deglacial rise in atmospheric CO_2.

Results

Radiocarbon. To assess both glacial and deglacial ventilation changes, we used paired samples of *Globigerina bulloides* and mixed benthic foraminifers from seven new sediment cores, located south of the present Subtropical Front (STF) (Fig. 1a). The core locations have sedimentation rates between 2.5 cm per kyr and 22 cm per kyr (Supplementary Table 2). Before the Last Glacial Maximum (LGM), ~29 cal. ka, all deep-water masses (below ~2,000 m) show $\Delta^{14}C$-values ranging from +200‰ to −100‰ (Fig. 2). At the same time, AAIW $\Delta^{14}C$ is clearly elevated with values of ~360‰. The most obvious feature of our reconstructed $\Delta^{14}C$-values over the LGM is the large glacial range of $\Delta^{14}C$ between ~830 and ~4,300 m (400‰ to −550‰; Fig. 2). About 21 cal. ka, deep-water $\Delta^{14}C$ at 4,300 and 2,066 m increases, followed by increasing radiocarbon values at 2,500 m at the onset of the last deglaciation. Parallel to increasing deep-water radiocarbon concentrations, AAIW-$\Delta^{14}C$ slightly decreases. At ~14.7 cal. ka, the $\Delta^{14}C$-records of all water depths converge and continue to evolve parallel to each other throughout the deglaciation and into the Holocene (Fig. 2). In the discussion, we used $\Delta\Delta^{14}C$-records for our interpretations. These records represent the $\Delta^{14}C$ offset of our data to the $\Delta^{14}C$-value of the past atmosphere[20]. We also calculated the $\Delta\Delta^{14}C_{adj}$ according to Cook and Keigwin[21] (Supplementary Fig. 3). This method corrects the initial $\Delta^{14}C$ values to the modern, pre-industrial ^{14}C profile (ref. 21). As the trend in our data remains the same, regardless of the method used, we use $\Delta\Delta^{14}C$ for our discussion, in order to improve the comparability to other studies.

Box modelling. We used a 1-box model to investigate the hypothesis that hydrothermal CO_2-fluxes might have contributed to our reconstructed maximum depletion in $\Delta^{14}C$ in the glacial deep ocean (Methods). We simulated the single effect of hydrothermal CO_2 inflow on oceanic $\Delta^{14}C$, as well as two sensitivity runs in which hydrothermal CO_2 inflow is combined with either carbonate compensation, leading to sediment dissolution[22] or with CO_2 sequestration, potentially connected with deep-ocean volcanism[23]. Our model simulates for the single effect of different CO_2 flux rates F (in µmol per kg per year; Supplementary Fig. 4) a drop in $\Delta^{14}C$ by −240‰ (F = 0.3), −380‰ (F = 0.6), −480‰ (F = 0.9) and −550‰ (F = 1.2). Only if the response of the marine carbonate system to this CO_2 flux (by carbonate compensation and the dissolution of sediments) is considered[22] (Supplementary Fig. 4), we calculate $\Delta^{14}C$ amplitudes in agreement with our maximally depleted data (approximately −500‰ to −600‰) for a hydrothermal flux of 0.6 µmol kg per year or larger. However, as hot rocks interact with seawater, MOR volcanism is also discussed as a potential CO_2 sink[23]. If we implement this process of similar size of the hydrothermal CO_2 flux (no net oceanic carbon change and therefore no carbonate compensation) we need a hydrothermal CO_2 flux F of 1.2 µmol kg per year to meet the maximum $\Delta^{14}C$ depletion of ~ −500‰, as observed in our data (Fig. 3a). To convert the CO_2 fluxes to gross carbon fluxes (PgC per year) we estimated the minimum area covered by our sediment cores as a ~1,000-m-thick water mass (2,500–3,600 m as covered

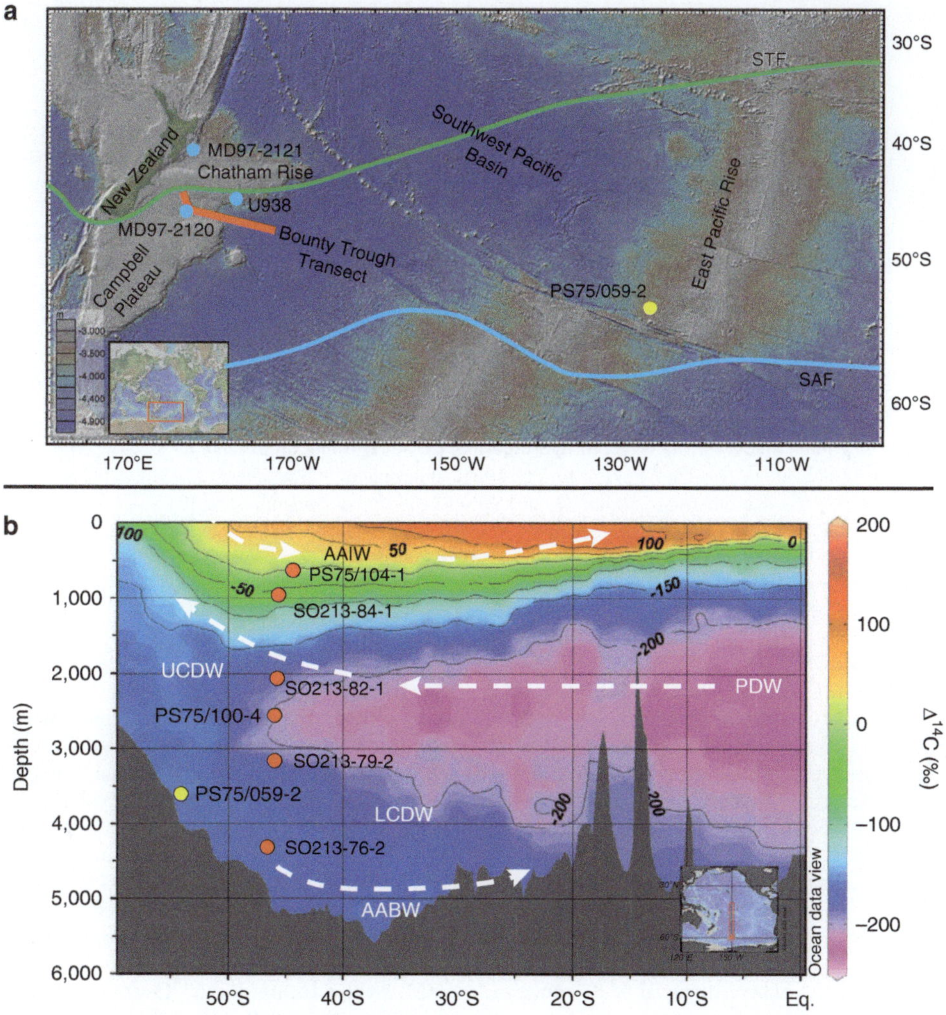

Figure 1 | Overview of the NZM showing core locations and water masses. (**a**) Map of the Southwest Pacific. The red bar represents the lateral area covered by our sediment core transect at the NZM. Yellow circle—position of our sediment core from the EPR. Blue circles—previous studies. U938 (ref. 5), MD97-2120 (ref. 15), MD97-2121 (ref. 6). Green line—Subtropical Front. Blue line—Subantarctic Front[66]. Map created using GeoMapApp. (**b**) Water mass section of modern South Pacific Δ^{14}C-concentrations. Sediment cores (red and yellow dots) projected into WOCE line P16 (ref. 67). AABW, Antarctic Bottom Water; WOCE, World Ocean Circulation Experiment. Section generated using ODV 4.7.2 (ref. 68).

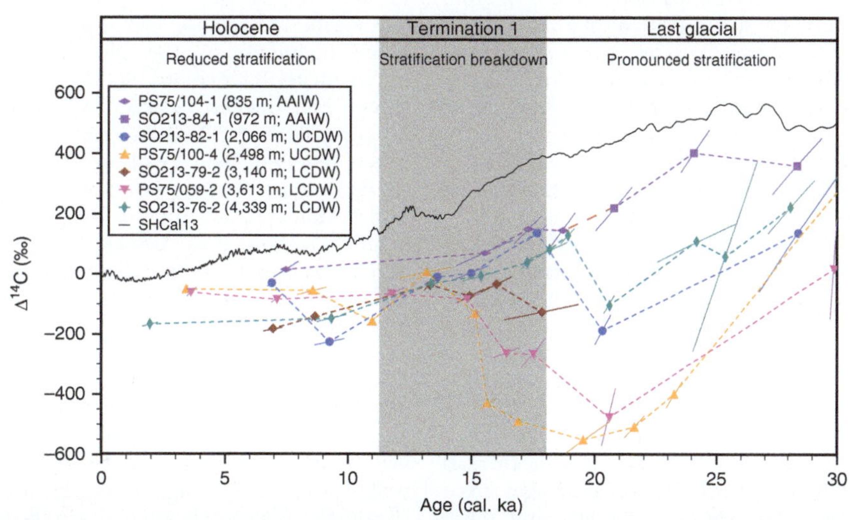

Figure 2 | Δ^{14}C changes of the major South Pacific water masses. The large glacial range in Δ^{14}C indicates a pronounced water mass stratification, followed by a progressive stratification breakdown during Termination 1 (grey shaded area). The broken red line indicates the area, where we spliced PS75/104-1 and SO213-84-1.

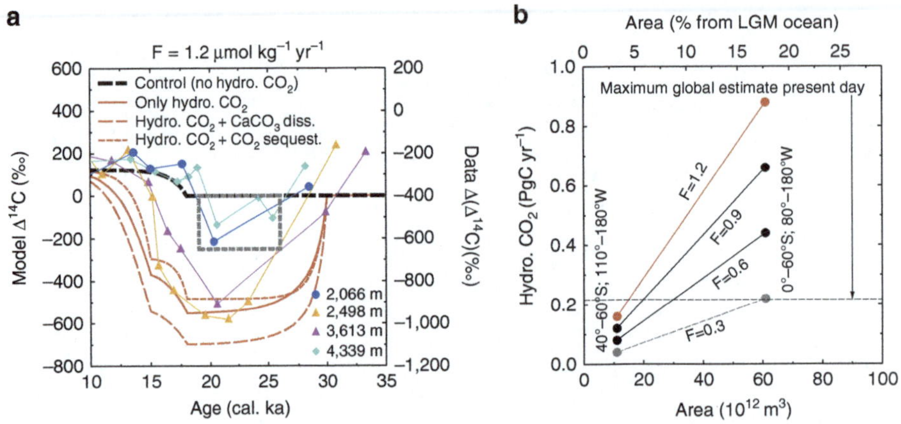

Figure 3 | Simulation results of a 1-box model for the effect of hydrothermal CO_2-outgassing on oceanic $\Delta^{14}C$. (**a**) Comparison of model-based $\Delta^{14}C$ with data-based $\Delta\Delta^{14}C$. SO213-82-1 (2,066 m; blue line); PS75/100-2 (2,498 m; orange line); PS75/059-2 (3,613 m; pink line); and SO213-76-2 (4,339 m; green line). The impact of our best-guess hydrothermal CO_2 flux (F) of 1.2 μmol kg^{-1} yr^{-1} between 30 and 15 cal. ka (red solid line) is compared with a control run (black broken line). In the control run, we only decreased the turnover time at 18 cal. ka according to Skinner *et al.*[6] from 2,700 to 1,500 years. In two sensitivity runs, we estimated the influence of $CaCO_3$ dissolution or CO_2 sequestration (red broken lines). The grey box indicates the $\Delta\Delta^{14}C$-area covered by previous Southern Ocean studies[4,6,8]. (**b**) Upscaling of our localized results for different hydrothermal CO_2 fluxes (F) to regional carbon fluxes as a function of area. Present day maximum global estimate of hydrothermal CO_2 fluxes (0.22 PgC yr^{-1}) (ref. 24) indicated by the black broken line.

by PS75/100-4 and PS75/059-2) ranging from 40°S to 60°S and 110°W to 180°W. The fluxes necessary to influence such a water mass would lead to a hydrothermal injection of CO_2 of 0.08–0.16 PgC per year (modern global flux is estimated to up to ~0.22 PgC per year)[24]. Upscaling to a larger water mass, spanning most of the South Pacific (1,000 m; 0°–60°S; 80°–180°W) would imply that the hydrothermal CO_2 flux might be as large as 0.44–0.88 PgC per year (Fig. 3b).

Discussion

Throughout the water column, the observed glacial $\Delta^{14}C$-range in our cores exceeds the modern and Holocene values by a factor of ~5 and indicates strong age differences and therefore enhanced stratification of the intermediate and deep glacial South Pacific. Along its pathway in the global thermohaline circulation PDW is fed into circumpolar waters and constitutes today's oldest water mass. The ^{14}C-depleted PDW presently extends to 2,000–2,500 m north of the Chatham Rise close to New Zealand[25] (Fig. 1b). From our transect, we are able to show that during the LGM, significantly ^{14}C-depleted and aged water masses occupied depths between ~2,000 and ~4,300 m in the Southern Westerly (SW) Pacific (Fig. 2). Our data locate the core of the ^{14}C-depleted water mass at the NZM in a of ~2,500 m (PS75/100-4; modern depth of Upper Circumpolar Deep Water; Fig. 4), yielding a maximum deep water to atmosphere offset in radiocarbon activities ($\Delta\Delta^{14}C$) of approximately −1,000‰. Analysing $\Delta\Delta^{14}C$ corrects for any impacts of changes in ^{14}C production[26] as well as for variable ocean-atmosphere exchange rates. A corresponding apparent ventilation age, based on benthic minus reservoir-corrected planktic ^{14}C ages would equate to ~8,000 years (Supplementary Fig. 3d). A similar glacial $\Delta\Delta^{14}C$ depletion of about −870‰ was reported from sediment core U938, which was recovered at the NZM in a water depth of 2,700 m (ref. 5) (Figs 1a and 4). We hypothesize that these extremely low $\Delta\Delta^{14}C$-values might be the result of the admixture of ^{14}C-dead hydrothermal CO_2 into a water mass with an initial high ventilation age, which was estimated to at least 2,700 years[6]. The upper and lower boundary of this old water mass are marked by higher $\Delta\Delta^{14}C$-values of −550‰ to −600‰ indicating a highly stratified water column (Fig. 4b). Similar $\Delta\Delta^{14}C$-values were reported at the NZM north of Chatham Rise at 2,314 m (ref. 6). This confines the most radiocarbon-depleted waters to a depth below ~2,300 m. The observed trend of ^{14}C between 830 and 4,300 m parallels the highest glacial nutrient concentrations off New Zealand, between 2,000 and 3,000 m (ref. 27) (Fig. 5), likewise indicative for the presence of aged, nutrient rich (low $\delta^{13}C$) and radiocarbon-depleted waters. Yet, the $\delta^{13}C$ reconstructions might yield a certain bias, as endobenthic (*Uvigerina*) instead epibenthic (*Cibicidoides*) foraminifera were used[27].

We traced the ^{14}C-depleted glacial carbon reservoir off New Zealand to the central South Pacific (EPR) 4,000 km east of the NZM. At this location, $\Delta\Delta^{14}C$-values are as negative as −900‰ at 3,600 m (PS75/059-2; Figs 1a and 4). Hence, we are confident that this water mass was not only restricted to the NZM but seems to have occupied large parts of the South Pacific. Further off, in the Drake Passage and the South Atlantic, glacial water masses have been identified in CDWs with $\Delta\Delta^{14}C$-values as low as −330‰ (ref. 8) and −540‰ (ref. 6), respectively (Fig. 4). In their timing and amplitudes, these records are similar to our Pacific radiocarbon signature characterizing the upper and lower boundary of the old carbon pool (Fig. 4b). Our intermediate-water record (SO213-84-1; modern depth of AAIW) shows the highest glacial $\Delta\Delta^{14}C$-values of our transect (approximately −90‰). We suggest that the ^{14}C-depleted deep waters represent the very old return flow from the North Pacific (PDW), similar to the modern circulation pattern (Fig. 1b). The distribution of radiocarbon in our reconstruction might indicate a floating carbon pool instead of a stagnant bottom layer. Several records from the North Pacific might corroborate this assumption. In the Gulf of Alaska[28] (MD02-2489) and off Kamchatka[29] (MD01-2416), as well, the glacial mid-depth water mass (Kamchatka) shows a considerable higher benthic to planktic ^{14}C offset than the deeper water mass off Alaska. Additional data from the northwest Pacific suggest the lowest glacial ^{14}C values in a water depth of ~2,300 m with better ventilated waters above and below[21]. In the Atlantic Ocean as well, Ferrari *et al.*[30] and Burke *et al.*[31] observed a mid-depth (floating) radiocarbon anomaly. A floating carbon pool might furthermore explain why no sign of old carbon was found in the deep equatorial Pacific below 4,000 m (ref. 32). Therefore, this record might have 'missed' the old, mid-depth carbon pool above.

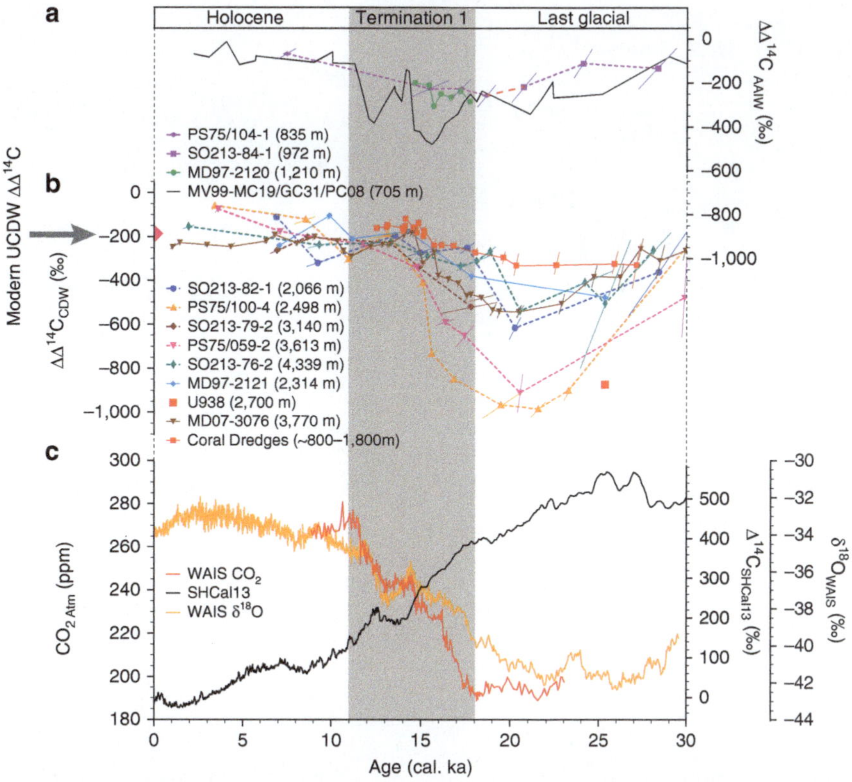

Figure 4 | Comparison of oceanic and atmospheric proxy records. $\Delta\Delta^{14}C$-values (ocean-atmosphere) of (**a**) the intermediate Pacific region; PS75/104-1 and SO213-84-1 (this study); MD97-2120 (ref. 15) (green line: Bounty Trough); MV99-MC19/GC31/PC08 (ref. 12) (black line: Northeast Pacific); (**b**) CDW $\Delta\Delta^{14}C$; PS75 and SO213 records (this study); MD97-2121 (ref. 6) (light blue line: north of Chatham Rise); U938[5] (red square: Bounty Trough); Coral dredges[8] (red line: Drake Passage); MD07-3076 (ref. 4 (brown line: South Atlantic); and Modern SW-Pacific UCDW $\Delta\Delta^{14}C$ (ref. 67) (pink triangle). (**c**) Atmospheric CO_2 concentrations[1] (red line); WAIS $\delta^{18}O$ record[69] (orange line) and atmospheric $\Delta^{14}C$-values[20] (black line). UCDW, Upper Circumpolar Deep Water.

Any explanation for the pronounced glacial radiocarbon-depletion of the deep SO has to involve a limited ocean–atmosphere exchange due to strengthened ocean stratification under glacial boundary conditions[3] (Fig. 6a). Northward-expanded Antarctic sea ice and SW Winds[33] contributed to reduced air–sea gas exchange and upwelling of deep-waters[30]. Surface freshening by melting sea-ice in the source regions of intermediate waters[34] and enhanced formation of highly saline Antarctic Bottom Water[35] may have set a density structure that led to reduced mixing and the encasement of old PDW (Fig. 6a). In addition, the shoaling of North Atlantic Deep Water[30] might have reduced the contribution of freshly ventilated waters into South Pacific CDW below ~2,000 m. These interacting key processes may have significantly contributed to the low radiocarbon values and are consistent with an enhanced glacial storage of carbon in the deep ocean.

Old water masses of 5,000–8,000 years are expected to be strongly oxygen depleted[36]. According to Sarnthein *et al.*[7], water masses with $\Delta^{14}C$ values lower than − 350‰ would be completely anoxic. However, pronounced anoxia have not been documented in the deep South Pacific between 2,500 and 3,600 m. Therefore, an admixture of ^{14}C-dead carbon via submarine tectonic activity along MOR[18] into a an old water mass in the deep South Pacific might have contributed to the extremely low radiocarbon values of the water mass at ~2,500–3,600 m. During the LGM, sea floor eruption rates along tectonically active plate boundaries may have intensified due to the lower glacial sea level[17–19]. This process might have released significant amounts of ^{14}C-dead CO_2 into the water column. Using a simple 1-box tested our hypothesis and calculated if

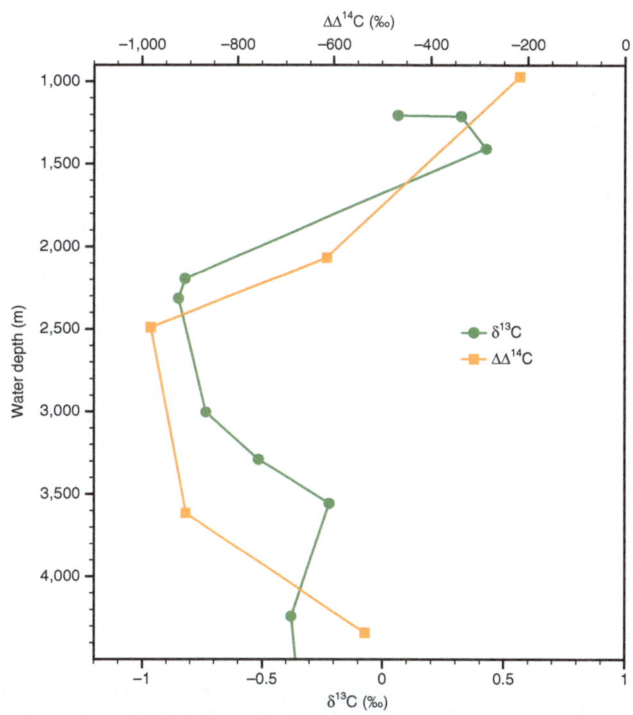

Figure 5 | Vertical distribution of carbon isotopes off New Zealand throughout the LGM. Orange line—$\Delta\Delta^{14}C$ (this study). Green line $\delta^{13}C_{Uvigerina}$[27].

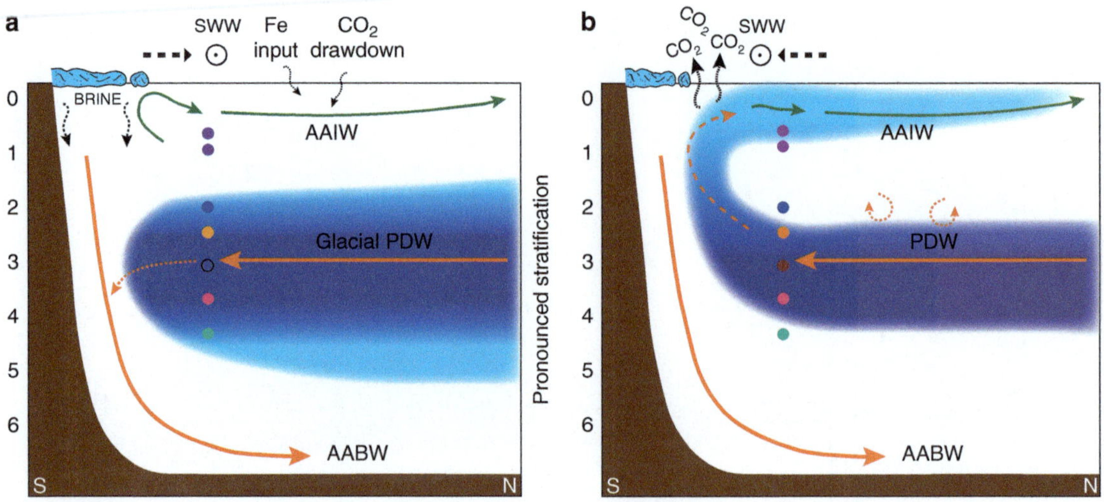

Figure 6 | Schematic representation of South Pacific overturning circulation. (**a**) Glacial pattern: northernmost extent of sea ice and SWW. Increased AABW-salinity by brine rejection favours stratification. Increased dust input promotes primary production and drawdown of CO_2. (**b**) Deglacial pattern: upwelling induced by southward shift of Antarctic sea ice and SWW. The erosion of the deep-water carbon pool releases ^{14}C-depleted CO_2 towards the atmosphere. Following air-sea gas exchange, the outgassing signal is incorporated into newly formed AAIW (light blue shading). Blue shading: poorly ventilated old and CO_2-rich waters; Darkest shading 2,500–3,600 m: water level influenced by hydrothermal CO_2. Green arrows: intermediate water; orange arrows: deep-water; light-blue areas: sea ice; SWW: Southern Westerly Winds; coloured circles: sediment cores (colour coding according to Fig. 2); black circle: SO213-79-2—no glacial data; and circular arrows: diffusional and diapycnal mixing.

the injection of hydrothermal CO_2 into the deep Pacific has the potential to amplify the $\Delta\Delta^{14}C$ minimum throughout the LGM (Fig. 3). To overcome the influence of the variable atmospheric ^{14}C-levels[26,37], we compared our simulated $\Delta^{14}C$ to our reconstructed $\Delta\Delta^{14}C$ values (deep ocean-to-atmosphere offset). The probably time-delayed response of submarine volcanism to changes in sea level complicate our flux calculations[19]. A crucial prerequisite for our hypothesis of the admixture of hydrothermal CO_2 is the presence of an already aged water mass with high nutrient concentrations and low $\Delta^{14}C$ levels (Fig. 7). According to the record of MD97-2121 (ref. 6) (2,314 m), the glacial ventilation age off New Zealand is at least 2,700 years. However, radiocarbon values might have been even lower as the MD97-2121 record lacks any data points between ~25 kyr and ~18 cal. ka (Fig. 4b). Our $\Delta\Delta^{14}C$ record of SO213-82-1 (2,066 m; Fig. 4b) is −600‰ at ~20kyr, ~100‰ lower than the minimum observed in MD97-2121 ~25 cal. ka, potentially indicating even higher turnover times. When we combine the radioactive decay, caused by an estimated water mass age of 2,700 years for the time of 35–18 cal. ka, with submarine ^{14}C-free volcanic CO_2 influx, our model calculates a decrease in $\Delta^{14}C$ for the corresponding water mass by additional −500‰ to −600‰ (Fig. 3a). This hydrothermal CO_2 outgassing (potentially accompanied by carbonate compensation and/or CO_2 sequestration) would lead to a maximum atmosphere-to-deep ocean offset of −800‰ to −1,000‰ $\Delta\Delta^{14}C$ (Supplementary Fig. 4e–g), comparable to the maximum depletion observed in PS75/059-2 and PS75/100-4. Today, in a water depth between ~2,500 and ~3,500 m, pronounced volcanic outgassing occurs along the southern EPR[38,39]. The resulting hydrothermal plume spreads towards the west and can be traced by the ^3He-signal in the broader western Pacific (Fig. 8)[40] and off northern New Zealand, right in the water depth under debate of ~2,500 m (ref. 41). Therefore, we argue that increased glacial outgassing of ^{14}C-dead volcanic CO_2 into a stratified ocean has the potential to significantly lower the $\Delta^{14}C$-content of an old (at least 2,700 years) water mass. The prominent Chatham Rise (Fig. 1a) might have acted as a physical barrier, blocking MD97-2121 (ref. 6) (~2,300 m) from the volcanic plume. As MD97-2121 lacks data for most of the last glacial (~18–25 cal. ka), we cannot fully exclude that this core might have been affected by volcanic CO_2 to some extent. Nevertheless, the ^{14}C-data of MD97-2121 are already significantly higher at ~18 cal. ka compared to the values of PS75/100-4. Therefore, we argue that the influence (if any) of hydrothermal activity must have been lower to the north of the Chatham Rise and/or at 2,300 m water depth. Despite the in detail unknown processes accompanying such a hydrothermal carbon flux, its admixture might add additional carbon to the ocean–atmosphere–biosphere system (0.08–0.16 PgC per year; Fig. 3b). However, the net carbon injection depends in detail on the strength of the additional processes carbonate compensation and CO_2 sequestration and might also be zero. Once the glacial processes, favouring stratification, are reversed, any net injected carbon might eventually be released to the atmosphere along with the carbon already stored within the deep ocean (Fig. 5b). A further quantification of the net carbon injection and its contribution to atmospheric CO_2 is not yet possible, since future investigations with process-based models are necessary. Furthermore, as the distribution of MOR is inhomogeneous in the world ocean, it is difficult to compare our local results to global CO_2 flux estimates[24]. In Fig. 3b, we illustrate that the water mass affected by hydrothermal CO_2 might span an area of between 3 and 17% of the global glacial ocean. While the results for the minimum area, covered by our sediment cores, are below the maximum estimate of present day global estimate of hydrothermal outgassing, the results for an area representative for most of the South Pacific are a factor of 2–4 times higher (Fig. 3b). This suggests that if the admixture of hydrothermal CO_2 is the process that can explain the minimum $\Delta\Delta^{14}C$ values recorded in the mid-depth South Pacific (PS75/100-4; PS75/059-2; U938 (ref. 5)) the global CO_2-fluxes from MOR throughout the LGM might have been much larger than today. However, if the initial water mass was older than the 2,700 years assumed in our model, the resulting fluxes might also have been smaller than stated here. Although mantle-CO_2 is depleted in both, $\Delta^{14}C$ and also in $\delta^{13}C$ (−5 ± 3‰ (refs 42 and 43)), its

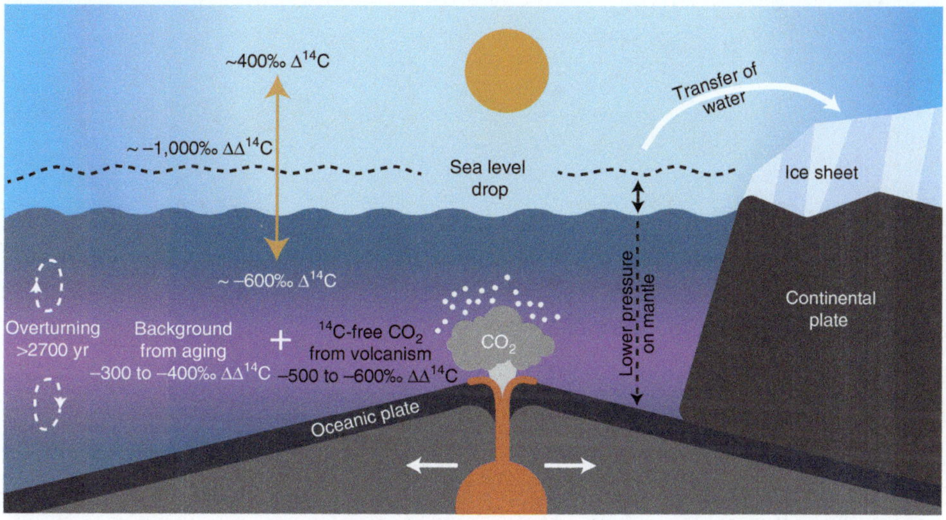

Figure 7 | Processes linking the glacial release of hydrothermal CO_2 and water mass $\Delta^{14}C$. The drop in global sea level triggers increased volcanic activity at MORs. The plume of ^{14}C-dead hydrothermal CO_2 is mixed into an aged water mass, in which the combined effects of surface reservoir age and deep-ocean turnover time, of ~2,700 years, led already to a background ocean-to-atmosphere offset in $\Delta\Delta^{14}C$ of −300‰ to −400‰. This admixture of hydrothermal CO_2 further lowers the water masses $\Delta\Delta^{14}C$ by another −500‰ to −600‰, yielding a total $\Delta\Delta^{14}C$ of about −1,000‰ (purple layer). Modified after Hand[70]. Reprinted with permission from AAAS.

influence might be stronger on radiocarbon. As ^{14}C is by far less common in the ocean than ^{13}C, it can be diluted (lowered) more easily than its non-radiogenic counterpart.

We assume that the previously outlined glacial/interglacial changes in the SO climate system (position of sea ice and westerlies; changes in water mass densities; changes in upwelling and circulation) are the major factors influencing the spatio-temporal evolution of the oceanic carbon pool. However, the extreme minima in $\Delta\Delta^{14}C$ between 2,500 and 3,600 m water depth in the South Pacific are most plausibly explained by the hypothesized admixture of hydrothermal CO_2 into an old and already ^{14}C-depleted water mass. Admittedly, this process would complicate the use of ^{14}C as a ventilation proxy for the water masses affected by hydrothermal CO_2. However, as the presence of an already existing ^{14}C-depleted glacial water mass is a crucial prerequisite for our model, our theory does not interfere with the concept of stratification, decreased ventilation and the presence of a glacial oceanic carbon pool.

In the Pacific, other processes affect the radiocarbon inventory of water masses as well. Stott and Timmermann[44] already suggested the release of ^{14}C-depleted carbon from gas clathrates. As the stability of such clathrates is located in shallow waters ~400 m (ref. 44), this process is not applicable for our mid-depth anomaly below ~2,500 m. Therefore, we reject the possibility of any large influence of ^{14}C-depleted CO_2 and CH_4 clathrates.

At the end of the LGM and during the transition into the Holocene (~20–11.5 cal. ka), converging $\Delta\Delta^{14}C$-values argue for a progressive destratification (Fig. 4). During this interval, the deep-water-to-atmosphere offset in radiocarbon between ~2,000 and ~4,300 m decreases significantly (Fig. 4b). Although the resolution of our deep-water sediment cores is rather low, their $\Delta\Delta^{14}C$-values increase within error synchronous to the rise in atmospheric CO_2 rise (Fig. 4).

During Termination 1, when the most radiocarbon-depleted deep waters rejuvenate, no pronounced depletion in AAIW $\Delta\Delta^{14}C$ is recorded (Fig. 4a). However, the intermediate-water $\Delta\Delta^{14}C$-values remain low from ~18–15 cal. ka. The decrease in the deep water-to-atmosphere $\Delta\Delta^{14}C$-offset and the abrupt drop in $\delta^{13}C$ of atmospheric CO_2 (ref. 45) suggests increased air–sea ... release of upwelled old CO_2

(Fig. 5b). The intermediate-waters from the NZM (this study) and the Chile Margin[14] significantly deviate from the (sub)tropical East Pacific, which shows two prominent drops in $\Delta\Delta^{14}C$ at the intermediate-water level during Termination 1 (refs 12, 13) (Fig. 4a). Therefore, it seems unlikely that southern sourced AAIW represents the source for the deglacial radiocarbon signals in the (sub)tropical East Pacific.

As we mentioned before, because of the close proximity, similar reservoir ages for all sediment cores are a requirement for our interpretations. However, the Holocene reservoir ages of two of our cores differ by ~1,000 years (Supplementary Fig. 5). Although this offset does not change the overall story, it prevents us from discussing the data of this time interval in more detail.

Our reconstructions provide new insights into the evolution and dynamic of the marine carbon inventory, its aging and the process of CO_2 release to the atmosphere. Growing evidence throughout large parts of the glacial South Pacific (this study; Skinner *et al.*[6]), the North Pacific[21,28,29], the Drake Passage[8] and the South Atlantic[4] suggest the existence of a floating body of very old, ^{14}C-depleted water between 2,000 and 4,300 m water depth, particularly in the tectonically active Pacific Ocean, where the glacial admixture of volcanic CO_2 might have influenced the ^{14}C-signature of this carbon pool from ~2,500 to ~3,600 m water depth. The effect of such volcanic CO_2 flux on the global carbon cycle as a whole and on atmospheric CO_2 in particular needs to be assessed in future studies, using more sophisticated models. During Termination 1, the atmosphere to deep-water $\Delta\Delta^{14}C$ of the carbon reservoir was reduced from about −1,000‰ to about −200‰ (Fig. 4b), indicating the erosion of the glacial carbon pool. This erosion is in accordance with the concept of a deglacial breakdown of SO stratification[3,45,46] and intensified deglacial wind-driven SO upwelling[2]. These processes ultimately culminated in the release of ^{14}C-depleted CO_2 from the deep ocean reservoir to the atmosphere (Fig. 5b), although in detail, the interaction of processes affecting the biological and physical pumps—and thus atmospheric CO_2—was probably more complex[47].

Methods

Sediment core details and sample treatment. The water mass transect that forms the backbone of this study consists of seven sediment cores retrieved during

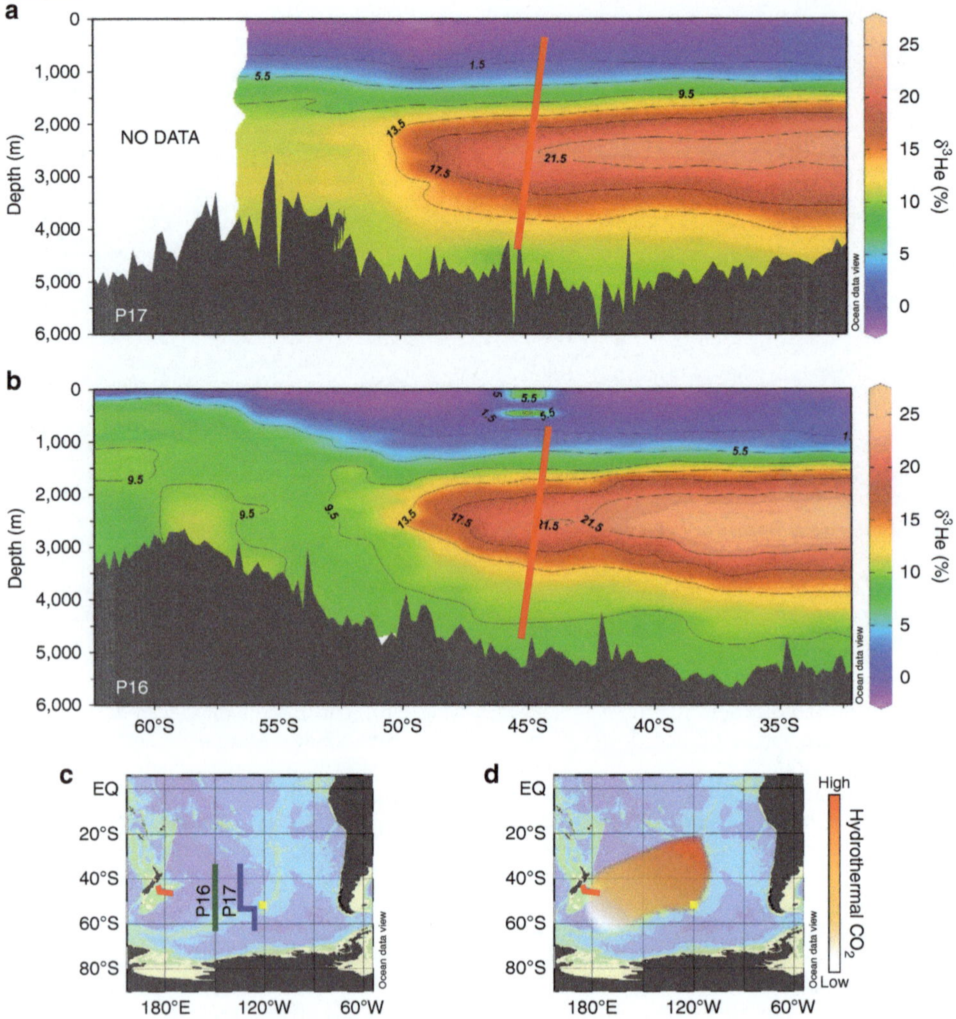

Figure 8 | Dispersal of hydrothermal ^{3}He in the Southwest Pacific. The modern hydrothermal ^{3}He plume, emanating at the southern EPR in a water depth of ~2,500 m (refs 38 and 39), can be traced throughout the southern Pacific towards New Zealand[41]. ^{3}He distribution along (**a**) WOCE line P17 (ref. 67) (~135° W) and (**b**) WOCE line P16 (ref. 67) (150° W). (**c**) Locations of ^{3}He sections P16 and P17. (**d**) Hypothesized dispersal of the glacial hydrothermal plume emanating from the EPR. Core locations and depths indicated by red bars and yellow squares. Panels generated using ODV 4.7.2 (ref. 68). WOCE, World Ocean Circulation Experiment.

the ANTXXVI/2 and SO213/2 cruises from the Bounty Trough off New Zealand (NZM) and from the EPR (Supplementary Figs 1 and 2). Collectively, these sediment cores record all water masses between 835 and 4,339 m, thus the modern water depths of the AAIW down to the Lower Circumpolar Deep Water with an imprint of Antarctic Bottom Water[27]. Positions, water depth, modern water masses and average sedimentation rates are reported in Supplementary Table 2.

An advantage of the NZM core transect is that it was not affected by potential glacial/interglacial shifts of the STF or the SAF[48,49]. The STF is bathymetrically fixed by the Chatham Rise (Supplementary Fig. 2), while the SAF is topographically steered by the submerged Campbell-Bounty Plateau[48,49]. As all sediment cores were located south of the STF and because of their close proximity to each other, we assume that changes in surface reservoir ages would affect all locations in a similar way. An exception is core PS75/059-2, which was retrieved ~4,200 km east of the Bounty Trough, at the western flank of the EPR (Supplementary Fig. 1) ~2°N of the SAF.

All sediment cores were split to form working and archival halves. The working half was sampled at 2 cm intervals. Depending on their water content, all samples were freeze dried for 2–3 days. Subsequently, the samples were wet sieved, using a 63 µm mesh sieve and dried at 50 °C for 2 days. As a last step, all samples were subdivided into the size fractions >400 µm, 315–400 µm, 250–315 µm, 125–250 µm and <125 µm. Benthic and planktic foraminifera were picked from 250 to 315 µm and 315 to 400 µm size fractions, taking great care to group similar sized individuals into samples for isotope analyses.

For the analysis of AAIW ventilation, we spliced sediment cores PS75/104-1 (835 m) and SO213-84-1 (972 m). We were forced, to combine both records as PS75/104-1 did not yield a sufficient amount of benthic foraminiferal fauna below a core depth of ~100 cm (LGM and older), while SO213-84-1 was significantly disturbed above ~50 cm core depth (Termination 1 and younger).

Radiocarbon measurements. For the reconstruction of radiocarbon activities, corresponding pairs of planktic (monospecific *G. bulloides*) and benthic (mix of *Cibicidoides wuellerstorfi* and *Uvigerina peregrina*) foraminifera were picked. To minimize the effect of contamination on our analyses, we paid special attention not to pick any broken, discoloured or filled tests. Radiocarbon measurements were performed at the National Ocean Science Accelerator Mass Spectrometer (NOSAMS) facility in Woods Hole, USA, the W. M. Keck Carbon Cycle AMS Laboratory at the University of California in Irvine, USA and at the Laboratory for Ion Beam Physics at the Eidgenössische Technische Hochschule in Zurich, Switzerland.

We calculated the difference in benthic and reservoir-corrected planktic (B-P) ^{14}C-ages to reconstruct the apparent ventilation ages of different water masses and compared our reconstructions with the method proposed by Cook and Keigwin[21] (Supplementary Fig. 4).

The equation of Adkins and Boyle[50] was used to determine the initial (paleo) radiocarbon activity (Δ^{14}C) of the benthic samples.

The difference of deep-water Δ^{14}C to the contemporaneous past atmospheric Δ^{14}C (called $\Delta\Delta^{14}$C) is the offset of our data from the IntCal13 reference curve[20].

Despite a potential bias by uncertainties in our age models and the IntCal13 (ref. 20) curve, we show in Supplementary Fig. 6 that the trend in $\Delta\Delta^{14}$C remains the same, regardless of the method used.

The radiocarbon dates for all cores are stored in the PANGAEA-database. Ventilation age errors were calculated from combined errors in ^{14}C ages and calibrated calendar ages. The error in $\Delta^{14}C$ was calculated from ^{14}C and calibrated age errors. All ^{14}C-data can be found in Supplementary Table 3 and on the PANGAEA-database.

Age control. For all sediment cores, an initial radiocarbon chronology was obtained from planktic ^{14}C-datings. Planktic radiocarbon ages were calibrated to calendar ages, using the calibration software Calib 7.0 (refs 51 and 52) with the embedded SHCal13 calibration curve[20]. To account for surface reservoir effects, we corrected all ^{14}C-ages according to reservoir age estimates by Skinner *et al.*[6]. However, to allow the use of ^{14}C as an unbiased proxy for deep-water ventilation, we fine-tuned our records to the nearby reference core MD97–2120 (ref. 53) (Supplementary Figs 7–9). We chose MD97–2120 (1,210 m water depth) as a reference core because of its position close to our sediments cores in the Bounty Trough. The original stratigraphy of MD97–2120 (refs 53 and 54) is based on nine ^{14}C measurements in the time interval between 0–35 kyr. Following the method applied by Rose *et al.*[15], we correlated the planktic $\delta^{18}O$ and Mg/Ca-derived SST records of MD97–2120 (ref. 54) with the EDC ice core δD record[55,56] (Supplementary Fig. 7). Correlating MD97–2120 via surface temperatures to Antarctic ice cores enables us to obtain a ^{14}C-independent stratigraphy that we can use as the baseline for the X-ray fluorescence (XRF) core-to-core correlation.

At the Alfred Wegener Institute in Bremerhaven, Germany, all sediment cores were analysed for their specific element abundances using an Avaatech XRF core-scanner with a preset step width of 1 cm. The resulting element content records (Sr, Ca, Fe and Sr/Fe) were then correlated to respective age-scaled XRF records of the sediment core MD97–2120 (refs 53 and 54) from the southern Chatham Rise (Supplementary Fig. 8) using the computer program AnalySeries 2.0.4.2. In particular, the Sr-counts were useful for the core-to-core correlation, as Sr represents $CaCO_3$, but is barley affected by differences in grain sizes or the sediments water-content[57]. Furthermore, we were able to utilize a distinctive vitric-rich tephra layer as a radiocarbon-independent stratigraphic marker within two sediment cores (SO213-76-2 and SO213-79-2). The tephra layer in both cores was geochemically characterized and identified as the widespread Kawakawa/Oruanui tephra (KOT; for analytical details see the section on tephra analyses). The KOT is a widespread silicic tephra erupted from Taupo Volcanic Centre in the central North Island of New Zealand with an age of ~25.36 cal. ka (ref. 58) and is the most important isochronous tephra marker erupted in the SW-Pacific region during the past 30,000 years[59]. This tephra was likewise found in reference core MD97–2120 (ref. 54). We adjusted the KOT age used by Pahnke *et al.*[54] to the revised age by Vandergoes *et al.*[58].

The age model of the EPR-core PS75/059-2 is based on the original approach of Lamy *et al.*[60]. However, as this age model lacks detailed tie-points between 0 and 30 cal. ka, we fine-tuned PS75/059-2 to the age-scaled record PS75/100-4, using their Sr XRF element counts. Despite a certain lack in the LGM variability of most XRF-records, we consider our age models as relatively robust. In particular the comparison of our records SO213-82-1 and SO213-76-2 to records from similar water depths in the South Pacific[6] (MD97–2121), the Drake Passage[8] (coral dredges) and the South Atlantic[4] (MD07–3076), reveals a similar timing as well as comparable $\Delta\Delta^{14}C$-values (Fig. 4) for all records.

The errors for the correlated age models were estimated from the offset to the ^{14}C-derived age model plus ^{14}C-errors.

Our XRF-based correlation method creates new surface reservoir ages for our sediment records. These differ from the surface reservoir age record of MD97–2121 (ref. 6), but are still in good agreement with these reconstructions (Supplementary Fig. 5). However, owing the XRF-correlation method, our reservoir ages show a slightly higher scatter than the records of Sikes *et al.*[5] and Skinner *et al.*[6]

Tephra analyses. In cores SO213-76-2 (507–508 cm) and SO213-79-2 (150–151 cm) a distinctive vitric-rich tephra layer was identified. Morphological expression, grain size and thickness of this tephra were indistinguishable and hence, most likely to represent the same eruptive event. Glass shard major element chemistry of this tephra layer was then determined by electron microprobe analysis and the results were compared with selected onshore and offshore KOT correlatives (Supplementary Fig. 10). The glass shard geochemistries were consequently indistinguishable which (a) affirms correlation to KOT and (b) augments the overall chronology of this study. All major element determinations were made on a JEOL Superprobe (JXA-8230) housed at Victoria University of Wellington, using the ZAF correction method. Analyses were performed using an accelerating voltage of 15 kV under a static electron beam operating at 8 nA. The electron beam was defocused between 10 and 20 µm. All elements calculated on a water-free basis, with H_2O by difference from 100%. Total Fe expressed as FeO_t. Mean and ±1 s.d. (Supplementary Table 4; in parentheses), based on *n* analyses. All samples normalized against glass standards VG-568 and ATHO-G.

Box modelling. Here we make some first-order estimates investigating the hypothesis of an impact of a potential hydrothermal CO_2 flux on the marine [illegible] and on $\Delta^{14}C$ in the South Pacific. We simulate carbon cycle [illegible] mid-depth South Pacific waters ~2,500–3,500 m (Supplementary Fig. 4a). We start our carbon cycle simulations from an LGM state that was previously simulated with the BICYCLE model[46]. We thus perturb a water mass with dissolved inorganic carbon (DIC) concentration of ~2,600 µmol kg^{-1}, with a total alkalinity of 2,667 µmol kg^{-1}. For LGM conditions (temperature ~0 °C; salinity = 35.8 PSU), these factors would lead to a pH of 7.7 and a carbonate ion concentration of ~70 µmol kg^{-1}.

We base our calculations on changes in concentrations for an undefined volume of the water mass and increase the amplitude of hydrothermal CO_2 until the ^{14}C-anomaly, seen in our data, is reproduced by our most simplistic model.

Similar to our data from 2,066 m (SO213-82-1), Skinner *et al.*[6] provide an independent glacial ventilation age of ~2,700 years, which decreases to ~1,500 years at 18 cal. ka. To obtain a stable $\Delta^{14}C$ of 0‰ in our water mass ~30 cal. ka (Fig. 3a, black broken line), the incoming carbon flux from X_i from oceanic mixing processes has a $\Delta^{14}C$ signature of +328‰. Corresponding to the decrease in ventilation age from 2,700 to 1,500 years, this flux increases at 18 cal. ka from 1 to 1.7 µmol per kg per year (Supplementary Fig. 4c). The outgoing carbon flux X_o, leaving our simulated 1-box water mass via oceanic transport processes, is determined by the turnover/ventilation time τ. X_o might change over time in size due to hydrothermal CO_2 injection, causing a rise in DIC within the water parcel. For the time window of minimum sea level[61] (Supplementary Fig. 4b) between 30 and 15 cal. ka, we assume a hydrothermal CO_2 flux (F) of 0, 0.3, 0.6, 0.9 or 1.2 µmol per kg per year (Supplementary Fig. 4d–g). Please note that this approach assumes an instantaneous feedback of the hydrothermal CO_2 outgassing rate to the removed load from the drop in global sea level and neglects any potential time delay in the solid earth response. Time delay of this process were briefly discussed previously[18] but might also be more complex[19].

On millennial timescales, the injection of hydrothermal CO_2 might lead to a partly dissolution of $CaCO_3$ in oceanic sediments[22,62]. Via this process of carbonate compensation, carbonate ions are brought into solution and added to the DIC pool. In this dissolution flux (D), the ratio of alkalinity and DIC is 2:1. If enough $CaCO_3$ is available for dissolution, the additional dissolution flux from carbonate compensation is on the order of 88% of the initial CO_2 injection[63]. In case of insufficient amounts of available $CaCO_3$, carbonate ion concentration and pH would fall. So far, it is estimated that the amount of dissolvable sediments is restricted to ~1,600 PgC (ref. 64), although simulation studies show that the actual dissolution for large CO_2-injections is smaller than this[22]. While the actual ^{14}C signature of dissolved $CaCO_3$ is not readily known, we assume the dissolution of ^{14}C-free sediment as the upper limit. This assumption is supported by the existence of very old surface sediments in the pelagic Southeast Pacific off Chile[65]. For simplicity, we here calculate an instantaneous additional impact of carbonate compensation. However, since this process is in reality time delayed with an e-folding time of a few thousand years[22], the most likely solution lies between both scenarios, with and without carbonate compensation.

In addition, we estimate the impact on $\Delta^{14}C$, if CO_2, of similar size (S) as the hydrothermal injection flux (F), is directly sequestered via magmatic processes[23], leading to stable DIC concentrations.

Please note that the three processes considered here (hydrothermal CO_2 flux F; carbonate dissolution D; CO_2 sequestration S) are roughly estimated due to their potential impact on oceanic $\Delta^{14}C$. In detail, these processes might be time-delayed or offset from each other.

Code availability. The computer code for the 1-box model used to calculate the simulation results shown in Fig. 3 and Supplementary Fig. 4 is available upon request from one of the co-authors (PK; peter.koehler@awi.de).

References

1. Marcott, S. A. *et al.* Centennial-scale changes in the global carbon cycle during the last deglaciation. *Nature* **514,** 616–619 (2014).
2. Anderson, R. F. *et al.* Wind-driven upwelling in the southern ocean and the deglacial rise in atmospheric CO_2. *Science* **323,** 1443–1448 (2009).
3. Sigman, D. M., Hain, M. P. & Haug, G. H. The polar ocean and glacial cycles in atmospheric CO_2 concentration. *Nature* **466,** 47–55 (2010).
4. Skinner, L. C., Fallon, S., Waelbroeck, C., Michel, E. & Barker, S. Ventilation of the deep Southern Ocean and deglacial CO_2 rise. *Science* **328,** 1147–1151 (2010).
5. Sikes, E. L., Samson, C. R., Gullderson, T. P. & Howard, W. R. Old radiocarbon ages in the Southwest Pacifc Ocean during the last glacial period and deglaciation. *Nature* **405,** 555–559 (2000).
6. Skinner, L. C. *et al.* Reduced ventilation and enhanced magnitude of the deep Pacific carbon pool during the last glacial period. *Earth Planet. Sci. Lett.* **411,** 45–52 (2015).
7. Sarnthein, M., Schneider, B. & Grootes, P. M. Peak glacial ^{14}C ventilation ages suggest major draw-down of carbon into the abyssal ocean. *Clim. Past* **9,** 929–965 (2013).
8. Chen, T. *et al.* Synchronous centennial abrupt events in the ocean and atmosphere during the last deglaciation. *Science* **349,** 1537–1541 (2015).
9. DeVries, T. & Primeau, F. An improved method for estimating water-mass ventilation age from radiocarbon data. *Earth Planet. Sci. Lett.* **295,** 367–378 (2010).

10. Toggweiler, J. R., Russell, J. L. & Carson, S. R. Midlatitude westerlies, atmospheric CO_2, and climate change during the ice ages. *Paleoceanography* **21,** PA2005 (2006).
11. Bostock, H. C., Sutton, P. J., Williams, M. J. M. & Opdyke, B. N. Reviewing the circulation and mixing of Antarctic intermediate water in the South Pacific using evidence from geochemical tracers and Argo float trajectories. *Deep-Sea Res. I* **73,** 84–98 (2013).
12. Marchitto, T. M., Lehman, S. J., Ortiz, J. D., Flückinger, J. & van Geen, A. Marine radiocarbon evidence for the mechanism of deglacial atmospheric CO_2 rise. *Science* **316,** 1456–1459 (2007).
13. Stott, L. D., Southon, J., Timmermann, A. & Koutavas, A. Radiocarbon age anomaly at intermediate water depth in the Pacific Ocean during the last deglaciation. *Paleoceanography* **24,** PA2223 (2009).
14. De Pol-Holz, R., Keigwin, L. D., Southon, J., Hebbeln, D. & Mohtadi, M. No signature of abyssal carbon in intermediate waters off Chile during deglaciation. *Nat. Geosci.* **3,** 192–195 (2010).
15. Rose, K. A. *et al.* Upper-ocean-to-atmosphere radiocarbon offsets imply fast deglacial carbon dioxide release. *Nature* **466,** 1093–1097 (2010).
16. Lindsay, C. M., Lehman, S. J., Marchitto, T. M. & Ortiz, J. D. The surface expression of radiocarbon anomalies near Baja California during deglaciation. *Earth Planet. Sci. Lett.* **422,** 67–74 (2015).
17. Lund, D. C. & Asimow, P. D. Does sea level influence mid-ocean ridge magmatism on Milankovitch timescales? *Geochem. Geophys. Geosyst.* **12,** Q12009 (2011).
18. Tolstoy, M. Mid-ocean ridge eruptions as a climate valve. *Geophys. Res. Lett.* **42,** 1346–1351 (2015).
19. Crowley, J. W., Katz, R. F., Huybers, P., Langmuir, C. H. & Park, S.-H. Glacial cycles drive variations in the production of oceanic crust. *Science* **347,** 1237–1240 (2015).
20. Reimer, P. J. *et al.* IntCal13 and Marine13 radiocarbon age calibration curves 0-50,000 years Cal BP. *Radiocarbon* **55,** 1869–1887 (2013).
21. Cook, M. S. & Keigwin, L. D. Radiocarbon profiles of the NW Pacific from the LGM and deglaciation: evaluating ventilation metrics and the effect of uncertain surface reservoir ages. *Paleoceanography* **30,** 174–195 (2015).
22. Archer, D., Kheshgi, H. & Maier-Reimer, E. Multiple timescales for neutralization of fossi fuel CO_2. *Geophys. Res. Lett.* **24,** 405–408 (1997).
23. Alt, J. C. & Teagle, D. A. H. The uptake of carbon during alteration of ocean crust. *Geochim. Cosmochim. Acta* **63,** 1527–1535 (1999).
24. Burton, M. R., Sawyer, G. M. & Granieri, D. Deep carbon emissions from volcanoes. *Rev. Mineral. Geochem.* **75,** 323–354 (2013).
25. Bostock, H. C., Hayward, B. W., Neil, H. L., Currie, K. I. & Dunbar, G. B. Deep-water carbonate concentrations in the Southwest Pacific. *Deep-Sea Res. I* **58,** 72–85 (2011).
26. Köhler, P., Muscheler, R. & Fischer, H. A model-based interpretation of low-frequency changes in the carbon cycle during the last 120,000 years and its implications for the reconstruction of atmospheric $\Delta^{14}C$. *Geochem. Geophys. Geosyst.* **7,** Q11N06 (2006).
27. McCave, I. N., Carter, L. & Hall, I. R. Glacial-interglacial changes in water mass structure and flow in the SW Pacific Ocean. *Quat. Sci. Rev.* **27,** 1886–1908 (2008).
28. Rae, J. W. B. *et al.* Deep water formation in the North Pacific and deglacial CO_2 rise. *Paleoceanography* **29,** 645–667 (2014).
29. Sarnthein, M., Grootes, P. M., Kennet, J. P. & Nadeau, M.-J. in *Ocean Circulation: Mechanisms and Impacts—Past and Future Changes of Meridional Overturning* Vol. 173 (eds Schmittner, A., Chiang, J. C. H. & Hemming, S. R.) 175–196 (AGU, 2007).
30. Ferrari, R. *et al.* Antarctic sea ice control on ocean circulation in present and glacial climates. *Proc. Natl Acad. Sci. USA* **111,** 8753–8758 (2014).
31. Burke, A. *et al.* The glacial mid-depth radiocarbon bulge and its implications for the overturning circulation. *Paleoceanography* **30,** 1021–1039 (2015).
32. Broecker, W. & Clark, E. Search for a glacial-age ^{14}C-depleted ocean reservoir. *Geophys. Res. Lett.* **37,** L13606 (2010).
33. Kohfeld, K. E. *et al.* Southern Hemisphere westerly wind changes during the Last Glacial Maximum: paleo-data synthesis. *Quat. Sci. Rev.* **68,** 76–95 (2013).
34. Ronge, T. A. *et al.* Pushing the boundaries: Glacial/Interglacial variability of intermediate- and deep-waters in the southwest Pacific over the last 350,000 years. *Paleoceanography* **30,** 23–38 (2015).
35. Adkins, J. F. The role of deep ocean circulation in setting glacial climates. *Paleoceanography* **28,** 539–561 (2013).
36. Hain, M. P., Sigman, D. M. & Haug, G. H. Shortcomings of the isolated abyssal reservoir model for deglacial radiocarbon changes in the mid-depth Indo-Pacific Ocean. *Geophys. Res. Lett.* **38,** L04604 (2011).
37. Hain, M. P., Sigman, D. M. & Haug, G. H. Distinct roles of the Southern Ocean and North Atlantic in the deglacial atmospheric radiocarbon decline. *Earth Planet. Sci. Lett.* **394,** 198–208 (2014).
38. Lupton, J. Hydrothermal helium plumes in the Pacific Ocean. *J. Geophys. Res.* **103,** 15853–15868 (1998).
39. Lupton, J. & Craig, H. A major Helium-3 source at 15°S on the East Pacific Rise. *Science* **214,** 13–18 (1981).
40. Talley, L. D. *Hydrographic Atlas of the World Ocean Circulation Experiment (WOCE). Vol. 2: Pacific Ocean* (eds Sparrow, M., Chapman, P. & Gould, J.) (International WOCE Project Office, Southampton, UK, 2007).
41. de Ronde, C. E. J. *et al.* Intra-oceaniic subduction-related hydrothermal venting, Kermadec volcanic arc, New Zealand. *Earth. Planet. Sci. Lett.* **193,** 359–369 (2001).
42. Sano, Y. & Marty, B. Origin of carbon in fumarolic gas from island arcs. *Chem. Geol.* **119,** 265–274 (1995).
43. Deines, P. The carbon isotope geochemistry of mantle xenoliths. *Earth Sci. Rev.* 247–278 (2002).
44. Stott, L. D. & Timmermann, A. in *Abrupt Climate Change: Mechanisms, Patterns, and Impacts* (eds Harunur, R., Polyak, L. & Mosley-Thompson, E.) 123–138 (American Geophysical Union, 2011).
45. Schmitt, J. *et al.* Carbon isotope constraints on the deglacial CO_2 rise from ice cores. *Science* **336,** 711–714 (2012).
46. Köhler, P., Fischer, H., Munhoven, G. & Zeebe, R. E. Quantitative interpretation of atmospheric carbon records over the last glacial termination. *Global Biochem. Cycles* **19,** GB4020 (2005).
47. Völker, C. & Köhler, P. Responses of ocean circulation and carbon cycle to changes in the position of the Southern Hemisphere westerlies at Last Glacial Maximum. *Paleoceanography* **28,** 726–739 (2013).
48. Sikes, E. L., Howard, W. R., Neil, H. L. & Volkman, J. K. Glacial-interglacial sea surface temperature changes across the subtropical front east of New Zealand based on alkenone unsaturation ratios and foraminiferal assemblages. *Paleoceanography* **17,** PA000640 (2002).
49. Hayward, B. W. *et al.* The effect of submerged plateaux on Pleistocene gyral circulation and sea-surface temperatures in the Southwest Pacific. *Global Planet. Change* **63,** 309–316 (2008).
50. Adkins, J. F. & Boyle, E. A. Changing atmospheric $\Delta^{14}C$ and the record of deepwater paleoventilation ages. *Paleoceanography* **12,** 337–344 (1997).
51. Stuiver, M. & Reimer, P. J. Extended 14C Data Base and Revised Calib 3.0 ^{14}C Age Calibration Program. *Radiocarbon* **35,** 215–230 (1993).
52. Calib 7.0 v. 7.0. Available at: http://calib.qub.ac.uk (2014).
53. Pahnke, K. & Zahn, R. Southern Hemisphere water mass conversion linked with North Atlantic climate variability. *Science* **307,** 1741–1746 (2005).
54. Pahnke, K., Zahn, R., Elderfield, H. & Schulz, M. 340,000-Year centennial-scale marine record of Southern Hemisphere climatic oscillation. *Science* **301,** 948–952 (2003).
55. EPICA Comminity Members. One-to-one coupling of glacial climate variability in Greenland and Antarctica. *Nature* **444,** 195–198 (2006).
56. Veres, D. *et al.* The Antarctic ice core chronology (AICC2012): an optimized thousand years multi-parameter and multi-site dating approach for the last 120. *Clim. Past* **9,** 1773–1748 (2013).
57. Tjallingii, R., Kölling, M. & Bickert, T. Influence of the water content on X-ray fluorescence core- scanning measurements in soft marine sediments. *Geochem. Geophys. Geosyst.* **8,** Q02004 (2007).
58. Vandergoes, M. J. *et al.* A revised age for the Kawakawa/Oruanui tephra, a key marker for the Last Glacial Maximum in New Zealand. *Quat. Sci. Rev.* **74,** 195–201 (2013).
59. Alloway, B. *et al.* Towards a climate event stratigraphy for New Zealand over the past 30 000 years (NZ-INTIMATE project). *J. Quat. Sci.* **22,** 9–35 (2007).
60. Lamy, F. *et al.* Increased dust deposition in the Pacific Southern Ocean during glacial periods. *Science* **343,** 403–407 (2014).
61. Bintanja, R., van de Wal, R. S. W. & Oerlemans, J. Modelled atmospheric temperatures and global sea levels over the past million years. *Nature* **437,** 125–128 (2005).
62. Zeebe, R. E. & Caldeira, K. Close mass balance of long-term carbon fluxes from ice-core CO_2 and ocean chemistry records. *Nat. Geosci.* **1,** 312–315 (2008).
63. Zeebe, R. E. & Wolf-Gladrow, D. in *CO2 in Seawater: Equilibrium, Kinetics, Isotopes Vol. 65 Elsevier Oceanography Series.* (eds Zeebe, R. E. & Wolf-Gladrow, D.Ch. 3, 141–250 (Elsevier, 2001).
64. Archer, D. An Atlas of the distribution of calcium carbonate in sediments of the deep sea. *Global Biochem. Cycles* **10,** 159–174 (1996).
65. Tiedemann, R. *FS Sonne Fahrtbericht/Cruise Report SO213* (Alfred Wegener Institute, 2012).
66. Orsi, A. H., Whitworth, III T. & Nowlin, Jr W. D. On the meridional extent and fronts of the Antarctic circumpolar current. *Deep Sea Res. I* **42,** 641–673 (1995).
67. Key, R. M. *et al.* A global ocean carbon climatology: results from Global Data Analysis Project (GLODAP). *Global Biochem. Cycles* **18,** GB4031 (2004).
68. Ocean Data View v. 4.7.2. Available at: http://owa.awi.de (2015).
69. WAIS Divide Project Members. Onset of deglacial warming in West Antarctica driven by local orbital forcing. *Nature* **500,** 440–444 (2013).
70. Hand, E. Seafloor grooves record the beat of the ice ages. *Science* **347,** 593–594 (2015).

Acknowledgements
We thank the crews of *R/V Sonne* and *R/V Polarstern*; M. Sarnthein for helpful comments; R. Fröhlking, B. Glückselig, N. Lensch, L. Ritzenhofen, L. Schönborn, M. Seebeck, R. Sieger, M. Warnkroß and S. Wiebe for technical support. T.A.R. was funded by the Federal Ministry of Education and Research (BMBF; Germany) project 03G0213A—SOPATRA. BVA received funded support from an internal Victoria University of Wellington Science Faculty Research Grant. R.D.P.-H. was funded by Chilean FONDAP 15110009 and Iniciativa Científica Milenio NC120066. Data are available at http://doi.pangaea.de/10.1594/PANGAEA.833663.

Author contributions
T.A.R., R.T. and F.L. designed the study and wrote most of the manuscript; T.A.R. performed XRF measurements; T.A.R., R.D.P.-H., L.W. and J.S. performed accelerator mass spectrometry measurements; P.K. constructed the box model to calculate the effect of hydrothermal CO_2-injection on $\Delta^{14}C$; K.P. provided XRF measurements on reference core MD97-2120; B.V.A. performed tephra analyses; all authors contributed to the written manuscript.

Additional information

Competing financial interests: The authors declare no competing financial interests.

16

Observing mesoscale eddy effects on mode-water subduction and transport in the North Pacific

Lixiao Xu[1], Peiliang Li[1], Shang-Ping Xie[1,2], Qinyu Liu[1], Cong Liu[1] & Wendian Gao[1]

While modelling studies suggest that mesoscale eddies strengthen the subduction of mode waters, this eddy effect has never been observed in the field. Here we report results from a field campaign from March 2014 that captured the eddy effects on mode-water subduction south of the Kuroshio Extension east of Japan. The experiment deployed 17 Argo floats in an anticyclonic eddy (AC) with enhanced daily sampling. Analysis of over 3,000 hydrographic profiles following the AC reveals that potential vorticity and apparent oxygen utilization distributions are asymmetric outside the AC core, with enhanced subduction near the southeastern rim of the AC. There, the southward eddy flow advects newly ventilated mode water from the north into the main thermocline. Our results show that subduction by eddy lateral advection is comparable in magnitude to that by the mean flow—an effect that needs to be better represented in climate models.

[1]Physical Oceanography Laboratory/Qingdao Collaborative Innovation Center of Marine Science and Technology, Ocean University of China, Qingdao 266100, China. [2]Scripps Institution of Oceanography, University of California San Diego, La Jolla, California 92093, USA. Correspondence and requests for materials should be addressed to S.-P.X. (email: sxie@ucsd.edu).

A universal feature of the global subtropical gyres is the presence of mode water, a thick layer of water with homogeneous properties within the main thermocline[1,2] (Supplementary Note 1). Mode water is important in the climate system as 'memories' of climate variability[3] and by 'breathing in' anthropogenic carbon dioxide[4,5]. Current climate models suffer large biases in representing mode-water subduction[6,7]. Mode-water subduction takes place near the western boundary current extensions with high eddy activities[8,9]. Previous studies propose the poor representation of mesoscale eddy effects as the major cause of the biases[7,10]. Results from eddy-resolving models suggest that mesoscale eddy contribution to the mode-water subduction is on the same order of magnitude as that by the mean flow[11–13].

Physical processes by which mesoscale eddies affect subduction remain unclear, and are challenging to observe in the field. ACs have been observed to trap mode water while migrating equatorwards[14–17]; however, previous data did not resolve eddies to provide a quantitative estimate of eddy subduction and transport. The transport due to eddy-trapping is much weaker in the meridional than the zonal direction[18], given the predominant westward propagation of mesoscale eddies[19]. In addition to the eddy-trapping transport, we propose a larger, complementary eddy lateral advective effect. Specifically, in the North Pacific, the southward eddy flow on the eastern rim of the AC advects south the dense, winter deep-mixed layer water from the north across large-scale density gradient into the permanent thermocline[7]. The cross-frontal eddy advection leads to an equatorward 'bolus' transport[20] of the newly ventilated mode water. Although this eddy-advective effect on mode-water subduction has been suggested from eddy-resolving model simulations[7,13], it has never been verified in field observations.

The present study investigates this eddy lateral advective effect on subduction and transport of the North Pacific subtropical mode water (STMW) through a targeted field experiment: the Pacific Mode-Water Ventilation Experiment (P-MoVE; ref. 21) that made the first direct observations of eddy subduction processes in the western North Pacific. We deploy 17 Argo-profiling floats with enhanced daily sampling in an AC south of the Kuroshio Extension (KE) in late March 2014 (See Methods). On the basis of more than 3,000 hydrographic Argo profiles following the AC, we find that the southward eddy flow transports relatively low potential vorticity (PV) and apparent oxygen utilization (AOU) from the northern winter deep-mixed layer into the permanent thermocline, suggesting a 'sweet spot' of subduction outside the eddy core on the southeastern rim of the AC. Together with results from the eddy-resolving model, we quantify that the southward eddy PV flux in the study region is comparable in magnitude to that by the mean flow, with the lateral advection dominating over the eddy-trapping effect. We conclude that the eddy lateral advection plays a key role in ventilating the interior of the upper ocean and increasing the net subduction of mode water.

Results

Observing eddy-induced subduction and transport. During the P-MoVE cruise, 17 Argo-profiling floats with enhanced daily sampling were deployed in an AC in the STMW-formation region in late March 2014 (Methods; Fig. 1a). The AC trajectory (red line in Fig. 1a) is determined by tracking the time-varying AC centre from satellite altimetry analysis, cross-checked against the geometry of velocity streamlines around the sea-level anomaly (SLA) maximum. Several Argo floats followed the westward-migrating AC through summer until the eddy disintegrated on encountering the Izu Ridge in late September (Fig. 1b). The floats

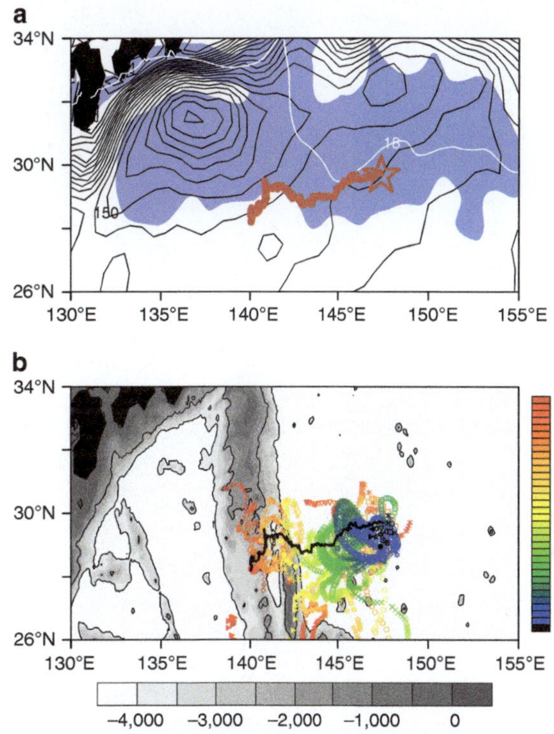

Figure 1 | Trajectories of the AC and Argo-profiling floats. (**a**) Trajectory of the target AC centre from 27 March to 10 September 2014 (thick red line). The star sign denotes where the 17 Argo floats are deployed. The black contours denote the mean dynamical topography (contour interval (CI) = 10 cm) to illustrate the large-scale circulation, and the blue shading highlights the STMW-formation region with SST of 17–19 °C (18 °C in white contour) on 27 March. (**b**) Trajectories of the 17 Argo floats (symbols with colours indicating the time). The trajectories end on 10 September 2014 when the AC dissipated. The symbol intervals for each float are daily. The grey shading indicates water depth (CI = 500 m; the −2,000 and −4,000 m contours are highlighted in thin black lines). A major bathymetric feature in the region is the Izu Ridge along 140°E.

provide more than 3,000 hydrographic profiles that offered an unprecedented detailed view of STMW subduction and dissipation. This section reports results from the field campaign.

We use an eddy-following coordinate system—(Δx, Δy) for the relative position of the Argo floats to the AC centre—to construct the composite AC fields based on the data of the 17 Argo-profiling floats (Methods). Figure 2a–i shows in raw data dots the mixed layer depth, AOU and PV around the AC on the core density surface of STMW ($25.3\sigma_\theta$) for March to August. Here PV is calculated as $q = -\frac{f}{\rho_0}\frac{\partial\rho}{\partial z}$, where ρ is potential density, f is the Coriolis parameter and ρ_0 is a reference density (1024 kg m^{-3}). Oxygen sensors have been deployed on the 17 Argo floats, and AOU is defined as the difference between the saturated and observed dissolved oxygen (DO) concentrations[22]. Similar patterns are obtained when the Argo data are interpolated on a 0.1×0.1 grid (Supplementary Fig. 1). During March–June, the data display east–west asymmetries (Fig. 2 and Supplementary Fig. 1), with larger mixed layer depth and lower values of PV and AOU in the southeast than in the northwest part of the AC (Supplementary Fig. 2a–c).We compare the vertical sections along the southeast and northwest tracks of the AC for PV (Fig. 3) and AOU (Fig. 4). The PV and AOU in the STMW layer show asymmetry between the southeast and northwest rims of the AC. Along the southeastern track of the

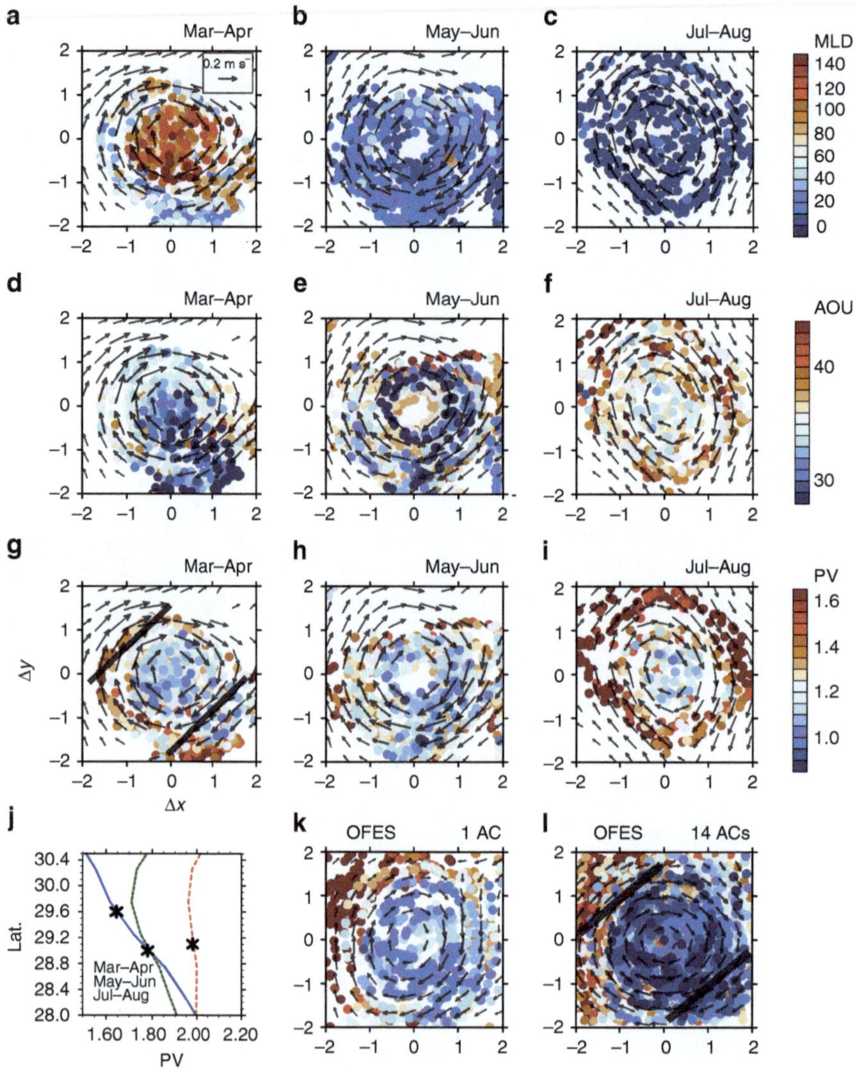

Figure 2 | The AC fields in the core layer of STMW. Observed raw data dots are displayed in the top three rows: (**a-c**) mixed layer depth (MLD; m), (**d-f**) AOU (ml kg^{-1}) and (**g-i**) PV ($10^{-10} m^{-1} s^{-1}$) on the core layer of STMW ($25.3\sigma_\theta$). A coordinate system (Δx, Δy) is used relative to the AC centre (the red line in Fig. 1a or the black line in Fig. 1b). The outer boundary of the eddy core, defined as the zero relative vorticity contour, is normalized here as between [−1 1]. Arrows denote geostrophic currents in m s^{-1}. The bottom left plot (**j**) denotes the zonally averaged (135°-150°E) PV on the core layer of STMW ($25.3\sigma_\theta$) as a function of latitude based on the Roemmich-Gilson Argo climatology: the solid blue line for March-April, dotted green line for May-June and dash red line for July-August. The latitude of AC is marked in black * in **j**. (**k-l**) Data dots of PV (colour shading) and geostrophic current (vectors) for March to April, sampled by synthetic Argo profiles deployed in OFES to mimic the sampling of the field campaign (Methods); (**k**) one AC in OFES and (**l**) 14 ACs in OFES. The slanted straight black lines in **g**,**l** are the positions of the transections in Figs 3 and 4.

AC, the STMW is ventilated and renewed with lowest PV ($<1.5\times10^{-10} m^{-1} s^{-1}$) and AOU ($<25$ ml kg^{-1}) during March–April (Figs 3a,b and 4a). After earlier April, the STMW pycnostad loses contact with the atmosphere, but the PV (AOU) lower than $1.5\times10^{-10} m^{-1} s^{-1}$ (25 ml kg^{-1}) persists until June (Figs 3e and 4c), implying the eddy southward advection of low PV and AOU from the northern STMW-formation region. By contrast, on the northwest rim of the AC (Figs 3c,d,f and 4b,d), the PV and AOU in STMW are much higher ($PV>1.5\times10^{-10} m^{-1} s^{-1}$; $AOU>35$ ml kg^{-1}). No ventilation occurs there, and the STMW pycnostad is sheltered from the surface even in March.

The subduction of low PV and AOU only happens on the southeast rim of the AC. There, the southward eddy flow carries low PV water from the deep-mixed layer into the permanent thermocline, suggesting a 'sweet spot' of subduction by eddies (Figs 2a,g,3a,b and 4a and Supplementary Fig. 1a,g). Owing to the southward intrusion of the deep-mixed layer and the thick mode-water layer underneath, the vertical density gradient is low on the eastern rim of the AC (Supplementary Fig. 3). In May–June, the STMW is capped by the seasonal thermocline that forms under surface warming (Fig. 3e). The southward advection of low PV and AOU (Figs 2e,h,3e and 4c) by the eddy flow from the STMW-formation region in the north continues as the large-scale, meridional PV gradient remains strong (green dot line in Fig. 2j). From late June to August, by contrast, the newly formed STMW in the north erodes steadily as the seasonal thermocline develops, weakening the background meridional PV gradient (red dash line in Fig. 2j). The southward advection of STMW on the eastern rim of the AC weakens and eventually ceases (Fig. 2f,i).

Thus, our field observations have captured eddy effects on STMW subduction and ventilation. The PV asymmetry between the southeast and northwest rims of the AC (Figs 2g,h,3e,f and Supplementary Fig. 2a–c) is a result of eddy advection across the

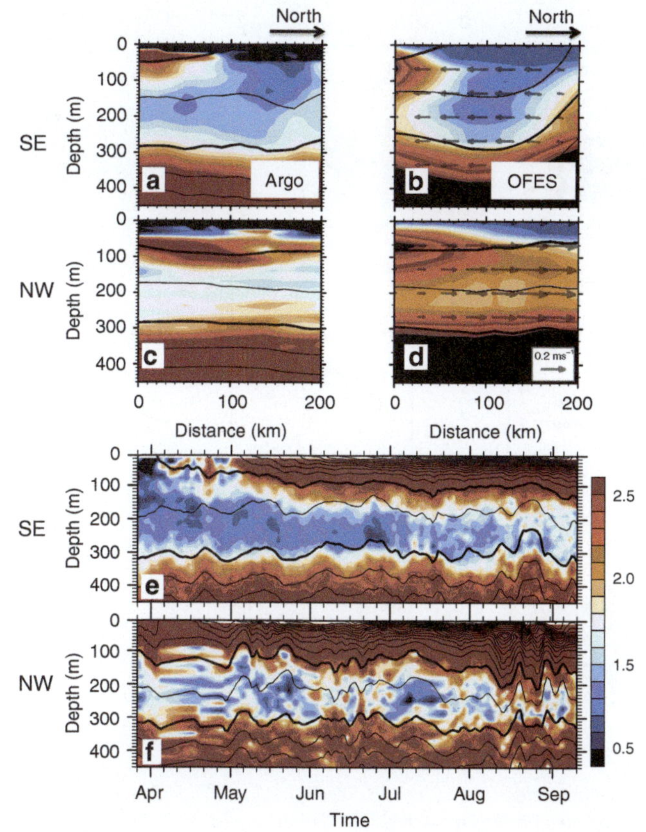

Figure 3 | Vertical sections on southeastern and northwestern flanks of the AC. The southeastern and northwestern sections (thick black lines in Fig. 2g,l) are in the circumferential tangent direction of the AC. Vertical sections along (**a**,**b**) the southeast (SE) and (**c**,**d**) the northwest (NW) tracks for March to April, based on (**a**,**c**) the 17 Argo floats and (**b**,**d**) the eddy-resolving OFES (Methods). The horizontal ordinate is the distance from the southernmost point. The time-depth sections averaged along the (**e**) southeastern and (**f**) northwestern tracks are shown in the lower tow panels. Black contours denote potential density, thickened to highlight the 25.0 and 25.5 kg m^{-3} isopycnals that correspond to the top and bottom of the STMW layer, respectively. Coloured shades denote PV (10^{-10} m^{-1}s^{-1}), and vectors in **b**,**d** show the along-track flow.

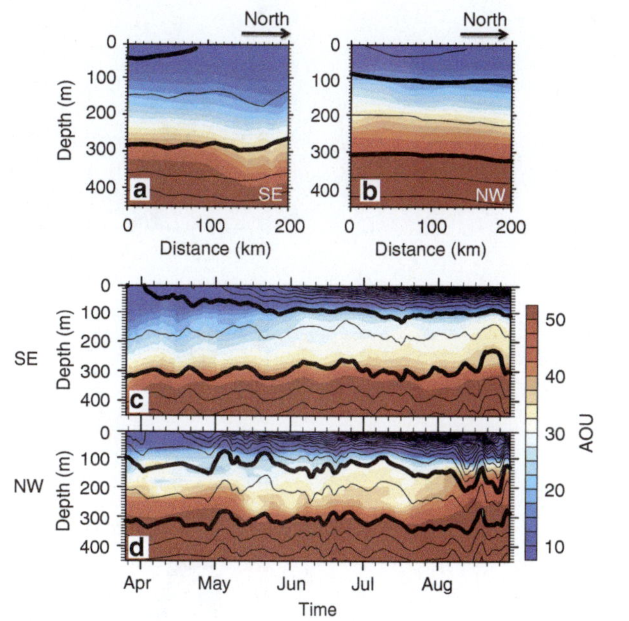

Figure 4 | The AOU cross-sections along southeastern and northwestern flanks of the AC based on data from the 17 Argo floats. The southeastern and northwestern sections are denoted in thick black lines in Fig. 2g. Vertical sections along (**a**) the SE and (**b**) the NW flank of the AC during March to April; the horizontal ordinate is the distance from the southernmost point. Time-depth sections averaged along the (**c**) southeastern and (**d**) northwestern tracks. Colour shades denote AOU (ml kg^{-1}). Black contours denote potential density fields; with thick curves represent the isopycnal 25.0 and 25.5 kg m^{-3}, corresponding to the top and bottom of the STMW layer, respectively.

background PV gradient. By advecting south the denser low PV water from the winter deep-mixed layer, the southward eddy flow acts to ventilate the interior of the upper ocean.

Data model comparison. Does the current eddy-resolving model faithfully represent the observed eddy-advective effect on STMW subduction? To address this question, we deploy synthetic Argo floats in the Ocean General Circulation Model (OGCM) for the Earth Simulator (OFES) model, and use the 'offline particle-tracking method' to track the floats. The Methods section (*Track ACs in the model*) describes how we deploy synthetic Argo floats in the eddy-resolving model OFES to mimic the sampling in the field. We tracked 14 ACs located in the study region (140°E–150°E, 28°–33°N) in late March. On the basis of the samples taken by the synthetic Argo floats, we construct the composite AC fields in OFES, and compare with observations.

We find that the eddy-resolving model successfully captures the eddy-advective effect. The east–west asymmetry in PV during March–April is confirmed in the OFES composites (Fig. 2k,l). As in observations, low (high) values of PV are found in the southeast (northwest) part of the AC (Supplementary Fig. 2d). The sampling of the east–west PV asymmetry can be improved by a more widely spread Argo array for the first few days; however, such an array is unstable on the outer rim of the eddy. In the model, the synthetic Argo floats away from the AC core cannot follow the AC movement and leave the AC soon.

Figure 5 provides a closer look into the eddy-advective effect in both the model and observations. PV is nearly homogeneous inside the AC core that traps winter deep-mixed layer water from the AC origin, but is asymmetric outside the AC core between the west and east rims where the outer ring of the eddy stirs the ambient fluid. There, the southward (northward) eddy flow advects relatively low (high) PV into (from) the permanent thermocline (Figs 3a–d and 5a,b,d,e,g,h), with a net effect to strengthen the ventilation of the interior ocean. The majority of the eddy PV flux occurs outside the eddy core as detailed in the next section.

We chose an AC from the model that has the eddy PV flux close to that of the AC in the field to compare the zonal asymmetry of PV (q') and tangential velocity (v_r') in Fig. 5. In OFES, the east–west PV asymmetry is sharper and the scatters of PV from the composite mean are narrower (Fig. 5e) than those in our observations (Fig. 5b), possibly because of insufficient horizontal dissipation[23].

For comparison with the observed AC that trapped low PV water in the eddy centre (Fig. 2g–i), Fig. 2k,l shows results for 'mode-water-trapping' ACs in the OFES model. Not all ACs trap low PV in their eddy centres (Supplementary Fig. 4a,b). However, both types of ACs show east–west PV asymmetry in the study region during March–April, illustrating the eddy lateral advective effect. Furthermore, cyclonic eddies (CCs) near the STMW ventilation region show similar eddy lateral advective effects

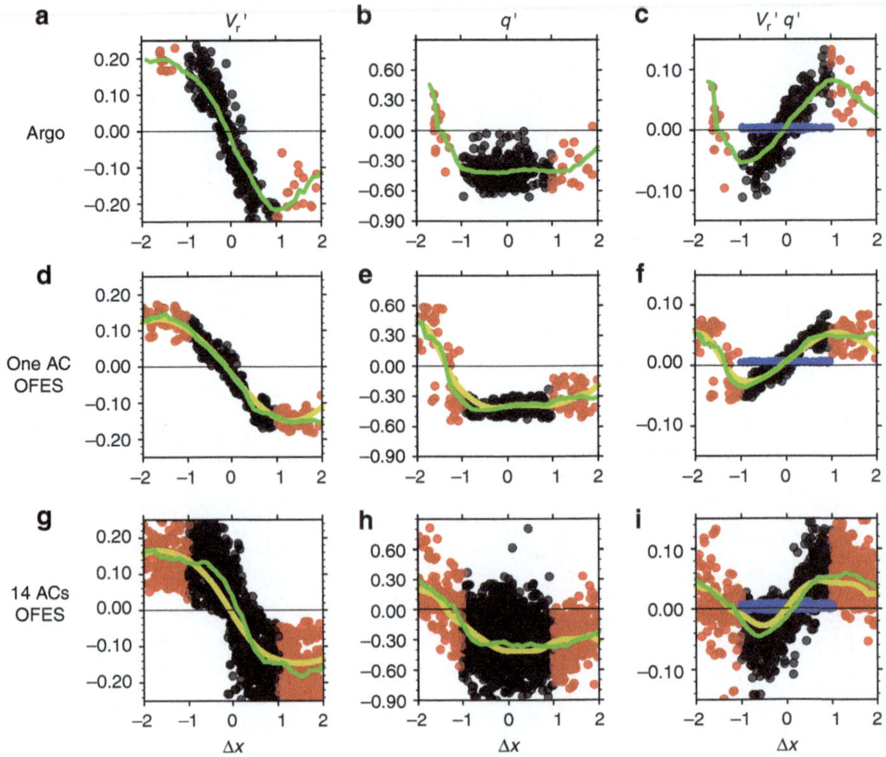

Figure 5 | The meridional PV advection by ACs. The abscissa represents zonal sections across the AC centre. Recall that [−1, 1] defines the eddy core. (**a,d,g**) The tangential velocity of AC (v_r' in $m\,s^{-1}$); (**b,e,h**) the PV anomaly from the zonal-averaged background climatology (q' in $10^{-10}\,m^{-1}\,s^{-1}$); and (**c,f,i**) the eddy-advective transport ($v_r'q'$ in black and red dots in $10^{-10}\,s^{-2}$). The black dots are within the eddy core [−1, 1] where the integrated eddy flow transport is offset, and the red dots are in ±[1, 2]. The eddy-trapping transport (C_y in blue dots) is superimposed in the right panels. (**a-c**) for Argo observations, (**d-f**) for one AC and (**g-i**) for 14 ACs in OFES. Here we choose one AC from the model (**d-f**) that has the eddy PV flux close to that of the AC in the field to compare the zonal asymmetry of PV and tangential velocity. The green lines denote the average for each $\Delta x = 0.1$ bin calculated with the Argo sampling. The yellow lines in (**d-i**) are the exact calculation of the time average based on the full model data (Methods).

Table 1 | The time-averaged meridional PV transport by a single eddy during March and April.

	Lateral advective transport	Trapping transport
Argo observations	8.24 ± 0.82	0.73 ± 0.07
OFES	7.09 ± 0.71	0.72 ± 0.07

AC, anticyclonic eddy; OFES, OGCM for the Earth Simulator; PV, potential vorticity.
On the basis of the Argo sampling (Methods), the eddy lateral advective and trapping transports (in $10^{-12}\,s^{-2}$) are estimated for the observed AC, and the average of 14 ACs in OFES (Fig. 5i). The range of value includes the ±10% error due to limited Argo sampling.

(Supplementary Fig. 4c). The east–west PV asymmetry is comparable in magnitude between the two types of ACs and CCs (Supplementary Fig. 4d).

Estimate the eddy-induced PV transport. On the basis of the Argo sampling, we estimate the time-averaged PV flux by a single AC during March and April on the core density surface of STMW, including the eddy-trapping and lateral advective transports (Table 1). The relationship between the eddy PV flux and the mode-water subduction rate is discussed in Methods. The Methods section (*Estimate the time-averaged PV transport by a single AC*) also describes in detail how we integrate the spatially and temporally varying Argo samples for the time-averaged eddy PV flux by individual ACs. We compare the estimates between the model and observations, and with the exact calculation from full model data (Fig. 5 and Table 1). The Argo sampling errors are less than 10%.

We compare the eddy-trapping and the eddy-advective PV transport based on the observed AC and the mode-water-trapping ACs we tracked in the model (Table 1). The net eddy-advective transport is large around the rim of the AC ($1 < |\Delta x| < 2$), but it is almost zero within the eddy core (($|\Delta x| < 1$), where the PV is well mixed and PV asymmetry is weak (Fig. 5c,f,i). The zonally integrated eddy PV flux by the observed AC with a normalized eddy radius is estimated at $8.24 \pm 0.82 \times 10^{-12}\,s^{-2}$ because of lateral advection, and $0.73 \pm 0.07 \times 10^{-12}\,s^{-2}$ because of the eddy-trapping effect. The eddy-trapping transport is smaller by one order of magnitude than the eddy lateral advective transport (Fig. 5c and Table 1). This result is confirmed by analysis of 14 ACs from the eddy-resolving model (Fig. 5i and Table 1).

The eddy lateral advective transport is comparable in magnitude to that by the mean flow. Directly calculated from OFES, the zonal-integrated PV transport on the core density surface of STMW across 30°N (140–150°E) by the mean flow is at $2.92 \times 10^{-6}\,m\,s^{-2}$, while the eddy-induced PV transport is at $2.88 \times 10^{-6}\,m\,s^{-2}$. Typically, there are 2.26 AC–CC pairs passing this section with an averaged eddy radius of 98.67 km. If we assume that eddy transport is symmetric between ACs and CCs, the equivalent PV flux by one single eddy (AC or CC) with a normalized eddy radius would be $6.45 \times 10^{-12}\,s^{-2}$. This is in broad agreement with the estimate by the synthetic Argo data in the model ($7.09 \pm 0.71 \times 10^{-12}\,s^{-2}$). These results support the importance of the eddy lateral advection for STMW subduction.

We address the estimate errors associated with limited Argo samples by comparing the exact eddy PV transport based on the full model data with the estimates based on the synthetic Argo samples. We find that the sampling of 17 Argo profiles per day captures the sub-eddy-scale east–west PV asymmetry and the time-averaged meridional eddy PV transport during March and April (green and yellow lines in Fig. 5d–i). The error caused by the Argo sampling is less than 10%, compared with perfect sampling in the model (Supplementary Fig. 5). This lends some credibility to the estimates from our Argo observations. Although 17 Argo floats could not fully resolve the instantaneous eddy PV flux that varies in space and time (as indicated by scatters from the composite mean in Fig. 5), our enhanced daily sampling enables a time-averaged estimate with errors less than 10%.

Quantitatively, our observations of one single eddy captured eddy-advective effects in the fields but only allow for order of magnitude estimates of these effects. Our estimate of the eddy PV flux is subject to several sources of uncertainty due to the limited sample size (only one AC observed in the field) and spatiotemporal variability among mesoscale eddies. These uncertainties are difficult to quantify at the current stage. The eddy PV fluxes, especially the long-term mean, need to be better quantified in the future with more observations.

Variations in mixing with the AC. The isopycnal PV is well mixed within the AC core but displays a strong asymmetry outside the core in the east–west direction (Fig. 5). The enhanced temporal and vertical sampling of our Argo observations allows us to examine the difference in vertical mixing between inside and outside the AC core (Figs 2i and 6). As the season progresses, the low PV trapped in the AC core seems less dissipated than that in the surroundings (Fig. 2i), perhaps because the PV barrier reduces lateral mixing[18]. In many of our Argo profiles for March–May, the STMW displays a 'multicore structure'[24,25], with more than one PV minima in the vertical (Fig. 6c) both in and outside the AC core. Remarkably, the PV minima prefer to form at discrete densities of 25.0, 25.2 and 25.4σ_θ (Fig. 6a,b). In summer, pycnostads with multiple PV minima in STMW become rare outside the AC and gradually mixed into one single broad dense core (Fig. 6b,e). In the AC core, by contrast, the multicore structure of STMW persists even in summer (Fig. 6a,d), indicating a weak vertical mixing there. The summer histogram retains the characteristic tri-modal structure in the AC core, while it transforms into a smooth distribution outside that shifts towards higher density due to the seasonal erosion of the light STMW.

Our enhanced vertical sampling yielded a large number of high-resolution profiles that enables studying the seasonal evolution of the STMW multicore structure for the first time. Such studies will shed light on how the multicore structure forms and why it persists for a longer time inside the AC core.

Discussion

We have observed the eddy-advective effect on mode-water subduction using a specially designed eddy-resolving array of 17 Argo-profiling floats deployed in an AC in the mode-water formation region. Our analysis of over 3,000 hydrographic profiles reveals that the PV and AOU are asymmetric about the AC centre, a structure indicative of lateral advection by the eddy. We further estimated that the southward eddy PV flux in the study region is comparable in magnitude with that by the mean flow, with the lateral advection dominating over the eddy-trapping effect. More eddy-resolving observations are needed to improve the estimates of long-term mean eddy fluxes. We found that the eddy lateral advection has significant seasonal variations, being strongest in winter and gradually weakening after June. The seasonal variations are closely related to the strength of the

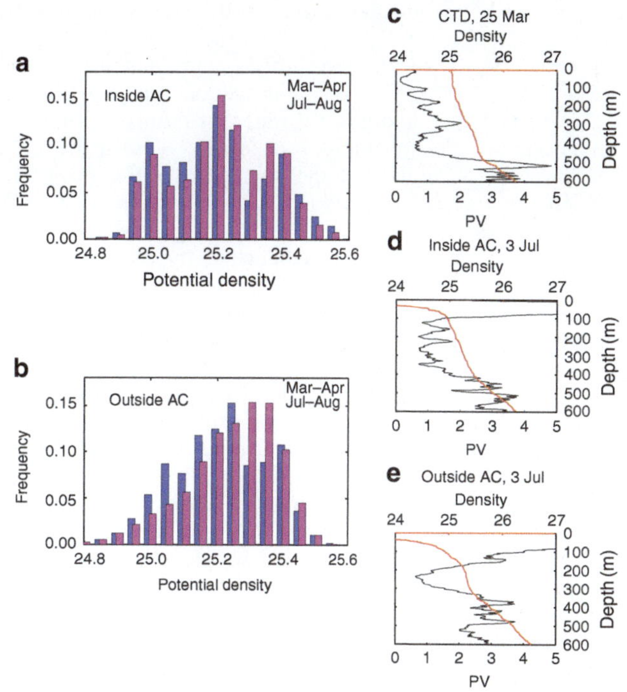

Figure 6 | Seasonal variations of the multicore structure in and outside the AC core. The AC core is defined as the interior region enclosed by the maximum of Azimuthal speed. Proportion of profiles with vertical PV minimum in 0.05σ_θ bins based on Argo floats (**a**) in and (**b**) outside the AC core, with blue (magenta) bars for March–April (July–August). Results for May–June are similar to those for July–August. Vertical profiles of PV (10^{-10} m^{-1}s^{-1}) and potential density (kg m^{-3}) for (**c**) at CTD station at 148°E and 29.6°N on 25 March 2014, from Argo floats on 3 July, (**d**) in (World Meteorological Organization (WMO) ID 2901559) and (**e**) outside (WMO ID 2901563) the AC core. Vertical profiles in **d**,**e** are typical of multiple (single) core STMW in (outside) the AC in summer.

background PV gradient that forms between the newly ventilated STMW to the north and the old STMW to the south. As the seasonal thermocline develops, the background meridional PV gradient weakens, and the southward PV advection eventually ceases.

The maximum eddy subduction takes place around the eastern rim of the AC ($1<\Delta x<2$; Figs 2g and 5). There, the subducted fluids may escape from the eddy into the interior ocean. Indeed, some of the floats left the eddy and moved southwards from the eastern rim of the AC (Fig. 1b). Besides, PV is not conservative along a trajectory of eddy flow as clear in the east–west asymmetry in Fig. 2g because of dissipation (for example, submesoscale processes). Thus, the eddy lateral advection causes a net PV flux to ventilate the interior of the upper ocean and to increase the net subduction of mode water. As long as meridional background PV gradients are maintained, the eddy advection is active, even if a fluid particle released in early winter circulates around the eddy and re-enters the mixed layer before the winter ends. This is different from Stommel's Ekman demon[26].

While eddy tracer fluxes have been estimated with satellite altimetry and surface drifters[27–31], previous studies mainly focused on the sea surface. Recent studies[32,33] using subsurface Argo data investigated an important but different eddy effect of mixing high PV from polewards of the KE (a damping effect on mode-water formation). Our field campaign targets at the subduction process south of the KE. Our subsurface data resolve the AC in a statistical sense, permitting observational estimates of eddy subduction effects in the region for the first time.

Eddy advection significantly affects the subduction and transport of STMW. In coarse-resolution climate models (Fig. 7a), the subduction of low PV water is confined to a narrow point, and the southward transport of low PV is along the mean streamline (Fig. 7a), indicating the dominance of the mean flow advection. By contrast, in gridded Argo observations, STMW subducts into the thermocline in a broad region of the winter deep-mixed layer (Fig. 7b). After subduction, the PV minimum is trapped near the outcrop within a narrow band. The width of the low PV band is ~200–300 km, corresponding to the length scale of mesoscale eddies. Even within the climatological low PV band, ~49% of historical Argo profiles do not contain STMW (Fig. 7c,d and Supplementary Fig. 6), another fact that underlines the importance of eddies for the subduction and transport of STMW.

A popular parameterization of geostrophic eddies widely used in climate models is the Gent and McWilliams scheme[34,35]. Coupled Model Intercomparison Project Phase 5 (CMIP5)-class models typically use Gent and McWilliams diffusion coefficients less than 1,000 $m^2 s^{-1}$ (refs 36,37). The values seem to be too small. Our Argo observations imply an eddy diffusivity on the order of ($10^4 m^2 s^{-1}$) in the study region (see Methods). This is consistent with several recent modelling studies that have advocated increasing the near-surface diffusivity to several thousand $m^2 s^{-1}$, but with a regional structure that peaks at O ($10^4 m^2 s^{-1}$) in the most energetic area of the ocean[10].

The value of our study is in the field campaign that was designed to test the hypothesis that the lateral advection by eddies enhances mode-water subduction. To our knowledge, it is the first time that this eddy effect has been detected in the field. We have successfully deployed an eddy-resolving array of 17 Argo floats inside an AC. Setting the parking depth at 500 dbar helped to keep many Argo floats trapped within the eddy, and the enhanced daily sampling yielded enough data for eddy-resolving composites. Reduced temporal sampling sharply reduces the number of profiles on the AC rim, deteriorating the representation of the sub-eddy-scale east–west asymmetry in PV structure (Supplementary Fig. 7). The enhanced vertical sampling, on the other hand, enables a large-scale study of seasonal evolution of the STMW multicore structure, a project that is underway. Additional eddy-resolving field experiments are necessary to better quantify the large-scale effects of mesoscale eddies.

The KE is the window where the winter-mixed layer ventilates the main thermocline of the North Pacific to the south. The ventilation makes the KE an important carbon sink[38–40]. Results from our field campaign call for improved representation of eddy effects on the subduction and transport of STMW to achieve more realistic simulations of the regional biogeochemical cycle and reliable projections of its change.

Methods

The field experiment. To investigate the eddy effect on subduction and transport of the STMW, the P-MoVE cruise deployed 17 Argo-profiling floats (World Meteorological Organization IDs 2901550–2901566) inside an AC. We considered ACs located west of 150°E (A and B in Supplementary Fig. 8) in late March 2014 for deployment. Since the AC A was close to the Izu Ridge, it might dissipate over the rough topography. Thus, we chose the AC B near 147.5°E, 29.5°N, which is a typical eddy with a radius of 71.5 km and rotational speed of 29 $cm s^{-1}$ (Supplementary Fig. 9).

The P-MoVE was conducted on the Research Vessel (R/V) *Dongfanghong 2* from 17 March to 23 April 2014. During the cruise, the target AC position was identified in advance from the near-real-time SLA field and communicated to the vessel each day. Conductivity, temperature, and depth (CTD), expendable bathythermograph (XBT) and shipboard acoustic Doppler current Profiler measurements (Supplementary Fig. 10a) were first conducted around the target AC to identify the AC centre before deploying the Argo-profiling floats. A significant difference in the vertical temperature profiles existed between the western (black line in Supplementary Fig. 10b) and eastern (red line in Supplementary Fig. 10b)

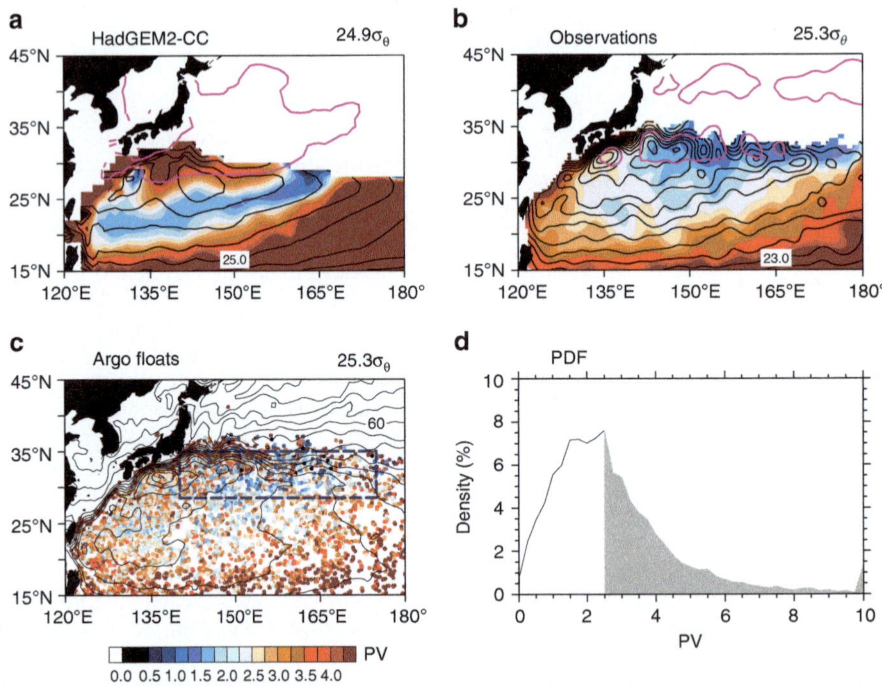

Figure 7 | Isopycnal PV on the core layer of STMW in March for climate model and gridded and ungridded raw data of Argo. (**a**) A randomly selected CMIP5 climate model (HadGEM2-CC), where eddies are parameterized. (**b**) Gridded observations based on Argo-profiling floats; the streamlines (acceleration potential, CI = 0.5 $m^2 s^{-2}$) are superimposed as black contours, and the winter MLD front (MLD = 150 m contour) in thick magenta line. (**c**) The ungridded raw data of Argo, along with the mean dynamical topography (CI = 10 cm). (**d**) PDF (%) of March isopycnal PV based on the Argo data for 2004–2014. Region for the calculation is denoted by the thick dashed rectangle in **c**. If we define the STMW using the critical PV value of $2.5 \times 10^{-10} m^{-1} s^{-1}$, ~49% of Argo-profiling floats do not contain STMW in its formation region in March (grey shading in **d**).

sides of the AC. Vertical temperature gradient was much weaker on the eastern side of the AC, suggesting fresh STMW formation there. Seventeen Argo floats were deployed on 27 March 2014 inside the AC (see deployment locations in Supplementary Fig. 10c). We deployed the Argo floats with an emphasis on the southeast part of the AC.

To keep the floats trapped in the migrating AC, we set the parking depth at 500 dbar. The floats dive to 1,000 dbar from the parking depth before the ascent, during which they conduct temperature, salinity and DO measurements. Each float makes observations daily as it rises from 1,000 dbar to near the sea surface, with a 2-dbar sampling interval above 600 m and 10 dbar below it. Compared with the conventional Argo sampling intervals of 10 days and 10 dbar, our enhanced time sampling proves crucial for our Argo array to maintain eddy-resolving capability, while the enhanced vertical sampling is instrumental in identifying reduced mixing of mode water within the AC core as detailed in Results.

Other data. To complement the P-MoVE Argo data, we include the following data: the historical ungridded raw data of Argo in the domain of interest (15–45°N, 120°–180°E) from 2004 to 2014, the daily SLA data from AVISO[41] on a horizontal grid of 1/3°, the eddy-tracking data set of Chelton *et al.*[19] and the daily outputs from OFES with a horizontal grid of 1/10° (ref. 42). All the Argo data passed the Argo real-time quality control, and those flagged as bad are eliminated. For the comparison of observational and simulated quantities, we use the gridded monthly Roemmich–Gilson Argo Climatology[43], the mean dynamical topography from the CNES-CLS09 product[44] and a randomly selected climate model—HadGEM2-CC from CMIP5 under the historical scenario.

Composite AC fields. On the basis of the daily SLA data from AVISO and the hydrographic profiles of the 17 Argo-profiling floats, we construct composites referenced to the AC centre (the red line in Fig. 1a or the black line in Fig. 1b). An eddy-following coordinate system (Δx, Δy) is used for the relative position of the Argo floats to the AC centre. The each-day AC centre is first identified based on the geometry of velocity streamlines around the SLA maximum. We interpolate each T–S and oxygen profile vertically on 1-dbar intervals using the Akima spline. Argo DO profile data are checked with shipboard CTDO and DO in the water samples at the time of launch (not shown here). Argo DO measurements failed the global range test after August 2014. The DO data after that are excluded.

Track ACs in the model. On the basis of the daily outputs of OFES, we use the 'offline particle-tracking' method to deploy synthetic Argo floats to track ACs in the model. We choose 14 ACs located within 140°E–150°E, 28°–33°N in March for 1994–2007. Seventeen synthetic Argo floats were deployed for each AC, and the deployment locations are similar as in Supplementary Fig. 10c. In line with the parking depth of 500 m, the synthetic Argo float's movement was determined by the model's daily velocity at 500 m. We record the float trajectory every day and take a vertical profile. Since the target AC is a 'mode-water eddy' where low PV water is trapped in the AC centre, we only track ACs with a low PV core in the model for one-to-one data-model comparison.

Estimate the time-averaged PV transport by a single AC. We define the time-averaged meridional PV (q) transport on the core density surface of STMW, $25.3\sigma_0$, across a zonal section ($-2<x<2$; the unit is the normalized distance from the eddy centre to the outer boundary of the eddy core) that the AC occupies as:

$$J = \left[\overline{vq}\right]_{|x|<2}, \tag{1}$$

where the bracket denotes zonal integration and the over bar denotes time mean. Note that equation (1) is under the eddy-following coordinate system. We can divide v (the meridional flow speed) and q into the climatological zonal average and the deviation from the climatological zonal average, $v = [\overline{v}] + v'$ and $q = [\overline{q}] + q'$. Then, the time mean meridional PV transport is $J = [\overline{v}][\overline{q}]_{|x|<2} + \left[\overline{v'q'}\right]_{|x|<2}$. We further define the eddy-induced velocity as:

$$v' = v'_r + C_y, \tag{2}$$

where v'_r is the relative velocity to the moving frame and C_y is the AC propagation velocity. For the 'mode-water-trapping' AC (Fig. 5), C_y is constant but v'_r is wholly antisymmetric about the AC centre, while q' is well mixed within the eddy core ($|\Delta x|<1$) but antisymmetric around the rim of the AC ($1<|\Delta x|<2$). Thus, $\left[\overline{C_y q'}\right]_{1<|x|<2} = 0$ and $\left[\overline{v'_r q'}\right]_{|x|<1} = 0$. The total meridional transport can be decomposited as follows:

$$\begin{aligned} J &= J_{mn} + J_{eddy\_trap} + J_{eddy\_adv} \\ &= [\overline{v}][\overline{q}]_{|x|<2} + \left[\overline{C_y q'}\right]_{|x|<1} + \left[\overline{v'_r q'}\right]_{1<|x|<2}. \end{aligned} \tag{3}$$

Here the eddy effects are divided into two parts: the Eddy-trapping effect $J_{eddy\_trap} = \left[\overline{C_y q'}\right]_{|x|<1}$ and the non-eddy-trapping lateral advective effect $J_{eddy\_adv} = \left[\overline{v'_r q'}\right]_{1<|x|<2}$.

Using equation (3), we estimate the time-averaged eddy PV flux for March–April based on the Argo sampling. We first calculate $v'_r q'$ ($C_y q'$) directly for each Argo sample ($-0.5<y<0.5$), and then obtain the time mean $\overline{v'_r q'}$ per $\Delta x = 0.1$ bin (solid green line in Fig. 5), and finally integrate $\overline{v'_r q'}$ zonally for the total eddy PV flux by one AC (Table 1).

This estimate (here after $J^{E}_{eddy\_adv}$) contains errors due to limited Argo sampling. To quantify the error, we compare the estimate $J^{E}_{eddy\_adv}$ to the exact calculation ($J^{T}_{eddy\_adv}$) from the full model data. For the full model data, we first calculate the instantaneous $v'_r q'$ ($C_y q'$) for each grid point around the AC, and then integrate temporally and zonally. We calculate the estimate error, $U = \frac{|J^{E}_{eddy\_adv} - J^{T}_{eddy\_adv}|}{J^{E}_{eddy\_adv}}$, for each AC we tracked in OFES and find that the error due to limited Argo sampling is less than 10% (Supplementary Fig. 5). Including the $\pm 10\%$ error due to limited Argo sampling, we add a value range for the estimate of the eddy PV flux in Table 1.

Relation between eddy PV flux and subduction rate. The meridional PV flux (T) on the core density surface of STMW is the product of velocity (v) and PV (q),

$$T = vq, \tag{4}$$

and the time mean PV flux is

$$\overline{T} = \overline{v}\overline{q} + \overline{v'q'}. \tag{5}$$

The time mean PV transport can be rewritten as a 'transport velocity' after dividing by the mean PV,

$$\frac{\overline{T}}{\overline{q}} = \overline{v} + \frac{\overline{v'q'}}{\overline{q}} = (\overline{v} + v*), \tag{6}$$

where the eddy-induced velocity in equations (6) is $v* = \frac{\overline{v'q'}}{\overline{q}}$, known as the 'bolus velocity'[20].

Following Marshall[11], the water mass subduction rate ($\overline{S}^{WM}$) in an eddying ocean can be described in terms of transport velocities

$$\overline{S}^{WM} = -\left\{\frac{\partial \overline{h}}{\partial t} + (\overline{v} + v^*)\frac{\partial \overline{h}}{\partial y} + (\overline{w} + w^*)\right\}, \tag{7}$$

where $[(\overline{v} + v^*), (\overline{w} + w^*)]$ is the transport velocity at the base of the mixed layer, $z = -h$. In particular, the subduction by mesoscale eddies is

$$S_{eddy} = -\left\{v^*\frac{\partial \overline{h}}{\partial y} + w^*\right\}. \tag{8}$$

From equations (6) and (7), we obtain the relation between the subduction rate and PV flux that transports low PV water from mixed layer into the permanent thermocline:

$$\overline{S}^{WM} = -\left\{\frac{\partial \overline{h}}{\partial t} + \frac{\overline{v}\overline{q} + \overline{v'q'}}{\overline{q}}\frac{\partial \overline{h}}{\partial y} + (\overline{w} + w^*)\right\}. \tag{9}$$

The comparison between the mean and the eddy PV flux in the present study reflects the relative contribution of the mean and the eddy flow to the subduction rate by lateral induction.

Estimate the eddy diffusivity coefficients. According to previous studies[6,35], the eddy PV flux is related to the mean PV gradient by an eddy closure hypothesis,

$$J = \overline{q'v'} = -\kappa\frac{\partial \overline{q}}{\partial y}, \tag{10}$$

where κ is the eddy diffusivity. Here we use equation (10) to diagnose κ.

The averaged eddy PV flux for the target AC $J_{obs} = \left[\overline{q'v'_r}\right]$ is estimated at $8.24 \pm 0.82 \times 10^{-12}\,s^{-2}$ (Table 1). This calculation for zonal integration is within $[-2R, 2R]$, where R is the eddy radius. As the target AC is a typical eddy in the study region (Supplementary Fig. 9), we extend the estimate to a broad region by considering the frequency of eddy occurrence

$$J = \alpha J_{obs}, \tag{11}$$

where α is the frequency of the eddy occurrence in the study region (135–175°E, 28–32°N). Here α is defined as:

$$\alpha = \frac{N*(4R)^2*T_{eddy}}{T*L_{lon}*L_{lat}}, \tag{12}$$

where T is the total length of observations, L_{lon} and L_{lot} are the zonal and meridional distances of the study region, respectively, N is the total number of mesoscale eddies (both ACs and CCs, if we assume that eddy PV transport is symmetric between them), $4R$ represents the extent for integration for J_{obs} and T_{eddy} is the averaged lifetime of mesoscale eddies.

On the basis of the eddy-tracking data set of Chelton *et al.*[19] for 1992–2012, in the study region (135–175°E, 28–32°N), the total number of mesoscale eddies $N = 1,147$ (eddies near the boundaries are excluded), the average eddy radius $R = 89.5$ km and the averaged lifetime of mesoscale eddies $T_{eddy} = 11$ weeks. Together with the length of observations $T = 1,095$ weeks, and the zonal and meridional distances of the study region $L_{lon} = 3,849.3$ km and $L_{lat} = 444.5$ km, respectively, α is estimated as 86.3%. We obtain a long-term mean eddy PV flux in the study region, $J = 7.11 \pm 0.71 \times 10^{-12}\,s^{-2}$.

The corresponding mean PV gradient $\frac{\partial \bar{q}}{\partial y}$ is estimated to be $-1.64 \times 10^{-16}\,m^{-2}\,s^{-1}$ based on the Roemmich-Gilson Argo climatology (blue line in Fig. 2j). Correspondingly, the eddy diffusivity is in the range of $(4.34 \pm 0.43) \times 10^{4}\,m^{2}\,s^{-1}$. Our Argo observations imply an eddy diffusivity on the order of $10^{4}\,m^{2}\,s^{-1}$ in the study region.

References

1. Hanawa, K. & Talley, L. D. Mode waters. *Int. Geophys.* **77,** 373–386 (2001).
2. Oka, E. & Qiu, B. Progress of North Pacific mode water research in the past decade. *J. Oceanogr.* **68,** 5–20 (2012).
3. Gu, D. & Philander, S. G. H. Interdecadal climate fluctuations that depend on exchanges between the tropics and extratropics. *Science* **275,** 805–807 (1997).
4. Bates, N. R., Pequignet, A. C., Johnson, R. J. & Gruber, N. A variable sink for atmospheric CO_2 in subtropical mode water of the North Atlantic Ocean. *Nature* **420,** 489–493 (2002).
5. Sallée, J. B., Matear, R. J., Rintoul, S. R. & Lenton, A. Localized subduction of anthropogenic carbon dioxide in the Southern Hemisphere oceans. *Nat. Geosci.* **5,** 579–584 (2012).
6. Lee, M. M., Nurser, J. G., Sallée, I. & Stevens, J. B. Subduction over the southern Indian Ocean in a high-resolution atmosphere-ocean coupled model. *J. Climate* **24,** 3830–3849 (2011).
7. Xu, L., Xie, S. P., McClean, J. L., Liu, Q. & Sasaki, H. Mesoscale eddy effects on the subduction of North Pacific mode waters. *J. Geophys. Res. Oceans* **119,** 4867–4886 (2014).
8. Oka, E. & Suga, T. Formation region of North Pacific subtropical mode water in the late winter of 2003. *Geophys. Res. Lett.* **30,** 2205 (2003).
9. Naveira-Garabato, A. C., Allen, J., Leach, H., Strass, V. & Pollard, R. Mesoscale subduction at the Antarctic Polar Front driven by baroclinic instability. *J. Phys. Oceanogr.* **31,** 2087–2107 (2001).
10. Sallée, J. B. & Rintoul, S. R. Parameterization of eddy-induced subduction in the Southern Ocean surface-layer. *Ocean Model.* **39,** 146–153 (2011).
11. Marshall, D. Subduction of water masses in an eddying ocean. *J. Mar. Res.* **55,** 201–222 (1997).
12. Qu, T., Xie, S. P., Mitsudera, H. & Ishida, A. Subduction of the North Pacific mode waters in a global high-resolution GCM. *J. Phys. Oceanogr.* **32,** 746–763 (2002).
13. Nishikawa, S., Tsujino, H., Sakamoto, K. & Nakano, H. Effects of mesoscale eddies on subduction and distribution of subtropical mode water in an eddy-resolving OGCM of the western North Pacific. *J. Phys. Oceanogr.* **40,** 1748–1765 (2010).
14. Uehara, H., Suga, T., Hanawa, K. & Shikama, N. A role of eddies in formation and transport of North Pacific subtropical mode water. *Geophys. Res. Lett.* **30,** 1705 (2003).
15. Pan, A. & Liu, Q. Mesoscale eddy effects on the wintertime vertical mixing in the formation region of the North Pacific Subtropical Mode Water. *Chin. Sci. Bull.* **50,** 1949–1956 (2005).
16. Kouketsu, S. *et al.* The role of meso-scale eddies in mixed layer deepening and mode water formation in the western North Pacific. *J. Oceanogr.* **68,** 63–77 (2011).
17. Liu, C. & Li, P. The impact of meso-scale eddies on the subtropical mode water in the western North Pacific. *J. Ocean Univ. China* **12,** 230–236 (2013).
18. Zhang, Z., Wang, W. & Qiu, B. Oceanic mass transport by mesoscale eddies. *Science* **345,** 322–324 (2014).
19. Chelton, D. B., Schlax, M. G. & Samelson, R. M. Global observations of nonlinear mesoscale eddies. *Prog. Oceanogr.* **91,** 167–216 (2011).
20. Lee, M.-M., Marshall, D. P. & Williams, R. G. On the eddy transfer of tracers: advective or diffusive? *J. Mar. Res.* **55,** 483–505 (1997).
21. Xie, S. P. Advancing climate dynamics toward reliable regional climate projections. *J. Ocean Univ. China* **12,** 191–200 (2013).
22. Ito, T., Follows, M. J. & Boyle, E. A. Is AOU a good measure of respiration in the oceans? *Geophys. Res. Lett.* **31,** L17305 (2004).
23. Nakano, H., Tsujino, H. & Sakamoto, K. Tracer transport in cold-core rings pinched off from the Kuroshio extension in an eddy-resolving ocean general circulation model. *J. Geophys. Res. Oceans* **118,** 5461–5488 (2013).
24. Taneda, T., Suga, T. & Hanawa, K. Subtropical mode water variation in the southwestern part of the North Pacific subtropical gyre. *J., Geophys. Res.* **105,** 19591–19598 (2000).
25. Oka, E. *et al.* 'Eddy resolving' observation of the North Pacific subtropical mode water. *J. Phys. Oceanogr.* **41,** 666–681 (2011).
26. Stommel, H. Determination of water mass properties of water pumped down from the Ekman layer to the geostrophic flow below. *Proc. Natl Acad. Sci. USA* **76,** 3051–3055 (1979).
27. Owens, W. B. A synoptic and statistical description of the Gulf Stream and subtropical gyre using SOFAR floats. *J. Phys. Oceanogr.* **14,** 104–113 (1984).
28. Holloway, G. Estimation of oceanic eddy transports from satellite altimetry. *Nature* **323,** 243–244 (1986).
29. Davis, R. Lagrangian ocean studies. *Annu. Rev. Fluid Mech.* **23,** 43–64 (1991).
30. Stammer, D. On eddy characteristics, eddy transports, and mean flow properties. *J. Phys. Oceanogr.* **28,** 727–739 (1998).
31. Abernathey, R. P. & Marshall, J. Global surface eddy diffusivities derived from satellite altimetry. *J. Geophys. Res. Ocean* **118,** 901–916 (2013).
32. Qiu, B. & Chen, S. Decadal variability in the formation of the North Pacific subtropical mode water: oceanic versus atmospheric control. *J. Phys. Oceanogr.* **36,** 1365–1380 (2006).
33. Qiu, B., Chen, S. & Hacker, P. Effect of mesoscale eddies on subtropical mode water variability from the Kuroshio Extension System Study (KESS). *J. Phys. Oceanogr.* **37,** 982–1000 (2007).
34. Redi, M. H. Oceanic isopycnal mixing by coordinate rotation. *J. Phys. Oceanogr.* **12,** 1154–1158 (1982).
35. Gent, P. R. & McWilliams, J. C. Isopycnal mixing in ocean circulation models. *J. Phys. Oceanogr.* **20,** 150–155 (1990).
36. Danabasoglu, G. & Marshall, G. Effects of vertical variations of thickness diffusivity in an ocean general circulation model. *Ocean Model.* **18,** 122–141 (2007).
37. Griffies, S. *et al.* Coordinated ocean-ice reference experiments (Cores). *Ocean Model.* **26,** 1–46 (2009).
38. Cronin, M. F., Meinig, C., Sabine, C. L., Ichikawa, H. & Tomita, H. Surface mooring network in the Kuroshio extension. *Syst. J. IEEE* **2,** 424–430 (2008).
39. Nakano, H. *et al.* Uptake mechanism of anthropogenic CO2 in the Kuroshio extension region in an ocean general circulation model. *J. Oceanogr.* **67,** 765–783 (2011).
40. Ishii, M. *et al.* Air–sea CO2 flux in the Pacific Ocean for the period 1990–2009. *Biogeosciences* **11,** 709–734 (2014).
41. Dibarboure, G., Lauret, O., Mertz, F. & Rosmordu, V. *SSALTO/DUACS User Handbook: (M)SLA and (M)ADT Near-Real Time and Delayed Time Products*, Version 1, Revision 9, CLS 39 (AVISO, 2008).
42. Sasaki, H. *et al. in High Resolution Numerical Modelling of the Atmosphere and Ocean* 157–186 (Springer, 2008).
43. Roemmich, D. & Gilson, J. The 2004-2008 mean and annual cycle of temperature, salinity, and steric height in the global ocean from the Argo Program. *Prog. Oceanogr.* **82,** 81–100 (2009).
44. Rio, M. H., Guinehut, S. & Larnicol, G. New CNES-CLS09 global mean dynamic topography computed from the combination of GRACE data, altimetry, and in situ measurements. *J. Geophys. Res.* **116,** C07018 (2011).

Acknowledgements

We wish to thank the captain and crew of the 'R/V *Dongfanghong 2*' for conducting the P-MoVE and the reviewers for their constructive comments. This work was supported by the National Basic Research Program of China (2012CB955602), the Natural Science Foundation of China (41490643, 41476002, 41490641, 41506010, 41176006 and 41221063), the China Postdoctoral Science Foundation-funded project, the Fundamental Research Funds for the Central Universities (201503029) and the US National Science Foundation.

Author contributions

L.X. and S.-P.X. designed the study and wrote the manuscript. P.L. led the cruise. L.X. performed the data analysis, with contributions from C.L. and W.G. Q.L. contributed to improving the analysis. All authors participated in the discussions of the results.

Additional information

Competing financial interests: The authors declare no competing financial interests.

17

Tracking ocean heat uptake during the surface warming hiatus

Wei Liu[1], Shang-Ping Xie[1] & Jian Lu[2]

Ocean heat uptake is observed to penetrate deep into the Atlantic and Southern Oceans during the recent hiatus of global warming. Here we show that the deep heat penetration in these two basins is not unique to the hiatus but is characteristic of anthropogenic warming and merely reflects the depth of the mean meridional overturning circulation in the basin. We find, however, that heat redistribution in the upper 350 m between the Pacific and Indian Oceans is closely tied to the surface warming hiatus. The Indian Ocean shows an anomalous warming below 50 m during hiatus events due to an enhanced heat transport by the Indonesian throughflow in response to the intensified trade winds in the equatorial Pacific. Thus, the Pacific and Indian Oceans are the key regions to track ocean heat uptake during the surface warming hiatus.

[1] Climate, Atmospheric Sciences, and Physical Oceanography (CASPO), Scripps Institution of Oceanography, University of California, San Diego, 9500 Gilman Drive, La Jolla, California 92093, USA. [2] Pacific Northwest National Laboratory, 902 Battelle Blvd, Richland, Washington 99354, USA. Correspondence and requests for materials should be addressed to W.L. (email: wel109@ucsd.edu).

The global surface temperature increase has slowed over the past 16 years[1–3]. The warming hiatus is regarded as a result of natural variability against the centennial warming trend, while radiative forcing change also contributes[4–6]. Two leading theories have been proposed to explain the hiatus. One suggests that the hiatus is closely associated with the negative phase of the Interdecadal Pacific Oscillation (IPO)[7–9], manifested as a La Niña-like cooling and intensified easterly winds over the equatorial Pacific[10,11]. Consistent with this theory, the recent hiatus can be successfully simulated by nudging the sea surface temperature (SST) or trade winds to observations over the tropical Pacific[12–14]. The other idea is based on an energy view, focusing on the vertical energy redistribution in the ocean on the decadal timescale. The surface warming hiatus is accompanied by an excessive global deep ocean (>700 m) heat uptake[15], consistent with a positive radiative imbalance at the top of the atmosphere[16–20]. These two theories are not mutually exclusive. Indeed, the accelerated trade winds at the IPO-negative phase tilts the equatorial thermocline, causing an apparent vertical heat redistribution[13,21].

Recent studies[21–23] further attempted to infer mechanisms for the warming hiatus by tracking regional ocean heat uptake at different levels. Specifically, an earlier model study[21] found that surface warming hiatus is significantly correlated with accelerated warming in deep levels in all basins, below 750 m in the Atlantic and Southern Oceans and below 300 m in the Pacific and Indian Oceans. Observational data analyses, however, showed different results, from models and between data sets. In the Ishii data (Methods), the increase of ocean heat content (OHC) between 300 and 1,500 m accelerates in the Atlantic and Southern Oceans but changes little in the Pacific and Indian Oceans[22]. In the World Ocean Atlas (WOA) data (Methods), on the other hand, the recent surface warming hiatus is associated with a heat rearrangement within the upper 300 m between the Pacific and Indian Oceans[23], along with an increased heat transport from the Pacific to the Indian Ocean by the Indonesian throughflow (ITF)[24]. These debates highlight a gap between models and observations, large uncertainties in OHC observations[19,25–28], and a complex relationship between global mean surface temperature (GMST) and OHC[29].

In this study, we revisit the energy theory and combine multiple observational data sets and state-of-art model simulations to examine the relationship between the GMST and regional OHC on decadal timescales. We show that the deep ocean heat uptake in the Atlantic and Southern Oceans primarily reflects the downward penetration of anthropogenic heat via the deep meridional overturning circulation (MOC), while OHC in the upper 350 m in the Pacific and Indian Oceans is closely tied to the surface warming hiatus.

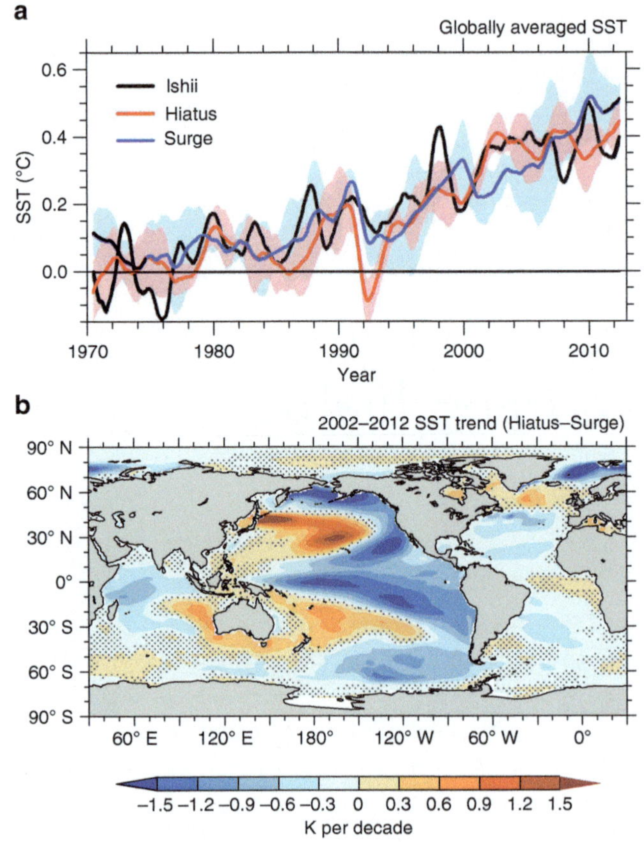

Figure 1 | Observed and simulated SST. (**a**) Monthly mean globally averaged SST from the Ishii data (black) and the CESM ensemble simulations. Two groups (named the Hiatus and Surge) of ensembles are shown, whose variations (one s.d.) are shaded in light pink and light blue and ensemble means are drawn in red and blue. All the curves are shown as a 12-month running mean by subtracting the annual mean value of the first year (year 1970). (**b**) The SST trend difference between the ensemble means of the CESM Hiatus and Surge groups during 2002–2012 (shading in *K* per decade). Stippling indicates region below 95% significance computed from a two-tailed *t*-test.

Results

Global and regional OHC changes. The observed GMST (defined as globally averaged SST in this study, not including land surface temperature) from the Ishii data[30] (Methods) shows a surface warming slowdown since 1998, with a near-zero warming trend during 2002–2012 (Fig. 1). Using the same data, ref. 22 shows that the OHC anomalies (from the 1970–2012 mean) increase during the hiatus period, and the warming penetrates deeper than 300 m only in the Atlantic and Southern Oceans (Supplementary Fig. 1). The large heat penetration was then taken as suggesting the importance of these basins for the hiatus. Here, instead of referencing to the 1970–2012 mean as in ref. 22, we compute ocean temperature anomalies as deviations from the 1970 values because referencing to 1970 facilitates an easy inspection of the OHC variation over the entire analysis period. From our rendering in Fig. 2, the most prominent feature of historical OHC variation is the signal of anthropologic warming: both global and regional OHCs continuously increase during 1970–2012. The interbasin difference of OHC increase is not unique to the hiatus period but common to the entire analysis period. Between 300 and 1,500 m, the OHC in the Atlantic, Southern Ocean, Pacific and Indian Ocean (Supplementary Fig. 1) exhibits a warming trend of 0.19, 0.23, 0.05 and 0.09×10^{23} J per decade during 1998–2012, which accounts for 33.8%, 40.7%, 8.2% and 15.8% of the global OHC warming trend. Such interbasin partition holds when we extend the analysis period to 1970–2012 (and to other observational data sets, see Supplementary Fig. 2). The Atlantic, Southern Ocean, Pacific and Indian Ocean explain 30.7%, 41.3%, 13.5% and 5.4% of the global warming within 300–1,500 m, respectively. This result demonstrates the dominance of the Atlantic and Southern Oceans in the deep ocean sequestration of heat, and more importantly, indicates that the penetration of heat below 700 m is not uniquely tied to interdecadal modulations of the surface warming rate.

Model simulations are consistent with the above result. From the Community Earth System Model (CESM) large ensemble simulations[31] (Methods), four members are selected each for the

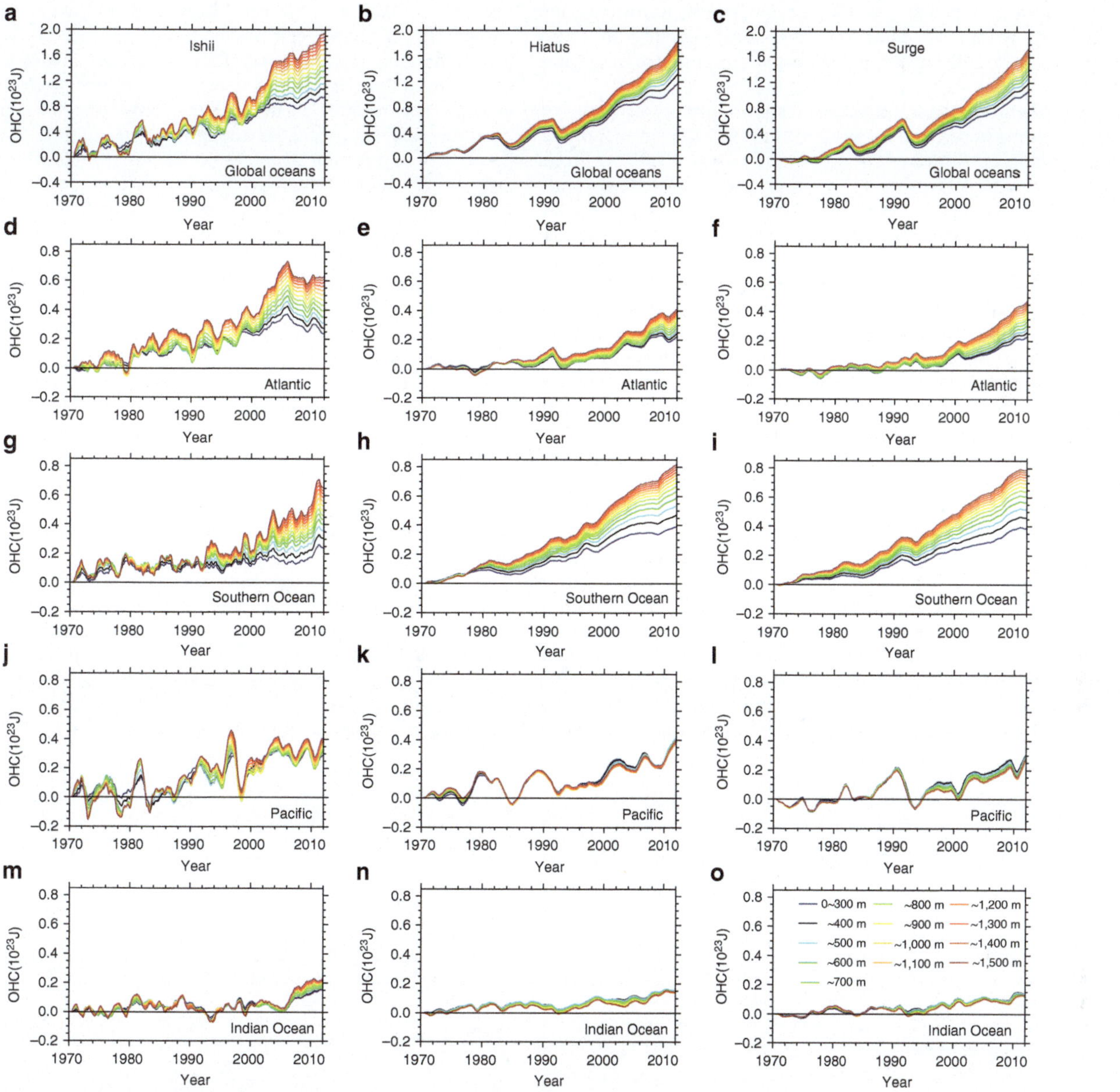

Figure 2 | Observed and simulated OHC. OHC integrated from the surface to indicated depths in global oceans, the Atlantic, Southern Ocean, Pacific and Indian Ocean from the Ishii data (**a,d,g,j,m**) and the ensemble means of the CESM Hiatus (**b,e,h,k,n**) and Surge (**c,f,i,l,o**) groups. All the curves are shown as a 12-month running mean by subtracting the annual mean value of the first year (year 1970).

Hiatus and Surge groups. The Hiatus group members are where the decadal trend of GMST is once negative during 2002–2012, whereas the Surge group are the runs corresponding to individual Hiatus members but with the largest warming trend during the same span (Supplementary Fig. 3; Supplementary Table 1). In Fig. 1a, the Hiatus group features a warming hiatus, with a near-zero warming rate since the early 2000s, while the Surge group shows a continuous surface warming, much like the ensemble mean of the Couple Model Intercomparison Project phase 5 (CMIP5) models[8,32]. The Hiatus and Surge groups share the same radiative forcing, but have different initial conditions. The difference between two groups is thus due to natural variability. The spatial pattern of the Hiatus minus Surge difference is characterized by negative SST trends in the tropical Pacific, and positive trends in the subtropical North and South Pacific (Fig. 1b), a pattern resembling the La Niña-like negative phase of the IPO.

Despite the differences in the GMST warming rate, subsurface OHCs (Methods) in the Ishii data, the Hiatus and Surge groups share a common evolution pattern. During 1970–2012, the global OHC keeps increasing, and the rate increases when it is integrated over a greater depth (Fig. 2). The Atlantic and Southern Oceans contribute most to OHC increase at depths >300 m, and this dominance in deep ocean contribution is not dependent on the GMST warming rate. During 1998–2012, the increase of 300–1,500 m OHC in the Atlantic and Southern Oceans accounts, respectively, for 30.7% and 50.1% of global OHC increase in the Hiatus group, and 43.2% and 46.3% in the Surge group. It is clear that results from either group are consistent with observations that show a deep ocean uptake in these basins. The deep heat storage takes place in the Atlantic and Southern Oceans over the entire 43 years, regardless whether the GMST warming slows down (in the Ishii data and the Hiatus group) or accelerates (in the Surge group). The deep heat penetration in the Atlantic and

Southern Oceans is the result of anthropogenic warming, not unique to the hiatus. Therefore, the deep warming in these two basins is not the basis to argue for their importance in driving the hiatus.

To further test the robustness of above results, we look for more warming hiatus and surge periods in the future events of the large ensembles. Running 10-year linear trends of GMST from 38-member simulations reveal another five hiatus events with negative decadal trends after 2012. There are nine hiatus decades in total. For each hiatus decade, we inspect the trends in the other 37 members and choose the one with the largest warming trend as the corresponding warming surge decade. The same span ensures the pair of hiatus and surge events share the identical radiative forcing. Thus, we expand the Hiatus and Surge groups to nine members each by including five pairs of hiatus (Supplementary Fig. 4) and surge decades (Supplementary Table 1) in the future simulations (the new groups are referred as the fHiatus and fSurge, see Methods). The decadal trend of composite GMST is -0.014 ± 0.023 K per decade in the fHiatus group and 0.376 ± 0.098 K per decade in the fSurge group, where the range ($\pm$ one s.d. of group $\times$ 1.86) represents the 95% range from a one-sided Student's t-test[21]. The trend spreads of two groups do not overlap, so these two groups are well separated regarding the surface warming rate.

Figure 3 compares the composite average trends of global and regional OHCs between the fHiatus and fSurge groups, where the error bar denotes one s.d. among members. To track the vertical heat redistribution over decades, we separate the ocean into four layers: 0–50 m, 50–350 m, 350–700 m and below 700 m. For the upper 350 m, the composite global OHC trend for nine hiatus decades is 0.39×10^{23}J per decade, a reduction of 31.6% compared with the composite trend in the fSurge group. This reduction is compensated in the deeper layers where the OHC trend for the hiatus decades is greater than the surge decades. For the 350–700 m layer, the composite global average in the fHiatus group is 32.0% larger (0.21 versus 0.15×10^{23}J per decade), and for the layer below 700 m, the composite global average in the fHiatus group is 5.6% larger (0.40 versus 0.37×10^{23}J per decade). The vertical heat redistribution is such that more heat is stored in the deep layers during hiatus decades[21].

Deep ocean below 700 m contributes ~38% of the whole ocean heat uptake. The difference in deep ocean heat uptake between the hiatus and surge groups is one order of magnitude smaller than their mean uptake (Fig. 3a). The s.d. within each group is also a few times smaller than the group mean. Thus, the deep ocean heat uptake observed during the hiatus decade reflects primarily anthropogenic warming, not the decadal variations that cause the hiatus. In support of observational results[23], our analyses of the large ensemble simulations show that statistically we do not expect to see a significant correspondence between decadal modulations of the GMST trend and global deep ocean heat uptake.

A clear pattern emerges from the regional ocean heat uptake below 700 m (Fig. 3b): it is large in the Atlantic and Southern Oceans (each making up ~36% of the global uptake) but small in the Pacific and Indian Oceans (~28% combined), a result consistent with observations[22]. Like the global mean, the deep ocean heat uptake in the Atlantic and Southern Oceans is dominated by anthropogenic warming, while the decadal

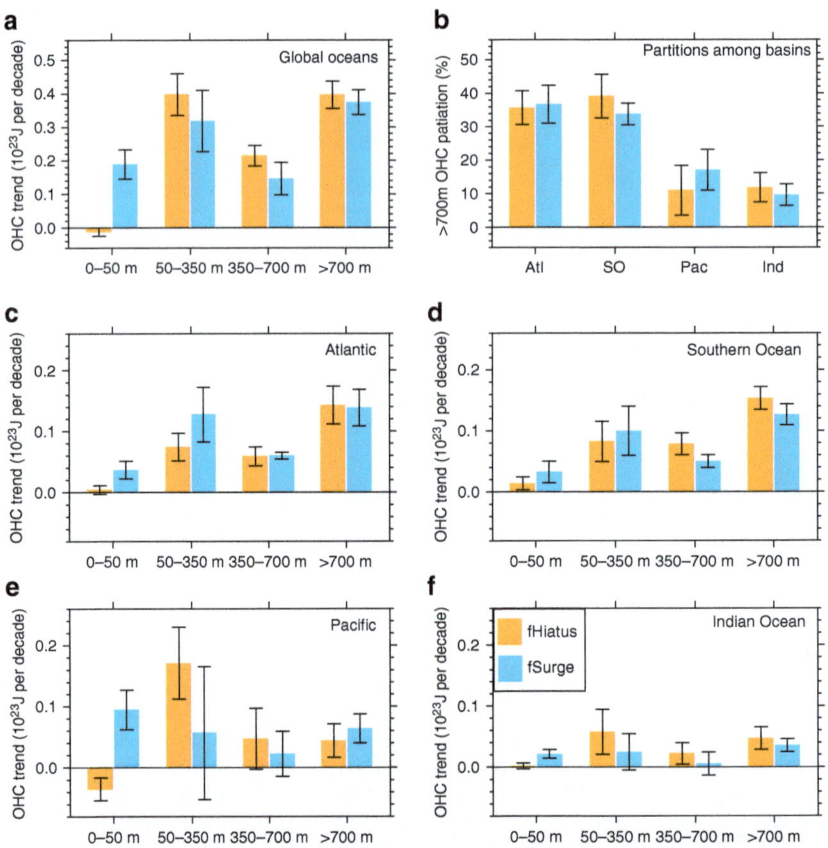

Figure 3 | OHC during hiatus and surge events. (**a**) Composite linear trends of global OHC during hiatus (orange bars) and surge (deep sky-blue bars) decades in the 21st century for four layers (surface to 50 m, 50-350 m, 350-700 m and below 700 m) with error bars showing one s.d. (**c-f**) Similar to **a** but for the Atlantic, Southern Ocean, Pacific and Indian Ocean. (**b**) The partition of individual basins for the OHC trend in oceans deeper than 700 m.

variability (including the fHiatus-fSurge difference) is an order of magnitude smaller. We can therefore apply the same conclusion made for global heat uptake below 700 m to these individual basins: the penetration of heat below 700 m in these basins does not make a major contribution to surface hiatus events or surges, either individually, or collectively.

Mechanisms for regional OHC variations. Physically, the disconnection between the decadal variations in GMST and OHC below 700 m reflects the fact that the variations of GMST and OHC below 700 m are governed by different mechanisms. GMST is affected by the atmosphere–ocean interaction and SST variability is organized into coherent patterns like the IPO[8,9,13]. Below 700 m, temperature variations are governed by distinct subsurface ocean dynamics, especially the MOCs in the Atlantic and Southern Oceans. MOCs prove to set the depth at which anthropogenic warming penetrates, which extend to great depths (>1,500 m) in the Atlantic and Southern Oceans. The former is known as the Atlantic meridional overturning circulation (AMOC) (Fig. 4c) and the latter is a residual circulation resulting from wind-driven and eddy-mediating mechanisms (Fig. 4f). The sinking motions of the deep MOCs (40–80°N in the Atlantic and 40–50°S in the Southern Ocean) sequester anthropogenic heat in the deep ocean below 700 m in these two regions, causing a pronounced abyssal warming (Fig. 4a–f). MOCs in the Pacific and Indian Oceans, however, are limited to shallow depths (< ~300 m). They appear as symmetric cells about the equator in the former while as a single anticlockwise cell in the latter, trapping anthropogenic heat mainly within the upper ~300 m in these two basins (Fig. 4g–l).

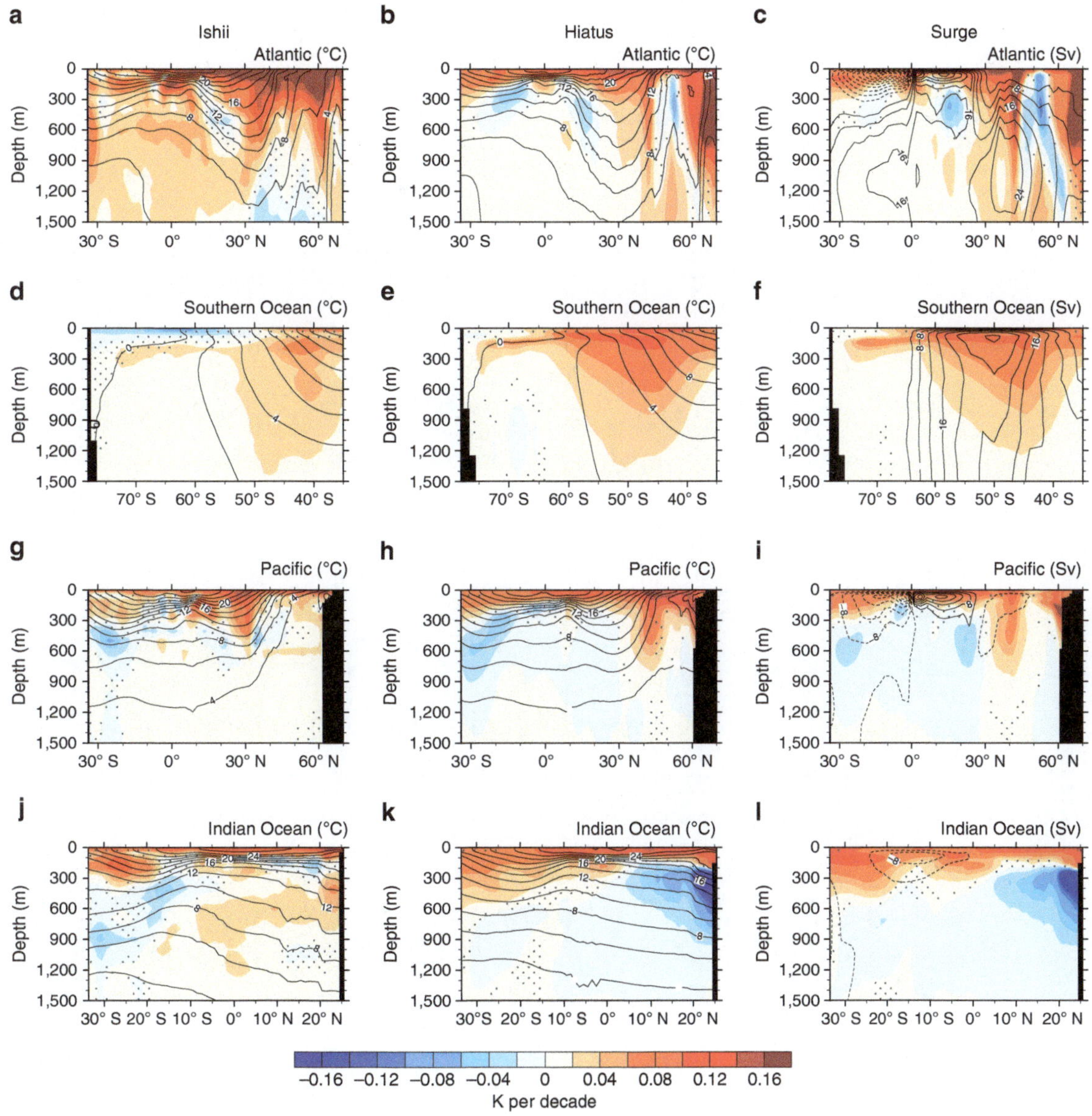

Figure 4 | Observed and simulated subsurface temperature trends. Zonal mean temperature trends during 1970-2012 (shading in *K* per decade) in the upper 1,500 m in global oceans, the Atlantic, Southern Ocean, Pacific and Indian Ocean from the Ishii data (**a**,**d**,**g**,**j**) and the ensemble means of the CESM Hiatus (**b**,**e**,**h**,**k**) and Surge (**c**,**f**,**i**,**l**) groups. Stippling indicates region below 95% significance computed from a two-tailed *t*-test. Contours show the 1970-2012 climatology of isotherm (°C) and meridional overturning stream-function (S_v, $1S_v = 1 \times 10^6\, m^3\, s^{-1}$). Since isotherms and meridional overturning stream-functions are highly similar between the Hiatus and Surge groups, for simplicity, the former is included in panels of the Ishii data and the hiatus group, while the latter is included in panels of the Surge group.

Heat uptake above 700 m, by contrast, shows certain correlation with GMST. Within the top three layers, differences in Indo-Pacific heat uptake between the fHiatus and fSurge groups are robust and on the same order of magnitude as their mean uptakes (Fig. 3e,f). During hiatus periods, anomalous cooling happens in the surface mixed layer (0–50 m) in all basins, especially in the Pacific. Over the Pacific and Indian Oceans, the 0–50 m cooling is compensated by warming within 50–350 m (Fig. 3e,f). This result is different from an earlier model study[21], which showed an overall cooling in the upper 300 m in the Indian Ocean but consistent with observations[23]. Here we include four observational data sets to compare with model results: the Ishii data, the EN4 data[33], the WOA data[34] and the latest European Centre for Medium-Range Weather Forecasts ocean reanalysis system 4 (ECMWF ORAS4) product[15,35] (Methods). All the observational data sets and CESM simulation consistently show an anomalous warming in the Indian Ocean below $\sim$50 m (Supplementary Fig. 5e) and an accelerated OHC increase (Fig. 5f) during hiatus periods. This Indian Ocean OHC increase corresponds to a concurrent Pacific OHC decrease in the 0–100 m layer (Fig. 5e), indicating heat redistribution between these two basins. The distribution patterns vary among data sets due to data uncertainties, while the model result well lies within the range of observational uncertainties (Fig. 5b). Our model result shows that the total Indo-Pacific OHC change is close to zero within the upper 350 m (Fig. 5b), meaning that most of the hiatus-related cooling in the surface and mixed layer is compensated by warming in 50–350 m. Therefore, enhanced heat uptake below 350 m, as suggested by ref. 21, is not required in these two basins.

Figure 6b,c and Supplementary Fig. 6d further show that warming in the Indian Ocean mostly happens in the tropics at the thermocline depth (70–150 m). This warming pattern appears related to a shift towards a La Niña-like state[23] and a change of the ITF[24]. To further investigate the La Niña-like shift and related Indo-Pacific heat rearrangement, we consider the difference of climate trend between the Hiatus and Surge groups during 2002–2012. Our results show a pattern in the Pacific consistent with the La Niña-like negative phase of the IPO as described by ref. 13. Specially, anomalous high sea level pressure (SLP) centres in the mid-latitudes and intensified trade winds reflect the accelerated Walker and Hadley cells (Supplementary Fig. 7a). Strengthened surface winds accelerate equatorial surface currents (Supplementary Fig. 7c), the Equatorial Undercurrent (EUC) (Supplementary Fig. 7d) and the Pacific shallow MOC (Fig. 6a and Supplementary Fig. 7e,f). The tropical thermocline deepens in the central and western Pacific (Fig. 6c; Supplementary Fig. 7b) with an anomalous subsurface warming maximum due to

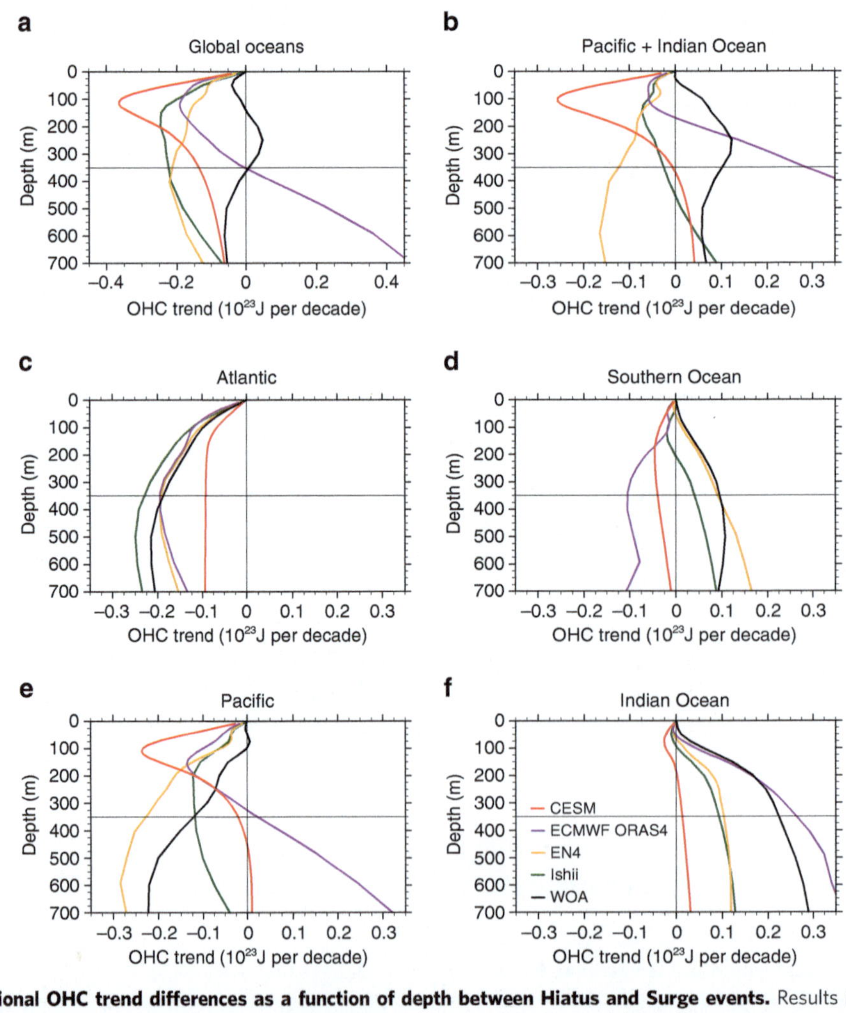

Figure 5 | Global and regional OHC trend differences as a function of depth between Hiatus and Surge events. Results in (**a**) global oceans, (**b**) the Pacific and Indian Oceans, (**c**) the Atlantic, (**d**) the Southern Ocean, (**e**) the Pacific and (**f**) the Indian Ocean are from observational data sets and the CESM simulations. OHC is integrated from the surface to indicated depths. Observations include the ECMWF ORAS4 reanalysis product (purple), the EN4 data (orange), the Ishii data (dark green) and the WOA data (blue). The CESM result (red) shows the OHC trend difference between the fHiatus and fSurge groups (the fHiatus ensemble mean minus the fSurge ensemble mean).

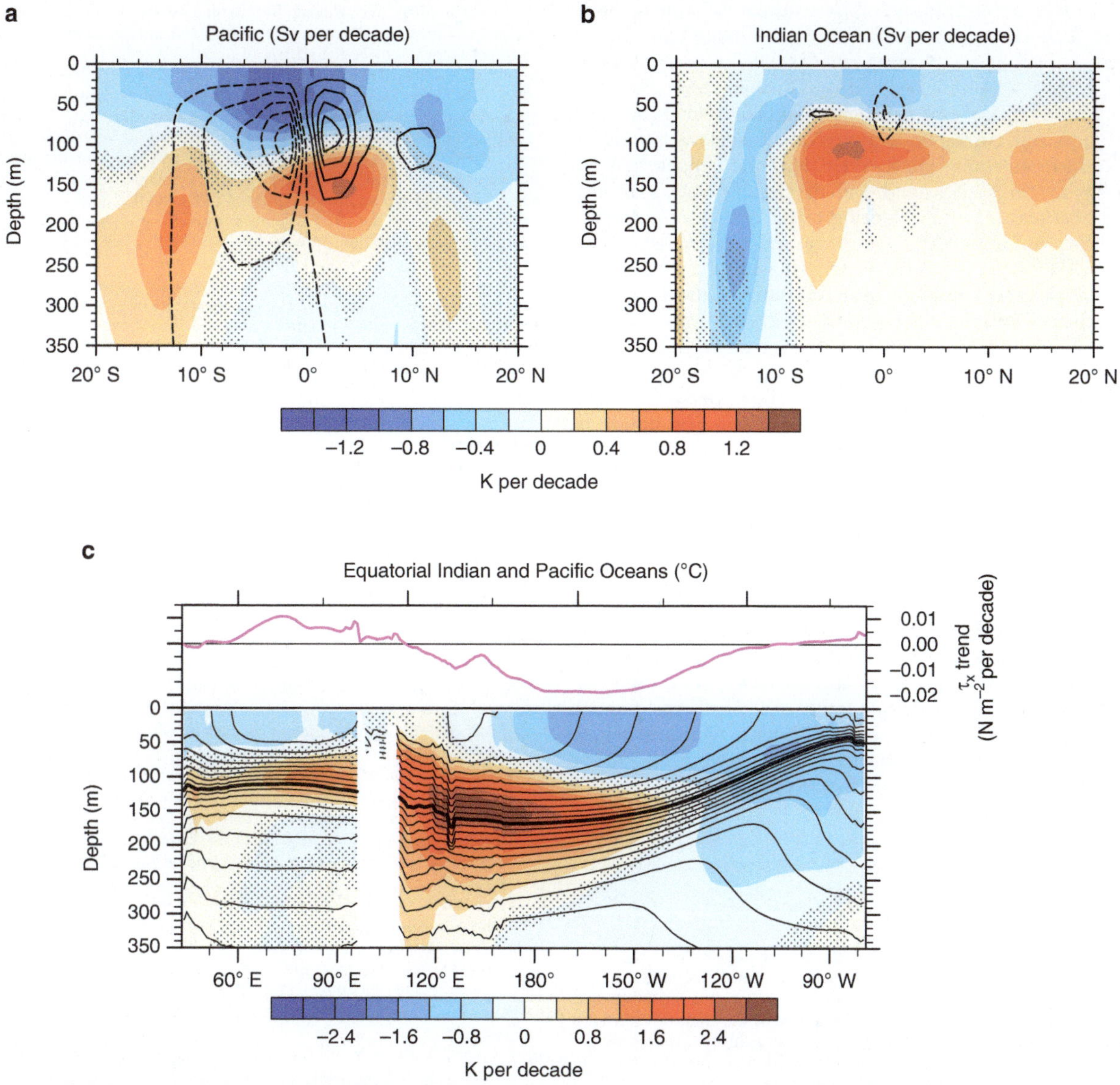

Figure 6 | Temperature trend differences in the tropical Pacific and Indian Oceans in the upper 350 m between the CESM Hiatus and Surge groups during 2002-2012. (**a**) The trend difference of zonal mean temperature (shading in *K* per decade) in the tropical Pacific, superposed by the trend difference of meridional overturning stream-function (contoured by $2S_v$ per decade, with zero contours omitted). (**b**) Similar to **a** but for the tropical Indian Ocean. (**c**) Trend differences of zonal wind stress (orchid) and temperature (shading in *K* per decade) along equatorial band (5° S-5° N) in the Indian and Pacific Oceans. The mean isotherms during 2002-2012 (38-ensemble mean) are also included as contours with an interval of 1 °C. The 20 °C contour is thickened to indicate the depth of thermocline.

increased equatorial pycnocline heat convergence, and shoals in the eastern Pacific with enhanced upwelling and surface cooling (Figs 6c; 1b).

The anomalous subsurface warming and accelerated OHC increase in the Indian Ocean is closely associated with the change in the Pacific. Anomalous warm water in the tropical western Pacific can be transported into the Indian Ocean via the Indonesian passages. During hiatus events, both the ITF volume transport and the ITF heat transport markedly increase[24], especially over the upper 350 m (Supplementary Fig. 8), which substantially contributes to the OHC increase in the Indian Ocean. Besides, local response in the Indian Ocean during the La Niña-like shift also facilitates the acceleration of the OHC increase in the Indian Ocean. Anomalous low SLP occurs in the subtropical southeastern Indian Ocean, with relaxed southeasterlies along Sumatra (Supplementary Fig. 7a). The weakened along-shore winds reduce the upwelling and deepen the thermocline off the coast. At the equator, anomalous westerlies pile up water eastwards and depress the thermocline in the eastern tropical Indian Ocean (Fig. 6c). Both processes contribute to the anomalous subsurface warming and OHC increase in the Indian Ocean.

Discussion

We show that the observed global OHC in the 300–1,500 m is dominated by a continuous warming trend over the past 43 years, and the penetration depth of ocean warming in individual basins is determined by the depth of the regional MOC. The deep (700–1,500 m) ocean heat uptake observed during the 21st century in the Atlantic and Southern Oceans primarily reflects anthropogenic heat entering the subsurface ocean via the deep

MOC. This deep penetration occurs at a similar rate during both hiatus and surge periods and is therefore not a major contributor to decadal changes in the warming rate of GMST. Our analyses of the CESM large ensemble simulations further show that the regional pattern of decadal ocean heat uptake below 700 m is indeed robust but corresponds poorly to decadal variations in global surface warming rate. OHC in the upper 350 m, particularly in the Pacific and Indian Oceans, shows robust features of decadal variations. During hiatus decades, the Indian Ocean shows an anomalous warming and accelerated OHC increase below 50 m, which is associated with a La Niña-like climate shift and an enhanced heat transport of the ITF. This Indian Ocean warming corresponds to a concurrent Pacific cooling in the 0–100 m layer, thus indicative of an Indo-Pacific heat rearrangement. We find that warming within 50–350 m mostly compensates the 0–50 m cooling in the Pacific and Indian Oceans so that enhanced deep-level (below 350 m) heat uptake is not required in these two basins. We therefore conclude that the ocean heat uptake in the upper 350 m in the Pacific and Indian Oceans is closely tied to the surface warming hiatus, whereas the regional pattern of deep ocean heat uptake (below 700 m) is not a valid basis to infer which basin drives the surface warming hiatus.

The vertical and interbasin heat redistribution in the ocean is not well understood. In addition to MOCs, the wind-driven decadal variability in ocean circulation is a plausible mechanism for heat redistribution[36]. For example, decadal variability of the Southern Ocean heat uptake is suggested to be caused by modulations in winds associated with the Southern Annual Mode[37]. The need to develop a rigorous energy view and test it for the hiatus[38] spotlights the important problem of three-dimensional redistribution of heat in the ocean.

Methods

The OHC calculation. In this study, OHC within a certain layer is calculated as

$$\mathrm{OHC} = \varrho_0 C_P \int_{z_1}^{z_2} T(z)\mathrm{d}z \qquad (1)$$

where ϱ_0 and C_P denote sea water density and specific heat capacity, respectively. z_1 and z_2 denote the upper and lower limits of the layer depth. $T(z)$ denotes temperature profile, a function of depth z. Because we focus on the variability over the decadal time scales, a 12-month running is applied to the OHC time series. We calculate the OHC anomalies to the first year (year 1970) to manifest the continuous subsurface warming (Fig. 2), and to the period of 1970–2012 to reproduce the results by ref. 22 (Supplementary Fig. 9).

The observational data sets. Four observational data sets are selected that cover the span of 1970–2012. They are composed of two categories: (1) objective analyses of *in-situ* observations (for example, expendable bathy-thermographs (XBTs), conductivity–temperature–depth (CTD) measurements from research ships and Argo floats) and (2) reanalysis products that assimilate a variety of available observations in an ocean model. The former includes the EN4 data[33], the WOA data[34] and the data from a research leading by Ishii[30] (called the Ishii data in this study). The latter includes the ECMWF ORAS4 product[15,35]. In particular, the EN4 data is monthly mean temperature at 42 levels extending to beyond 5,000 m from 1900 to the present. The WOA data is annual mean temperature at 16 levels in the upper 700 m during 1955–2012. The Ishii data is monthly mean temperature at 24 levels in the upper 1,500 m during 1945–2012. The ECMWF ORAS4 product is monthly temperature at 42 levels extending to beyond 5,000 m during 1958–2014. For the Ishii data, the annual mean anomalies for globally integrated upper 700 m OHC during 1950–2014 is available from Japan Meteorological Agency. Due to a sparse sampling and its induced large OHC error in the early era (Supplementary Fig. 10), the period of 1970–2012 is selected for the SST and OHC analyses.
In this study, to examine the changes in the OHC (Fig. 5) and temperature (Supplementary Fig. 5) between recent hiatus event and prior surge event, we consider the global and regional OHC (temperature) trends for the periods from 2003 to 2012 and from 1992 to 2001, and calculate the difference between the two periods.

The CESM large ensemble simulations. The CESM community provides a production of a large ensemble (38 members) using a 1-degree CESM-CAM5 with biogeochemistry (BGC)[31]. Ensemble members go from 1920–2100 using historical forcing (1920–2005) and Representative Concentration Pathway 85 (RCP 8.5) forcing (2005–2100). We calculate the running 10-year linear trends of globally averaged SST from the 38-ensemble members and choose the members simulating the recent hiatus event by the criteria of negative decadal trend appearing within the period of 2002–2012. As shown in Supplementary Fig. 3, four members (M09, M16, M19 and M31) happen to capture the recent hiatus event, while four (M13, M29 M37 and M07) simulate a warming surge during the same span. Thus we collect the former four as the 'Hiatus' group and the latter four as the 'Surge' group. For each group, the ensemble mean SST and OHC are calculated and compared in Figs 1 and 2. The model drift is estimated from a 500-year pre-industrial control run and removed in the SST and OHC analyses. Moreover, to test the robustness of the OHC difference between the hiatus and surge events, we expand the Hiatus and Surge groups by including five pairs of hiatus and surge decades (Supplementary Table 1) in the future simulations and name the new groups are as the fHiatus and fSurge. To calculate the ensemble mean or standard deviation of the OHC (Figs 3 and 5) and temperature (Supplementary Fig. 5) of the fHiatus and fSurge, values in future hiatus and surge events are scaled to 2002–2011 by multiplying by a ratio of 38-ensmble mean global OHC trend during 2002–2012 to 38-ensmble mean global OHC trend during the event period.

References

1. Easterling, D. R. & Wehner, M. F. Is the climate warming or cooling? *Geophys. Res. Lett.* **36,** L08706 (2009).
2. Katsman, C. A. & van Oldenborgh, G. J. Tracing the upper ocean's "missing heat". *Geophys. Res. Lett.* **38,** L14610 (2011).
3. Foster, G. & Rahmstorf, S. Global temperature evolution 1979–2010. *Environ. Res. Lett.* **6,** 044022 (2011).
4. Solomon, S. *et al.* Contributions of Stratospheric Water Vapor to Decadal Changes in the Rate of Global Warming. *Science* **32,** 1219–1223 (2010).
5. Kaufmann, R. K., Heikki, K., Michael, L. M. & James, H. S. Reconciling anthropogenic climate change with observed temperature 1998–2008. *Proc. Natl Acad. Sci. USA* **108,** 11790–11793 (2011).
6. Fyfe, J. C., von Salzen, K., Cole, J. N. S., Gillett, N. P. & Vernier, J.-P. Surface response to stratospheric aerosol changes in a coupled atmosphere–ocean model. *Geophys. Res. Lett.* **40,** 584–588 (2013).
7. Meehl, G. A., Hu, A., Arblaster, J. M., Fasullo, J. & Trenberth, K. E. Externally forced and internally generated decadal climate variability associated with the Interdecadal Pacific Oscillation. *J. Clim.* **26,** 7298–7310 (2013).
8. Meehl, G. A., Teng, H. & Arblaster, J. M. Climate model simulations of the observed early-2000s hiatus of global warming. *Nat. Clim. Change* **4,** 892–902 (2014).
9. Trenberth, K. E., Fasullo, J. T., Branstator, G. & Phillips, A. S. Seasonal aspects of the recent pause in surface warming. *Nat. Clim. Change* **4,** 911–916 (2014).
10. Zhang, Y., Wallace, J. M. & Battisti, D. S. ENSO-like interdecadal variability: 1900-93. *J. Clim.* **10,** 1004–1020 (1997).
11. Power, S., Casey, T., Folland, C., Colman, A. & Mehta, V. Inter-decadal modulation of the impact of ENSO on Australia. *Clim. Dyn.* **15,** 319–324 (1999).
12. Kosaka, Y. & Xie, S.-P. Recent global-warming hiatus tied to equatorial Pacific surface cooling. *Nature* **501,** 403–407 (2013).
13. England, M. H. *et al.* Recent intensification of wind-driven circulation in the Pacific and the ongoing warming hiatus. *Nat. Clim. Change* **4,** 222–227 (2014).
14. Watanabe, M. *et al.* Contribution of natural decadal variability to global warming acceleration and hiatus. *Nat. Clim. Change* **4,** 893–897 (2014).
15. Balmaseda, M. A., Trenberth, K. E. & Källén, E. Distinctive climate signals in reanalysis of global ocean heat content. *Geophys. Res. Lett.* **40,** 1754–1759 (2013).
16. Hansen, J. *et al.* Earth's energy imbalance: Confirmation and implications. *Science* **308,** 1431–1435 (2005).
17. Hansen, J., Sato, M., Kharecha, P. & von Schuckmann, K. Earth's energy imbalance and implications. *Atmos. Chem. Phys.* **11,** 13421–13449 (2011).
18. Trenberth, K. E., Fasullo, J. T. & Balmaseda, M. A. Earth's Energy Imbalance. *J. Clim.* **27,** 3129–3144 (2014).
19. Trenberth, K. E., Fasullo, J. T. & Kiehl, J. Earth's global energy budget. *Bull. Am. Meteorol. Soc.* **90,** 311–323 (2009).
20. Loeb, N. G. *et al.* Observed changes in top-of-the-atmosphere radiation and upper-ocean heating consistent within uncertainty. *Nat. Geosci* **5,** 110–113 (2012).
21. Meehl, G. A., Arblaster, J. M., Fasullo, J. T., Hu, A. & Trenberth, K. E. Model-based evidence of deep-ocean heat uptake during surface-temperature hiatus periods. *Nat. Clim. Change* **1,** 360–364 (2011).
22. Chen, X. & Tung, K. K. Varying planetary heat sink led to global-warming slowdown and acceleration. *Science* **345,** 897–903 (2014).
23. Nieves, V., Willis, J. K. & Patzert, W. C. Recent hiatus caused by decadal shift in Indo-Pacific heating. *Science* **349,** 532–535 (2015).
24. Lee, S.-K. *et al.* Pacific origin of the abrupt increase in Indian Ocean heat content during the warming hiatus. *Nat. Geosci* **8,** 445–449 (2015).
25. Lyman, J. M. *et al.* Robust warming of the global upper ocean. *Nature* **465,** 334-337 (2010).

26. Durack, P. J., Gleckler, P. J., Landerer, F. W. & Taylor, K. E. Quantifying underestimates of long-term upper-ocean warming. *Nat. Clim. Change* **4,** 999–1005 (2014).
27. Smith, D. M. *et al.* Earth's energy imbalance since 1960 in observations and CMIP5 models. *Geophys. Res. Lett.* **42,** 1205–1213 (2015).
28. Roemmich, D. *et al.* Unabated planetary warming and its ocean structure since 2006. *Nat. Clim. Change* **5,** 240–245 (2015).
29. Palmer, M. D., McNeall, D. J. & Dunstone, N. J. Importance of the deep ocean for estimating decadal changes in Earth's radiation balance. *Geophys. Res. Lett.* **38,** L13707 (2011).
30. Ishii, M., Shouji, A., Sugimoto, S. & Matsumoto, T. Objective analyses of SST and marine meteorological variables for the 20th century using COADS and the Kobe Collection. *Int. J. Climatol.* **25,** 865–879 (2005).
31. Kay, J. E. *et al.* The Community Earth System Model (CESM) Large Ensemble Project: A Community Resource for Studying Climate Change in the Presence of Internal Climate Variability. *Bull. Amer. Meteor. Soc.* **96,** 1333–1349 (2015).
32. Fyfe, J. C., Gillett, N. P. & Zwiers, F. W. Overestimated global warming over the past 20 years. *Nat. Clim. Change* **3,** 767–769 (2013).
33. Good, S. A., Martin, M. J. & Rayner, N. A. EN4: quality controlled ocean temperature and salinity profiles and monthly objective analyses with uncertainty estimates. *J. Geophys. Res.* **118,** 6704–6716 (2013).
34. Levitus, S. *et al.* World ocean heat content and thermosteric sea level change (0–2000 m), 1955–2010. *Geophys. Res. Lett.* **39,** L10603 (2012).
35. Balmaseda, M. A., Mogensen, K. & Weaver, A. T. Evaluation of the ECMWF ocean reanalysis system ORAS4. *Q.J.R. Meteorol. Soc* **139,** 1132–1161 (2013).
36. Huang, R. X. Heaving modes in the world oceans. *Clim. Dyn.* **45,** 3563–3591 (2015).
37. Drijfhout, S. S. *et al.* Surface warming hiatus caused by increased heat uptake across multiple ocean basins. *Geophys. Res. Lett.* **41,** 7868–7874 (2014).
38. Xie, S.-P., Kosaka, Y. & Okumura, Y. M. Distinct energy budgets for anthropogenic and natural changes during global warming hiatus. *Nat. Geosci* **9,** 29–33 (2016).

Acknowledgements

This work is supported by NSF AGS-1249145. Jian Lu was supported by the U.S. Department of Energy Office of Science Biological and Environmental Research (BER) as part of the Regional and Global Climate Modeling program. The CESM model output data are publicly available on the Earth System Grid (ESG). For observational and reanalysis data, the EN4 data are publicly available at http://hadobs.metoffice.com/en4/, the Ishii data are publicly available at http://rda.ucar.edu/data sets/ds285.3/ and http://www.data.jma.go.jp/gmd/kaiyou/english/ohc/ohc_data_en.html, the WOA data are publicly available at https://www.nodc.noaa.gov/OC5/3M_HEAT_CONTENT/ and the ECMWF ORAS4 data are publicly available at https://climatedataguide.ucar.edu/climate-data/oras4-ecmwf-oceanreanalysis-and-derived-ocean-heat-content and Integrated Climate Data Center -ICDC.

Author contributions

W.L. performed the analyses. W.L. and S.-P.X. wrote the paper. J.L. contributed to the interpretation of results.

Additional information

Competing financial interests: The authors declare no competing financial interests.

Mid Pleistocene foraminiferal mass extinction coupled with phytoplankton evolution

Sev Kender[1,2], Erin L. McClymont[3], Aurora C. Elmore[3,4], Dario Emanuele[5], Melanie J. Leng[1,2] & Henry Elderfield[4,‡]

Understanding the interaction between climate and biotic evolution is crucial for deciphering the sensitivity of life. An enigmatic mass extinction occurred in the deep oceans during the Mid Pleistocene, with a loss of over 100 species (20%) of sea floor calcareous foraminifera. An evolutionarily conservative group, benthic foraminifera often comprise >50% of eukaryote biomass on the deep-ocean floor. Here we test extinction hypotheses (temperature, corrosiveness and productivity) in the Tasman Sea, using geochemistry and micropalaeontology, and find evidence from several globally distributed sites that the extinction was caused by a change in phytoplankton food source. Coccolithophore evolution may have enhanced the seasonal 'bloom' nature of primary productivity and fundamentally shifted it towards a more intra-annually variable state at ~0.8 Ma. Our results highlight intra-annual variability as a potential new consideration for Mid Pleistocene global biogeochemical climate models, and imply that deep-sea biota may be sensitive to future changes in productivity.

[1] Centre for Environmental Geochemistry, School of Geography, University of Nottingham, Nottingham NG7 2RD, UK. [2] British Geological Survey, Keyworth, Nottingham NG12 5GG, UK. [3] Department of Geography, Durham University, Durham DH1 3LE, UK. [4] Department of Earth Sciences, University of Cambridge, Cambridge CB2 3EQ, UK. [5] DST—Università del Sannio, Via dei Mulini 59a, Benevento 82100, Italy. Correspondence and requests for materials should be addressed to S.K. (email: sev.kender@nottingham.ac.uk) or to (sesev@bgs.ac.uk).
‡Deceased

High-resolution continuous records, such as those found in the marine realm, provide an outstanding opportunity to pair fossil occurrences with geochemical environmental proxies to examine the possible causes of mass extinctions. The Mid Pleistocene transition (MPT, ~1.2–0.6 Ma) was characterized by global cooling, glacial stage lengthening, changing ocean circulation and evolution of terrestrial and marine biota[1–3]. During the MPT, a stepwise extinction of over 100 species of deep-water benthic foraminifera occurred, targeted to a specific morphological group (largely the elongated 'stilostomellids' with ornamented apertures, referred to here as the 'extinction group') within just three families[4–7]. Benthic foraminifera are an important group often comprising >50% of the total benthic eukaryote biomass on the deep-ocean floor[8]. The cause of this last mass extinction, their only significant extinction in the last 15 Myr, is still unknown[6,7,9–11]. The deep ocean is the largest and most stable habitat on Earth[10], and the only other two mass extinctions of benthic foraminifera in the Cenozoic occurred over the Palaeocene–Eocene and Eocene–Oligocene boundaries[10]. Understanding the ecological sensitivity of the deep sea is important as it is a habitat increasingly under stress[12], and may be sensitive to current and future changes in productivity[13]. This last mass extinction in benthic foraminifera, completed by ~0.8–0.6 Ma, has been extensively studied and shown to be global in extent, although slightly diachronous and apparently stepwise during glacials[4,7,11,14]. Higher latitude localities (above ~60°) have not been studied in detail[7], but Bering Sea[15] and Kerguelen Plateau Southern Ocean[7] foraminiferal assemblages show very low abundances of the extinction group before the MPT. A lack of living representatives has hampered our understanding of their ecological preferences and therefore the cause of the extinction. Leading hypotheses include the following: changing deep-water physical properties, such as cooling and increased oxygenation[6,7,10,14]; a change in productivity or extinction of a particular phytoplankton food source[7,11]; and increased 'seasonality' or annual irregularity of organic carbon export[9–11]. Although bottom-water temperature, oxygenation and corrosiveness have been challenged as the cause, partly because of a global uniform change being unlikely[7,10], these properties are yet to be directly tested with environmental parameters reconstructed alongside extinction group data[7].

In this study, we address these hypotheses by examining the first records of extinction group abundance generated together with proxies for bottom-water temperature, carbonate saturation ($\Delta[CO_3^{2-}]$, related to corrosiveness), sea-surface temperature (SST) and productivity. We also generated records of phytoplankton assemblages from two sites to compare with a published South Pacific record[11]. Here we provide direct evidence from benthic foraminiferal Mg/Ca which disproves the hypothesis that changing bottom-water temperature controlled foraminiferal decline[6]. However, we do find compelling micropalaeontological evidence for a change in phytoplankton food source as the cause of the extinction. We show that a global dominance of the nannoplankton 'small' *Gephyrocapsa* occurred at ~0.8 Ma, coinciding with the largest decline in benthic foraminifera. We conclude that small *Gephyrocapsa* evolution may have enhanced intra-annual variability of primary productivity and carbon flux reaching much of the mid-low latitude ocean floor where the extinction was centred, and that the extinction group may have been unable to cope with long periods of seasonally reduced organic carbon flux during parts of the year.

Results

Benthic extinction in the Tasman Sea. We reconstructed environmental parameters using a wide range of proxies at Deep Sea Drilling Project (DSDP) Site 593 over the period ~0.75–1 Ma, covering the benthic extinction (Fig. 1). Site 593 is situated in 1,063 m of water on the Challenger Plateau of the Tasman Sea (SW Pacific Ocean, Supplementary Fig. 1), lying to the north of the modern Subtropical Front, which is a complex zone delineated by large gradients in SST and salinity[16]. SSTs in the Tasman Sea are considered to be sensitive to glacial–interglacial displacement of the Subtropical Front[7,17]. Site 593 is bathed by Antarctic Intermediate Water, which is broadly characterized by low salinity (34.3–34.5 PSU), low temperatures (3.5–10 °C) and high dissolved oxygen (200–250 μmoles kg^{-1}; refs 18,19).

Previous low-resolution (~22 kyr) benthic foraminiferal data from Site 593 (ref. 5) indicated that the extinction occurred between 0.8 and 0.9 Ma. Our higher-resolution (~3.5 kyr) results indicate the extinction group declined in the Tasman Sea, irrespective of size fraction, throughout the study interval with an initial overall reduction in abundance at ~0.95 Ma, and a major decline towards very low abundance at ~0.83 Ma (Fig. 1g). Our data are consistent with other studies that conclude that the architecture of the extinction is captured in all size fractions[7], and suggest that it was not associated with an early shortening of the life cycle, which might be apparent with an increased proportion of small specimens. Sites below ~1 km water depth typically have the highest extinction group diversity[7], and species richness at Site 593 is relatively low at ~10–12 species per ~10-cc sample (Fig. 1h). Site 593 is dominated by *Strictocostella scharbergana* and *Siphonodosaria lepidula*, which decline abruptly at 0.85 and 0.8 Ma, respectively (Fig. 1i,j). Different species are dominant in extinction assemblages in the Pacific, Atlantic and Mediterranean, although they are morphologically (and thus possibly ecologically) related and also became extinct during the MPT[7].

Bottom-water temperature and corrosiveness. Decreased intermediate/deep-water temperature is a hypothesized cause of the extinction[6], possibly due to increased oxygenation and its impact on an inferred microbial food source[6,10,14]. Bottom-water temperature at intermediate-depth Site 593, reconstructed from Mg/Ca of infaunal *Uvigerina peregrina* (see Methods), ranges from ~3 to 8 °C (modern temperature is 4.5 °C), with warmer interglacials (Fig. 1e). Seawater absolute magnesium (17 Ma residence time) and calcium (1 Ma residence time) concentrations would have been slightly different during the Mid Pleistocene[20], thus having an impact on the accuracy of the temperature estimates based on modern calibration (possibly by up to ~1 °C), although the overall trends will be unchanged. There is no apparent correlation between bottom-water temperature and faunal abundances during the pre-extinction period before ~0.83 Ma (Fig. 2a), nor any secular change over the extinction itself, and we conclude that the benthic extinction at Site 593 was not caused by temperature changes. Increased bottom-water corrosiveness is another physical property that has been proposed to have an impact on benthic foraminifera[21,22]. The living position of the extinction group is unknown, but has been inferred as infaunal[7,9,11]. If this were the case, pore water $[CO_3^{2-}]$ might be more relevant regarding the extinction, although pore waters have lower $[CO_3^{2-}]$ and changes in bottom-water $[CO_3^{2-}]$ would have influenced pore water values (in addition to other factors such as organic carbon flux and sedimentation rate). Bottom-water $\Delta[CO_3^{2-}]$ at Site 593, reconstructed from B/Ca of epifaunal *Cibicidoides wuellerstorfi* (see Methods), shows relatively low values ranging from 10 to 25 μmol kg^{-1}, with highest values recorded during cooler glacials and some potential negative outliers before 0.9 Ma (Fig. 1f). Similar to Mg/Ca, the absolute B/Ca of seawater would have been slightly different from

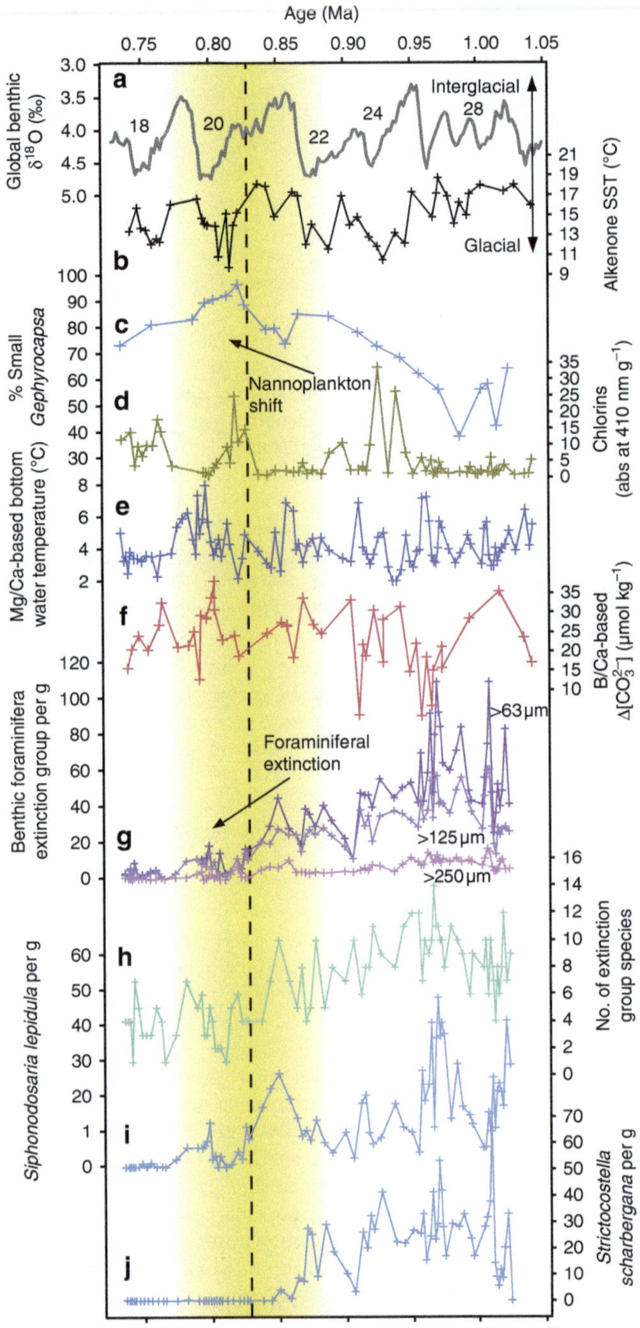

Figure 1 | Reconstructed environmental proxies for Tasman Sea DSDP Site 593. (**a**) Global benthic foraminifera $\delta^{18}O$ composite[52] showing colder glacial (positive) and interglacial cycles. (**b**) Sea-surface temperature reconstructed from alkenones. (**c**) Small *Gephyrocapsa* as a % of total nannofossil assemblage. (**d**) Concentration of chlorin pigments used here as a proxy for photosynthetic material related to primary productivity[24]. (**e**) Bottom-water temperature reconstructed from benthic foraminiferal *U. peregrina* Mg/Ca ratios. (**f**) Bottom-water $\Delta[CO_3^{2-}]$ reconstructed from benthic foraminiferal *C. wuellerstorfi* B/Ca ratios. (**g-j**) Abundance per g and number of species of benthic foraminifera from the extinction group. Note that none of the environmental proxies (**d-f**) follows abundance of the foraminiferal extinction group (**g-j**). The vertical yellow bar indicates the interval over which the foraminiferal extinction occurs, and the vertical dashed line indicates where small *Gephyrocapsa* dominates the assemblage.

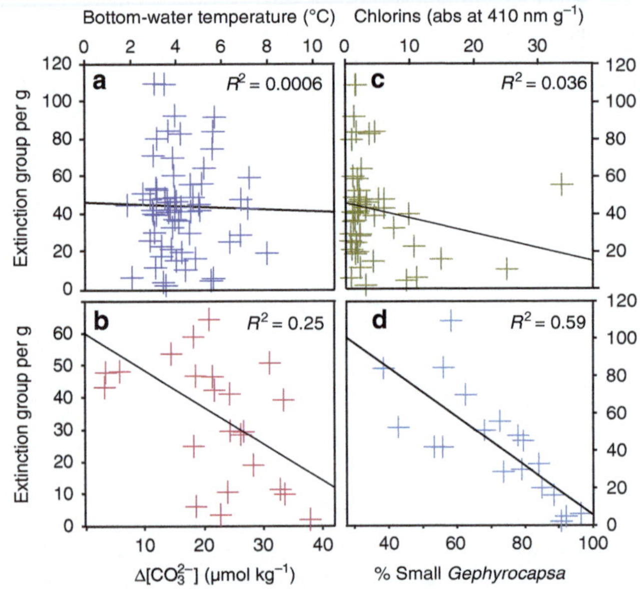

Figure 2 | Foraminiferal extinction group against other parameters for the pre-extinction interval ~0.8–1 Ma. (**a**) Bottom-water temperature derived from foraminiferal Mg/Ca, (**b**) bottom-water $\Delta[CO_3^{2-}]$ derived from foraminiferal B/Ca, (**c**) chlorin P410 from bulk sediment and (**d**) % small *Gephyrocapsa*.

modern values, thus potentially having an impact on the absolute values of our calculated $\Delta[CO_3^{2-}]$ even though the overall trends should be considered accurate[23]. Since all the values are oversaturated with respect to *in situ* $[CO_3^{2-}]$ (that is, all have positive $\Delta[CO_3^{2-}]$ values), and the extinction group abundance does not co-vary with $\Delta[CO_3^{2-}]$ during the pre-extinction interval (Fig. 2b), we conclude that the extinction group was tolerant to values in this range and could not have become globally extinct because of increased bottom-water corrosiveness.

Carbon flux. Benthic foraminifera are influenced by several ecological factors, which include bottom-water oxygenation, bottom-water sediment heterogeneity and hydrodynamics, temperature, corrosiveness and organic carbon type and flux[21,22]. However, above the lysocline (below which calcite dissolution occurs), typical open-ocean benthic ecology is primarily affected by the organic carbon flux—which is related to primary productivity (quantity, type and duration) and remineralization of particulate organic carbon as it is transported to depth[10]. A change in organic carbon export, linked to primary productivity, has been hypothesized as an alternative cause of the extinction[7,9,10]. SST at Site 593 (Fig. 1b), reconstructed using the alkenone proxy $U^{K}_{37}{}'$ (see Methods), shows significant variability over the study interval, ranging from 10 to 18 °C, with cooler temperatures during glacials after ~0.95 Ma. Since modern SSTs in the Tasman Sea are tightly coupled to the position of the productive Subtropical Front, we anticipate that shifts to Subtropical Front position might have influenced the organic matter flux. Sedimentary chlorin (Fig. 1d) is derived from photosynthetic material and specifically from chlorophyll pigments[24]. At this distal oceanic location they are likely to have originated from a proximal phytoplankton source, although a terrestrial contribution cannot be discounted by this study. Chlorin concentrations are typically higher during colder glacials at 0.75, 0.82 and 0.93 Ma, possibly indicating enhanced productivity[24] at times of Subtropical Front northward migration (cooler SSTs). However, despite significant variations, there is no prior correlation (Fig. 2c) nor secular change in chlorin concentration (and by extension phytoplankton carbon flux), which could account for the extinction in benthic foraminifera focused at ~0.83 Ma (Fig. 1).

Phytoplankton food source. The extinction group therefore appears to have been relatively tolerant to variations in physical water properties and potentially overall organic carbon flux changes at Site 593 before the ~0.83 Ma decline. Changing organic carbon *source* is another possible extinction mechanism[7,9–11]. The extinction group probably lived infaunally, according to their elongated morphology and lower shell $\delta^{13}C$ than epifaunal foraminifera[7,9,11], and preferred relatively high organic carbon flux at upper-abyssal to mid-bathyal depths; their global abundance reflects this distribution[7]. The specialized architectural function of extinction group apertures has been discussed at length[11] but remains unknown, and may have helped direct pseudopodial flow for detritus feeding[10,11], perhaps leaving them sensitive to a change in organic carbon supply.

Considering the strong benthic–pelagic coupling of benthic foraminifera[10], we compiled published high-resolution calcareous nannoplankton assemblage records to assess the potential for a global changing source of organic carbon causing the extinction. Coccolithophores are one of the major mid-low latitude phytoplankton groups contributing to the organic carbon pump, known to undergo rapid evolution in the Pleistocene[25]. From published records, we identified that a significant peak in the morphological genus 'small *Gephyrocapsa*' (<3 μm) occurred in the SE Atlantic[26], NW Pacific[27] and SW Pacific[11] centred at ~0.8 Ma (Fig. 3b). This peak in abundance is consistent with records from the Indian Ocean[25], and other nannoplankton records that do not differentiate this particular species from other small placoliths in the North Atlantic, South Atlantic and Mediterranean Sea (Supplementary Fig. 2). Although most of these sites do not have benthic foraminiferal data with which to directly compare the nannoplankton assemblages, Site MD97-2114 in the SW Pacific indicates that the benthic extinction coincided with this peak[11], and other sites[7] indicate that at a global scale the extinction had largely taken place by 0.8 Ma (Fig. 3c). To test the hypothesis that dominant small *Gephyrocapsa* could have been implicated in the benthic extinction, we paired records of nannoplankton and foraminiferal assemblage data in the Tasman Sea (Fig. 1c) and the North Atlantic (Supplementary Fig. 3). Both records show a consistently high % small *Gephyrocapsa* (>90%) during the extinction interval at ~0.83 Ma. Interestingly, the first step in the extinction at ~0.96 Ma at Site 593 (Fig. 1g) coincides with an initial increase in the % small *Gephyrocapsa* from ~50 to 65% (Fig. 1c), showing the possibility that % small *Gephyrocapsa* was already exerting some control over extinction group abundance. From our research, the abundance of small *Gephyrocapsa* is the only oceanic parameter that shows a correlation with the extinction group leading up to the extinction in the Tasman Sea (Fig. 2d). To address the possible significance of this correlation, we explore the various possible environmental changes controlling the two groups and describe a new conceptual model that can account for both data sets.

Discussion

Our new nannoplankton data (Fig. 1c and Supplementary Fig. 3), together with other compiled records (Fig. 3b and Supplementary Fig. 2), indicate that the global oceans became dominated by small *Gephyrocapsa* around ~0.8 Ma, a finding that has not previously been highlighted, perhaps because of the relatively

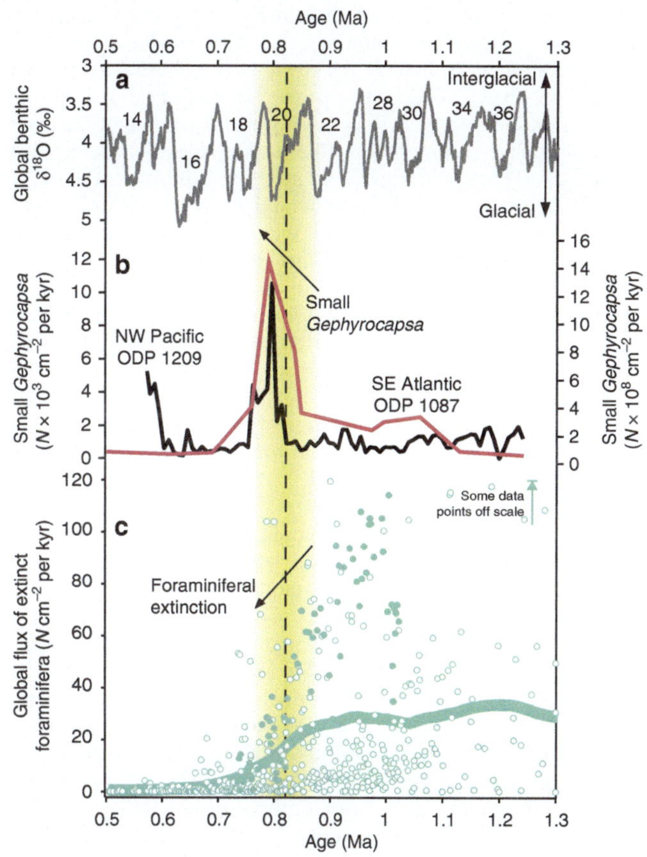

Figure 3 | Nannoplankton assemblages compared with extinct benthic foraminifera over the Mid Pleistocene. (**a**) Global deep-sea $\delta^{18}O$ composite[52]. (**b**) Accumulation rate of small *Gephyrocapsa* at Ocean Drilling Program (ODP) Site 1087 (ref. 26) and ODP Site 1209 (ref. 27). (**c**) Flux (accumulation rate) of extinct foraminifera from 15 global sites compiled by ref. 7 (open symbols), and including new data from this study (solid symbols), with a 0.2-pt LOESS smoothing spline (bold). Some data points are off the scale; smoothed line takes into account all data. Note how the peak in small *Gephyrocapsa* dominance at ~0.8 Ma occurs in the NW Pacific and SE Atlantic (and SW Pacific and N Atlantic, Supplementary Fig. 3), and coincides with persistently low abundance of the extinction group thereafter. Vertical yellow and dashed lines as in Fig. 1.

short duration of this event compared with typical biostratigraphic sampling resolution. In some of the records, the dominance may have occurred sooner (for example, Site 980, N Atlantic) than others (for example, Site 593, Tasman Sea), such that there is diachroneity in the onset of this event; however, the crucial point is that all global records with coccolith data show high abundances of small *Gephyrocapsa* at ~0.8 Ma. In summary, evidence for a global peak in small *Gephyrocapsa* at ~0.8 Ma comes from sites in the SE Atlantic[26], SW Pacific[11], NW Pacific[27], Indian Ocean[25], N Atlantic and Tasman Sea (Supplementary Figs 2 and 3). The cause of the small *Gephyrocapsa* event remains unknown; however, it may have been an evolutionary adaptation[25,28] as the *Gephyrocapsa* lineage is thought to be relatively independent of temperature and nutrients, but dependent on light intensity and day length[25]. The well-known abundance peak of *G. caribbeanica* at ~0.6 Ma has been linked to increased blooms because of a sustained reduction in orbital eccentricity affecting day length and light intensity[25]; a sustained eccentricity minima is also observed during the benthic extinction event at ~0.8 Ma (Supplementary Fig. 2).

Regardless of the cause of the small *Gephyrocapsa* event, its correlation with the extinction at ~0.8 Ma provides a new piece of evidence when evaluating the cause of the benthic extinction. Correlation itself does not prove causation; however, as phytoplankton provide a food supply for marine benthos, how might this change in phytoplankton assemblages have had an impact on benthic foraminifera? We propose that the most plausible extinction mechanism would be a change towards a more enhanced variation in annual export production and delivery of organic carbon to the deep ocean[9–11]. There is micropalaeontological evidence for an increase in intra-annual variability of phytodetrital pulses during the extinction interval, with increased seasonal phytodetrital benthic foraminifera in the Indian Ocean[9] and SW Pacific[11] from ~0.8 Ma. Previous studies have suggested that this could not have caused the benthic extinction because physical processes that enhance seasonality could not have occurred throughout the global ocean at the same time, as the benthic extinction did[7,11]. However, if the ecology of small *Gephyrocapsa* was similar to its modern descendent, *Emiliania huxleyi*, a seasonal bloom species that dominates mid-low latitude nannoplankton assemblages during the Holocene[25,28], the increased global presence of small *Gephyrocapsa* may represent an early adaptation to more bloom-type ecology within nannoplankton, where intra-annual export production changed. Modern (interglacial) global productivity is dominantly annually uneven outside the oligotrophic ocean gyres (Fig. 4). Modelling studies indicate that mid-low latitude oceanic intra-annual variability of production may decrease with future projected warming[29], potentially implying that intra-annual production variability could have been higher (compared with modern variability) during colder glacials. Any increased intra-annual variability during MPT glacials would have reduced organic carbon flux during some seasons, while increasing flux at other times. In addition to preferring relatively high organic carbon flux[7,9], the unique apertural ornamentation of the extinction group may denote a different feeding strategy. We speculate that they required a more uniform supply of carbon throughout the year. This is in contrast to extant species, typically able to survive for years without fresh phytodetritus[30]. The phytoplankton change may have primed the group for complete extinction at ~0.8–0.6 Ma. There remains the possibility that both the small *Gephyrocapsa* event and the benthic extinction reflect an alternative, common cause. Although we consider it unlikely, as we have discounted various water property changes (Figs 1 and 2), perhaps an unconstrained factor such as a reduction in surface and deep-ocean CO_2 played a role.

Our proposed mechanism is consistent with the timing, bathymetric and geographic distribution of the extinction event. To summarize, our proposed ecology for the extinction group is that they preferred a relatively high flux of organic carbon *regularly* delivered to the sea floor such that there were no long periods within the year where flux was very low. This explains their distribution before the extinction[7], which was higher in mid-low latitude regions as opposed to high latitudes, as high latitudes would have had a more uneven supply of export production throughout the year (Fig. 4). It was also lower in oligotrophic regions such as the South Pacific Gyre, South China Sea and Mediterranean Sea, and higher in modern eutrophic regions[7]. Finally, it was higher at mid-bathyal to upper-abyssal depths compared with the oligotrophic lower abyssal ocean, where the carbon flux is very low. This proposed ecology also explains their disappearance first at deeper abyssal depths (oligotrophic), the Mediterranean (oligotrophic) and during glacials[7], which may have experienced more variability of seasonal export production, according to modelling studies[29].

Our proposed mechanism can explain the extinction because we invoke evolution within a phytoplankton group prevalent in the global oceans where the extinction group was most abundant before the extinction. Thus, although small *Gephyrocapsa* may not have been abundant at higher latitudes where siliceous phytoplankton predominate today, this was a region with low extinction group abundance before 0.8 Ma and thus not a suitable refuge for the group when mid-low latitude small *Gephyrocapsa* began to dominate nannoplankton assemblages.

The long-term decline of the extinction group during the Eocene–Oligocene, and increase in phytodetritus-exploiting species, has been suggested to be because of enhanced seasonality in primary productivity forced by cooling[10,31]. In this sense, the Cenozoic decline of the extinction group could be categorized as a 'slow' mass extinction[32], as export production may have exhibited gradually more intra-annual variability. However, the abrupt decline by ~0.8 Ma coinciding with the small *Gephyrocapsa* event would indicate a 'rapid' mass extinction with biological causes, perhaps with similarities to planktonic foraminiferal mass extinctions during the Palaeocene–Eocene and Eocene–Oligocene transitions[32]. The timing of our proposed increase in this mid-low latitude intra-annual variability, inferred from the global small *Gephyrocapsa* dominance and benthic extinction, may be important for the development of climate during the MPT as variability of primary production could have had an impact on carbon export and storage[33]. Further modelling studies are needed to ascertain the impact of an enhanced intra-annual variability of production on climate (during MPT cooling), forced by evolutionary changes within the biosphere rather than physical processes alone.

Methods

DSDP Site 593 stratigraphy and age model. DSDP Site 593 (40°30.47′S, 167°40.47′E, 1,063 m water depth) was cored on the Challenger Plateau of the Tasman Sea in the SW Pacific Ocean. The upper 393 m of recovered sediment is foraminiferal-bearing nannofossil ooze, hosting very abundant and visually well-preserved benthic foraminifera[34]. A low-resolution, orbitally tuned stratigraphy was available to guide sampling based on a shipboard bio- and magnetostratigraphy and coarse-resolution benthic $\delta^{18}O$ analyses on infaunal *Uvigerina* spp.[35]. The updated age model here has been generated after re-assigning this to the GTS2012 timescale[36]. Samples for the new isotope stratigraphy were analysed at 10–20-cm resolution in cores 593Z-1H-1 through 593Z-5H-2 (~0–36.3 m depth) to yield a mean sample resolution of ~5 kyr for the time period 0–1.1 Ma. The revised glacial–interglacial stratigraphy for the interval 0–0.4 Ma, based on *C. wuellerstorfi* oxygen isotopes and the Brunhes/Matuyama palaeomagnetic reversal, is presented in ref. 37. Here we created a revised isotope stratigraphy for the period 0.4–1.1 Ma using new $\delta^{18}O$ analyses of the epifaunal species *C. wuellerstorfi* (Supplementary Fig. 4 and Supplementary Data 1). The tuning targets are shown in Supplementary Table 1. The Potaka tephra (1 Ma (ref. 38)) is clearly identified at 21.50 mbsf, and lies above a distinct benthic $\delta^{18}O$ minimum, which is thus aligned to MIS 31. The top of the Olduvai magnetochron is not well represented, but the base of the Olduvai and the Gauss/Matuyama boundary were used to guide identification of the key isotope stages. A linear sedimentation rate is assumed between all tie points.

Micropalaeontology analysis. Benthic foraminiferal analyses were carried out at the British Geological Survey, Keyworth, UK, and at the University of Leicester, UK, using an Olympus SZX10 binocular microscope. Seventy-eight sediment samples of ~10 cc were oven dried at 40 °C and washed with deionized water over a 63-μm sieve to remove clays (at Durham University). The >63-μm fraction was then dried, and individual species of foraminifera were counted under a binocular microscope and transferred into cardboard reference slides. Residues were sieved into size fractions, and picked separately, according to a procedure outline in ref. 7, so that the >250-μm fraction had 100% picked, the >125-μm fraction had an average of 37% picked and the >63-μm fraction had an average of 21% picked (Supplementary Data 2). Fraction splits were used in order to increase the number of samples analysed because of the high volume of sediment. Abundance in splits was then normalized for dry sediment weight to estimate a total number of extinction group specimens per g of dried sediment for each sample[7]. A total of 31 species and taxonomic groups were identified using the taxonomic work of Hayward *et al.*[7], and photographed using a Hitachi S-3600N scanning electron microscope in the Geology Department at the University of Leicester (Supplementary Fig. 5). The BFAR (benthic foraminiferal accumulation rate,

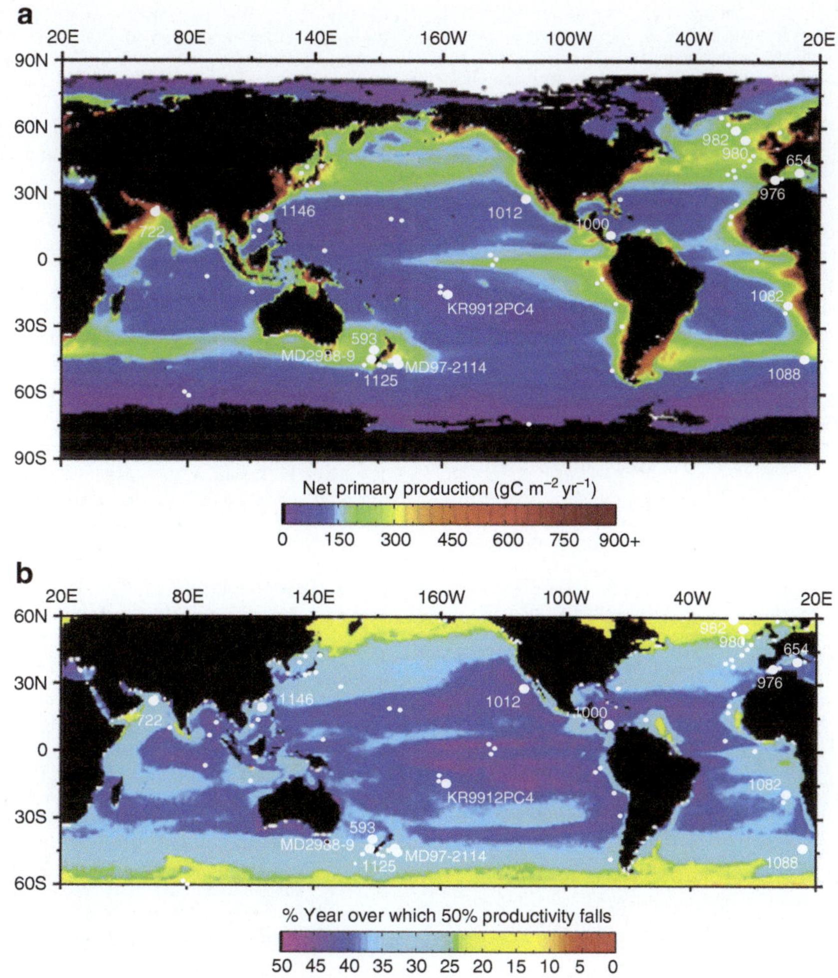

Figure 4 | Distribution of extinction group study sites and modern productivity. Location of core sites on maps showing the mathematical estimates of modern net primary production, and modern unevenness of annual primary production (figure adapted from ref. 53). Sites shown indicate where the Mid Pleistocene benthic foraminiferal extinction group has been studied; the DSDP/ODP site numbers are given for higher-resolution studies[7] that correspond to the sites used to construct Fig. 3. (**a**) Global average net primary productivity[53]. (**b**) The geographic distribution of the seasonal variation of net primary production (seasonality) for the years 1998-2007. The colours refer to the % of the year over which 50% productivity takes place, such that purple indicates no seasonality (colours modified from ref. 53). We propose that glacial MPT seasonality may have been higher than modern seasonality. Note that seasonality is generally lowest in regions where primary productivity is low.

specimens $cm^2 kyr^{-1}$) was calculated by using sediment dry bulk density (ρ, in $g\,cm^{-3}$), sedimentation rate (v, in $cm\,kyr^{-1}$) and number of specimens per g (x, specimens g^{-1}) as follows:

$$\mathrm{BFAR} = \rho v x \qquad (1)$$

Calcareous nannofossil species were analysed at the British Geological Survey, Keyworth, UK. A total of 33 smear slides were prepared from the central part of fresh core sediment samples and were analysed using a Zeiss cross-polarizing light microscope at × 1,000 magnification. A counting phase was performed to obtain a qualitative and quantitative description of the assemblages based on percentage abundances of each recognized taxon (Supplementary Data 3). Nannofossil counts were performed in random visual fields on slides where the nannofossils were homogeneously distributed. At least 300 specimens $>3\,\mu m$ were counted per slide in a varying number of fields of view. Specimens $<3\,\mu m$ were included in a separate set of counts to quantify their dominance. As a result, two counts were obtained: the total assemblage and the subdominant one excluding specimens $<3\,\mu m$.

Most of the species identified in this study belong to the family Noelaerhabdaceae and are recognized at generic and specific levels: *G. oceanica*, *G. caribbeanica*, *G. muellerae* recognized during this interval as *G. margereli*[39], *G. omega* and small *Gephyrocapsa* (specimens $<3\,\mu m$, mainly constituted of *G. aperta* and *G. ericsonii*). Other taxa are present within the assemblages recognized at generic, specific and subspecific levels. At the generic level, minor taxa are represented by the genres *Syracosphaera*, *Pontosphaera*, *Reticulofenestra* and *Rhabdosphaera*. At the specific level, minor taxa are represented by the species *Pseudoemiliania lacunosa*, *R. asanoi*, *Helicosphaera carteri*, *H. hyalina* and *Umbilicoshpaera sibogae*. Finally, *Coccolithus pelagicus* and *Calcidiscus leptoporus* were distinguished at subspecific and submorphotype levels on the basis of their coccolith length. Refs 40,41 have documented the existence of two *C. pelagicus* subspecies: *C. pelagicus pelagicus*, the cold form, and *C. pelagicus braarudii*, the temperate and upwelling waters form[42]. Three *C. leptoporus* morphotypes have been distinguished: *C. leptoporus* type small ($<5.5\,\mu m$ (refs 41,43)); *C. leptoporus* type medium (5.5–8 μm; (refs 41,43)); and *C. leptoporus* type large ($>8\,\mu m$ (refs 41,43)).

Chlorin and alkenone analyses. Separate original samples were freeze-dried, homogenized, and alkenones and chlorins (diagenetic transformation products of chlorophyll)[24,44] were extracted with an organic solvent mixture of dichloromethane/methanol (Supplementary Data 4), following the microwave-assisted protocol of ref. 45. The microwave temperature was increased from room temperature to 70 °C over 2 min, held at this temperature for 5 min and then allowed to cool. Extracts were subsequently centrifuged in test tubes, and the supernatant dried by rotary evaporation before being divided into two aliquots for analysis. All analyses were carried out in the Geography Department at Durham University. Chlorins were analysed by ultraviolet-visible spectrophotometry, quantified at the 410 and 665 nm wavelengths, and normalized for extracted sample weight in g (ref. 45). One aliquot was dissolved again in a known volume of acetone, and analysed using a dionex photodiode array detector coupled to a quaternary pump[45]. Absorbance across the 350–850-nm wavelength was recorded and quantified at 410 and 665 nm, which corresponds to the diagenetic transformation products of chlorophyll[24,44]. Samples were analysed in triplicate, and the means are reported here (Supplementary Data 4). The average s.d. within samples was 0.44 units (410 nm) and 0.08 units (665 nm).

Alkenones were isolated from the second aliquot of the lipid extract using silica column chromatography, eluting with *n*-hexane (for hydrocarbons), dichloromethane (for ketones) and methanol (for polar compounds). Alkenones were quantified by Thermo Scientific Trace 1310 gas chromatograph fitted with a flame ionization detector. Separation was achieved with a fused silica column (30 m × 0.25 mm inner diameter) coated with 0.25 µm of 5% phenyl methyl siloxane (HP-5MS) and using He as the carrier gas. After injection, the following oven temperature programme was used: 60–200 °C at 20 °C min^{-1}, 200–320 °C at 6 °C min^{-1}, and then held at 320 °C for 35 min. SST was calculated using the $U^{K}_{37}{}'$ index[46] and the global mean annual SST calibration[47].

Oxygen isotope analysis. Approximately four individual tests of *C. wuellerstorfi* from the >250-µm fraction of each sample were analysed using an IsoPrime dual inlet mass spectrometer plus Multiprep device at the NERC Isotope Geosciences Laboratory. $\delta^{18}O$ is reported as per mille (‰) deviations of the isotopic ratios ($^{18}O/^{16}O$), and is calculated on the V-PDB scale using an internal laboratory standard that is routinely calibrated against NBS-19 standards (Supplementary Data 1). Average analytical reproducibility for $\delta^{18}O$ of the calcite standard is <0.1‰.

Trace elemental analysis. Approximately 10 pristine individuals (with white calcite, no visible clay and no visible recrystallization) per sample of benthic foraminifera *U. peregrina* and *C. wuellerstorfi* were selected from the >250-µm fraction. Particular care was taken to select the correct *U. peregrina* and *C. wuellerstorfi* morphotypes, as *Planulina/Cibicidoides* have been shown to have significant species-specific elemental fractionation[48,49]. Benthic foraminifera were cleaned following the oxidative procedure of ref. 50, and analysed first for calcium concentration using inductively coupled plasma optical emission spectrometer (ICP-OES) (Varian Vista). *U. peregrina* were then re-analysed at 100 p.p.m. [Ca] for Mg/Ca by ICP-OES and the *C. wuellerstorfi* samples were re-analysed at 10 p.p.m. [Ca] by ICP-MS (Element) for B/Ca. All trace element sample preparation and analysis occurred at the Godwin Laboratory for Palaeoclimate Research, Cambridge University (Supplementary Data 5). All analysed tests appeared visually well preserved, and low Fe/Ca and Mn/Ca values recorded simultaneously by ICP-OES indicate that samples were effectively cleaned and were devoid of diagenetic effects. Estimations of past IWT (intermediate water temperature) were then generated using the most recently published calibration from the Mg/Ca ratio of *Uvigerina*[2,51] (in mmol mol^{-1}), which has been shown to be suitable for this site for the interval from 0.4–0 Ma (ref. 37).

$$Mg/Ca_{U.peregrina} = 0.1IWT + 1.0 \quad (2)$$

Bottom-water $\Delta[CO_3^{2-}]$ (defined as the difference between $[CO_3^{2-}]_{in\ situ}$ and $[CO_3^{2-}]_{saturation}$) was calculated from the B/Ca ratio of epifaunal foraminifera *C. wuellerstorfi* (in µmol mol^{-1}) using the following calibration[49]. This proxy has also previously been used at this site for the interval from 0 to 0.4 Ma (ref. 37).

$$B/Ca_{C.wuellerstrorfi} = 1.14 \pm 0.048\Delta[CO_3^{2-}] + 177.1 \pm 1.41 \quad (3)$$

References

1. Clark, P. U. *et al.* The middle Pleistocene transition: characteristics, mechanisms, and implications for long-term changes in atmospheric pCO_2. *Quat. Sci. Rev.* **25,** 3150–3184 (2006).
2. Elderfield, H. *et al.* Evolution of ocean temperature and ice volume through the mid-Pleistocene climate transition. *Science* **337,** 704–709 (2012).
3. McClymont, E. L., Sosdian, S. M., Rosell-Melé, A. & Rosenthal, Y. Pleistocene sea-surface temperature evolution: early cooling, delayed glacial intensification, and implications for the mid-Pleistocene climate transition. *Earth-Sci. Rev.* **123,** 173–193 (2013).
4. Weinholz, P. & Lutze, G. F. in *Proceedings of the Ocean Drilling Program, Scientific Results* (eds Ruddiman, W. *et al.*) **108,** 113–117 (the Ocean Drilling Program, 1989).
5. Hayward, B. W. Late Pliocene to middle Pleistocene extinctions of deep-sea benthic foraminifera ('Stilostomella extinction') in the south-west Pacific. *J. Foraminiferal Res.* **32,** 274–306 (2002).
6. Hayward, B. W. *et al.* Last global extinction in the deep sea during the mid-Pleistocene climate transition. *Paleoceanography* **22,** PA3103 (2007).
7. Hayward, B. W. *et al. The last global extinction (Mid-Pleistocene) of deep sea benthic foraminifera (Chrysalogoniidae, Ellipsoidinidae, Glandulonodosariidae, Plectofrondiculariidae, Pleurostomellidae, Stilostomellidae), their Late Cretaceous-Cenozoic history and taxonomy* (Cushman Foundation for Foraminiferal Research Special Publication 43 408, 2012).
8. Gooday, A. J. *et al.* in *Deep-Sea Food Chains and the Global Carbon Cycle* (eds Rowe, G. T. & Pariente, V.) 63–91 (NATO ASI Series, Springer, Dordrecht, The Netherlands, 1992).
9. Hess, S. & Kuhnt, W. Neogene and quaternary paleoceanographic changes in the southern South China Sea (Site 1143): the benthic foraminiferal record. *Mar. Micropaleontol.* **54,** 63–87 (2005).
10. Thomas, E. Cenozoic mass extinctions in the deep sea: what perturbs the largest habitat on Earth? *Geol. Soc. Am.* **424,** 1–23 (2007).
11. Mancin, N., Hayward, B. W., Trattenero, I., Cobianchi, M. & Lupi, C. Can the morphology of deep-sea benthic foraminifera reveal what caused their extinction during the mid-Pleistocene climate transition? *Mar. Micropaleontol.* **104,** 53–70 (2013).
12. Mengerink, K. J. *et al.* A call for deep-ocean stewardship. *Science* **344,** 696–698 (2014).
13. Behrenfeld, M. J. *et al.* Climate-driven trends in contemporary ocean productivity. *Nature* **444,** 752–755 (2006).
14. Hayward, B. W., Grenfell, H. R., Carter, R. & Hayward, J. J. Depth distribution of Recent deep-sea benthic foraminifera east of New Zealand, and their potential for improving palaeobathymetric assessments of Neogene microfaunas. *N. Z. J. Geol. Geophys.* **44,** 555–587 (2001).
15. Takahashi, K., Ravelo, A. C. & Alvarez Zarikian, C. A. & the Expedition 323 Scientists. in *Proceedings of the International Ocean Drilling Program* **323** (2011).
16. Hamilton, L. J. Structure of the subtropical front in the Tasman Sea. *Deep Sea Res. Part I Oceanogr. Res. Pap.* **53,** 1989–2009 (2006).
17. Carter, R. M., Gammon, P. R. & Millwood, L. Glacial–interglacial (MIS 1–10) migrations of the subtropical front across ODP Site 1119, Canterbury Bight, Southwest Pacific Ocean. *Mar. Geol.* **205,** 29–58 (2004).
18. Bostock, H. C., Sutton, P. J., Williams, M. J. M. & Opdyke, B. N. Reviewing the circulation and mixing of Antarctic intermediate water in the South Pacific using evidence from geochemical tracers and Argo float trajectories. *Deep Sea Res. Part I Oceanogr. Res. Pap.* **73,** 84–98 (2013).
19. Talley, L. D. in *Mechanisms of Global Climate Change at Millennial Time Scales.* (eds Clark, P. U., Webb, R. S. & Keigwin, L. D.) 1–22 (AGU, Geophys. Mono. Series., 1999).
20. Evans, D. & Müller, W. Deep time foraminifera Mg/Ca paleothermometry: nonlinear correction for secular change in seawater Mg/Ca. *Paleoceanography* **27,** PA4205 (2012).
21. Levin, L. A. *et al.* Environmental influences on regional deep-sea species diversity. *Annu. Rev. Ecol. Syst.* **32,** 51–93 (2001).
22. Jorissen, F. J., Fontanier, C. & Thomas, E. in *Proxies in Late Cenozoic Palaeoceanography: Pt. 2: Biological tracers and biomarkers* (eds Hillaire-Marcel, C. & de Vernal, A.) 263–326 (Elsevier, 2007).
23. Kender, S., Yu, J. & Peck, V. L. Deep ocean carbonate ion increase linked to Miocene CO_2 decline. *Sci. Rep.* **4,** 4187 (2014).
24. Harris, P. G. *et al.* Chlorin accumulation rate as a proxy for quaternary marine primary productivity proxy. *Nature* **383,** 63–65 (1996).
25. Rickaby, R. E. M. *et al.* Coccolith chemistry reveals secular variations in the global ocean carbon cycle? *Earth Planet. Sci. Lett.* **253,** 83–95 (2007).
26. McClymont, E. L. *et al.* Alkenone and coccolith records of the mid-Pleistocene in the south-east Atlantic: implications for the $U^{K}_{37}{}'$ index and South African climate. *Quat. Sci. Rev.* **24,** 1559–1572 (2005).
27. Lupi, C., Bordiga, M. & Cobianchi, M. *Gephyrocapsa* occurrence during the Middle Pleistocene transition in the Northern Pacific Ocean (Shatsky Rise). *Geobios* **45,** 209–217 (2012).
28. Bollmann, J., Baumann, K. H. & Thierstein, H. R. Global dominance of *Gephyrocapsa coccoliths* in the late Pleistocene: selective dissolution, evolution, or global environmental change? *Paleoceanography* **13,** 517–529 (1998).
29. Henson, S., Cole, H., Beaulieu, C. & Yool, A. The impact of global warming on seasonality of ocean primary productivity. *Biogeosciences* **10,** 4357–4369 (2013).
30. Alve, E. Benthic foraminiferal responses to absence of fresh phytodetritus: a two-year experiment. *Mar. Micropaleontol.* **76,** 67–65 (2010).
31. Thomas, E. & Gooday, A. J. Deep-sea benthic foraminifera: tracers for Cenozoic changes in oceanic productivity? *Geology* **24,** 355–358 (1996).
32. Molina, E. Evidence and causes of the main extinction events in the Paleogene based on extinction and survival patterns of foraminifera. *Earth-Sci. Rev.* **140,** 166–181 (2015).
33. Lutz, M. J., Caldeira, K., Dunbar, R. B. & Behrenfeld, M. J. Seasonal rhythms of net primary production and particulate organic carbon flux to depth describe the efficiency of biological pump in the global ocean. *J. Geophys. Res.* **112,** C10011 (2007).
34. Shipboard Scientific Party in *Initial Reports of the Deep Sea Drilling Project* (eds Kennett, J. P., von der Borch, C. C. *et al.*) **90,** 551–651 (Deep Sea Drilling Project, 1996).
35. Head, P. S. & Nelson, C. S. in *Evolution of the Tasman Sea Basin* (eds van der Lingen, G. J., Swanson, K. M. & Muir, R. J.) 159–179 (A.A. Balkema, 1994).
36. Gradstein, F. M., Ogg, J. G., Schmitz, M. D. & Ogg, G. M. *The Geologic Time Scale 2012* (Elsevier, 2012).

37. Elmore, A. C. *et al.* Antarctic intermediate water properties since 400ka recorded in infaunal (*Uvigerina peregrina*) and epifaunal (*Planulina wuellerstorfi*) benthic foraminifera. *Earth Planet. Sci. Lett.* **428,** 193–203 (2015).
38. Shane, P. A. R. A widespread, early Pleistocene tephra (Potaka tephra, 1 Ma) in New Zealand: character, distribution, and implications. *N. Z. J. Geol. Geophys.* **37,** 25–35 (1994).
39. Bréhéret, J. Formes nouvelles quaternaires et actuelles de la famille des Gephyrocapsaceae (Coccolithophorides). *C. R. Acad. Sci.* **287,** 447–449 (1978).
40. Baumann, K. H., Böckel, B. & Frenz, M. in *Coccolithophores—from Molecular Processes to Global Impact* (eds Thierstein, H. R. & Young, J. R.) 367–402 (Springer, 2004).
41. Geisen, M. *et al.* Lifecycle associations involving pairs of holococcolithophorid species- intraspecific variation or cryptic speciation? *Eur. J. Phycol.* **37,** 531–550 (2002).
42. Marino, M., Maiorano, P. & Lirer, F. Changes in calcareous nannofossil assemblages during the Mid-Pleistocene revolution. *Mar. Micropaleontol.* **69,** 70–90 (2008).
43. Sáez, A. G., Probert, I., Young, J. R. & Medlin, L. K. in *Coccolithphores - from Molecular Processes to Global Impact* (eds Thierstein, H. R. & Young, J. R.) 251–270 (Springer, 2004).
44. Baker, E. W. & Louda, J. W. in *Biological Markers in the Sedimentary Record* (ed. Johns, R. B.) 125–226 (Elsevier, 1986).
45. Kornilova, O. & Rosell-Melé, A. Application of microwave-assisted extraction to the analysis of biomarker climate proxies in marine sediments. *Org. Geochem.* **34,** 1517–1523 (2003).
46. Prahl, F. G. & Wakeham, S. G. Calibration of unsaturation patterns in long-chain ketone compositions for palaeotemperature assessment. *Nature* **320,** 367–369 (1987).
47. Müller, P. J., Kirst, G., Ruhland, G., Storch, I. V. & Rosell-Melé, A. Calibration of the alkenone paleotemperature index $U^{k}_{37}{}'$ based on core-tops from the eastern South Atlantic and the global ocean (60°N–60°S). *Geochim. Cosmochim. Acta* **62,** 1757–1772 (1998).
48. Yu, J. & Elderfield, H. Benthic foraminiferal B/Ca ratios reflect deep water carbonate saturation state. *Earth Planet. Sci. Lett.* **258,** 73–86 (2007).
49. Rae, J. W. B., Foster, G. L., Schmidt, D. N. & Elliot, T. Boron isotopes and B/Ca in benthic foraminifera: proxies for the deep ocean carbonate system. *Earth Planet. Sci. Lett.* **302,** 403–413 (2011).
50. Barker, S., Greaves, M. & Elderfield, H. A study of cleaning procedures used for foraminiferal Mg/Ca paleothermometry. *Geochem. Geophys. Geosyst.* **4,** 1–20 (2003).
51. Elderfield, H. *et al.* A record of bottom water temperature and seawater d18O for the Southern Ocean over the past 440kyr based on Mg/Ca of benthic foraminiferal *Uvigerina spp. Quat. Sci. Rev.* **29,** 160–169 (2010).
52. Lisiecki, L. E. & Raymo, M. E. A Pliocene-Pleistocene stack of 57 globally distributed benthic $\delta^{18}O$ records. *Paleoceanography* **20,** PA1003 (2005).
53. Brown, C. W., Uz, S. S. & Corliss, B. H. Seasonality of oceanic primary production and its interannual variability from 1998 to 2007. *Deep Sea Res. Part I Oceanogr. Res. Pap.* **90,** 166–175 (2014).

Acknowledgements

We thank Mervyn Greaves for analytical assistance, Adeyinka Aturamu for SEM assistance, and the International Ocean Discovery Program for providing samples. This work was supported by the UK Natural Environment Research Council (awards NE/I024372/1 to S.K., NE/I027703/1 to E.L.M. and IP-1339-1112 to E.L.M.), a Philip Leverhulme Prize (awarded to E.L.M.) and The European Research Council (award 2010-NEWLOG ADG-267931 to H.E.).

Author contributions

S.K. and E.L.M. devised the study. A.C.E. produced the Mg/Ca and B/Ca data and constructed the age model, S.K. produced the foraminiferal assemblage data, E.L.M. produced the organic geochemistry data, D.E. and E.L.M. produced the nannofossil assemblage data, A.C.E. and M.J.L. produced the oxygen isotope data. All authors contributed to the ideas and writing of the manuscript.

Additional information

Competing financial interests: The authors declare no competing financial interests.

Cyanobacterial symbionts diverged in the late Cretaceous towards lineage-specific nitrogen fixation factories in single-celled phytoplankton

Francisco M. Cornejo-Castillo[1], Ana M. Cabello[1], Guillem Salazar[1], Patricia Sánchez-Baracaldo[2], Gipsi Lima-Mendez[3,4,5], Pascal Hingamp[6], Adriana Alberti[7], Shinichi Sunagawa[8], Peer Bork[8,9], Colomban de Vargas[10,11], Jeroen Raes[3,4,5], Chris Bowler[12], Patrick Wincker[7,13,14], Jonathan P. Zehr[15], Josep M. Gasol[1], Ramon Massana[1] & Silvia G. Acinas[1]

The unicellular cyanobacterium UCYN-A, one of the major contributors to nitrogen fixation in the open ocean, lives in symbiosis with single-celled phytoplankton. UCYN-A includes several closely related lineages whose partner fidelity, genome-wide expression and time of evolutionary divergence remain to be resolved. Here we detect and distinguish UCYN-A1 and UCYN-A2 lineages in symbiosis with two distinct prymnesiophyte partners in the South Atlantic Ocean. Both symbiotic systems are lineage specific and differ in the number of UCYN-A cells involved. Our analyses infer a streamlined genome expression towards nitrogen fixation in both UCYN-A lineages. Comparative genomics reveal a strong purifying selection in UCYN-A1 and UCYN-A2 with a diversification process ~91 Myr ago, in the late Cretaceous, after the low-nutrient regime period occurred during the Jurassic. These findings suggest that UCYN-A diversified in a co-evolutionary process, wherein their prymnesiophyte partners acted as a barrier driving an allopatric speciation of extant UCYN-A lineages.

[1] Department of Marine Biology and Oceanography, Institut de Ciències del Mar (ICM), CSIC, Pg. Marítim de la Barceloneta 37-49, 08003 Barcelona, Spain. [2] School of Geographical Sciences, University of Bristol, Bristol BS8 1SS, UK. [3] Department of Microbiology and Immunology, Rega Institute KU Leuven, Herestraat 49, 3000 Leuven, Belgium. [4] VIB Center for the Biology of Disease, VIB, Herestraat 49, 3000 Leuven, Belgium. [5] Department of Applied Biological Sciences (DBIT), Vrije Universiteit Brussel, Pleinlaan 2, 1050 Brussels, Belgium. [6] Aix Marseille Université CNRS IGS UMR 7256, 13288 Marseille, France. [7] CEA-Institut de Génomique, Genoscope, Centre National de Séquençage, 2 rue Gaston Crémieux, CP5706, F-91057 Evry, France. [8] European Molecular Biology Laboratory, Structural and Computational Biology Unit, Meyerhofstrasse 1, 69117 Heidelberg, Germany. [9] Max-Delbrück-Centre for Molecular Medicine, 13092 Berlin, Germany. [10] CNRS, UMR 7144, Station Biologique de Roscoff, Place Georges Teissier, 29680 Roscoff, France. [11] Sorbonne Universités, UPMC Université Paris 06, UMR 7144, Station Biologique de Roscoff, Place Georges Teissier, 29680 Roscoff, France. [12] Ecole Normale Supérieure, PSL Research University, Institut de Biologie de l'Ecole Normale Supérieure (IBENS), CNRS UMR 8197, INSERM U1024, 46 rue d'Ulm, F-75005 Paris, France. [13] Université d'Evry, UMR 8030, CP5706 Evry, France. [14] Centre National de la Recherche Scientifique (CNRS), UMR 8030, CP5706 Evry, France. [15] Department of Ocean Sciences, University of California, Santa Cruz, California 95064, USA. Correspondence and requests for materials should be addressed to S.G.A. (email: sacinas@icm.csic.es).

Symbiotic relationships involving diazotrophic microorganisms, that is, those capable of converting dissolved dinitrogen gas into ammonia, are of relevant interest in marine biogeochemistry because they represent major sources of fixed nitrogen, a limiting nutrient for primary production in the world's oceans[1]. As such, identifying these interactions is essential for understanding the role of symbiosis in biogeochemical cycles. Fortunately, the application of novel approaches such as high-throughput sequencing and single-cell genomics has greatly accelerated the pace of microbial symbiosis research[2,3]. This is notable in the case of *Candidatus Atelocyanobacterium thalassa* (UCYN-A), a unicellular diazotrophic cyanobacterium, and its partner, a single-celled eukaryotic alga of the class Prymnesiophyceae[4]. Prymnesiophytes as well as UCYN-A are abundant and widely distributed members of the marine plankton and represent ecologically relevant players in carbon and nitrogen cycles[5–9]. The streamlined genome of UCYN-A and the striking lack of genes encoding the photosystem II complex, the Calvin/Benson/Bassham cycle for carbon fixation, as well as other essential pathways such as the tricarboxylic acid cycle, hinted at a symbiotic lifestyle[10–12]. UCYN-A is now known to establish a mutualistic relationship based on the exchange of fixed carbon and nitrogen with two different cell-sized prymnesiophyte partners, the unicellular alga *Braarudosphaera bigelowii* (7–10 μm)[13,14] and an uncultured closely related prymnesiophyte (1–3 μm)[4,15].

Phylogenomic analyses have demonstrated the monophyly of UCYN-A within the marine cyanobacteria clade that includes *Crocosphaera* sp. and *Cyanothece* sp. clades[12]. Phylogenetic analysis of the UCYN-A nitrogenase gene (*nifH*) sequences, a common marker used to address the diversity of N_2-fixing microorganisms, distinguished at least three distinct UCYN-A clades: UCYN-A1, UCYN-A2 and UCYN-A3 (ref. 14). Comparative genomics revealed that UCYN-A1 and UCYN-A2 lineages share largely syntenic genomic structures, suggesting that both lineages diverged after genome reduction from a common ancestor[12]. Yet, their time of evolutionary divergence and evolutionary pressures remain unknown. It has been suggested that these two variants could be adapted to different niches, that is, coastal waters (*B. bigelowii*) and open ocean (its closely related prymnesiophyte)[14], but this ecological differentiation was recently ruled out[9]. Although the two prymnesiophyte partners could follow different ecological strategies[9], the partner fidelity has never been tested in this symbiotic system, and therefore we cannot assume a similar ecological niche for their symbionts. Comparative gene expression studies could help to disentangle the ecological distinction of these two UCYN-A lineages but they are scarce and solely focused on the *nifH* gene expression without showing a clear differentiation in lineage-specific patterns[14].

By designing and applying new probes in double catalysed reporter deposition fluorescence *in situ* hybridization (CARD-FISH), we identified the specific symbiotic associations at the UCYN-A lineage level in samples from South Atlantic waters from the *Tara* Oceans expedition, where we had previously verified significant abundances of the prymnesiophyte partners. The new probes allowed us to differentiate both symbiotic systems that resulted to vary in the number of UCYN-A cells involved. The coupled analyses of metagenomes and metatranscriptomes from surface and deep chlorophyll maximum (DCM) depths that encompassed four different plankton size fractions distinguish different prymnesiophyte partners based on difference in cell sizes captured in different size fractions, complementing and extending the results obtained by CARD-FISH. Gene expression was explored in the two UCYN-A lineages to decipher whether distinct lineages, in association with distinct partners, exhibit different expression patterns. Finally, we investigated the evolutionary pressures acting on UCYN-A1 and UCYN-A2 lineages by comparative genomic analyses and performed phylogenomic analyses to estimate the age divergence of the two symbiotic lineages. Our findings support a symbiont–host co-evolutionary scenario in the marine environment originating from a single ancestral symbiotic event in the late Cretaceous from which at least two different UCYN-A lineages diversified to become lineage-specific nitrogen fixation factories in their prymnesiophyte partners. Together, these investigations improve our understanding of the relevance of co-evolutionary processes in marine ecosystems and the ecological significance of N_2-fixing symbiosis in the marine biogeochemical cycles.

Results and Discussion

Partner fidelity of the two UCYN-A lineages. Microscopic *in situ* identification of different UCYN-A lineages as well as their prymnesiophyte partners by specific CARD-FISH staining is crucial to determine the specificity of their relationships. The CARD-FISH method has been successfully applied to identify unicellular diazotrophic cyanobacteria[16] as well as specifically targeting the UCYN-A clade[15,17]. However, to our knowledge there was not any reported probe to distinguish UCYN-A at the lineage level. We designed a competitor probe to be used with the UCYN-A732 probe[15] to distinguish UCYN-A1 and UCYN-A2 lineages (Fig. 1a–c; Supplementary Table 1). Similarly, we designed two probes to distinguish the two prymnesiophyte partners, *B. bigelowii* (UBRADO69 probe) and the closely related prymnesiophyte (UPRYM69 probe) (Fig. 1a–c; Supplementary Table 1). The UCYN-A732 probe, in the absence of its competitor, labelled UCYN-A cells inside either *B. bigelowii* or the closely related prymnesiophyte partner (Fig. 1a,c). However, when the UBRADO69 probe was applied with the UCYN-A732 probe together with its competitor, UCYN-A cells were unlabelled or labelled when accompanying *B. bigelowii* or the closely related prymnesiophyte partner, respectively (Fig. 1b). It has been proposed that smaller UCYN-A cells are associated with smaller prymnesiophyte cells and vice versa, indicating different growth stages[17]. However, those findings were interpreted from microscopic observations of the UCYN-A symbiosis detected with the general prymnesiophyte PRYM02 and UCYN-A732 (without its competitor) probes, that is, without the ability to distinguish UCYN-A1 and UCYN-A2 cells. The results presented here show that both prymnesiophyte partners are phylogenetically closely related but distinct species, and therefore we suggest that the observed differences in cell sizes of prymnesiophyte partners reflect distinct species rather than different growth stages of the same species. These results demonstrate that UCYN-A lineages display partner fidelity with their prymnesiophyte partners, being *B. bigelowii* and the closely related prymnesiophyte in specific association with UCYN-A2 and UCYN-A1 lineages, respectively.

The number of UCYN-A cells per partner is lineage specific. Previous studies have shown that the prymnesiophyte partners can harbour one or two UCYN-A cells[4,9,13,15], pointing to a coupling between the prymnesiophyte cell division and the number of symbiotic cells, at least for UCYN-A1 (ref. 9). In our samples, only one UCYN-A1 cell per prymnesiophyte cell was detected (Fig. 1a,b). By contrast, *B. bigelowii* carried a symbiosome-like compartment with a variable but higher number of UCYN-A2 cells (~3–10 cells) (Fig. 1b,c). This structure was observed both attached to the host and in a free state, as an entity composed by several UCYN-A2 cells enclosed by a common envelope (Fig. 1c). In a previous study, the UCYN-A2 cells found

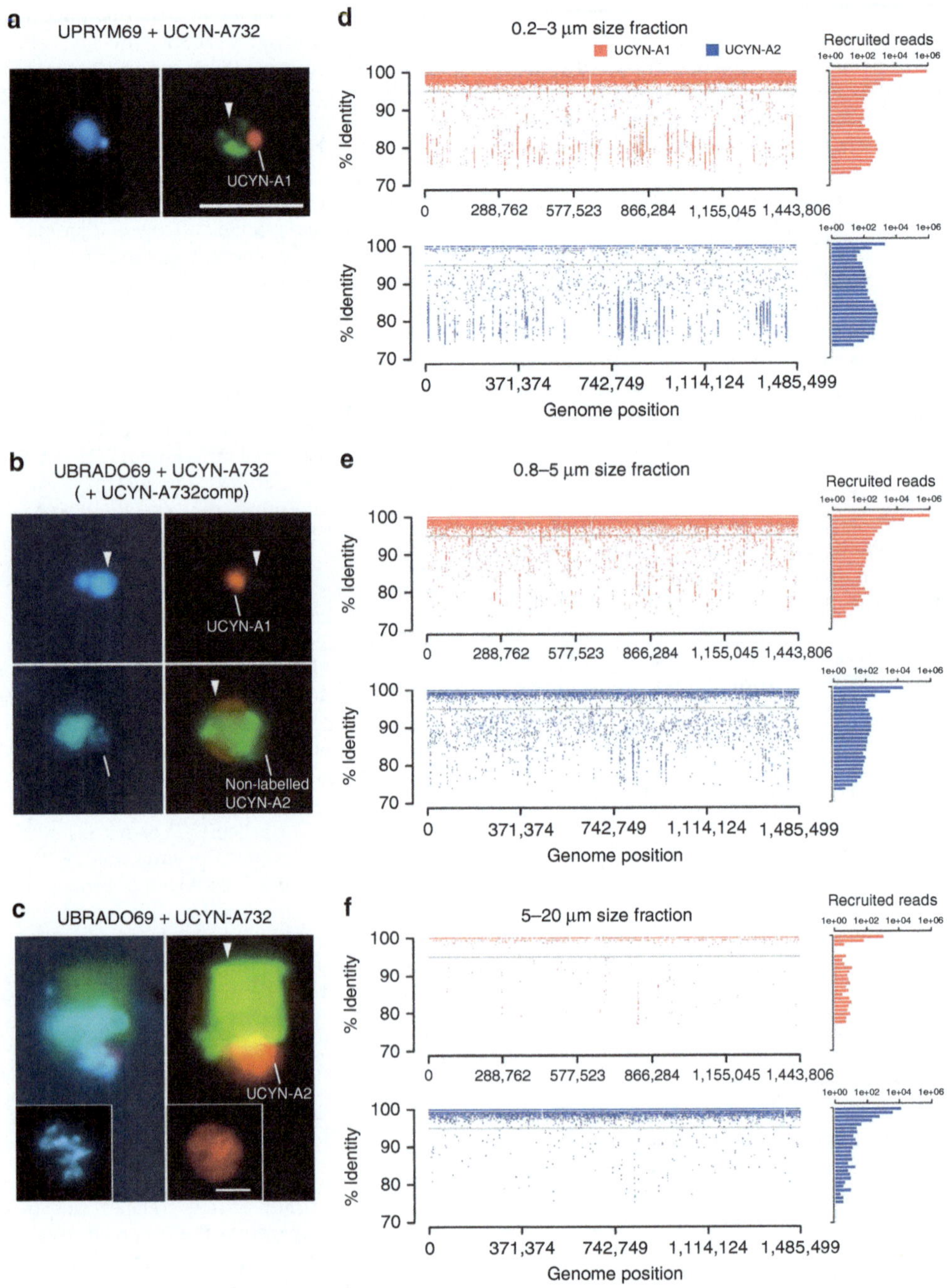

Figure 1 | Partner specificity and variation of UCYN-A lineages with plankton size fraction. (**a–c**) Epifluorescence microscopy images with the double-CARD-FISH assay showing the specificity of symbiont-host pairs and (**d–f**) fragment recruitment of UCYN-A lineages in size-fractionated metagenomes from surface waters collected in station TARA_078. (**a–c**) Left panels correspond to the 4′,6-diamidino-2-phenylindole signal (blue-labelled DNA); right panels correspond to the combined signal of the prymnesiophyte-specific probes (green-labelled host under blue light excitation) and the UCYN-A probe (red-labelled symbiont under green-light excitation). (**a**) UCYN-A1 with its prymnesiophyte partner; (**b**) the two UCYN-A symbiotic pairs, indicating the specific labelling of UCYN-A1 (upper) and *B. bigelowii* (lower) with their specific partners, the small prymnesiophyte closely related to *B. bigelowii* and UCYN-A2 respectively; (**c**) *B. bigelowii* with UCYN-A2. The inset in **c** shows the detail of non-associated UCYN-A2 cells within a common symbiotic structure. Prymnesiophyte partners are indicated by arrow heads. Scale bar in **a** represents 5 μm and this scale is shared in **a–c** except in the inset of **c** where it indicates 2 μm. (**d–f**) On the left side, recruitment of metagenomic reads using UCYN-A1 and UCYN-A2 genomes as reference. Reads are plotted as red (UCYN-A1) or blue (UCYN-A2) dots depending on the closest hit genome, representing the covered genome positions (*x* axis) and the % of identity with the closest reference (*y* axis). A horizontal grey line set at 95% indicates the threshold for reads representing members of the same population as the reference genome. On the right side, histograms represent the number of recruited reads, in logarithmic scale, by UCYN-A1 (red) or UCYN-A2 (blue) genomes in intervals of 1% identity, from 100 to 70% identity.

in *B. bigelowii* were separated from the *B. bigelowii* cytoplasm by a single membrane, likely a perisymbiont membrane, and the envelope of the UCYN-A2 itself consisted of three layers, possibly an outer membrane, a peptidoglycan wall and a plasma membrane[13]. Although UCYN-A1 and UCYN-A2 are very similar in terms of gene content, the genes involved in cell wall biogenesis and cell shape determination appear to be only present in UCYN-A2, suggesting clear structural differences associated with its host[12]. Therefore, our observations hint at different symbiotic organizations: while the UCYN-A1 lineage has one or two separated cells per host, the UCYN-A2 lineage may harbour up to 10 cells per prymnesiophyte partner cell within a common symbiotic structure.

UCYN-A lineages vary in different plankton size fractions. A total of eight marine metagenomes from stations TARA_078 and TARA_076 were analysed to assess the distribution of UCYN-A lineages in several plankton size fractions (0.2–3, 0.8–5, 5–20 and >0.8 μm) of the microbial assemblages in surface and DCM waters (Table 1). We used the two UCYN-A genomes sequenced to date as reference genomes[11,12] in the fragment recruitment of these metagenomic samples (Table 1). Because of the UCYN-A partner fidelity displayed by double CARD-FISH (see above), metagenomic sequence reads from UCYN-A lineages should vary with size fraction as predicted by the different cell sizes of the prymnesiophyte partners. The sequence reads from the UCYN-A1 lineage were primarily present in surface waters within the size fraction range of the small prymnesiophyte partner (0.2–3, 0.8–5 and >0.8 μm; Table 1). Almost 100% of the UCYN-A1 genome was recovered in each of the metagenomes from surface of these size fractions in the two stations. Likewise, UCYN-A1 sequence reads were poorly represented in the 5–20 μm size fraction (~10% of genome recovery; Fig. 1d–f; Table 1). On the other hand, in TARA_078, the UCYN-A2 sequence read distribution in surface waters was consistent with the *B. bigelowii* cell size, that is, UCYN-A2 reads were nearly absent in the 0.2–3 μm size fraction metagenomes, but were more abundant in the 0.8–5, 5–20 and >0.8 μm fractions. In all these larger fractions, the UCYN-A2 reached high genome recovery values (90%, 76% and 99%, respectively), except for the >0.8 μm fraction in TARA_076 where UCYN-A2 was virtually absent (Fig. 1d–f, Table 1). In the >0.8 μm size fraction, UCYN-A1 was approximately nine times more abundant than UCYN-A2 in TARA_078 (Table 1). Likewise, in the same station, the small prymnesiophyte partner was more abundant than *B. bigelowii* based on V9 18S ribosomal RNA (rRNA) tags[9]. In the DCM samples, both UCYN-A lineages were poorly represented in the metagenome sequences, accounting for <14% and 1% of genome recovery for UCYN-A1 and UCYN-A2, respectively (Table 1). The same vertical distribution has been observed for their prymnesiophyte partners that were found preferentially in surface layers, while the rest of the prymnesiophyte assemblage peaked at the DCM[9]. Therefore, although the UCYN-A1 lineage was in general more abundant than UCYN-A2, a transition from the UCYN-A1 to UCYN-A2 lineage was observed from smaller to larger size fractions, likely explained by the partner fidelity and the difference in cell size of their prymnesiophyte partners.

Another interesting finding was that most of the metagenomic (and metatranscriptomic) reads mapping to the UCYN-A1 or UCYN-A2 genomes had very high sequence identities (>99% to their respective reference genome; Fig. 1d–f), which suggests an extremely low microdiversity within populations that were sampled from geographically distant regions in the Pacific (ALOHA and SIO) and South Atlantic Oceans (this study). The size-fractionated sampling strategy combined with the metagenomic analyses reported in this study will be also important to uncover the genomic pool of new UCYN-A lineages, such as UCYN-A3, to identify the lineage-specific distribution of UCYN-A populations and to set the cell size range of their partners, a first step for their identification.

UCYN-A expression is streamlined to fuel nitrogen fixation. The analyses of seven size-fractionated metatranscriptomes from two stations (TARA_078 and TARA_076) and depths (surface and DCM) allowed for the first time a whole-genome transcription profiling of these widely distributed diazotrophic cyanobacteria (Table 1). In surface waters, UCYN-A1 transcripts were in general more abundant than those from UCYN-A2, except in the 5–20 μm size fraction (TARA_078) in which the latter were dominant (Table 1). The gene expression of 1,131 and 1,179 protein-coding genes in UCYN-A1 (Supplementary Data 1) and UCYN-A2 (Supplementary Data 2), respectively, were examined. In both lineages, the nitrogen fixation operon, including the *nifH* gene, was the most highly expressed gene-cluster accounting for a quarter of the total transcripts (Fig. 2a,b). In the >0.8 μm size fraction (TARA_078), despite UCYN-A1 being more abundant than UCYN-A2, the expressed *nifH* transcripts per cell were almost two times higher for UCYN-A2 (648.33) than for UCYN-A1 (396.60; Supplementary Data 1 and 2). It is well known that biological nitrogen fixation

Table 1 | Fragment recruitment (FR) of UCYN-A lineages.

Station	Depth	Sample	Fraction (μm)	Sequencing depth (reads)	FR (reads) UCYN-A1	FR (reads) UCYN-A2	Genome recovery (%) UCYN-A1	Genome recovery (%) UCYN-A2
76	SRF	MG	0.2-3	177,019,968	188,088	26	99.30	0.14
76	SRF	MT	0.2-3	18,908,305	25,340	137	21.35	0.37
76	SRF	MG	>0.8	73,651,199	54,776	147	98.61	1.35
76	SRF	MT	>0.8	10,283,396	12,143	322	15.01	0.59
76	DCM	MG	>0.8	115,099,936	848	3	9.00	0.03
76	DCM	MT	>0.8	12,998,358	76	3	0.49	0.02
78	SRF	MG	0.2-3	155,580,203	842,234	2,395	99.94	13.95
78	SRF	MT	0.2-3	13,151,362	133,693	453	46.61	0.99
78	SRF	MG	0.8-5	105,731,269	980,895	24,021	99.81	90.24
78	SRF	MG	5-20	139,070,786	1,182	17,028	10.14	76.47
78	SRF	MT*	5-20	97,646,287	292	17,862	1.76	34.69
78	SRF	MG	>0.8	163,575,710	719,803	81,528	99.32	99.03
78	SRF	MT	>0.8	9,966,043	44,613	9,415	30.77	11.51
78	DCM	MG	>0.8	86,446,300	1,358	45	13.32	0.48
78	DCM	MT	>0.8	10,659,304	82	10	0.71	0.07

DCM, deep chlorophyll maximum; MG, metagenome; MT, metatranscriptome; SRF, surface.
*A protocol that selectively sequenced RNA sequences with poly(A) tails was conducted.

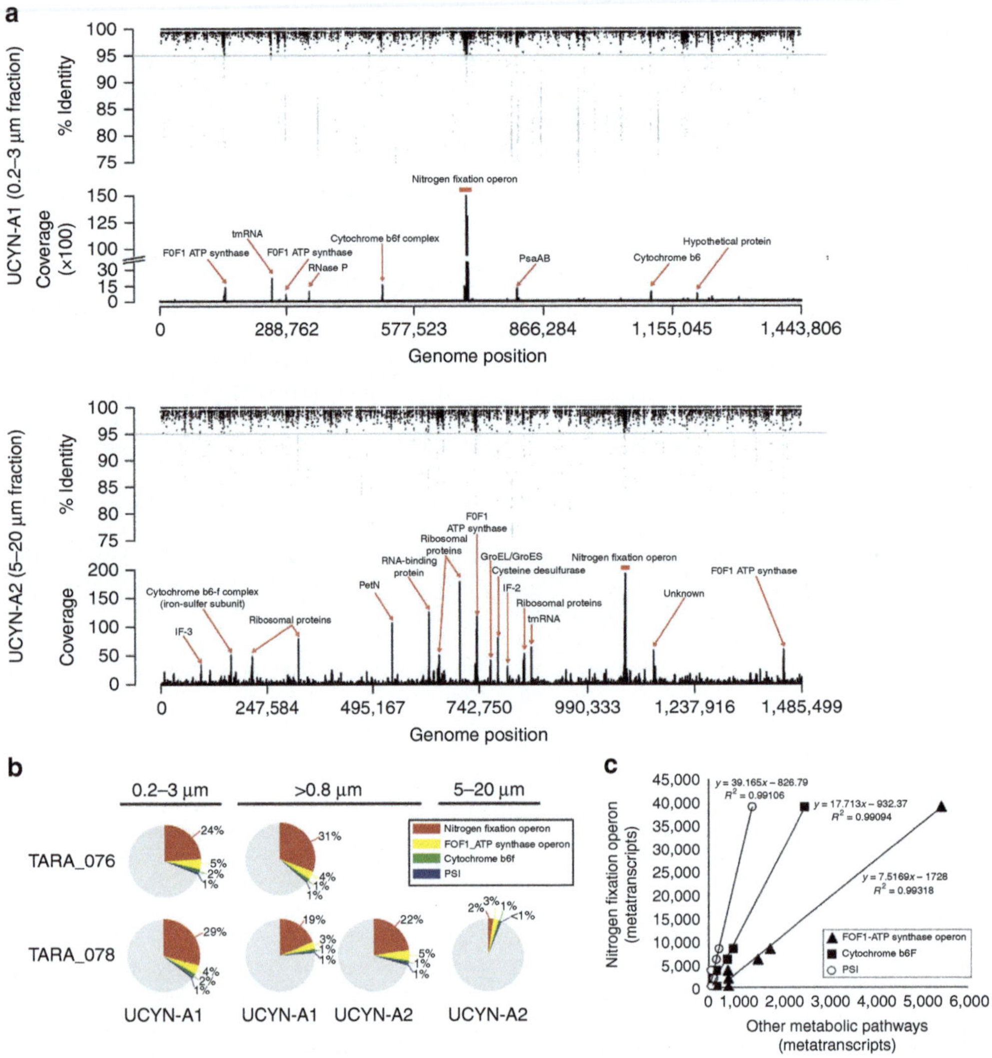

Figure 2 | Genome expression in UCYN-A1 and UCYN-A2 lineages. (**a**) Metatranscriptome recruitment at the surface of the TARA_078 station of UCYN-A1 (0.2-3 μm) and UCYN-A2 (5-20 μm) transcripts. Transcripts are plotted as black dots representing the covered genome positions and the % of identity with the closest reference. A horizontal grey line set at 95% identity shows the threshold used to count the number of times, or coverage, that a gene was expressed. The most expressed genes in both lineages are highlighted. (**b**) Relative contribution of nitrogen fixation operon, FOF1-ATP synthase operon, cytochrome b_6f and PSI genes to the total UCYN-A transcripts contribution in surface samples; percentages are indicated. (**c**) Transcript counts of nitrogen fixation operon versus those of ATP synthase (triangle), cytochrome b_6f (square) and PSI (open circle) transcripts. All of these transcripts were significantly correlated ($P<10^{-5}$) and regression lines, regression equations and R^2 values are indicated in the figure.

has a high energetic cost (16 mol of ATP to generate 2 mol of ammonia). Notably, the F0F1-ATP synthase operon and genes encoding for the cytochrome b_6f complex and photosystem I complex (PSI) were highly transcribed and positively correlated ($P<10^{-5}$, $N=6$, linear regression analysis) with the nitrogen fixation operon transcript abundances (Fig. 2c). These findings suggest that the generation of reducing power and the ATP synthesis could be coupled to fuel the nitrogen fixation process in UCYN-A. Likewise, UCYN-A2 might have higher nitrogen fixation rates per cell than UCYN-A1 based on the higher number of *nifH* transcripts per cell. It is reasonable to assume that the differences in *nifH* gene expression between the UCYN-A lineages could simply reflect the differences in the cell size of their partners with differential nutrient requirements for growth. In addition, it has been indirectly demonstrated that the nitrogen fixation of UCYN-A supports the CO_2 fixation of its prymnesiophyte partner[18]. Therefore, we hypothesize that the larger *B. bigelowii* host cell would meet its larger N nutrient requirements by partnering with a larger number of UCYN-A2 symbiotic cells.

Nitrogen-fixing microorganisms, and particularly cyanobacteria, should protect their nitrogenase from inactivation by oxygen. The absence of the ability to use photosystem II that evolves O_2 explains why UCYN-A appears to fix N_2 and express the nitrogenase genes during the day[19]. However, its association with an oxygen-evolving partner could make the nitrogenase enzyme in UCYN-A not completely safe from oxygen. We observed that the *sufB* gene (cysteine desulferase), involved in the assembly or repair of oxygen-labile iron–sulfur clusters under oxidative stress, was highly transcribed (Supplementary Data 1 and 2). It may be that UCYN-A requires high expression level of *sufB* genes to repair the nitrogenase enzyme from oxygenic inactivation, suggesting then a similar role than for the peroxidase genes found in their

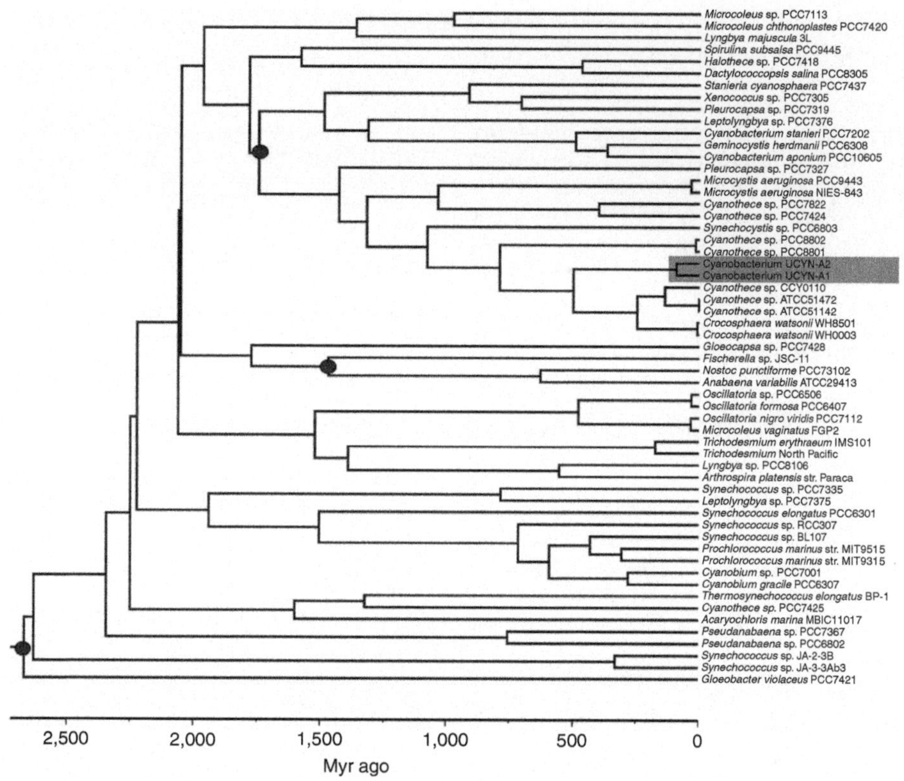

Figure 3 | Time-calibrated cyanobacteria tree. The phylogeny shown was estimated based on 135 proteins from 57 taxa. Three calibration points (black circles) were used for the tree presented and were treated as soft bounds. The root of the tree was set with a maximum age of 2,700 Myr ago and a minimum age of 2,320 Myr ago. Divergence time for the ancestor of cyanobacteria UCYN-A1 and UCYN-A2 (highlighted with a grey box) is given with the corresponding values for the posterior 95% confidence intervals in Supplementary Table 2.

genomes[11,12]. Our findings reveal that UCYN-A lineages dedicate a large transcriptional investment to fix nitrogen representing the first whole-genome expression profiling in environmental UCYN-A populations.

UCYN-A diverged during the late Cretaceous. Our findings on partner fidelity in UCYN-A point to the hypothesis of symbiont–host co-evolution[14]. To analyse the selection pressure and evolution of the protein-coding genes, we calculated the number of synonymous or silent (Ks) and non-synonymous (Ka, inducing amino-acid change) nucleotide substitutions[20,21] for 887 protein-coding genes shared by the UCYN-A1 and UCYN-A2 genomes (Supplementary Data 3). The Ka/Ks ratio may offer important clues about the selection pressure where ratios <1 indicate purifying selection and ratios >1 point to positive selection[22]. We found that 873 out of the 887 protein-coding genes were under purifying selection ($P<0.05$, codon-based Z-test) (Supplementary Data 3). The 14 remaining genes also presented $Ka/Ks<1$ but were not statistically well supported ($P>0.05$). Purifying selection means that synonymous mutations are maintained, while non-synonymous mutations are continuously removed from the population. We did not detect signs of large-scale positive selection, that is, no apparent strong adaptation to novel niches in UCYN-A lineages, suggesting that the evolutionary forces for niche adaptation would act on the prymnesiophyte partners rather than on UCYN-A. Our results are consistent with the fact that UCYN-A2 lacks the same major pathways and proteins that are absent in UCYN-A1 (ref. 12), indicating then that the symbionts were genetically adapted to their hosts before they were separated by speciation.

The age of divergence for UCYN-A1 and UCYN-A2 lineages was calculated by phylogenomic and Bayesian relaxed molecular clock analyses (Fig. 3; Supplementary Table 2). Our results indicate that UCYN-A1 and UCYN-A2 lineages diverged around 91 Myr ago, that is, during the late Cretaceous. In agreement, *B. bigelowii* has a fossil record extending back to the late Cretaceous (ca. 100 Myr ago)[23], reported from neritic and pelagic sediments, for example, in lower Paleogene sediments immediately above the K/Pg mass extinction level as well as in the Oligocene Diversity Minimum[24,25]. In the Jurassic, between 190 and 100 Myr ago, nutrient availability in the ocean was lower than at any point during the last 550 Myr ago[26]. It is therefore likely that the symbiotic relationship between the common ancestor of UCYN-A1 and UCYN-A2, and a *Braarudosphaera*-related species was established by the late Cretaceous to cope with extremely low-nutrient conditions and a generalized oligotrophy in marine surface waters, as it has been recognized for other symbiotic system such as the Acantharia–*Phaeocystis* symbiosis[27]. UCYN-A then underwent purifying selection, progressively reducing its genome to the point that it became an obligate symbiont. An analogous discovery was the case of the two Rhopalodiaceae freshwater diatom species, *Rhopalodia gibba* and *Epithemia turgida* having acquired N_2-fixing endosymbionts[28,29]. Similar to the two UCYN-A partnerships described here, phylogenies of these two diatoms species and their intracellular symbionts were found to be congruent and, concordantly, a single symbiotic event has been proposed[29]. Probably, a similar scenario can be envisioned here for the two UCYN-A partnerships.

Taking into account that the number of symbiotic cells harboured by distinct prymnesiophyte partners is different and phylogenetically dependent, that is, the larger *B. bigelowii* can harbour a variable number (up to 10) of UCYN-A2 cells, while the small prymnesiophyte partner harboured only one or two UCYN-A1 cells, it is reasonable to think that a larger nutrient acquisition could be linked to a larger number of symbionts. Indeed, the whole-genome expression patterns suggested that the metabolic investment in UCYN-A1 and UCYN-A2 is mainly focused on the nitrogen fixation machinery. Our evolutionary analysis revealed that UCYN-A1 and UCYN-A2 were genetically adapted to their prymnesiophyte partners before UCYN-A speciation (purifying selection) but, on the contrary, the prymnesiophyte partners seem to follow different ecological strategies[9], suggesting a speciation process under positive selection. Our results suggest that the partner fidelity shown by UCYN-A lineages together with the speciation in the common ancestor of *B. bigelowii* and its closely related prymnesiophyte may have forced an allopatric speciation of UCYN-A1 and UCYN-A2 populations in the late Cretaceous. Comparative genome analysis of the two prymnesiophyte partners would clarify whether these two algal species underwent positive selection through evolution by adaptation to novel niches. As revealed by *nifH* phylogenetic analysis, it seems that novel UCYN-A lineages, such as UCYN-A3, and prymnesiophyte (or not prymnesiophyte) partners, will help to understand the evolutionary relationships of N_2-fixing cyanobacterial symbionts and the extent of their ecological relevance on marine biogeochemical cycles.

The present study offers new insights into the marine nitrogen-fixing UCYN-A symbiosis by disentangling the partner fidelity, host–symbiont organization and size distribution, gene expression and evolution of UCYN-A1 and UCYN-A2 lineages. These results demonstrate that specific UCYN-A symbiotic pairs co-exist without cross-symbiotic partnerships. The fact that its distribution occupies new plankton size fractions accordantly to the host size should be considered in global nitrogen fixation models. The number of UCYN-A1 and UCYN-A2 cells involved in this symbiosis differs and appears to be a conserved phylogenetic trait. Remarkably, about a quarter of the UCYN-A transcripts were from nitrogen fixation genes, highlighting the importance of nitrogen fixation in this symbiosis. Our results present further evidences of a host and symbiont co-evolution scenario in the marine environment, probably derived from a single ancestral symbiotic event wherein at least two different lineages diversified in the late Cretaceous. Investigation of N_2-fixing cyanobacterial symbionts and their partners should provide clues for discovering new ecological compartments for nitrogen fixation that would increase our understanding of the nitrogen cycle in the ocean.

Methods

Sample choice. From a total of 243 metagenomes from 68 globally distributed stations from *Tara* Oceans expedition[30], the abundance of UCYN-A based on 16S $_{mi}$TAGs[31,32] and their corresponding prymnesiophyte partners evaluated by V9 18S iTAGS[9,33], pointed out to a couple of stations, that is, TARA_078 (30° 8′ 12.12″ S, 43° 17′ 23.64″ W) and TARA_076 (20° 56′ 7.44″ S, 35° 10′ 49.08″ W) in the South Atlantic Ocean in which this symbiotic system were significantly abundant[9], and therefore these two stations were chosen to further explore the UCYN-A symbiotic system.

Sample collection. For the whole-cell CARD-FISH, 10 ml of surface seawater (pre-filtered with 20-μm pore-size mesh) was fixed with paraformaldehyde (1.5% final concentration) at 4 °C overnight and gently filtered through 0.2-μm pore-size polycarbonate filters (Millipore, GTTP, 25 mm diameter). For nucleic acid extractions and sequencing, surface seawater was collected and subsequently separated into four size fractions (0.2–3, 0.8–5, 5–20 and >0.8-μm pore-size filters)[34,32]. After filtration, filters were kept for ~4 weeks at −20 °C on the schooner and then at −80 °C in the laboratory until processed for hybridization or nucleic acid extraction.

Design of CARD-FISH probes. For the design of specific oligonucleotide probes targeting *B. bigelowii* and the closely related prymnesiophyte partner, a total of 580 sequences, 18S rRNA gene sequences, belonging to the class Prymnesiophyceae were retrieved from the PR2 database[35], aligned using MAFFT[36] and the alignment was verified manually to remove chimeras and sequences with ambiguities (466 sequences were kept). A maximum likelihood phylogenetic tree was built using RAxML[37] with 100 trees for both topology and bootstrap analyses, and visualized with iTol[38,39] (Supplementary Fig. 1). The newly designed probe UBRADO69 targeted *B. bigelowii*, while probe UPRYM69 targeted the closely related prymnesiophyte partner (Supplementary Table 1). UBRADO69 and UPRYM69 probes differed in only one position, and required a competitor to avoid unspecific hybridizations. Therefore, the labelled probe UBRADO69 was used in combination with the unlabelled UPRYM69 oligonucleotide for the detection of *B. bigelowii*, and vice versa for the detection of the closely related prymnesiophyte partner (Supplementary Table 1). Two helpers, helper-A PRYM and helper-B PRYM, were designed to improve the hybridization process for both probes (Supplementary Table 1). The UCYN-A732 probe designed against UCYN-A by targeting the 16S rRNA[15] has only one mismatch with the UCYN-A2 sequence and a competitor was designed to distinguish specifically UCYN-A1 and UCYN-A2 clades with high specificity (Supplementary Table 1). The specificity of the new probes was checked with the online tool ProbeCheck (http://www.cme.msu.edu/RDP/) and by searching in the GenBank database (http://www.ncbi.nlm.nih.gov/index.html) to detect potential matching sequences in non-target groups.

CARD-FISH assay and epifluorescence microscopy. A preliminary double-hybridization assay using the universal haptophyte PRYM02 probe[40] and UCYN-A732 was first applied to check whether the partner of UCYN-A in our sample belong to class Prymnesiophyceae. To specifically target the different UCYN-A lineages and their prymnesiophyte hosts, a double-CARD-FISH assay was performed for each partnership (according to the multi-colour CARD-FISH protocol[41]). For the first hybridization step, the specific probe for one of the prymnesiophyte partners (UBRADO69 or UPRYM69) was used and, for the second step, the UCYN-A732 probe was used. To check the specificity of symbiont pairs, an additional double-CARD-FISH assay was carried out with the UBRADO69 probe and the UCYN-A732 as described before with the addition of the UCYN-A732 competitor to the hybridization buffer (probe, helpers and competitor at 0.16 ng μl^{-1}). Filters were embedded in low-gelling-point agarose 0.1% (w/v) to minimize cell loss, and cell walls were permeabilized with lysozyme (37 °C, 1 h) and acromopeptidase solutions (37 °C, 0.5 h). For the first CARD-FISH step (described in more detail in Cabello *et al.*[9]), filters were hybridized overnight at 46 °C in 40% formamide (FA) hybridization buffer containing a mixture of the HRP (horseradish peroxidase)-labelled probe, helpers and competitor oligonucleotides. Filters were then rinsed in washing buffer at 48 °C and tyramide signal amplification was performed for 40 min at room temperature in the dark in a buffer containing 4 μg ml^{-1} Alexa 488-labelled tyramide. Before the second hybridization, the HRP from the first probe was inactivated with 0.01 M HCl for 10 min at room temperature in the dark[41]. The second CARD-FISH used the probe UCYN-A732 and its corresponding helpers and was applied according to Krupke *et al.*[15]. UCYN-A cells were hybridized for 3 h at 35 °C in 50% FA hybridization buffer, rinsed in washing buffer for 15 min at 37 °C and tyramide signal amplification was done as before but using 1 μg ml^{-1} Alexa 594-labelled tyramide. Preparations were counterstained with 4′,6-diamidino-2-phenylindole at 5 μg ml^{-1}, mounted in antifading reagent (77% glycerol, 15% VECTASHIELD and 8% 20× PBS) and kept frozen until microscopic analysis. A no-probe control showed that there was no signal coming from endogenous peroxidases. Filters were observed by epifluorescence microscopy (Olympus BX61) at 1,000× under ultraviolet (4′,6-diamidino-2-phenylindole signal of the nucleous), blue-light (green-labelled host cells with Alexa 488) or green-light (red-labelled symbionts with Alexa 594) excitations. Micrographs were taken using an Olympus DP72 camera (Olympus America Inc.) attached to the microscope.

Hybridization conditions for the UPRYM69 and UBRADO69 probes were optimized testing different FA concentrations in the hybridization buffer and varying the hybridization temperature. The UPRYM69 probe (together with the competitor oligonucleotide) was tested in NE Atlantic surface samples, where UCYN-A1 host cells were ~86% of prymnesiophytes (~550 cells per ml). Initially, we tried 20–30–40–50% FA in the buffer and the temperature of 35 °C for hybridization. At 20% FA, host cells carrying UCYN-A (n = 89) displayed a faint fluorescent signal (90%) or were not labelled (10%), whereas above 40% FA, no hybridized cells were detected. Signal was improved using helper oligonucleotides and host cells displayed a bright homogeneous signal at all FA concentrations, but we observed cross-hybridization (observed as fluorescent dots all over the cells) in larger prymnesiophyte-like cells not associated to UCYN-A even at 50% FA. Thus, we tested 40 and 50% FA in a hybridization temperature of 46 °C. The 40% FA showed optimal signal intensity, labelling small prymnesiophytes cells (~2.5 μm) always carrying UCYN-A and no cross-hybridization was observed. We applied these conditions to hybridize the surface sample TARA_078. In this sample, in addition to the labelled small host cells observed in the NE Atlantic, we observed larger host cells not labelled by the UPRYM69 probe. To verify that these cells were the UCYN-A2 host, we applied the UBRADO69 probe with the same conditions (as both probes differ in only 1 position) and we found the complementary result: the larger host was labelled but not the smaller one. With the optimized conditions (40% FA, 46 °C), the probes were labelling specifically the target host without cross-hybridization.

Nucleic acid extractions and sequencing. Surface and DCM seawater samples collected by *Tara* Oceans' station 76 and 78 in the South Atlantic Ocean (TARA_076 and TARA_078) for metagenomic sequencing were size fractionated.

For surface samples, metagenomes from two and four fractions were analysed in TARA_076 (0.2–3 and >0.8 μm) and TARA_078 (0.2–3, 0.8–5, 5–20 and >0.8 μm), respectively. For DCM samples, metagenomes from one fraction were analysed in TARA_076 (>0.8 μm) and TARA_078 (>0.8 μm). Seawater samples for metatranscriptomic sequencing used also several size fractions. For surface samples, metatranscriptomes from two and three fractions were analysed in TARA_076 (0.2–3 and >0.8 μm) and TARA_078 (0.2–3, 5–20 and >0.8 μm), respectively. For DCM samples, metatranscriptomes from one fraction were analysed in TARA_076 (>0.8 μm) and TARA_078 (>0.8 μm). DNA and RNA extraction protocols for the different size fractions and metagenome sequencing are described in refs 31–33.

cDNA synthesis and sequencing. For 0.2–3 μm and >0.8 μm filters, bacterial rRNA depletion was carried out on 240–500 ng total RNA using Ribo-Zero Magnetic Kit for Bacteria (Epicentre, Madison, WI). The Ribo-Zero depletion protocol was modified to be adapted to low RNA input amounts[42]. Depleted RNA was used to synthetize complementary DNA (cDNA) with SMARTer Stranded RNA-Seq Kit (Clontech, Mountain View, CA)[42]. For 5–20-μm filter from TARA_078, cDNA was synthetized starting from 50 ng total RNA using SMARTer Ultra Low RNA Kit (Clontech) by oligodT priming, following the manufacturer protocol. Full-length double-stranded cDNA was fragmented to a 150–600-bp size range using the E210 Covaris instrument (Covaris Inc., USA). Then, fragments were end-repaired and 3′-adenylated, and ligated to Illumina adaptors using NEBNext Sample Reagent Set (New England Biolabs, Ipswich, MA). Fragments were PCR-amplified using Illumina adapter-specific primers and purified. All metatranscriptomic libraries were quantified by qPCR using the KAPA Library Quantification Kit for Illumina Libraries (KapaBiosystems, Wilmington, MA) and library profiles were assessed using the DNA High Sensitivity LabChip kit on an Agilent Bioanalyzer (Agilent Technologies, Santa Clara, CA). Libraries were sequenced on Illumina HiSeq2000 instrument (Illumina, San Diego,CA) using 100 base-length read chemistry in a paired-end mode. Sequencing depth for each sample is detailed in Table 1.

Nucleotide data deposition. Nucleotides data used in this study have been deposited in the European Nucleotide Archive (ENA; www.ebi.ac.uk/ena) under the following accession numbers: ERR1001626-27, ERR1007415-18, ERR1013384-85, ERR599006, ERR599010, ERR599022, ERR599126, ERR599237, ERR599240, ERR599250, ERR599253, ERR599275 and ERR599311.

Fragment recruitment analysis from *omics* data sets. BLAST+ v2.2.25 was used to recruit metagenomic and metatranscriptomic reads similar to the two UCYN-A genomes sequenced up to date[11,12] using default parameter values, except for the following: -perc_identity 50, -evalue 0.0001. Metagenomic/metatranscriptomic reads belonging to 23S, 16S and 5S rRNA genes or Internal Transcribed Spacer (ITS) regions as well as those aligned along <90% of its length were excluded (Table 1). The genome recovery was calculated as the percentage of nucleotide positions within the reference genomes aligned with metagenomic or metatranscriptomics reads >95% identity, threshold used for representing members of the same population as the reference genome[43] (Table 1). To assess the gene expression at the genome level, we first used the gene positions to count the number of metatranscripts covering each gene. Then, we normalized these counts using two approaches (i) by UCYN-A single-copy house-keeping genes (*recA* and *gyrB* metatranscript counts), and (ii) by metagenomic read counts for each UCYN-A gene (in this case, we also normalized by sequencing depth; Supplementary Tables 2 and 3).

Phylogenomic and relaxed molecular clock analyses. Sequence data for 57 cyanobacterial genomes were used to estimate the phylogenetic relationships of UCYN-A1 (ref. 11) and UCYN-A2 (ref. 12). We analysed 135 protein sequences that have shown to be highly conserved, to have undergone a minimum number of gene duplications and also to represent a wide diversity of cellular functions[44]. Maximum likelihood analyses and bootstrap values were performed using RAxML 7.4.2 (ref. 37). Bayesian relaxed molecular clock analyses as implemented in MCMCTree[45] and PhyloBayes 3.3b[46] were performed to estimate divergence times of UCYN-A1 and UCYN-A2 (Supplementary Table 2). We applied the uncorrelated gamma multipliers model[47], as this model seems to fit better cyanobacteria nucleotide data sets based on Bayes factors[48]. Age divergences for UCYN-A1 and UCYN-A2 were estimated based on three genes: LSU (3,002 characters), SSU (1,546 characters) and *rpoC1* (1,887 characters). In PhyloBayes[46], we implemented the CAT-GT replacement model of nucleotide evolution. For all non-calibrated nodes, we used a birth–death prior[49] on divergence times.
A permissive gamma distributed root prior of 2,500 Myr ago was also implemented (s.d. = 200 Myr ago, which allowed the 95% credibility interval of the root node to range between 2,300 and 2,700 Myr ago). We treated all calibrations as soft allowing for 2.5% on each side for an upper and lower bound. In MCMCTree, LSU, SSU and *rpoC1* were treated as separate loci and branch lengths were estimated in BASEML[45]. We used the HKY85 (ref. 50) model of nucleotide evolution based on Bayes factor analyses[48]. We used 1 billion years per unit time for all analyses. The gamma prior G (α and β) used to describe how variable rates are across branches was specified as follows G (1, 7). The mean and s.d. was specified as $m = \alpha/\beta$. The gamma priors for the substitution model parameters κ (transition/transversion rate ratio) and α (gamma shape parameter for variable rates among sites) were all specified by gamma distributions. Respective means and s.d.'s were (6, 2) for κ and (1, 1) for α. For all analyses, we used fixed values for the birth–death process $\lambda = \mu = 1$ and $\rho = 0$. Analyses were performed at least twice to ensure convergence of the MCMC, although only one analysis is reported. For all age calibrations, both minimum and maximum bounds were soft and specified by uniform distributions between the maximum/minimum time constraints with 2.5% tail probabilities above/below these limits allowing for molecular data to correct for conflicting fossil information[51]. To check whether analyses had converged, we used Tracer v1.5.0 (http://beast.bio.ed.ac.uk/Tracer). For the cyanobacterial root, 2,700 Myr ago[52] and 2,320 Myr ago[53] (the rise in atmospheric oxygen) were set as the maximum and minimum age, respectively. Other fossils exhibiting unique morphological features were assigned to well-supported groups such as the Nostocales[54] and the clade containing two *Pleurocapsa* genomes (PCC 7319 and PCC 7327) in the Pleurocapsales[55].

References

1. Karl, D. *et al.* Dinitrogen fixation in the world's oceans. *Biogeochemistry* **57/58,** 47–98 (2002).
2. McFall-Ngai, M. Are biologists in 'future shock'? Symbiosis integrates biology across domains. *Nat. Rev. Microbiol.* **6,** 789–792 (2008).
3. Martinez-Garcia, M. *et al.* Unveiling in situ interactions between marine protists and bacteria through single cell sequencing. *ISME J.* **6,** 703–707 (2012).
4. Thompson, A. W. *et al.* Unicellular cyanobacterium symbiotic with a single-celled eukaryotic alga. *Science* **337,** 1546–1550 (2012).
5. Montoya, J. P. *et al.* High rates of N_2 fixation by unicellular diazotrophs in the oligotrophic Pacific Ocean. *Nature* **430,** 1027–1032 (2004).
6. Jardillier, L., Zubkov, M. V., Pearman, J. & Scanlan, D. J. Significant CO_2 fixation by small prymnesiophytes in the subtropical and tropical northeast Atlantic Ocean. *ISME J.* **4,** 1180–1192 (2010).
7. Goebel, N. L. *et al.* Abundance and distribution of major groups of diazotrophic cyanobacteria and their potential contribution to N_2 fixation in the tropical Atlantic Ocean. *Environ. Microbiol.* **12,** 3272–3289 (2010).
8. Zehr, J. P. & Kudela, R. M. Nitrogen cycle of the open ocean: from genes to ecosystems. *Ann. Rev. Mar. Sci.* **3,** 197–225 (2011).
9. Cabello, A. M. *et al.* Global distribution and vertical patterns of a prymnesiophyte-cyanobacteria obligate symbiosis. *ISME J.* **10,** 693–706 (2015).
10. Zehr, J. P. *et al.* Globally distributed uncultivated oceanic N_2-fixing cyanobacteria lack oxygenic photosystem II. *Science* **322,** 1110–1112 (2008).
11. Tripp, H. J. *et al.* Metabolic streamlining in an open-ocean nitrogen-fixing cyanobacterium. *Nature* **464,** 90–94 (2010).
12. Bombar, D. *et al.* Comparative genomics reveals surprising divergence of two closely related strains of uncultivated UCYN-A cyanobacteria. *ISME J.* **8,** 2530–2542 (2014).
13. Hagino, K., Onuma, R., Kawachi, M. & Horiguchi, T. Discovery of an endosymbiotic nitrogen-fixing cyanobacterium UCYN-A in Braarudosphaera bigelowii (Prymnesiophyceae). *PLoS ONE* **8,** e81749 (2013).
14. Thompson, A. *et al.* Genetic diversity of the unicellular nitrogen-fixing cyanobacteria UCYN-A and its prymnesiophyte host. *Environ. Microbiol.* **16,** 3238–3249 (2014).
15. Krupke, A. *et al.* In situ identification and N_2 and C fixation rates of uncultivated cyanobacteria populations. *Syst. Appl. Microbiol.* **36,** 259–271 (2013).
16. Le Moal, M., Collin, H. & Biegala, I. C. Intriguing diversity among diazotrophic picoplankton along a Mediterranean transect: a dominance of rhizobia. *Biogeosciences* **8,** 827–840 (2011).
17. Krupke, A. *et al.* Distribution of a consortium between unicellular algae and the N_2 fixing cyanobacterium UCYN-A in the North Atlantic Ocean. *Environ. Microbiol.* **16,** 3153–3167 (2014).
18. Krupke, A. *et al.* The effect of nutrients on carbon and nitrogen fixation by the UCYN-A-haptophyte symbiosis. *ISME J.* **9,** 1635–1647 (2015).
19. Zehr, J. P. Nitrogen fixation by marine cyanobacteria. *Trends Microbiol.* **19,** 162–173 (2011).
20. Li, W. H. Unbiased estimation of the rates of synonymous and nonsynonymous substitution. *J. Mol. Evol.* **36,** 96–99 (1993).
21. Hurst, L. D. The Ka/Ks ratio: diagnosing the form of sequence evolution. *Trends Genet.* **18,** 486–487 (2002).
22. McDonald, J. H. & Kreitman, M. Adaptive protein evolution at the Adh locus in Drosophila. *Nature* **351,** 652–654 (1991).
23. Bown, P. R. (ed). *Calcareous Nannofossil Biostratigraphy* 328 (Kluwer Academic, 1998).
24. Peleo-Alampay, A. M., Mead, G. A. & Wei, W. Unusual Oligocene Braarudo- sphaera-rich layers of the South Atlantic and their palaeoceanographic implications. *J. Nannoplankton Res.* **21,** 17–26 (1999).
25. Bown, P., Lees, J. & Young, J. in *Coccolithophores: From Molecular Process to Global Impact* (eds Thierstein, H. R. & Young, J. R.) 481–508 (Springer, 2004).

26. Cardenas, A. L. & Harries, P. J. Effect of nutrient availability on marine origination rates throughout the Phanerozoic eon. *Nat. Geosci.* **3**, 430–434 (2010).
27. Decelle, J. *et al.* An original mode of symbiosis in open ocean plankton. *Proc. Natl Acad. Sci. USA* **109**, 18000–18005 (2012).
28. Kneip, C., Lockhart, P., Voss, C. & Maier, U.-G. Nitrogen fixation in eukaryotes - new models for symbiosis. *BMC Evol. Biol.* **7**, 55 (2007).
29. Nakayama, T. *et al.* Spheroid bodies in rhopalodiacean diatoms were derived from a single endosymbiotic cyanobacterium. *J. Plant Res.* **124**, 93–97 (2011).
30. Karsenti, E. *et al.* A holistic approach to marine eco-systems biology. *PLoS Biol.* **9**, e1001177 (2011).
31. Sunagawa, S. *et al.* Structure and function of the global ocean microbiome. *Science* **348**, 1261359 (2015).
32. Logares, R. *et al.* Metagenomic 16S rDNA Illumina tags are a powerful alternative to amplicon sequencing to explore diversity and structure of microbial communities. *Environ. Microbio.* **16**, 2659–2671 (2013).
33. De Vargas, C. *et al.* Ocean plankton. Eukaryotic plankton diversity in the sunlit ocean. *Science* **348**, 1261605 (2015).
34. Pesant, S. *et al.* Open science resources for the discovery and analysis of Tara Oceans Data. *Sci. Data* **2**, 150023 (2015).
35. Guillou, L. *et al.* The Protist Ribosomal Reference database (PR2): a catalog of unicellular eukaryote Small Sub-Unit rRNA sequences with curated taxonomy. *Nucleic Acids Res.* **41**, D597–D604 (2013).
36. Katoh, K., Misawa, K., Kuma, K. & Miyata, T. MAFFT: a novel method for rapid multiple sequence alignment based on fast Fourier transform. *Nucleic Acids Res.* **30**, 3059–3066 (2002).
37. Stamatakis, A. RAxML-VI-HPC: maximum likelihood-based phylogenetic analyses with thousands of taxa and mixed models. *Bioinformatics* **22**, 2688–2690 (2006).
38. Letunic, I. & Bork, P. Interactive Tree Of Life (iTOL): an online tool for phylogenetic tree display and annotation. *Bioinformatics* **23**, 127–128 (2007).
39. Letunic, I. & Bork, P. Interactive Tree Of Life v2: online annotation and display of phylogenetic trees made easy. *Nucleic Acids Res.* **39**, W475–W478 (2011).
40. Simon, N. *et al.* Oligonucleotide probes for the identification of three algal groups by dot blot and fluorescent whole-cell hybridization. *J. Eukaryot. Microbiol.* **47**, 76–84 (2000).
41. Pernthaler, A., Pernthaler, J. & Amann, R. Sensitive multi-color fluorescence in situ hybridization for the identification of environmental microorganisms. *Mol. Microb. Ecol. Man.* **3**, 2613–2627 (2004).
42. Alberti, A. *et al.* Comparison of library preparation methods reveals their impact on interpretation of metatranscriptomic data. *BMC Genomics* **15**, 912 (2014).
43. Caro-quintero, A. & Konstantinidis, K. T. Bacterial species may exist, metagenomics reveal. *Environ. Microbiol.* **14**, 347–355 (2011).
44. Blank, C. E. & Sánchez-Baracaldo, P. Timing of morphological and ecological innovations in the cyanobacteria - a key to understanding the rise in atmospheric oxygen. *Geobiology* **8**, 1–23 (2010).
45. Yang, Z. PAML 4: phylogenetic analysis by maximum likelihood. *Mol. Biol. Evol.* **24**, 1586–1591 (2007).
46. Lartillot, N., Lepage, T. & Blanquart, S. PhyloBayes 3: a Bayesian software package for phylogenetic reconstruction and molecular dating. *Bioinformatics* **25**, 2286–2288 (2009).
47. Drummond, A. J., Ho, S. Y. W., Phillips, M. J. & Rambaut, A. Relaxed phylogenetics and dating with confidence. *PLoS Biol.* **4**, 699–710 (2006).
48. Sánchez-Baracaldo, P., Ridgwell, A. & Raven, J. A. A neoproterozoic transition in the marine nitrogen cycle. *Curr. Biol.* **24**, 652–657 (2014).
49. Lepage, T., Bryant, D., Philippe, H. & Lartillot, N. A general comparison of relaxed molecular clock models. *Mol. Biol. Evol.* **24**, 2669–2680 (2007).
50. Hasegawa, M., Kishino, H. & Yano, T. Dating of the human-ape splitting by a molecular clock of mitochondrial DNA. *J. Mol. Evol.* **22**, 160–174 (1985).
51. Yang, Z. & Rannala, B. Bayesian estimation of species divergence times under a molecular clock using multiple fossil calibrations with soft bounds. *Mol. Biol. Evol.* **23**, 212–226 (2006).
52. Brocks, J. J., Buick, R., Summons, R. E. & Logan, G. A. A reconstruction of Archean biological diversity based on molecular fossils from the 2.78 to 2.45 billion-year-old Mount Bruce Supergroup, Hamersley Basin, Western Australia. *Geochim. Cosmochim. Acta* **67**, 4321–4335 (2003).
53. Bekker, A. *et al.* Dating the rise of atmospheric oxygen. *Nature* **427**, 117–120 (2004).
54. Tomitani, A., Knoll, A. H., Cavanaugh, C. M. & Ohno, T. The evolutionary diversification of cyanobacteria: molecular-phylogenetic and paleontological perspectives. *Proc. Natl Acad. Sci. USA* **103**, 5442–5447 (2006).
55. Zhang, Y. & Golubic, S. Endolithic microfossils (cyanophyta) from early Proterozoic stromatolites, Hebei, China. *Acta Micropaleontol. Sin.* **4**, 1–3 (1987).

Acknowledgements

We are indebted to the scientists and crew on board of *Tara* Oceans expedition. We thank V. Balagué and I. Forn for technical support, P. Sánchez and R. Logares for providing access to bioinformatic resources and for bioinformatic advice, and I. Ferrera for critical reading. This work was supported by the Spanish Ministry of Science and Innovation grant CGL2011-26848/BOS MicroOcean PANGENOMICS, and TANIT (CONES 2010-0036) from the Agència de Gestió d'Ajuts Universitaris i Recerca to S.G.A. This research was further supported by European Molecular Biology Laboratory (EMBL), Investissements d'avenir grants Oceanomics (ANR-11-BTBR-OOO8) and France Génomique (ANR-10-INBS-09-08). F.M.C.-C. and A.M.C. held FPI (MICINN) fellowships and G.S. held PhD JAE-Predoc (CSIC) fellowship. S.G.A. was supported by FP7-OCEAN-2011 project 'Micro B3'. This article is contribution number 37 of Tara Oceans.

Author contributions

F.M.C.-C. and S.G.A designed the study. F.M.C.-C. and A.M.C. designed the CARD-FISH probes and optimized the CARD-FISH assay. A.A. and P.W. performed the metagenome and metatranscriptome sequencing. A.M.C., G.S. and P.S.-B. contributed to the data analyses. F.M.C.-C. and S.G.A. wrote the manuscript and all authors contributed to improve the manuscript.

Additional information

Competing financial interests: The authors declare no competing financial interests.

20

North Atlantic Deep Water Production during the Last Glacial Maximum

Jacob N.W. Howe[1], Alexander M. Piotrowski[1], Taryn L. Noble[1,2], Stefan Mulitza[3], Cristiano M. Chiessi[4] & Germain Bayon[5]

Changes in deep ocean ventilation are commonly invoked as the primary cause of lower glacial atmospheric CO_2. The water mass structure of the glacial deep Atlantic Ocean and the mechanism by which it may have sequestered carbon remain elusive. Here we present neodymium isotope measurements from cores throughout the Atlantic that reveal glacial–interglacial changes in water mass distributions. These results demonstrate the sustained production of North Atlantic Deep Water under glacial conditions, indicating that southern-sourced waters were not as spatially extensive during the Last Glacial Maximum as previously believed. We demonstrate that the depleted glacial $\delta^{13}C$ values in the deep Atlantic Ocean cannot be explained solely by water mass source changes. A greater amount of respired carbon, therefore, must have been stored in the abyssal Atlantic during the Last Glacial Maximum. We infer that this was achieved by a sluggish deep overturning cell, comprised of well-mixed northern- and southern-sourced waters.

[1] Department of Earth Sciences, University of Cambridge, Cambridge CB2 3EQ, UK. [2] Institute for Marine and Antarctic Studies (IMAS), University of Tasmania, Hobart, Tasmania 7001, Australia. [3] MARUM-Center for Marine Environmental Sciences, University of Bremen, Leobener Strasse, D-28359 Bremen, Germany. [4] School of Arts, Sciences and Humanities, University of São Paulo, Av. Arlindo Bettio 1000, CEP03828-000 São Paulo SP, Brazil. [5] Institut Français de Recherche pour l'Exploitation de la Mer (IFREMER), Unité de Recherche Géosciences Marines, F-29280 Plouzané, France. Correspondence and requests for materials should be addressed to J.N.W.H. (email: jacob.howe@cantab.net).

Changes in Atlantic Meridional Overturning Circulation are important in controlling glacial–interglacial climatic shifts due to their role in regulating heat transport in the surface ocean and carbon storage in the deep ocean[1,2]. Nutrient-based proxy reconstructions suggest that overturning circulation in the Atlantic during the Last Glacial Maximum (LGM) was different to that of the modern ocean[3–5]. These nutrient proxy reconstructions are often interpreted as indicating that during the LGM North Atlantic Deep Water (NADW) shoaled to form Glacial North Atlantic Intermediate Water (GNAIW) and was completely replaced by southern-sourced water in the deep Atlantic (>2.5 km)[6,7]. However, elucidation of the water mass distribution in such reconstructions is complicated by the fact that the nutrient content of deep waters may vary independently of their source and dynamics[8]; meanwhile, modelling studies of glacial Atlantic overturning have produced conflicting results[9].

In the modern ocean Antarctic Bottom Water (AABW), which is formed in the high latitude Southern Ocean, represents a major inefficiency in the biological pump—the name given to the biologically mediated processes which sequester carbon in the deep ocean. In this region, deep waters are upwelled to the surface, allowing dissolved carbon to be outgassed to the atmosphere, but these waters sink again before their nutrient load can be fully consumed by marine organisms, thus they are said to have a high-preformed nutrient concentration[10]. The formation of NADW and AABW, and their resultant proportions in the deep ocean thus controls the preformed nutrient concentration of the deep ocean, which, in part, controls atmospheric pCO_2 (ref. 11). Ice core records show that atmospheric CO_2 concentrations rose by ~90 p.p.m. between the LGM and the Holocene[12]. However, considered in isolation, the replacement of high-preformed nutrient AABW with low-preformed nutrient NADW, which has been suggested to have occurred across the last deglaciation, would have increased the efficiency of the biological pump and thereby decreased atmospheric CO_2 (refs 11,13). If our current understanding of glacial–interglacial circulation changes is correct then other parts of the carbon cycle must have counteracted the effects of switching from the GNAIW to NADW mode of deep water formation across the deglaciation. Alternatively, nutrient proxy reconstructions of the Atlantic may be recording carbon cycle changes rather than differences in water mass sourcing.

Neodymium isotopes are a water mass tracer independent of biological processes, allowing deconvolution of changes in water mass sourcing and nutrient regeneration[14]. Modern NADW has a characteristic ε_{Nd} value (normalized $^{143}Nd/^{144}Nd$ ratio in parts per ten thousand) of −13.5 (ref. 15) while seawater in the deep Southern Ocean shows ε_{Nd} values around −8.5 (ref. 16), because it has a greater proportion of Pacific Deep Water (PDW) which has an ε_{Nd} of −3.5 (ref. 17). Away from continental margins water masses throughout the deep ocean reflect the quasi-conservative mixing of these deep water masses[14]. Planktic foraminifera from sea-floor sediment cores have been shown to successfully preserve bottom water ε_{Nd}, thereby offering an archive of past seawater ε_{Nd} (ref. 18).

Here we present Holocene and LGM seawater ε_{Nd} reconstructions, based on foraminiferal ε_{Nd} measurements from 24 cores (Supplementary Table 1, Supplementary Fig. 1) spanning from 46° S to 40° N in the Atlantic Ocean. These reconstructions are used to decipher the nature of changes in Atlantic water mass distributions between glacial and interglacial conditions. Although the LGM reconstruction shows more radiogenic neodymium isotope values than the Holocene profile, the observation of less radiogenic values in the deep North Atlantic than the deep South Atlantic indicates that NADW was produced under glacial conditions. When compared with benthic foraminiferal carbon isotope values, the neodymium isotope measurements reveal that more respired organic carbon was stored in the deep Atlantic Ocean during the LGM than in the Holocene.

Results

Core top foraminiferal ε_{Nd} versus seawater ε_{Nd}. Figure 1 cross plots the Holocene foraminiferal ε_{Nd} values (Supplementary Tables 2 and 3) against the nearest available published seawater ε_{Nd} measurements. Data points were only included in the cross plot if the seawater measurements were within both 10° of latitude and longitude and within 500 m depth of the core site. Most of the data points on the cross plot are within error of the 1:1 line, indicating that foraminifera are faithfully preserving seawater ε_{Nd}. Despite the distance criteria outlined above being applied, the three data points which are outside of error of the 1:1 line are likely due to the seawater data not being of sufficiently close proximity to represent the water mass bathing the core sites. This effect is likely to be particularly prominent in site locations that are near water mass boundaries.

Holocene Atlantic ε_{Nd} reconstruction. We gridded our Holocene foraminiferal ε_{Nd} measurements (Supplementary Table 2) with suitable published data (Fig. 2c). The published data includes results from measurements made on foraminifera, fish debris, high resolution crusts and leachates (Supplementary Tables 3 and 4). As some measurements made on leachates have been shown to be susceptible to contamination by the detrital fraction[19,20], results from sites where the core top leachate ε_{Nd} values deviated significantly from nearby seawater values were excluded from the reconstruction. The data points included from the crusts BM1969.05 and TR079 D-14 (Supplementary Table 4) were dated outside of the definition of the LGM used for selecting sediment core data (23–18 ka); however, the records from these crusts were used to argue for the stability of seawater ε_{Nd} across glacial–interglacial cycles in their respective locations[21]. The slightly older glacial data points from these crusts were therefore deemed appropriate for inclusion in the LGM time slice.

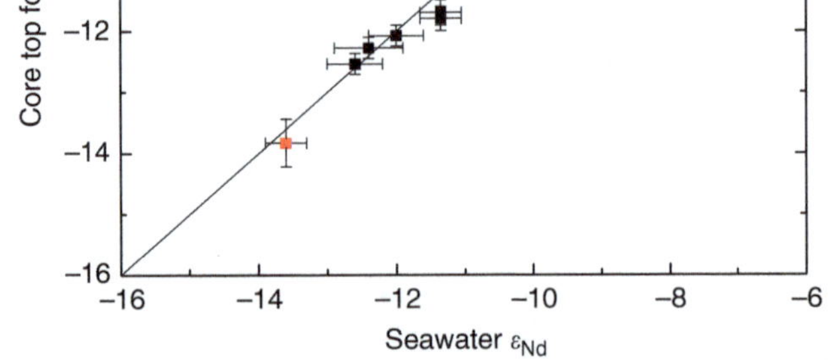

Figure 1 | Core top foraminiferal ε_{Nd} versus seawater ε_{Nd}. Modern open Atlantic seawater[15,16,55,56] ε_{Nd} cross plotted against Holocene Atlantic ε_{Nd} measurements made on uncleaned planktic foraminifera in this work (black squares; Supplementary Table 2) together with previously reported foraminiferal values used in this work (red squares; Supplementary Table 3)[18,56]. Error bars are 2σ external errors. A 1:1 line is also plotted.

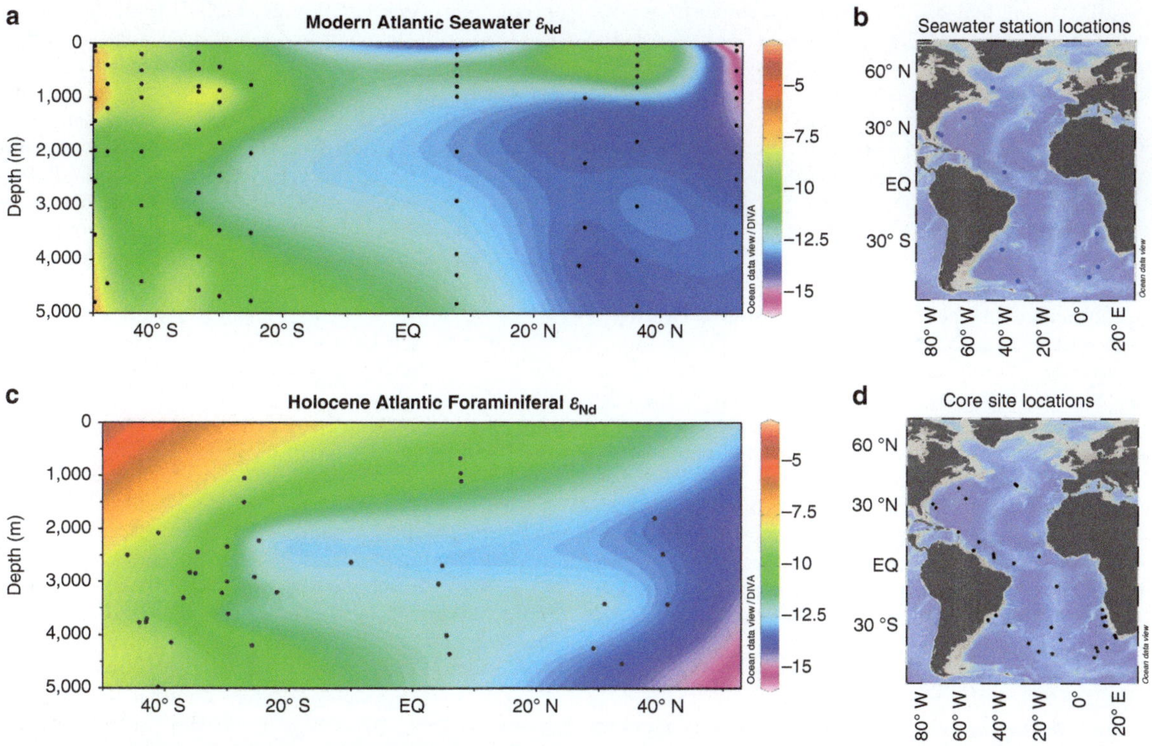

Figure 2 | Atlantic seawater ε_{Nd} and Holocene ε_{Nd} reconstruction. (**a**) Modern seawater ε_{Nd} profile from 50° S to 53° N for the western Atlantic and the eastern Atlantic Ocean south of the Walvis Ridge with data points indicated by large black dots[15,16,55]. (**b**) Map showing location of seawater ε_{Nd} profiles. (**c**) Reconstruction of Holocene Atlantic ε_{Nd} with data points indicated by large black dots (data is listed in Supplementary Tables 2,3 and 4). (**d**) Map showing location of sites used in the Holocene and LGM reconstructions. Figure created using Ocean Data View software[22].

DIVA gridding was performed using Ocean Data View[22] with the 'signal-to-noise ratio' variable set to a value of 25. This variable determines how much a single data point is able to influence the overall plot and this level was chosen to reduce the influence of single data points, as was done by Curry and Oppo[4].

The Holocene reconstruction shows the most unradiogenic ε_{Nd} values, around -13, below 1,500 m at the northernmost extent of the plot (Fig. 2c). These values become more radiogenic to the south, with the greatest propagation of this unradiogenic signal centred at 2,000–4,000 m water depth. This is surrounded by areas with more radiogenic ε_{Nd} values, from -8 to -10, at all depths south of 25° S; below 4,000 m from 25° S to the equator; and above 1,500 m at all latitudes. Although the data set is not as extensive as that used in $\delta^{13}C$ reconstructions[4], this reconstruction replicates the salient features of the seawater ε_{Nd} (Fig. 2a). The only significant mismatches are above 1,000 m at all latitudes and below 4,000 m north of 40° N; in these regions the gridding procedure extrapolates from the nearest data point to the edge of the profile due to a lack of data based constraints. These areas should therefore be interpreted with caution and for this reason the depth range of 0–1,000 m is excluded from later reconstructions.

Last Glacial Maximum Atlantic ε_{Nd} reconstruction. The good correlation between seawater ε_{Nd} and the salinity profile of the modern Atlantic[23] demonstrates that the modern seawater ε_{Nd} profile is the result of mixing between northern- and southern-sourced water masses. Collectively, these observations provide confidence that this selection of cores can be used to constrain past Atlantic water mass distributions below 1,000 m. The LGM ε_{Nd} reconstruction (Fig. 3b) shows the most unradiogenic values around -12.5 between 1,500 and 2,000 m extending from the North Atlantic to ~10° N which then transition to values around -8 at 40° S. In contrast, the deep North Atlantic is occupied by a homogeneous water mass with an ε_{Nd} around -10.5; ε_{Nd} values become more radiogenic to the south, reaching -5.5 at 45° S.

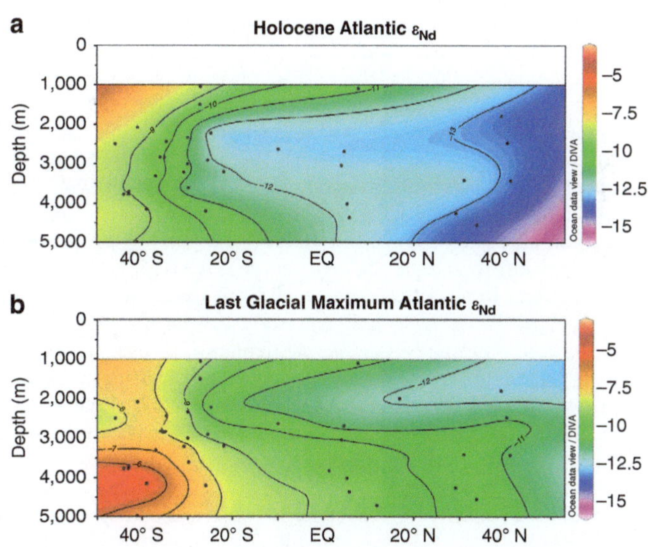

Figure 3 | Holocene and LGM Atlantic ε_{Nd} reconstructions. (**a**) Holocene and (**b**) LGM (23–18 ka) reconstructions of ε_{Nd} of the Atlantic Ocean from 50° S to 53° N measured on uncleaned foraminifera in this study and published authigenic data (Supplementary Tables 2,3 and 4)[18,21,25,26,32,56-59]. Core sites are given by the black dots; locations are plotted in Fig. 2d and Supplementary Fig. 1. Depths from 0 to 1,000 m have been left blank due to a lack of data. Figure created using Ocean Data View software[22].

The largest differences between the LGM and modern Atlantic profiles occur below 2,500 m (Fig. 3); the deep North Atlantic shifts from ε_{Nd} values around -10.5 in the glacial to -13.5 in the modern ocean, whilst the deep South Atlantic shifts from -5.5 to -8.5. This coherent spatial pattern of changes in ε_{Nd} values in the Atlantic between the LGM and the Holocene (Fig. 3b) is consistent with a change in water mass advection and indicates there was a greater influence of southern-sourced waters in the deep Atlantic under glacial conditions.

Discussion

The change observed in the deep South Atlantic between the LGM and Holocene (Fig. 3) is consistent with a lower flux of NADW into the Southern Ocean under cooler, glacial conditions[24,25]. This lower flux of NADW resulted in the deep Southern Ocean being filled with a greater proportion of water of Pacific and Indian Ocean origin that was more radiogenic in composition than Atlantic-sourced deep water[26–28]. The deep South Atlantic, however, remained less radiogenic than the deep Pacific[26,28], requiring a source of unradiogenic neodymium to the deep South Atlantic under glacial conditions. Furthermore, the deep North Atlantic was less radiogenic than the deep South Atlantic during the LGM (Fig. 3b); this implies that northern-sourced water with its unradiogenic ε_{Nd} must have been exported to abyssal depths under glacial climate conditions. The mixing of GNAIW with denser southern-sourced waters is deemed an unlikely explanation given the large amount of energy required to mix these intermediate waters down to abyssal depths in the North Atlantic. Rather, these observations suggest that two glacial North Atlantic-sourced water masses existed during the LGM: GNAIW at depths above 2,500 m and denser Glacial NADW (GNADW)[29] below 2,500 m.

The inference of a lower flux of GNADW under glacial conditions[24] than NADW in the Holocene resulting in a more radiogenic composition of Glacial AABW (GAABW) than AABW would cause the deep Atlantic to exhibit more radiogenic ε_{Nd} values without a change in the water mass mixing proportions in the deep Atlantic. By taking this end-member change of deep southern-sourced water ε_{Nd} into account, however, our results can be used to calculate the proportion of the Atlantic ventilated by NADW (%NADW) both in the Holocene and during the LGM (Supplementary Table 5), and thus elucidate changes in water mass mixing proportions from the effect of the change in the AABW end-member composition. In each case three end-members were required; namely NADW; AABW and Antarctic Intermediate Water, or their glacial counterparts (Supplementary Fig. 2). A binary mixing calculation was performed between a northern- and a southern-sourced water mass at each core site using equation (1). The choice of southern-sourced water varied with depth and latitude according to the observed boundary between Antarctic Intermediate Water and Lower Circumpolar Deep Water as defined by a salinity of 34.7 psu (ref. 30) and the gradient in neodymium concentration (but not ε_{Nd}) between them in the Atlantic sector of the modern Southern Ocean[16].

$$\%\,\mathrm{NADW} = \frac{[\mathrm{Nd}]_{\mathrm{SSW}}(\varepsilon_{\mathrm{Nd(NADW)}} - \varepsilon_{\mathrm{Nd(SSW)}})}{[\mathrm{Nd}]_{\mathrm{NADW}}\varepsilon_{\mathrm{Nd(NADW)}} - [\mathrm{Nd}]_{\mathrm{SSW}}\varepsilon_{\mathrm{Nd(SSW)}} - \varepsilon_{\mathrm{Nd}}\varepsilon_{\mathrm{Nd(NADW)}} + \varepsilon_{\mathrm{Nd(SSW)}}\varepsilon_{\mathrm{Nd}}} \times 100 \tag{1}$$

The modern seawater ε_{Nd} and [Nd] end-member values were taken from published seawater data (Supplementary Fig. 2). For the glacial ocean, the ε_{Nd} and [Nd] of NADW and PDW were kept constant; support for the stability of the ε_{Nd} of both comes from crust data[31] whilst there is, at present, no proxy for past neodymium concentration available. The more radiogenic ε_{Nd} of GAABW was taken from a South Atlantic record of benthic foraminifera[32] and the concentration calculated by conservative mixing of NADW and PDW (ref. 16). Unlike records from the deep South Atlantic[26,32], the new intermediate depth glacial foraminiferal ε_{Nd} data from the South Atlantic measured in this work exhibit no changes between the LGM and the Holocene (GeoB2107-3 and GeoB2104-3; Supplementary Table 2), so the neodymium composition of Glacial Antarctic Intermediate Water was assumed to be the same as in the modern ocean. This observation clearly demonstrates that the intermediate and deep South Atlantic were isotopically distinct in terms of neodymium during the LGM, reinforcing the need for two distinct southern-sourced end-member water masses in the mixing calculations.

Figure 4 shows the %NADW values calculated for the Holocene and LGM Atlantic gridded into meridional sections using the same technique employed for the ε_{Nd} data profiles plotted in Fig. 2. In the LGM ε_{Nd} profile (Fig. 3b), the most radiogenic values, around -5.5, are observed at 4,000 m depth in the South Atlantic, whereas less radiogenic values, around -6.5, are seen at $\sim$5,000 m at the same latitude. The more radiogenic values come from the Mid-Atlantic Ridge; the less radiogenic from the Cape Basin, which lies to the east of the Mid-Atlantic Ridge. The ε_{Nd} difference between these sites likely reflects a longitudinal mixing gradient between radiogenic, Pacific-derived, circumpolar waters[28] and less radiogenic North Atlantic derived waters[25]. A similar phenomenon is seen in modern seawater data from the South Atlantic although the offset is less pronounced[16]. The results from the deep Cape Basin[26] were, therefore, excluded from the mixing proportion plots in Fig. 4 as the longitudinal gradient in ε_{Nd} observed in the deep South Atlantic obscures the latitudinal mixing gradient which is the primary interest here.

The gridded profile of the Holocene %NADW results (Fig. 4a) shows the Atlantic north of 20° N and below 1,500 m depth is mostly (>90%) NADW. %NADW decreases at all depths to the

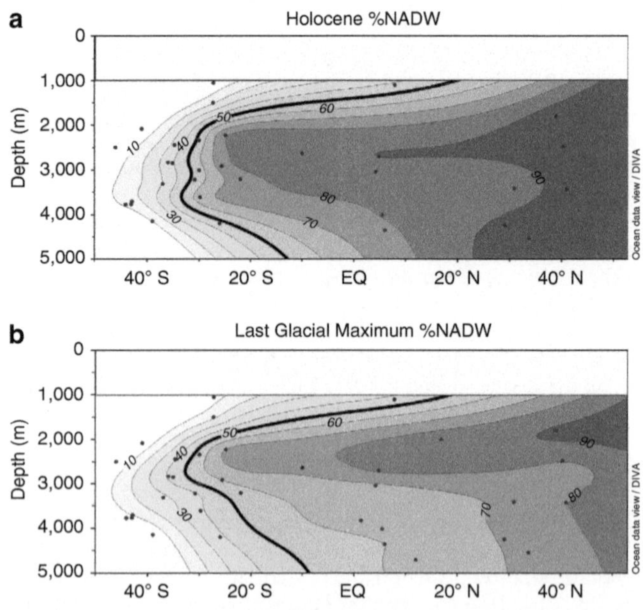

Figure 4 | Holocene and LGM NADW percentage. NADW percentage (%NADW) calculated for the Atlantic Ocean from 50° S to 53° N in the (**a**) Holocene and during the (**b**) LGM (23–18 ka) using authigenic ε_{Nd}. Depths from 0 to 1,000 m have been left blank due to a lack of data. Figure created using Ocean Data View software[22].

south, with the 50% mixing line occurring near 10° S at 5,000 m, but not until 30° S from 2,000 to 4,000 m. The LGM %NADW contours, including the 50% contour, (Fig. 4b) are similar to the Holocene profile above 2,500 m (Fig. 4a). The greatest difference between the two plots occurs below 2,500 m in the North Atlantic, which has significantly less NADW in the LGM profile (55–80% NADW) than in the Holocene profile (>90% NADW). From this observation it is clear that there was a greater proportion of southern-sourced waters in the deep Atlantic during the LGM than in the Holocene; however, it is also clear that the deep Atlantic was not occupied solely by southern-sourced waters during the LGM.

Although changes in the proportion of NADW in the deep Atlantic between glacial and interglacial conditions are the simplest explanation for the results in Figs 3 and 4, other alternatives must be considered. The sensitivity of the calculated %NADW for the glacial deep Bermuda Rise ε_{Nd} of -10.4 (OCE326-GGC6; 33.7° N, 57.6° W, 4,540 m)[18] to the neodymium composition of end-member water masses is shown in Fig. 5. The ε_{Nd} of AABW, considered to be known with reasonable confidence[32], was held constant. The ε_{Nd} of GNADW was varied from a near maximum possible value of -10.5 to a minimum of -16.5, similar to the least radiogenic value of -16 that was used in similar calculations performed for a nearby site[33]. As the %NADW result is dependent on the relative neodymium concentrations of the northern and southern end-members, the ratio of $[Nd]_{GNADW}$ to $[Nd]_{GAABW}$ was varied from 0.2 to 1. The latter value was used in similar calculations performed for a South Atlantic site[32] and is used as an upper limit in this work as ratios >1 would require more neodymium dissolved in the deep Atlantic than the deep Pacific, a situation that is not consistent with the observation that neodymium concentrations increase with water mass age in the modern ocean[34].

Using the glacial end-member values assigned in Fig. 4, the deep Bermuda Rise site was calculated to be bathed by 72% NADW during the LGM (black box, Fig. 5). In contrast an $\varepsilon_{Nd(GNADW)}$ of -10.5 and concentration ratio of 0.2 yields nearly 100% NADW, whilst an $\varepsilon_{Nd(GNADW)}$ of -16.5 and a concentration ratio of 1.0 means just 44% NADW is required to explain the ε_{Nd} observed at the Bermuda Rise during the LGM. Although the resultant uncertainty in the %NADW at the Bermuda Rise of ±28% is significant, all end-member configurations yield %NADW in the glacial deep North Atlantic values >44%. This is a robust finding that directly contradicts the notion that the deep Atlantic was dominated by southern-sourced waters during the LGM. Furthermore, the lower estimate of 44% is deemed unlikely to be realistic as less radiogenic $\varepsilon_{Nd(NADW)}$ values appear restricted to interstadials, and have been interpreted as pulses of Labrador Sea Water formation during these warm intervals[33] or unradiogenic weathering pulses during ice sheet retreat[35]. Rather, most evidence suggests that $\varepsilon_{Nd(GNADW)}$ values were similar to or slightly more radiogenic than NADW in the modern Atlantic[21] yielding higher %NADW values (Fig. 5). The influence of analytical error on the calculated %NADW values is small; using the external error bounds of 0.5 epsilon units[18] gives a range of 67–77% NADW at the deep Bermuda Rise during the LGM.

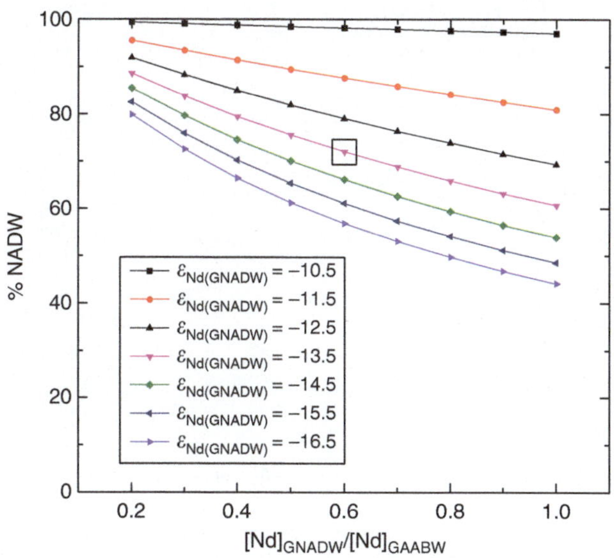

Figure 5 | Sensitivity test of NADW percentage calculation. Results of the sensitivity tests performed on the %NADW calculated for the Bermuda Rise ε_{Nd} of -10.4 during the LGM (OCE326-GGC6; 33.7° N, 57.6° W, 4,540 m)[18]. $\varepsilon_{Nd(GAABW)}$ was held constant at -5.5 (ref. 32), whilst $\varepsilon_{Nd(GNADW)}$ and the ratio of $[Nd]_{GNADW}$ to $[Nd]_{GAABW}$ were varied as shown. The result from end-members used in Fig. 4 ($\varepsilon_{Nd(GNADW)} = -13.5$ and $[Nd]_{GNADW}/[Nd]_{GAABW} = 0.60$) is highlighted by a black box.

Although these uncertainties do limit the certainty of the exact proportion of the deep Atlantic that was ventilated by northern-sourced waters during the LGM, the need for Glacial NADW appears inescapable. If localized processes, such as boundary exchange, were controlling the glacial ε_{Nd} profile, one would expect to see a heterogeneous profile controlled by regional detrital inputs. Furthermore, modelling studies have shown that boundary exchange processes are not able to explain the magnitude of the observed glacial–interglacial shifts in deep ocean ε_{Nd} and thus changes in water mass advection must be invoked to explain them[36]. Reconstructions of glacial deep water mass ventilation from radiocarbon[37], and carbonate ion concentration from B/Ca[38,39] also show a younger, better ventilated, water mass in the deep North Atlantic than the deep South Atlantic. These proxies, therefore, also provide evidence for a significant proportion of northern-sourced deep waters in the deep North Atlantic during the LGM, supporting the results shown in Fig. 4.

Reconciling the glacial ε_{Nd} values in the Atlantic with nutrient proxy based reconstructions requires a greater amount of respired organic carbon in the deep glacial Atlantic relative to the modern situation[4,8]. Circulation versus biological sources of carbon can be differentiated by cross plots of ε_{Nd} against benthic foraminiferal $\delta^{13}C$ for the Holocene and LGM Atlantic (Fig. 6). Deep water mass mixing end-members were ascribed to each plot as detailed in the Supplementary Information. The South Atlantic foraminiferal values (grey regions Fig. 6) were corrected for the Mackensen Effect, the phenomenon where benthic foraminifera display $\delta^{13}C$ values lower than that of the overlying deep water due to a phytodetrital layer of light organic carbon on the sea floor[40]. The composition of NADW is similar in the Holocene and LGM cross plots with $\delta^{13}C$ and ε_{Nd} values ~1.4‰ and -13.5, respectively, whereas glacial AABW $\delta^{13}C$ is more depleted (-0.5‰) and ε_{Nd} more radiogenic (-5.5) than its modern counterpart (0.4‰, -8.5).

The blue curves show the values expected for conservative mixing between these end-members for each time slice; those points falling along this mixing line have seen little input of biological carbon. During the LGM, data points lying close to the mixing line come from cores shallower than 3,000 m. Most of the glacial data points from cores located below 3,000 m sit well below the mixing curve which we interpret as being caused by the addition of respired organic matter with a low $\delta^{13}C$ (usually ~ -20‰)[41]. It is important to note that these cores represent locations within both northern- and southern- sourced glacial deep water, so this signal was not advected from either source but was acquired by biological processes along the advection pathway; this highlights the shortcomings of using $\delta^{13}C$ in isolation as a proxy for water mass mixing. The lower than expected $\delta^{13}C$ values could also in part be due to the use of mixed *Cibicidoides* species for some of the benthic foraminiferal $\delta^{13}C$ measurements. The variable depth habitats of different *Cibicidoides* species could lead to offsets of the measured foraminiferal calcite $\delta^{13}C$ from bottom water $\delta^{13}C$ (ref. 42). The inference of organic matter remineralization being the dominant cause of the offset, however, is supported by a recent study that modelled the glacial Atlantic and predicted a biological regenerative imprint in the glacial deep Atlantic $\delta^{13}C$ (ref. 8). The same study calculated that, after allowing for this remineralisation of organic matter, the glacial deep North Atlantic was ventilated by 50–80% NADW, which compares well with the ε_{Nd} derived values of 55–80% NADW produced in this work (Fig. 4b).

The accumulation of respired organic carbon in the deep Atlantic during the LGM could be due to higher surface productivity and thus export of organic carbon to the deep ocean under glacial climate conditions. Although glacial productivity reconstructions in the northwestern Atlantic are sparse, nearby regions show elevated export production during the LGM relative to the Holocene[43]. Higher glacial surface productivity, however, cannot explain the depth dependence of the $\delta^{13}C$ offset in the LGM (Fig. 6b), nor can it explain the chemocline seen at ~2,500 m in nutrient proxy reconstructions[3,4]. Thus it seems likely that the greater amount of respired organic matter in the deep Atlantic during the LGM must have also been at least partially due to a longer residence time of seawater in the glacial deep Atlantic than in the modern Atlantic. This conclusion is supported by a modelling study of a $^{231}Pa/^{230}Th$ data compilation which concluded that the glacial Atlantic had rapid overturning in the shallow cell but slower overturning at depth[44].

Our findings are summarized in hypothetical overturning schematics for the Atlantic Ocean during glacials and interglacials in Fig. 7. In the glacial scenario (Fig. 7b) there are two distinct northern-sourced water masses, separated by density differences. The glacial northern-sourced intermediate depth water mass was

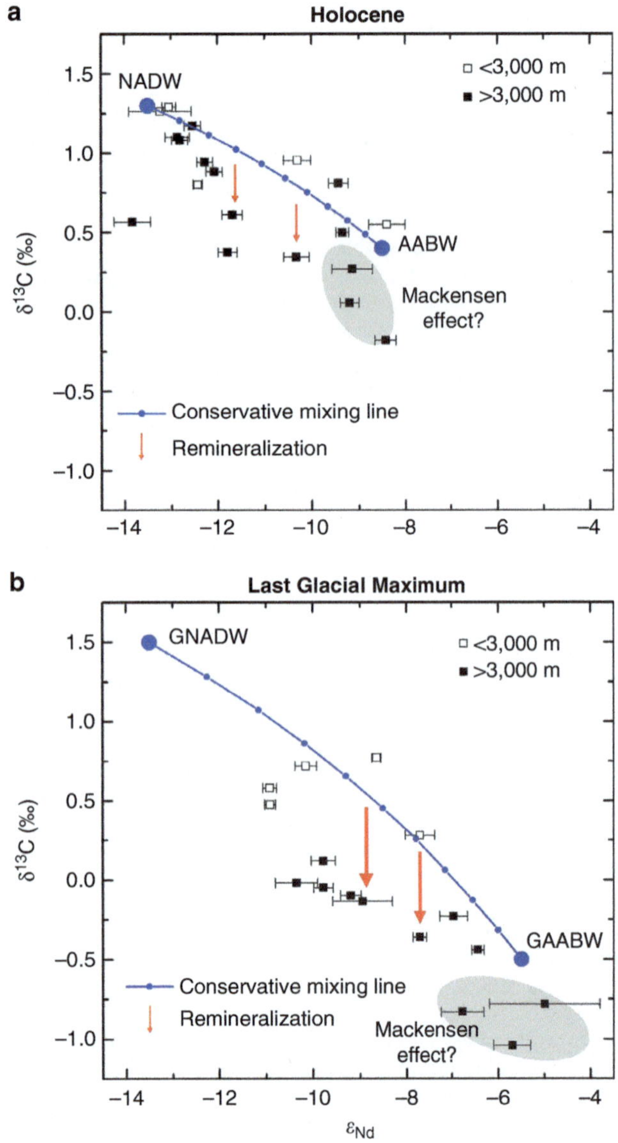

Figure 6 | Cross plots of Atlantic benthic foraminiferal $\delta^{13}C$ against foraminiferal ε_{Nd}. Cross plots of benthic foraminiferal $\delta^{13}C$ against foraminiferal ε_{Nd} for the deep Atlantic Ocean during the **(a)** Holocene and the **(b)** LGM. Water mass end-members labelled are NADW, AABW, GNADW and GAABW. The blue curve shows the values expected for conservative mixing between these water masses in the corresponding cross plot. Offsets from this line are attributed to either the remineralization of organic matter (red arrows) or a possible Mackensen Effect (shaded grey regions)[40]. Details of the data used are given in Supplementary Table 7, and how end-member values were assigned in Supplementary Note 1 and Supplementary Table 8. Error bars show the 2σ external error of the ε_{Nd} measurements.

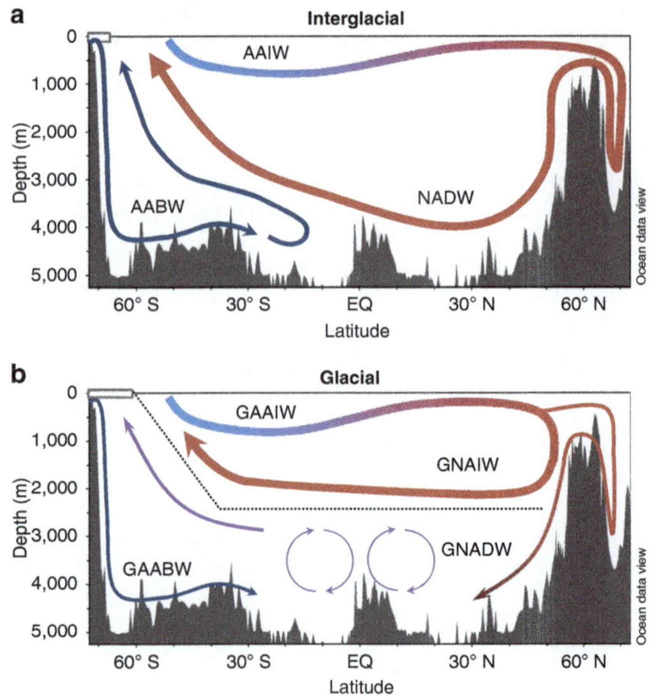

Figure 7 | Interglacial and Glacial Atlantic overturning schematics. Schematics of the structure of Atlantic Meridional Overturning Circulation during **(a)** interglacial and **(b)** glacial periods assuming the Holocene and LGM are typical of each. Water masses depicted are Antarctic Intermediate Water (AAIW), NADW and AABW and their glacial counterparts as well as GNAIW. Thin arrows associated with GNADW and GAABW in the glacial schematic represent the inferred lower flux of these water masses during glacial periods relative to interglacials. The dotted black line in the glacial schematic shows the location of the chemocline in nutrient proxy reconstructions at the interface between the two overturning cells. Figure partly created using Ocean Data View software[22].

likely formed by convection south of Iceland[45] and is inferred to have overturned rapidly[44]. The deep Atlantic cell is surmised to have overturned more slowly and may have been supplied by NADW formed in the Nordic Seas during seasonal ice-free periods coming over the Greenland-Scotland Ridge[46]. Brine rejection from sea ice may also have contributed to this process[47]. A lower flux of GNADW during the LGM than NADW in the Holocene in to the deep North Atlantic would then explain how AABW was able to penetrate much further north during the LGM (Fig. 4b) than in the modern Atlantic Ocean where it has little influence north of the equator (Fig. 4a). In contrast, the deep South Atlantic shows less of a change in water mass mixing proportions between the LGM and the Holocene (Fig. 4) because it is dominated by southern-sourced waters in both climate states (Fig. 7).

Many modelling studies and hypothesized glacial overturning schemes invoke GNAIW being incorporated into deep waters elsewhere in the glacial ocean to ventilate the deep ocean with low-preformed nutrient concentration water[10,13,48]. Here, we have presented evidence that instead the deep Atlantic was ventilated directly from the North Atlantic resolving the difficulty in reconciling glacial CO_2 drawdown with the observed proxy data. Another important prediction of our work is that switching from the glacial to interglacial mode of North Atlantic circulation (Fig. 7) would flush respired organic carbon from the deep Atlantic and would therefore be expected to raise atmospheric CO_2 without invoking changes in nutrient utilization in the high latitude Southern Ocean[13]. Although a full carbon cycle model would be required to quantify this effect, the resumption of strong NADW production at the start of the Bølling-Allerød inferred from ε_{Nd} and $^{231}Pa/^{230}Th$ records, coincides with a $\sim$12 p.p.m. increase in atmospheric CO_2 (Supplementary Fig. 3)[12,18,49]. This increase likely reflects the flushing of respired carbon from the deep Atlantic by strong NADW production during the deglaciation[50].

Methods

Site selection. ε_{Nd} measurements were made on uncleaned planktic foraminifera from Holocene and LGM samples of cores from throughout the western Atlantic basin or selected regions within the eastern Atlantic (Fig. 2d); all core names and locations are listed in Supplementary Table 1 and plotted in Supplementary Fig. 1. Eastern Atlantic cores were limited to sites south of the Walvis Ridge or sites on bathymetric rises above the sill depth of the Mid-Atlantic Ridge (3,750 m)[51]. This selection was based on the criteria outlined by Curry and Oppo[4], and was intended to avoid the regions of the eastern Atlantic which are ventilated through fracture zones in the Mid-Atlantic Ridge and thus do not display the same latitudinal water mass mixing gradient as the western Atlantic[52]. Cores north of 45° N were excluded as they have been shown to be susceptible to the influence of volcanic ash and IRD in the North Atlantic[19,53].

Age controls. Age controls for cores used in this work range from planktic or benthic foraminiferal $\delta^{18}O$ records to radiocarbon dates (Supplementary Table 1) and came from published age models, with the exception of the Ceara Rise cores. The radiocarbon-based age models developed in this work for the latter cores, ODP 925E, ODP 928B and ODP 929B, are presented in Supplementary Table 6. For all cores, depths which were assigned calendar ages between 23 and 18 ka were included in the LGM reconstruction.

Sample preparation. Samples were prepared for analysis following the methods of Roberts *et al.*[18] and references therein. In short, where possible, $\sim$80 mg of mixed planktic foraminifera were picked from the coarse fraction (>63 μm) for neodymium isotope measurements. After picking, foraminifera tests were broken open between two glass plates, rinsed, sonicated and any clays removed. The samples were then dissolved in 1 mol l^{-1} reagent grade acetic acid.
The REEs were extracted from the dissolved sample using Eichrom TRUspec resin in 100 ml Teflon columns. Neodymium was then separated from the other rare earth elements using Eichrom LNspec resin on volumetrically calibrated Teflon columns.

Neodymium isotopic measurements. Neodymium isotopes were analysed using the Nu Plasma HR or Neptune Plus multi-collector inductively coupled plasma mass spectrometers at the University of Cambridge. $^{146}Nd/^{144}Nd$ was normalized to 0.7219 and samples were bracketed with a concentration-matched solution of reference standard JNdi-1, the measured composition of which varied between runs but was corrected to the accepted value of $^{143}Nd/^{144}Nd = 0.512115$ (ref. 54). The ε_{Nd} of each sample is reported with the external error (2σ) of the bracketing standards from the corresponding measurement session, unless the internal error was larger than the external error, in which case the combined internal and external error (2σ) is reported.

Eight complete procedural blanks for the process from foraminiferal dissolution through to column chemistry were run on a TIMS Sector 54 at the University of Cambridge using a ^{150}Nd spike. For the typical sample, of at least 15 ng, the average blank of 66 pg of neodymium is <0.5% of the total neodymium.

Benthic foraminiferal stable isotopes. Benthic foraminifera (mixed *Cibicidoides* species; between 3 and 7 tests) were picked for stable isotope analysis from the coarse fraction (>125 μm) of samples without published $\delta^{13}C$ data. Samples were analysed by the Godwin Laboratory at the University of Cambridge using either a Micromass Multicarb Sample Preparation System attached to a VG SIRA or a Thermo Kiel device attached to a Thermo MAT253 Mass Spectrometer in dual inlet mode. Isotopic ratios are presented relative to standard Vienna PeeDee Belemnite; external precision was ±0.06‰ for $\delta^{13}C$.

References

1. Broecker, W. S. Glacial to interglacial changes in ocean chemistry. *Prog. Oceanogr.* **11**, 151–197 (1982).
2. Denton, G. H. *et al.* The last glacial termination. *Science* **328**, 1652–1656 (2010).
3. Lynch-Stieglitz, J. *et al.* Atlantic meridional overturning circulation during the Last Glacial Maximum. *Science* **316**, 66–69 (2007).
4. Curry, W. B. & Oppo, D. W. Glacial water mass geometry and the distribution of $\delta^{13}C$ of ΣCO_2 in the western Atlantic Ocean. *Paleoceanography* **20**, PA1017 (2005).
5. Marchal, O. & Curry, W. B. On the abyssal circulation in the glacial Atlantic. *J. Phys. Oceanogr.* **38**, 2014–2037 (2008).
6. Ferrari, R. *et al.* Antarctic sea ice control on ocean circulation in present and glacial climates. *Proc. Natl Acad. Sci. USA* **111**, 8753–8758 (2014).
7. Negre, C. *et al.* Reversed flow of Atlantic deep water during the Last Glacial Maximum. *Nature* **468**, 84–88 (2010).
8. Gebbie, G. How much did Glacial North Atlantic Water shoal? *Paleoceanography* **29**, 190–209 (2014).
9. Weber, S. L. *et al.* The modern and glacial overturning circulation in the Atlantic ocean in PMIP coupled model simulations. *Clim. Past* **3**, 51–64 (2007).
10. Sigman, D. M., Hain, M. P. & Haug, G. H. The polar ocean and glacial cycles in atmospheric CO_2 concentration. *Nature* **466**, 47–55 (2010).
11. Marinov, I. *et al.* Impact of oceanic circulation on biological carbon storage in the ocean and atmospheric pCO2. *Global Biogeochem. Cycles* **22**, GB3007 (2008).
12. Marcott, S. A. *et al.* Centennial-scale changes in the global carbon cycle during the last deglaciation. *Nature* **514**, 616–619 (2014).
13. Hain, M. P., Sigman, D. M. & Haug, G. H. Carbon dioxide effects of Antarctic stratification, North Atlantic Intermediate Water formation, and subantarctic nutrient drawdown during the last ice age: Diagnosis and synthesis in a geochemical box model. *Global Biogeochem. Cycles* **24**, GB4023 (2010).
14. Frank, M. Radiogenic isotopes: tracers of past ocean circulation and erosional input. *Rev. Geophys.* **40**, doi:10.1029/2000RG000094 (2002).
15. Piepgras, D. J. & Wasserburg, G. J. Rare earth element transport in the western North Atlantic inferred from Nd isotopic observations. *Geochim. Cosmochim. Acta* **51**, 1257–1271 (1987).
16. Stichel, T., Frank, M., Rickli, J. & Haley, B. A. The hafnium and neodymium isotope composition of seawater in the Atlantic sector of the Southern Ocean. *Earth Planet. Sci. Lett.* **317–318**, 282–294 (2012).
17. Amakawa, H., Sasaki, K. & Ebihara, M. Nd isotopic composition in the central North Pacific. *Geochim. Cosmochim. Acta* **73**, 4705–4719 (2009).
18. Roberts, N. L., Piotrowski, A. M., McManus, J. F. & Keigwin, L. D. Synchronous deglacial overturning and water mass source changes. *Science* **327**, 75–78 (2010).
19. Elmore, A. C., Piotrowski, A. M., Wright, J. D. & Scrivner, A. E. Testing the extraction of past seawater Nd isotopic composition from North Atlantic deep sea sediments and foraminifera. *Geochem. Geophys. Geosyst.* **12**, Q09008 (2011).

20. Wilson, D. J., Piotrowski, A. M., Galy, A. & Clegg, J. A. Reactivity of neodymium carriers in deep sea sediments: implications for boundary exchange and paleoceanography. *Geochim. Cosmochim. Acta* **109,** 197–221 (2013).
21. Foster, G. L., Vance, D. & Prytulak, J. No change in the neodymium isotope composition of deep water exported from the North Atlantic on glacial-interglacial time scales. *Geology* **35,** 37–40 (2007).
22. Schlitzer, R. Ocean Data View. http://odv.awi.de (2016).
23. Blanckenburg, F. V. Tracing past ocean circulation? *Science* **286,** 1862b–1863b (1999).
24. Rutberg, R. L., Hemming, S. R. & Goldstein, S. L. Reduced North Atlantic Deep Water flux to the glacial Southern Ocean inferred from neodymium isotope ratios. *Nature* **405,** 935–938 (2000).
25. Wei, R., Abouchami, W., Zahn, R. & Masque, P. Deep circulation changes in the South Atlantic since the Last Glacial Maximum from Nd isotope and multi-proxy records. *Earth Planet. Sci. Lett.* **434,** 18–29 (2016).
26. Piotrowski, A. M. *et al.* Reconstructing deglacial North and South Atlantic deep water sourcing using foraminiferal Nd isotopes. *Earth Planet. Sci. Lett.* **357–358,** 289–297 (2012).
27. Wilson, D. J., Piotrowski, A. M., Galy, A. & Banakar, V. K. Interhemispheric controls on deep ocean circulation and carbon chemistry during the last two glacial cycles. *Paleoceanography* **30,** PA2707 (2015).
28. Noble, T. L., Piotrowski, A. M. & McCave, I. N. Neodymium isotopic composition of intermediate and deep waters in the glacial southwest Pacific. *Earth Planet. Sci. Lett.* **384,** 27–36 (2013).
29. Völker, C. & Köhler, P. Responses of ocean circulation and carbon cycle to changes in the position of the Southern Hemisphere westerlies at Last Glacial Maximum. *Paleoceanography* **28,** 726–739 (2013).
30. Orsi, A. H., Whitworth, T. & Nowlin, W. D. On the meridional extent and fronts of the Antarctic Circumpolar Current. *Deep Sea Res. Part I Oceanogr. Res. Pap.* **42,** 641–673 (1995).
31. Pena, L. D. & Goldstein, S. L. Thermohaline circulation crisis and impacts during the mid-Pleistocene transition. *Science* **345,** 318–322 (2014).
32. Skinner, L. C. *et al.* North Atlantic versus Southern Ocean contributions to a deglacial surge in deep ocean ventilation. *Geology* **41,** 667–670 (2013).
33. Böhm, E. *et al.* Strong and deep Atlantic meridional overturning circulation during the last glacial cycle. *Nature* **517,** 73–76 (2015).
34. Elderfield, H., Whitfield, M., Burton, J. D., Bacon, M. P. & Liss, P. S. The oceanic chemistry of the rare-earth elements [and Discussion]. *Philos. Trans. R. Soc. A Math. Phys. Eng. Sci.* **325,** 105–126 (1988).
35. Crocket, K. C., Vance, D., Gutjahr, M., Foster, G. L. & Richards, D. A. Persistent Nordic deep-water overflow to the glacial North Atlantic. *Geology* **39,** 515–518 (2011).
36. Rempfer, J., Stocker, T. F., Joos, F. & Dutay, J.-C. Sensitivity of Nd isotopic composition in seawater to changes in Nd sources and paleoceanographic implications. *J. Geophys. Res.* **117,** C12010 (2012).
37. Skinner, L. C., Waelbroeck, C., Scrivner, A. E. & Fallon, S. J. Radiocarbon evidence for alternating northern and southern sources of ventilation of the deep Atlantic carbon pool during the last deglaciation. *Proc. Natl. Acad. Sci. USA* **111,** 5480–5484 (2014).
38. Yu, J., Elderfield, H. & Piotrowski, A. M. Seawater carbonate ion-$\delta^{13}C$ systematics and application to glacial–interglacial North Atlantic ocean circulation. *Earth Planet. Sci. Lett.* **271,** 209–220 (2008).
39. Yu, J. *et al.* Deep South Atlantic carbonate chemistry and increased interocean deep water exchange during last deglaciation. *Quat. Sci. Rev.* **90,** 80–89 (2014).
40. Mackensen, A., Hubberten, H., Bickert, T. & Ftitterer, D. K. The $\delta^{13}C$ in benthic foraminifera tests of *Fontbotia wuellerstorfi* (Schwager) relative to the $\delta^{13}C$ of dissolved inorganic carbon in Southern Ocean deep water: implications for glacial ocean circulation models. *Paleoceanography* **8,** 587–610 (1993).
41. Kroopnick, P. The distribution of ^{13}C of ΣCO_2 in the world oceans. *Deep. Sea Res. Part A Oceanogr. Res. Pap.* **32,** 57–84 (1985).
42. Hodell, D. A., Venz, K. A., Charles, C. D. & Ninnemann, U. S. Pleistocene vertical carbon isotope and carbonate gradients in the South Atlantic sector of the Southern Ocean. Geochemistry. *Geophys. Geosyst.* **4,** 1–19 (2003).
43. Kohfeld, K. E., Le Quéré, C., Le, Harrison, S. P. & Anderson, R. F. Role of Marine Biology in Glacial-Interglacial CO_2 Cycles. *Science* **308,** 74–78 (2005).
44. Lippold, J. *et al.* Strength and geometry of the glacial Atlantic Meridional Overturning Circulation. *Nat. Geosci.* **5,** 813–816 (2012).
45. Labeyrie, L. D. *et al.* Changes in the vertical structure of the North Atlantic Ocean between glacial and modern times. *Quat. Sci. Rev.* **11,** 401–413 (1992).
46. Millo, C., Sarnthein, M., Voelker, A. & Erlenkeuser, H. Variability of the Denmark Strait Overflow during the Last Glacial Maximum. *Boreas* **35,** 50–60 (2006).
47. Dokken, T. & Jansen, E. Rapid changes in the mechanism of ocean convection during the last glacial period. *Nature* **401,** 458–461 (1999).
48. Kwon, E. Y. *et al.* North Atlantic ventilation of 'southern-sourced' deep water in the glacial ocean. *Paleoceanography* **27,** 1–12 (2012).
49. McManus, J. F., Francois, R., Gherardi, J.-M., Keigwin, L. D. & Brown-Leger, S. Collapse and rapid resumption of Atlantic meridional circulation linked to deglacial climate changes. *Nature* **428,** 834–837 (2004).
50. Chen, A. T. *et al.* Synchronous centennial abrupt events in the ocean and atmosphere during the last deglaciation. *Science* **349,** 1537–1542 (2015).
51. Metcalf, W. G., Heezen, B. C. & Stalcup, M. C. The sill depth of the Mid-Atlantic Ridge in the equatorial region. *Deep Sea Res. Oceanogr. Abstr.* **11,** 1–10 (1964).
52. McCartney, M. S., Bennett, S. L. & Woodgate-Jones, M. E. Eastward flow through the Mid-Atlantic Ridge at 11° N and its influence on the abyss of the eastern basin. *J. Phys. Oceanogr.* **21,** 1089–1121 (1991).
53. Roberts, N. L. & Piotrowski, A. M. Radiogenic Nd isotope labeling of the northern NE Atlantic during MIS 2. *Earth Planet. Sci. Lett.* **423,** 125–133 (2015).
54. Tanaka, T. *et al.* JNdi-1: a neodymium isotopic reference in consistency with LaJolla neodymium. *Chem. Geol.* **168,** 279–281 (2000).
55. Jeandel, C. Concentration and isotopic composition of Nd in the South Atlantic Ocean. *Earth Planet. Sci. Lett.* **117,** 581–591 (1993).
56. Huang, K.-F., Oppo, D. W. & Curry, W. B. Decreased influence of Antarctic intermediate water in the tropical Atlantic during North Atlantic cold events. *Earth Planet. Sci. Lett.* **389,** 200–208 (2014).
57. Jonkers, L. *et al.* Deep circulation changes in the central South Atlantic during the past 145 kyrs reflected in a combined $^{231}Pa/^{230}Th$, Neodymium isotope and benthic $\delta^{13}C$ records. *Earth Planet. Sci. Lett.* **419,** 14–21 (2015).
58. Gutjahr, M., Frank, M., Stirling, C. H., Keigwin, L. D. & Halliday, A. N. Tracing the Nd isotope evolution of North Atlantic Deep and Intermediate Waters in the western North Atlantic since the Last Glacial Maximum from Blake Ridge sediments. *Earth Planet. Sci. Lett.* **266,** 61–77 (2008).
59. Lang, D. C. *et al.* Incursions of southern-sourced water into the deep North Atlantic during late Pliocene glacial intensification. *Nat. Geosci.* **9,** 375–379 (2016).

Acknowledgements

Sample material was provided by the Godwin Laboratory for Paleoclimate Research at the University of Cambridge, the International Ocean Discovery Program the GeoB Core Repository at the MARUM—Center for Marine Environmental Sciences, University of Bremen and Petrobras. We thank Jo Kerr and Aurora Elmore for providing additional samples. We thank Thiago Pereira dos Santos for providing the unpublished age model data for GL1090; we thank Jo Clegg and Vicky Rennie for technical support and Natalie Roberts for helpful discussions. Radiocarbon analyses were supported by NERC radiocarbon grant 1752.1013 and Nd isotope analyses by NERC grants NERC NE/K005235/1 and NERC NE/F006047/1 to A.M.P. J.N.W.H. was supported by a Rutherford Memorial Scholarship. S.M. was supported through the DFG Research Center/Cluster of Excellence 'The Ocean in the Earth System'. C.M.C. acknowledges financial support from FAPESP (grant 2012/17517-3).

Author contributions

J.N.W.H. and A.M.P. designed the study, J.N.W.H. performed the work, and J.N.W.H. and A.M.P. wrote the manuscript with contributions from all other authors. T.L.N. provided unpublished data from ODP 1088; S.M., C.M.C. and G.B. provided sample material and unpublished data.

Additional information

Competing financial interests: The authors declare no competing financial interests.

21

A global reanalysis of storm surges and extreme sea levels

Sanne Muis[1], Martin Verlaan[2,3], Hessel C. Winsemius[2], Jeroen C.J.H. Aerts[1] & Philip J. Ward[1]

Extreme sea levels, caused by storm surges and high tides, can have devastating societal impacts. To effectively protect our coasts, global information on coastal flooding is needed. Here we present the first global reanalysis of storm surges and extreme sea levels (GTSR data set) based on hydrodynamic modelling. GTSR covers the entire world's coastline and consists of time series of tides and surges, and estimates of extreme sea levels. Validation shows that there is good agreement between modelled and observed sea levels, and that the performance of GTSR is similar to that of many regional hydrodynamic models. Due to the limited resolution of the meteorological forcing, extremes are slightly underestimated. This particularly affects tropical cyclones, which requires further research. We foresee applications in assessing flood risk and impacts of climate change. As a first application of GTSR, we estimate that 1.3% of the global population is exposed to a 1 in 100-year flood.

[1] Institute for Environmental Studies (IVM), Vrije Universiteit Amsterdam, 1081 HV Amsterdam, The Netherlands. [2] Deltares, 2600 MH Delft, The Netherlands. [3] Mathematical Physics, Electrical Engineering, Mathematics and Computer Science (EEMCS), TU Delft, 2628 CD Delft, The Netherlands. Correspondence and requests for materials should be addressed to S.M. (email: sanne.muis@vu.nl).

Storm surge, a rise in sea level due to low atmospheric pressure and strong winds, is the main driver of coastal flood events[1]. The most extreme of these events are caused by tropical cyclones[2], but extra-tropical storms can also produce high sea levels, especially when they coincide with high tide[3]. With over 600 million people living in low-lying coastal areas[4], coastal floods can have devastating societal impacts. It is estimated that on average 0.8–1.1 million people per year are flooded globally[5]. This is reflected by disasters such as the flooding of The Netherlands and the United Kingdom in 1953, which resulted in over 2,000 fatalities[6], and led to the construction of a series of flood protection works along the Dutch coast and the Thames Barrier in London. Another more recent catastrophe was the flooding of New Orleans in 2005 due to tropical cyclone Katrina, which resulted in around 1,100 fatalities[7].

In recent years, coastal flood risk has increased due to population, economic growth[8] and land subsidence[9,10]. To date, socioeconomic development has been the main driver of increasing risk[11], but in the future sea level rise will be an important driver of increasing risk of coastal floods[5,12–14]. To analyse spatial patterns and temporal trends in coastal flood risk, several continental to global scale studies have been carried out[5,8,12,15–17] based on the extreme sea levels in the Dynamic Interactive Vulnerability Assessment (DIVA) input database[18,19]. The extreme sea levels in the DIVA database have been instrumental in many risk assessments, and have provided important insights into which areas face the highest risk, as well as the potential effects of sea level rise and adaptation. However, some applications require time series of sea levels (instead of extreme values), such as: assessing interannual variability; assessing the impact of changes in storm regimes; and the modelling of past events. Such time series can be obtained from tide gauge observations, but many regions at risk have insufficient numbers of tide gauges and/or record lengths available to reliably estimate extreme sea levels. Extreme sea levels vary significantly along the coast due to variability in storminess, coastline shape, and bathymetry[1]. Hence, interpolation between different stations will not accurately capture the spatial variability. Recently, more advanced techniques based on altimetry data have been developed[20], but the limited length of altimetry records prevent their application to low-probability extreme events[21,22]. Due to these data limitations, there is still limited understanding of the global coastal flood hazard, even under current (stationary) climate conditions.

At the regional scale, hydrodynamic modelling is the state-of-art approach to develop consistent and complete sea level reanalyses that can be applied to flood risk assessments[18,19]. As the modelling of surges in shallow coastal areas requires a high resolution, generally such a modelling approach has been computationally too costly to apply on the global-scale. The application of unstructured grids (or 'flexible mesh') in hydrodynamic models make it possible to have a sufficient resolution in shallow coastal areas[23], while maintaining computational efficiency[24]. These developments in hydrodynamic modelling combined with the increasing availability of global data sets on climate and elevation, make it possible to upscale the hydrodynamic modelling to the global scale.

Here we apply such a hydrodynamic approach for the first time at the global scale and present the Global Tide and Surge Reanalysis (GTSR) data set. To obtain the first near-coast global reanalysis of storm surges (1979–2014), we force the newly developed Global Tide and Surge Model (GTSM)[24] with wind speed and atmospheric pressure from the ERA-Interim global atmospheric reanalysis[25]. Tides are modelled separately using a recent update of the Finite Element Solution (FES2012) hydrodynamic model (Supplementary Fig. 1 and Supplementary Note 1)[26]. To validate the GTSR data set, we compare the results with observed sea levels from a global data set from University of Hawaii Sea Level Center (UHSLC). Results show that the GTSR time series agree very well with observed sea levels. Extreme sea levels are slightly underestimated, especially those induced by tropical cyclones; this is mainly due to the resolution of the meteorological forcing. To illustrate one of the potential applications of the GTSR data set, we assess global exposure to coastal flooding using a simple inundation model, and find that 1.3% of the global population, equal to 76 million people, is living in the 1 in 100-year floodplain. We foresee various other applications of the GTSR data set, such as: assessing flood risk at the global scale; assessing interannual variability of storminess; providing warning thresholds for operational forecasting models; and assessing the impacts of climate change.

Results

Validation of time series. For the total sea level, there is generally a good agreement between modelled and observed sea levels. The root mean square errors (RMSE), based on the time series with a 10 minute temporal resolution, are lower than 0.20 m for 80% of the stations (Fig. 1a). The average RMSE across all validation sites is 0.17 m (s.d. is 0.15 m). The performance of the surge levels is even better, with 95% of the stations having a RMSE lower than 0.2 m. The average RMSE is 0.11 m (s.d. is 0.05 m). For illustration, Fig. 2a– shows the modelled and observed surge levels for selected sites around the world. For tidal levels, about 85% of the stations have a RMSE lower than 0.2 m, but the maximum RMSE value is greater than 1 m. The average RMSE is 0.15 m (s.d. is 0.42 m). The large errors (RMSE >1 m) for the tide gauge stations Windham, Victoria and Puerto Montt (red dots in Fig. 1a) are caused by an over- or underestimation of the tidal amplitude. Supplementary Fig. 2a–b shows the RMSEs for the surge levels and tide levels for all observation stations.

To assess whether sea level extremes are also adequately represented in the GTSR time series, we calculate the performance based on daily maxima. Figure 3a–k shows scatter density plots for modelled and observed daily maxima for 12 selected locations. The majority of the daily maxima (orange to red areas in Fig. 3a–k) are close to the perfect-fit line, indicating a good performance. The performance for the majority of these 12 stations decreases for more extreme sea levels (the least-squares line diverges from the best-fit line). The underestimation of extreme sea levels is primarily an issue of resolution (see Discussion section). The average Pearson correlation coefficient is 0.83 (s.d. is 0.14), indicating a good representation of the maxima by the model. Over 75% of the stations have a correlation coefficient higher than 0.75. In regions prone to tropical cyclones (Supplementary Fig. 3a–b and Supplementary Note 2), such as the Caribbean Sea, we obtain correlation coefficients lower than 0.5. The average correlation coefficient in these tropical regions is 0.77, which is significantly lower than the average correlation coefficient of 0.87 in extra-tropical regions (two-tailed Student's t-test, $P<0.05$).

Validation of extremes. To obtain extreme sea levels for various return periods, we fit a Gumbel distribution to the annual maxima using the maximum-likelihood method. The Gumbel plot is shown in Fig. 4a–e for five selected stations. For all stations shown, the annual maxima follow a relatively straight line, indicating a good fit of the Gumbel distribution. Average relative errors for return periods from 5 to 100 years are in the range of 11–14%, increasing with higher return periods. As there is a limited number of stations with observation records longer than

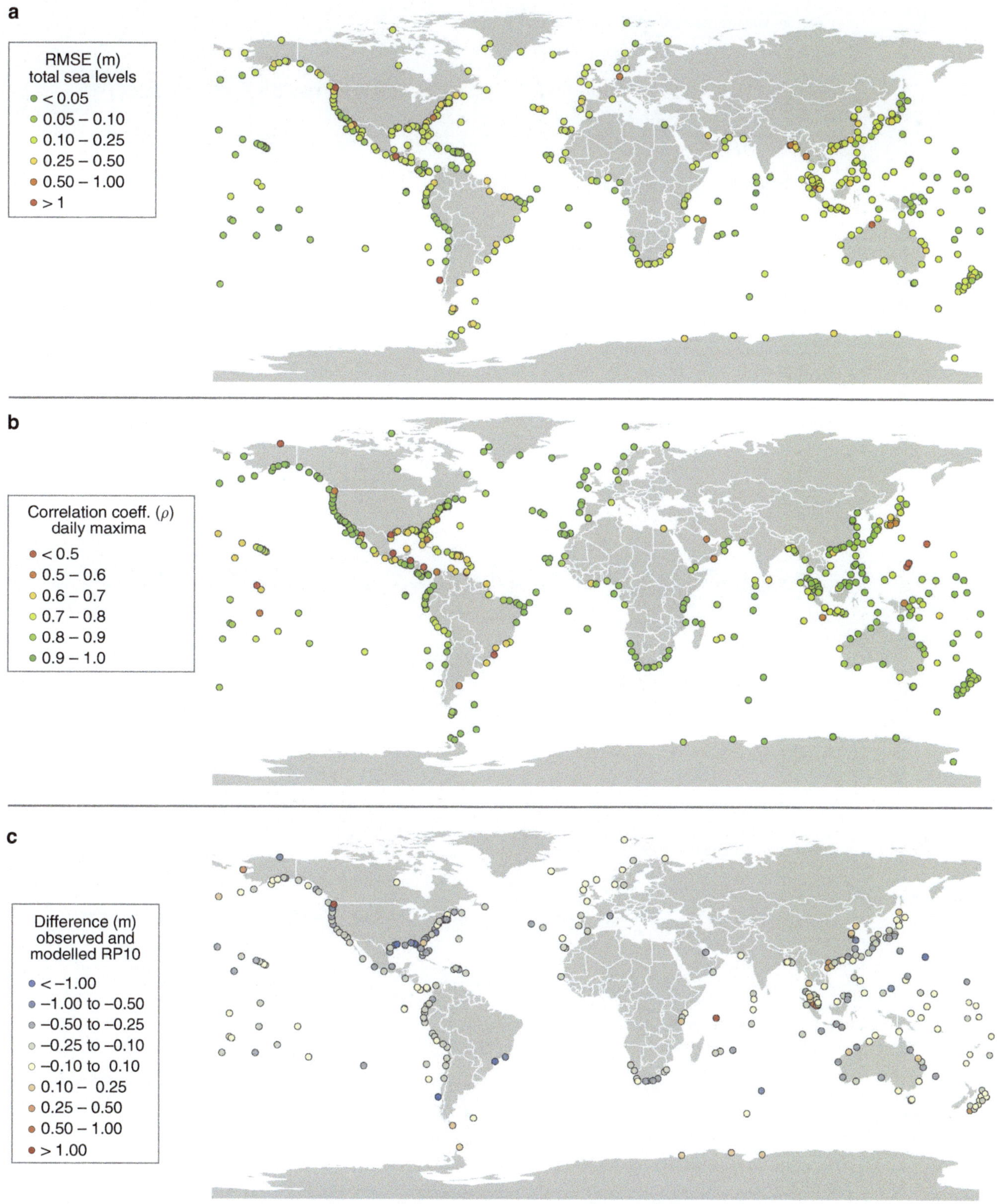

Figure 1 | Maps showing the performance of GTSR against observed sea levels. The performance of GTSR shown as (**a**) the RMSE (m) between modelled and observed sea level time series; (**b**) the Pearson correlation coefficient (ρ) between modelled and observed daily maximum sea levels; and (**c**) the bias (m) calculated by subtracting the modelled extreme sea levels from the observed extreme sea levels with a return period of 10 years.

the 30 years, we focus on 1 in 10-year extreme sea levels and use all stations that have an observation record longer than 10 years. For 75% of these 144 stations, the absolute error for the 1 in 10-year sea level is lower than 0.3 m (Fig. 1c). On average, the extremes are underestimated by −0.14 m (s.d. is 0.39). Supplementary Fig. 4 shows the results for the 1 in 100-year sea levels. The performance of GTSR is best in areas where extremes are dominated by large, extra-tropical storms, such as

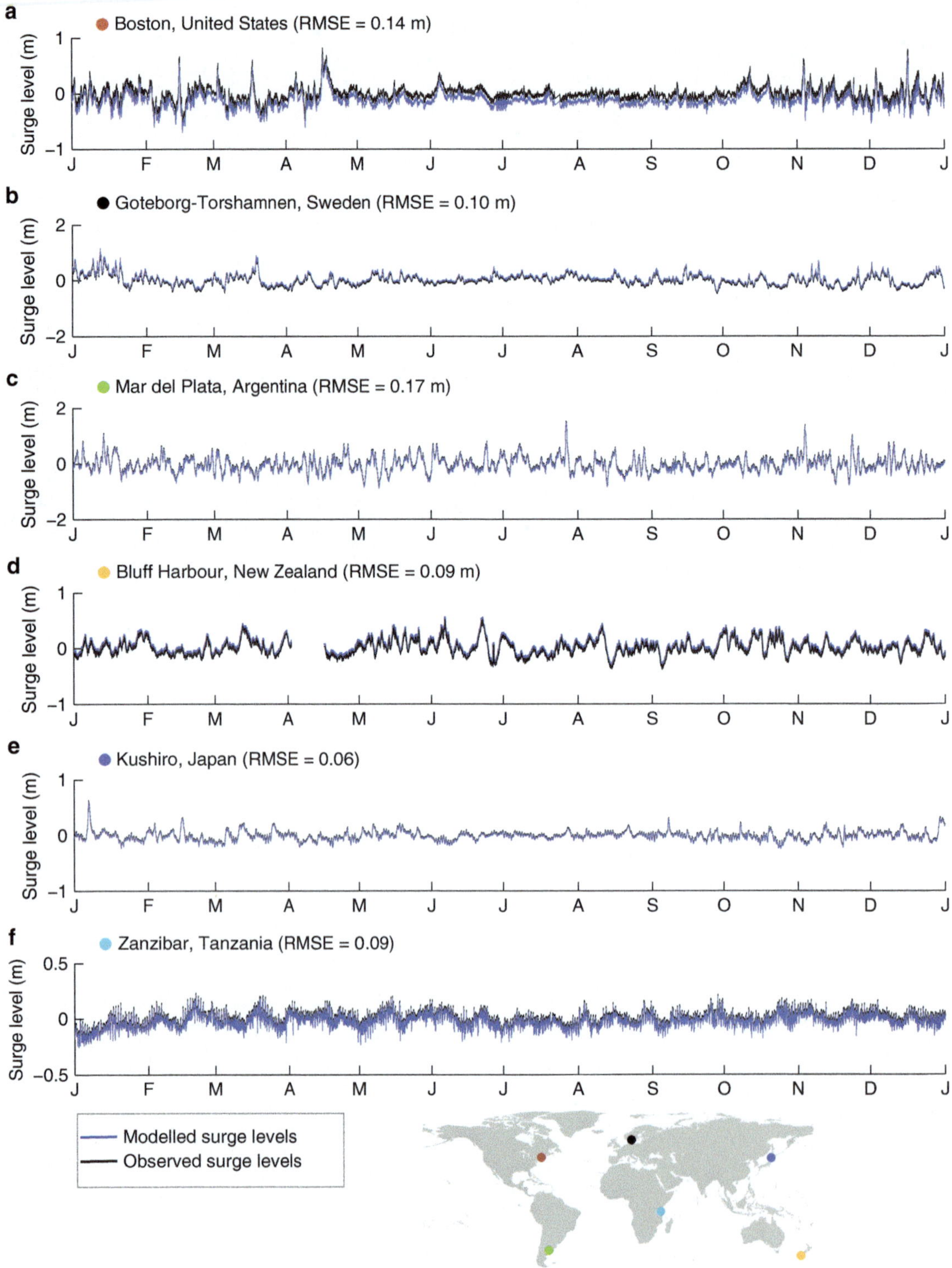

Figure 2 | Comparison of the modelled and observed surge levels at selected stations. Comparison of the modelled and observed surge levels for 2007 at six selected stations around the world, (**a**) Boston, United States; (**b**) Goteborg-Torshamn, Sweden; (**c**) Mar de la Plata, Argentina; (**d**) Bluff Harbour, New Zealand; (**e**) Kushiro, Japan; and (**f**) Zanzibar, Tanzania. The coloured dots in the world map indicate the location of the observation stations.

Europe, southeast Australia, eastern South-America and northwest North-America. In regions where storm surges are largely induced by tropical cyclones, the mean absolute difference between the extreme sea levels with a return period of 10 years is 0.23 m, which is significantly higher than in extra-tropical regions, where the mean absolute difference is 0.08 m (two-tailed Student's *t*-test, $P<0.05$).

Application to assess flood exposure. To illustrate a first application of the GTSR extremes, we calculate the flood hazard (that is, inundation extent) and flood exposure (that is, exposed people) based on 1 in 100-year extreme sea levels. This assessment is based on SRTM elevation[27] and GRUMP population density for the year 2000 (refs 28,29). For this first demonstration, we assume no protection from coastal flooding and planar flood levels.

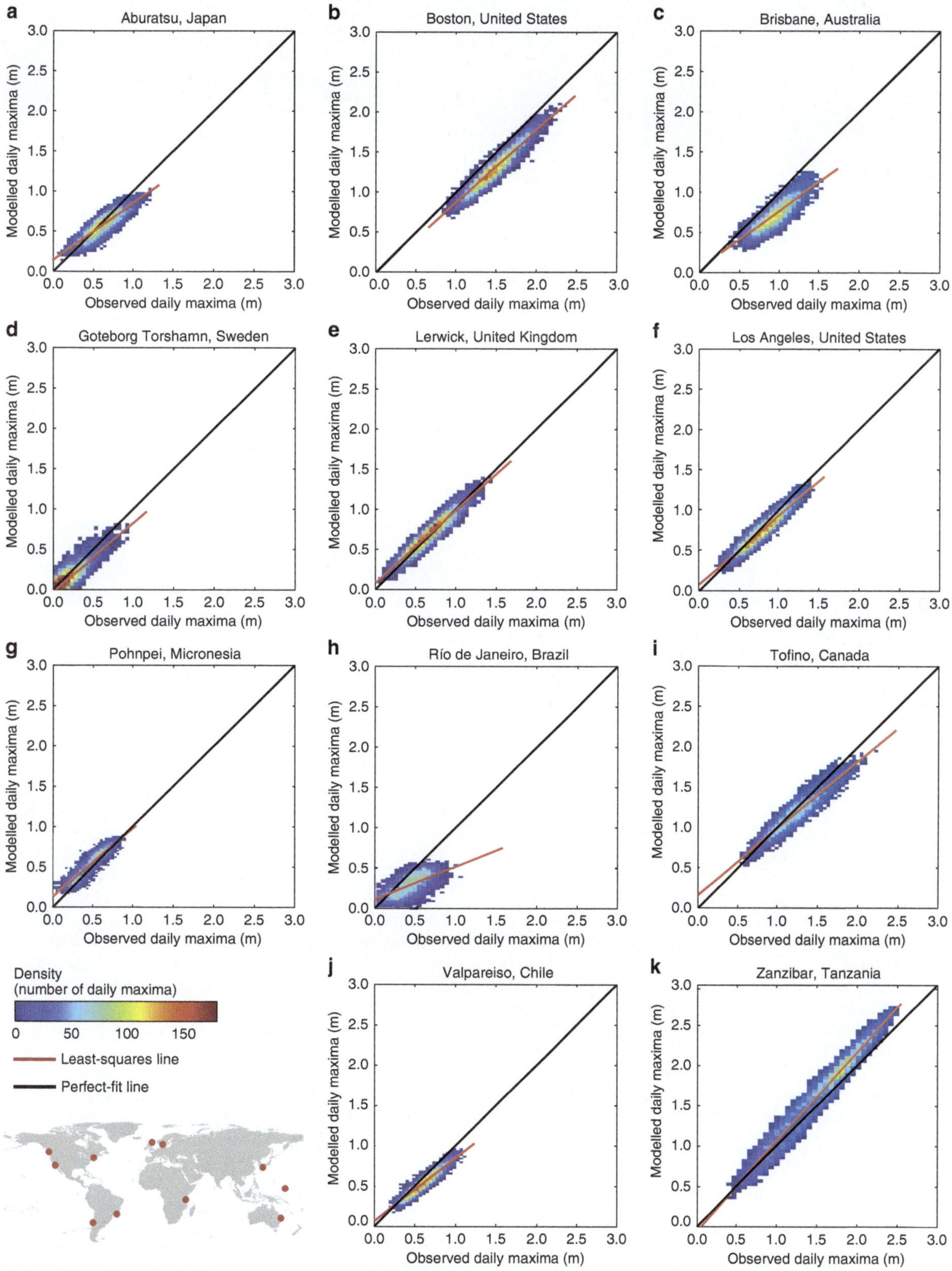

Figure 3 | Scatter plots of modelled and observed daily maxima at selected stations. Scatter plots for (**a**) Abaratsu, Japan; (**b**) Boston, United States; (**c**) Brisbane, Australia; (**d**) Goteborg-Torshamn, Sweden; (**e**) Lerwick, United Kingdom; (**f**) Los Angeles, United States; (**g**) Pohnpei, Micronesia; (**h**) Río de Janeiro, Brazil; (**i**) Tofino, Canda; (**j**) Valpareiso, Chile; and (**k**) Zanzibar, Tanzania. Colours indicate the data density for bins with a 5 cm × 5 cm size. The red dots in the world map indicate the location of the observations stations.

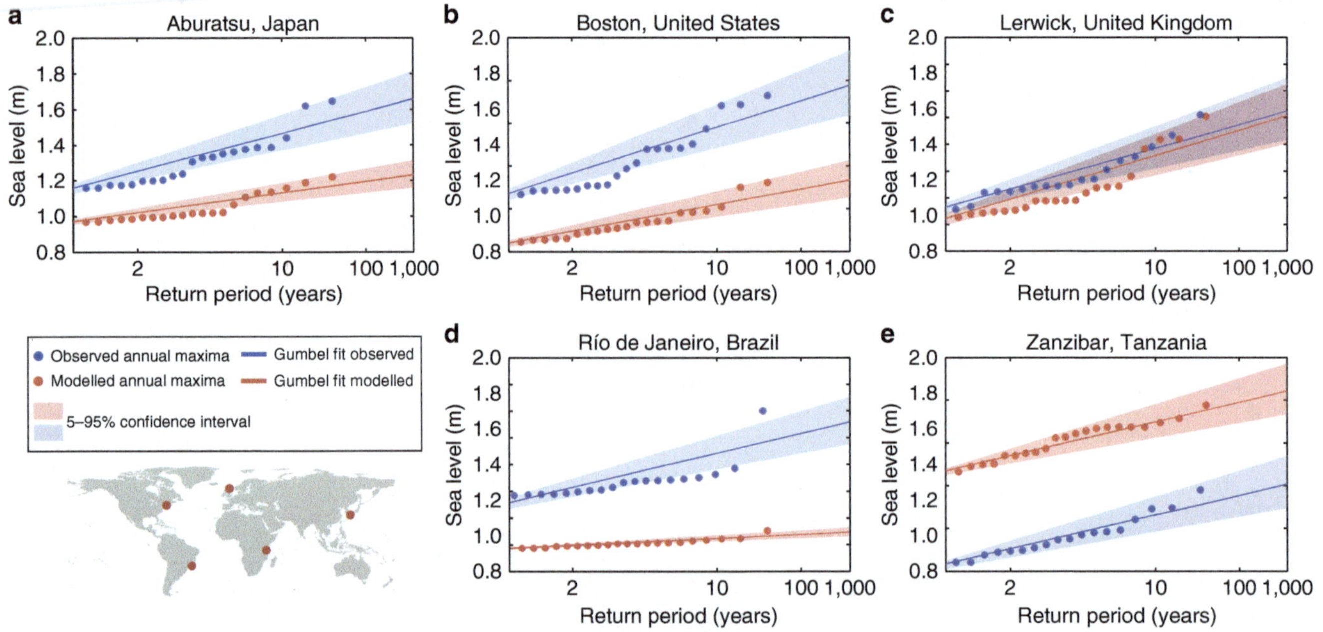

Figure 4 | Gumbel plot for the modelled and observed annual maxima at selected stations. Gumbel plots for five selected stations around the world, (**a**) Aburatsu, Japan; (**b**) Boston, United States; (**c**) Lerwick, United Kingdom; (**d**) Río de Janeiro, Brazil; and (**e**) Zanzibar, Tanzania. The annual maxima are plotted against the Gumbel variate, which is equal to $x = -\ln(-\ln(G(y)))$. The red dots in the world map indicate the location of the observation stations.

Results show that there are large regional variations in the 1 in 100-year extreme sea levels (Fig. 5a)[30]. Relatively high extreme sea levels are found in areas with high tidal amplitude, such as northwest Europe, south Argentina, China and Bangladesh. This spatial pattern agrees with other global data sets[19,20,31], and does not change for higher return periods (Supplementary Fig. 5a–b). Combining extreme sea levels with elevation, and assuming no flood protection, shows that major inundation occurs particularly in delta areas in Europe and Asia (Supplementary Fig. 6). The global population exposed to a 1 in 100-year flood is 76 million, which is equal to 1.3% of the total world population (Table 1). The flood exposure map in Fig. 5b shows that China has a particularly large exposure, with 37 million people, which is 3% of the country's population and 53% of the global exposure (Supplementary Fig. 7a shows the absolute exposure). Other countries that have a relatively high exposure include The Netherlands (8.4 million people, 53% of total population), Vietnam (4.7 million people, 6% of global exposure) and Egypt (4.3 million people, 6% of global exposure). Relative to the total national population, The Netherlands and Greenland stand out with 53% and 31% of the population exposed to a 1 in 100-year flood (Supplementary Fig. 7b), respectively; in both these countries the highest concentration of people is found near the coast[28].

We assess the uncertainty in the extreme value statistics by calculating the 5 and 95% confidence bounds of the Gumbel fit, and express the uncertainty as the percentage difference compared to extreme sea level based on the best-fit. For extreme sea levels with a return period of 100 years, the uncertainty of the Gumbel fit is below 10% for half of the world's coastline (Fig. 6a). The uncertainty is greater than 25% for only 4% of the world's coastline. The largest uncertainty is seen in regions where the extreme values are relatively low, such as the Mediterranean Sea and Caribbean Sea, where a small change in extreme sea level leads to a large increase in relative terms. To assess how sensitive the exposure estimates are to the uncertainty in extreme values statistics, we use the 5 and 95% confidence bounds as input to the inundation and impact model. Globally, the sensitivity to the uncertainty in extreme values is relatively small, with a range of −8 to +21% around the best-fit for exposed population (that is, exposure ranges between 70 and 92 million people). The results per country are shown in Fig. 6, which shows that the results for the USA, Thailand, Mauritania and several countries along the Baltic Sea are particularly sensitive, with uncertainty values greater than 50% of the best-fit. Of the top 10 countries listed in Table 1, the results for China and Vietnam have the largest uncertainty, with exposure estimates ranging from −34% to +39% and −28% to +4% respectively. While the largest uncertainty in extreme sea levels is found along the north coast of Russia, this does not lead to large uncertainty in exposure because the area has a very low population density. On the other hand, an uncertainty in extreme sea levels of 10–50% along the east coast of the USA and <10% along the west coast of the USA leads to uncertainty in the country's exposure estimates between −40% and +49%.

Discussion

The performance of the GTSR time series is similar to the performance of other hydrodynamic models that cover a large geographic domain. For example, Cid *et al.*[32] reported a mean RMSE of 0.08–0.10 m for surge levels in the Mediterranean Sea. For a hydrodynamic model covering the entire Australian coastline, Haigh *et al.*[33] reported mean RMSEs of 0.14 m and 0.05 m for total sea level and surge, respectively. The validation of GTSR shows that extreme sea levels are generally underestimated but that the differences with observed extreme sea levels are rather small (that is, <0.45 m for 90% of the observation stations for the 1 in 10-year sea level). For applications to risk assessment, which make use of data from global digital elevation models (DEMs), this underestimation is reasonable, since the vertical resolution of such DEMs is much greater. For example, the global SRTM DEM has a vertical resolution of 1 m, but its uncertainty is up to several metres[27]. Hence, while acknowledging that the

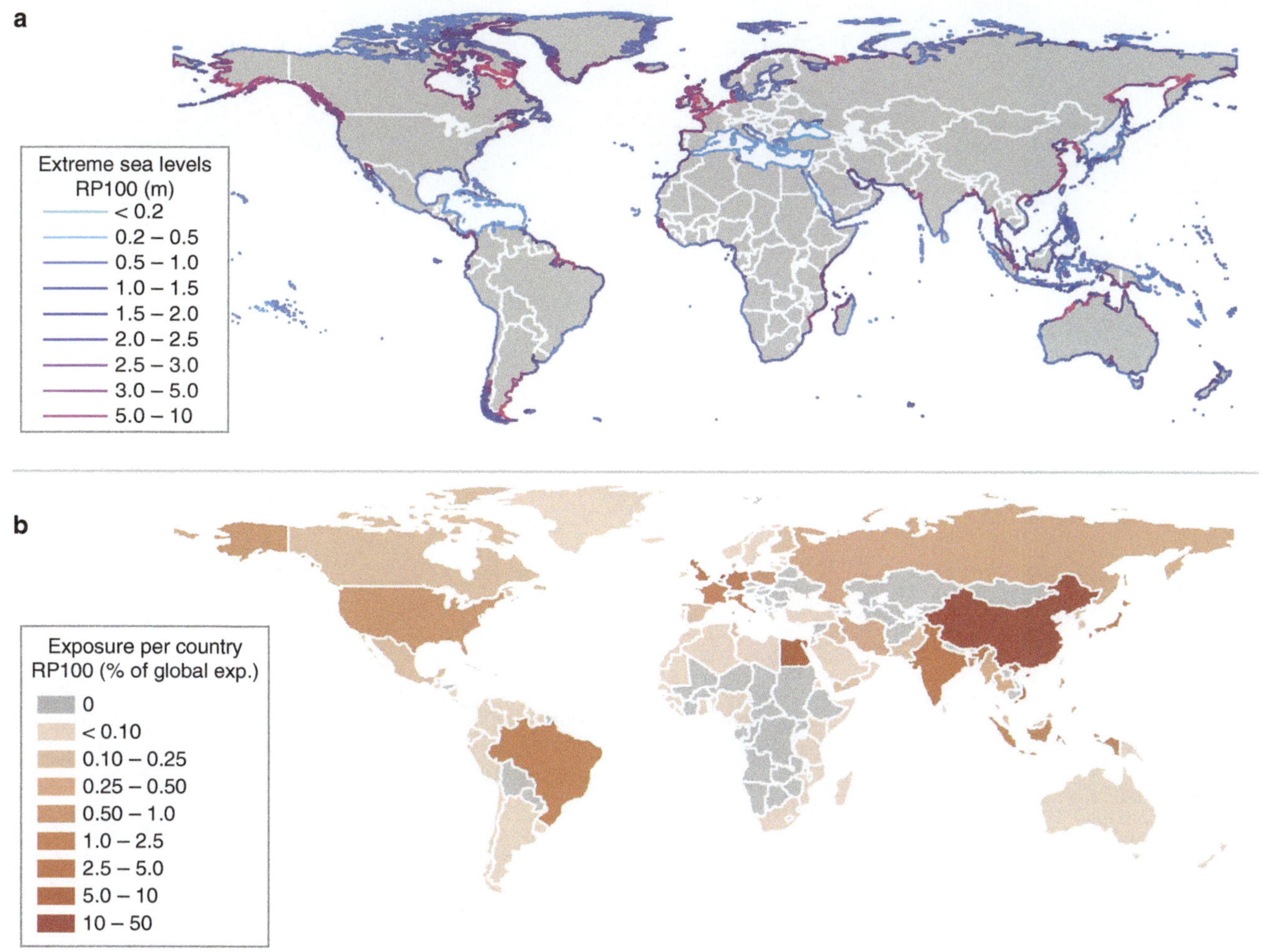

Figure 5 | Extreme sea levels with a return period of 100 years and the exposed population. Maps showing (**a**) the height of extreme sea levels with a return period of 100 years (based on the best Gumbel fit) around the entire world's coastline; and (**b**) the estimated exposed population estimates per country (relative to the global exposure) with return period of 100 years.

Table 1 | Absolute and relative exposed population to a 1 in 100-year flood for the 10 most exposed countries.

Country	Absolute exposure (in millions)	Relative exposure (% of population)
China	36	2.9
The Netherlands	8.4	53
Vietnam	4.7	6.0
Egypt	4.3	6.4
Germany	2.9	3.5
India	2.5	0.3
United Kingdom	2.4	4.1
Japan	2.0	1.6
Bangladesh	1.4	1.1
Belgium	1.1	11
World	**76**	**1.3**

errors can potentially be large in specific locations, we are confident that the GTSR time series (that is, surge, tide and total sea levels) and extremes are a very valuable addition to the global data sets that are currently available. Further benchmarking may be performed by comparing the GTSR extremes with other modelled data sets, such as extreme sea levels in the DIVA database[19] and regional hydrodynamic models[23,34].

There are several limitations to the GTSR time series and extremes. We aim to update the data set in the future, addressing some of the issues described here. First, the validation shows that extreme values are slightly underestimated. This is an inevitable result of the relatively coarse resolution of the model grid, bathymetry and meteorological forcing (compared with point observations). Extremes in wind speed and atmospheric pressure are smoothed in the ERA-Interim data set due to the temporal (6 h) and spatial (0.75°) resolution. This is particularly problematic for tropical cyclones, which are characterized by strong gradients in atmospheric pressures and wind fields both in time and in space. Our validation results show that the underestimation of extreme sea levels is more severe in tropical areas. However, because of the sparseness and shortness of the available records, the largest tropical cyclone-induced surges are not all included in the available observations[2]. Hence, if we compare the GTSR extremes against reported sea levels induced by tropical cyclones we see larger deviations. For example, the maximum sea levels during tropical cyclone Katrina in New Orleans (2005), and during Typhoon Haiyan (2013), in the Philippines exceeded 8 m (ref. 35) and 4–5 m (refs 36,37), respectively, while our extreme sea level estimates do not exceed 2–3 m for a return period of 1,000 year. This illustrates that accurately modelling these intense storms requires a much higher resolution than atmospheric

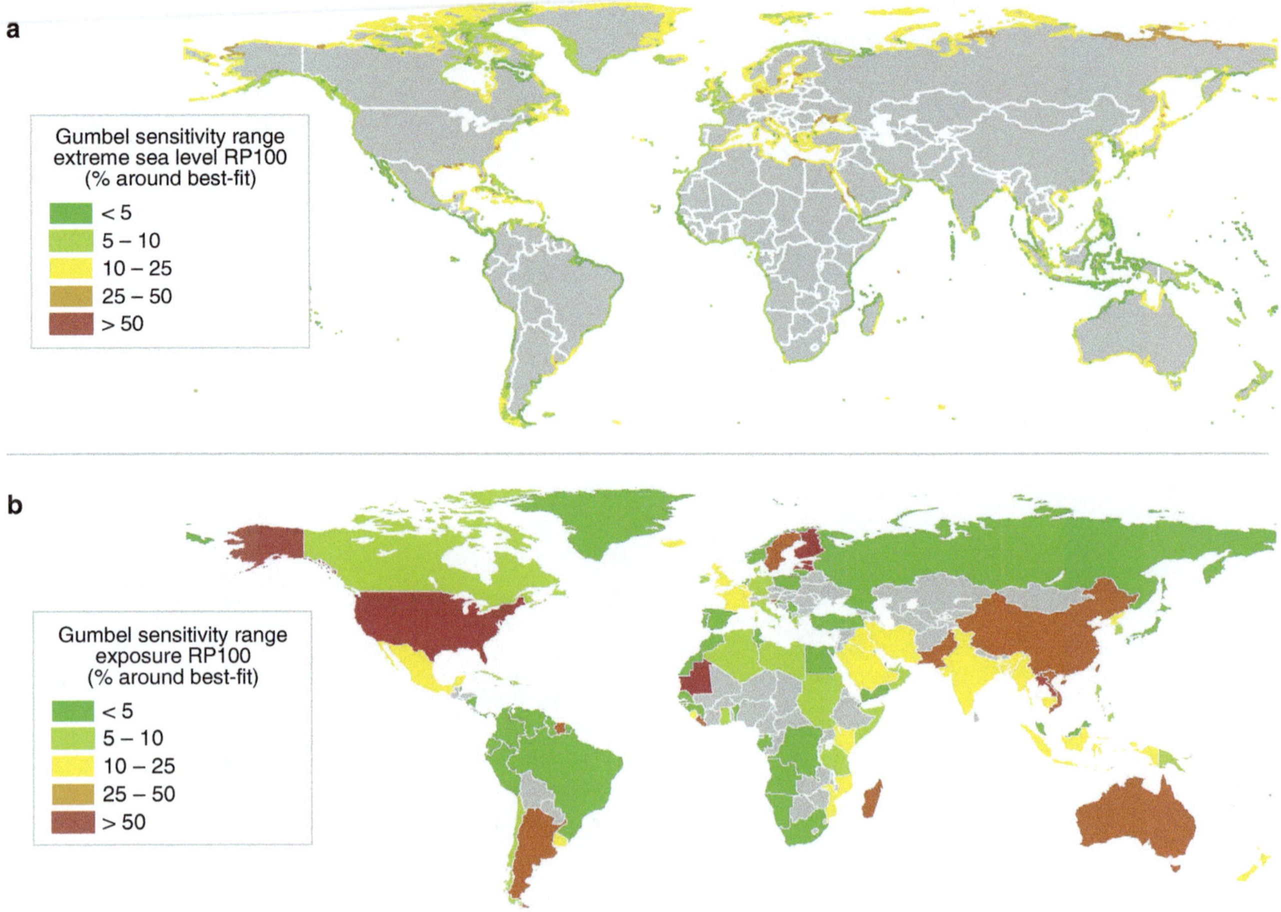

Figure 6 | Gumbel sensitivity range for extreme sea levels and flood exposure. Maps showing the uncertainty of the extreme values statistics for the 1 in 100-year return period. The values shown is the range for the 5-95% confidence bounds expressed as a percentage of the value for the best-fit for (**a**) the height of extreme sea around the entire world's coastline; and (**b**) the estimated exposed population estimates per country.

reanalysis data can deliver at present day. For updated versions of the GTSR data set, this issue could be resolved by generating localized wind fields based on storm track data and the parametric model of Holland[38]. This is expected to more accurately simulate surge levels resulting from tropical cyclones. However, even when the extreme sea levels are adequately modelled, time series of 36 years contain insufficient number of tropical cyclones to obtain reliable statistics of extreme values. Hence, synthetic resampling techniques are needed to extend the tropical cyclone record to a longer record[33,39,40]. Including tropical cyclones will lead to higher coastal flood risk, as the most damaging storm surges are often induced by tropical cyclones[41].

Second, we apply the annual maxima method to obtain the GTSR extremes. Although the method is widely applied, it neglects the fact that sea levels are composed of two independent processes, namely a tide-driven (deterministic) process and a surge-driven (stochastic) process. In addition, data are used inefficiently, as extreme values are estimated based on annual maxima only. To address these limitations, more sophisticated statistical methods, such as the *r* largest or joint probability, could be applied[42]. However, these methods are more sensitive to timing errors and/or temporal or spatial variation. Assuming that tides and storm surges behave independently, the estimates of extreme values could also be made more robust by resampling the surge and tide levels in time to obtain longer time series.

Third, the sea level variations in the GTSR time series are due to gravitational tides and barotropic changes (changes in wind and pressure). Baroclinic effects (density differences) are not considered. While in most parts of the world the generation of extreme sea levels is dominated by tide and surge, in some regions the variations in mean sea level are relatively large[43] and thus affect the total sea level. For example, this is the case in parts of the Australian coastline[23]. Also, non-linear interactions between storm surges and tides, the effects of waves, and precipitation and river flow are not considered. In this version of GTSR, tides and surges are modelled separately and surge-tide interactions are thus not included, although they are known to be important in shallow water areas with a large tidal range[44,45]. Including the surge-tide interaction has improved the model performance for regional hydrodynamic models[46]. For example, in the case of the North Sea, the RMSE was lowered by ca. 40% (ref. 47). However, at this stage this does not apply to GTSM (Supplementary Note 1), as the current version of GTSM is not capable of adequately reproducing tidal characteristics in all coastal regions. Wave setup may increase total sea levels considerably near the coast with the largest contribution in regions with steep slopes[48]. In deltas and estuaries, precipitation and river flow may also contribute to coastal flooding[49]. Including all these processes on a global scale is not feasible at present, but it is important to note that these processes may lead to significantly different extreme sea levels locally.

Also, the impact modelling has some limitations. Aside from limitations of the simple inundation model (see Methods section), the most important limitation is that flood protection is not included in this analysis, and many exposed countries are protected by dikes and storm surge barriers up to a certain design standard. For example, despite its high exposure rates, The Netherlands has very-low flood probabilities as all low-lying areas are protected by coastal defences, some with very high protection standards (1 in 10,000-year return period)[50]. However, Hallegatte *et al.*[12] estimated protection standards for major cities and found that a city such as Jakarta has a protection standard that equals a 10-year return period. Hence, for many countries listed in Table 1, the protection standard will be lower than a return period of 100 years. Using a similar methodology (planar inundation and no flood protection), but an alternative population database, Jongman *et al.*[8] estimated that 271 million people are exposed to 1 in 100-year coastal flooding in 2010. On the basis of the similar elevation and population data, but using a somewhat different methodology that includes flood protection, Hinkel *et al.*[5] estimated that between 160 million people are exposed to a 1 in 100-year flood. The differences in methodology and data make a direct comparison difficult, but our exposure estimate of 70–92 million people seems low compared with previous studies[5,8]. In countries prone to tropical cyclones, such as Bangladesh, exposure is underestimated as the tropical storms are responsible for the largest events[2]. Nevertheless, previous studies of global flood risk, including the studies of Jongman *et al.*[4] and Hinkel *et al.*[7], are based on the extreme sea levels in the DIVA database[19]. This data set has not been validated at the global scale, but compared to observed sea levels, Muis *et al.*[51] found that the DIVA 1 in 10-year extreme sea levels in Indonesia are overestimated with on average 0.8 m. Further research could assess differences in flood hazard between DIVA and GTSR and its influence on the estimated flood exposure.

In future research, the GTSR data set could be applied to assess global flood risk, both under current and future climate conditions. Coastal floods may become more severe due to sea level rise and changes in storminess[14,52]. While sea level rise may be directly added to the return levels[5], this is not valid for all regions[53], and the advantage of the physical based modelling approach is that the sensitivity of this assumption can be tested. Another advantage is that changes in storminess can be assessed using a dynamical approach by forcing the model with future global climate model simulations[54,55]. Furthermore, changes in storm duration can be propagated into the full dynamics of storm surge-induced flooding. Such climate change assessments can be used to identify areas that face rapidly increasing risks, which is important for planning disaster risk reduction efforts and to prioritize adaptation efforts[56]. Aside from flood risk applications, the GTSR data set can be used for a variety of other applications. For example, the GTSR time series may be used to: assess changes in storminess[57]; correct for meteorological effects in mean sea level[58]; disentangle the drivers of extreme events in different areas (tidal versus surge dominated)[20]; or assess the influence of interannual variability on risk[59]. As our new time series also includes the duration of flood events, another potential research direction may be the improvement of inundation modelling on a global scale by including flood duration in the inundation modelling. Recent research demonstrated the importance of assessing compound flooding in major US coastal cities[60]. In other delta regions, including compound flood events may also be critical for correctly assessing flood risk[61–64]. In combination with time series of precipitation or discharge, the GTSR time series could be used to assess compound floods on a global scale. Another application of the GTSM model that is currently being developed is the global operational forecasting system called GLOSSIS[65], which produces 10-day forecasts of coastal storm surges worldwide. The GTSR extremes could be used to raise warning flags and to identify potential flood hazards. In the future this should not only be based on exceeding a physical threshold, but also include potential impacts of a flood.

Methods

General approach. The method to develop the GTSR time series is based on two global hydrodynamic models: GTSM for storm surges, and FES2012 for tides (see Supplementary Fig. 8 for a flowchart). To simulate water levels in all coastal areas, while not generating huge amounts of data, model output is produced for 16,395 locations along the coastline based on the centroids of the DIVA segments database[19] and the locations of observation stations used for validation. Reduced data sets, useful for extreme value analyses, were generated by post-processing the results into daily maxima as well as annual maxima, for each of the output locations.

Modelling tides with FES2012. FES2012 is a global tidal model, which assimilates satellite altimetry data. The model is available for download at: http://www.aviso.altimetry.fr. Tidal elevations are distributed on a regular grid of 1/16°. Tides are simulated here with a 10-min time interval. A review on the performance of global tide models by Stammer[66] shows that FES2012 performs relatively well compared to other global tidal models in coastal areas.

Modelling storm surge with GTSM. GTSM is based on the Delft3D Flexible Mesh software developed by Deltares[24]. To accurately resolve hydrodynamic equations in topographically complex areas, such as coastal regions, while not decreasing the computational efficiency, it is desirable to locally refine the computational grid[24]. Delft3D FM enables this by allowing the use of unstructured grids. The cell size of the computational grid is dependent on the bathymetry and increases from 1/2° ($\sim$50 km) in deeper parts of the ocean towards 1/20° ($\sim$5 km) in shallow coastal areas (Supplementary Fig. 9). The bathymetric data with a resolution of 1/60° are collected from the General Bathymetric Chart of Oceans[67] and are interpolated onto the computational grid.

To simulate the sea levels resulting from storm surges, the model was forced with 10 m wind speed and atmospheric pressure obtained from the ERA-Interim data set developed by The European Centre For Medium-Range Weather Forecasts (ECMWF)[25]. The ERA-Interim data set is a global atmospheric reanalysis from 1979 to present day. The meteorological fields are available every 6 h and have a spatial resolution of $0.75 \times 0.75°$. The meteorological fields are temporally downscaled to 10 min using linear interpolation. The 10-m wind speed is translated into wind stress using a drag coefficient based on Charnock[68]. For consistency with the ECMWF climate model, we applied a Charnock parameter of 0.041. The sensitivity of applying different values for the Charnock parameter is discussed in the Supplementary Note 1.

The model simulations were carried out for each year separately, using a spin-up time of 11 days. Using a parallel setup with 16 cores, each simulation of one year takes approximately 30 h. The different runs were started in parallel, so in theory without other users on the computer cluster, the whole computation can be completed in 30 h.

Data used for validation. A global data set with observed sea levels is used to validate GTSR. Hourly levels from 472 stations (Supplementary Fig. 10) over the period 1980–2011 are obtained from the archives of the UHSLC (data set is available at http://uhslc.soest.hawaii.edu). All stations are quality-checked by tidal analysis using TideMAT 1.05 (ref. 69). The records of each station are only used when <20% of the data are missing. Each year is analysed separately and mean annual sea level is used as reference date; as such the sea levels are detrended. We subtract the tidal component from the total sea level to obtain the residual level. This component primarily contains the meteorological contribution to sea level (that is, surge level), but may also contain harmonic prediction errors or timing errors. However, we consider these errors as negligible as each station has been inspected visually. Decomposing the total sea level into a tide and surge component enables the separate validation of the two components. For each station with a record longer than 25 years, we also estimate the extreme sea levels for different return periods and compare those with modelled sea levels (see next paragraph).

Extreme value statistics. To estimate the probabilities of extreme sea levels, we apply extreme value statistics using the annual maxima method[70,71]. For each output location, we extract the annual maximum for the calendar years 1979–2014 and fit a Gumbel distribution using the maximum-likelihood method. From the parameterised distribution, we can obtain estimates of sea levels corresponding to selected return periods. We also obtain uncertainty estimates of the parameterised extreme value distribution (that is, 5 and 95% confidence bounds). The Gumbel distribution is often a good approximation of observed extreme sea levels and is frequently applied to estimate return periods[42]. While more advanced statistical methods are available, such as peak-over-threshold or joint probability, the annual maxima method is more robust to temporal and spatial variations, and is thus relatively easy to apply on a global scale.

Modelling inundation and flood exposure. On the basis of the extreme sea levels, coastal inundation is calculated using a GIS-based planar approach, which uses the extreme sea level and a DEM as input. Inundated areas are defined as areas that have an elevation lower than the water level, and have a direct connection to the sea[5,8]. The DEM is obtained during NASA's Shuttle Radar Topography Mission and its original resolution is $1'' \times 1''$ ($\sim 30 \times 30$ m at the equator)[27]. We use a 30" × 30" ($\sim 1 \times 1$ km at the equator) average to estimate inundation. Impacts of coastal floods are measured in terms of exposed population using the GRUMPv1 population counts maps for 2000 (refs 28,29). Flood protection is not included in the analysis and will lead to an overestimation of the flood hazard. Furthermore, the planar approach assumes that a maximum sea level during any event will travel infinitely far land inwards and will therefore overestimate the extent of the floodplain, in particular on coastlines with wide flat areas far land inwards. On the other hand, the spatial averaging of the DEM may result in smoothing of local depressions, causing underestimation of the flood extent in areas with relatively large upward gradients land inwards. Also, coastal cities are often in delta areas where the river may propagate the surge into the hinterland. In this sense, the approach may underestimate the flood extent.

References

1. Resio, D. T. & Westerink, J. J. Modeling the physics of storm surges. *Phys. Today* **61**, 33–38 (2008).
2. Needham, H. F., Keim, B. D. & Sathiaraj, D. A review of tropical cyclone-generated storm surges: Global data sources, observations, and impacts. *Rev. Geophys.* **53**, 545–591 (2015).
3. Pugh, D. T. *Tides, Surges and Mean Sea-Level* (John Wiley & Sons, 1987).
4. McGranahan, G., Balk, D. L. & Anderson, B. The rising tide: assessing the risks of climate change and human settlements in low elevation coastal zones. *Environ. Urban.* **19**, 17–37 (2007).
5. Hinkel, J. *et al.* Coastal flood damage and adaptation costs under 21st century sea-level rise. *Proc. Natl Acad. Sci. USA* **111**, 3292–3297 (2014).
6. Spencer, T., Brooks, S. M. & Möller, I. Storm-surge impact depends on setting. *Nature* **505**, 26 (2013).
7. Rappaport, E. N. Fatalities in the United States from Atlantic tropical cyclones: new data and interpretation. *Bull. Am. Meteorol. Soc.* **95**, 341–346 (2014).
8. Jongman, B., Ward, P. J. & Aerts, J. C. J. H. Global exposure to river and coastal flooding: long term trends and changes. *Glob. Environ. Chang.* **22**, 823–835 (2012).
9. Nicholls, R. J. & Cazenave, A. Sea-level rise and its impact on coastal zones. *Science* **328**, 1517–1520 (2010).
10. Syvitski, J. P. M. *et al.* Sinking deltas due to human activities. *Nat. Geosci.* **2**, 681–686 (2009).
11. Bouwer, L. M. Have disaster losses increased due to anthropogenic climate change? *Bull. Am. Meteorol. Soc.* **6**, 791–793 (2011).
12. Hallegatte, S., Green, C., Nicholls, R. J. & Corfee-Morlot, J. Future flood losses in major coastal cities. *Nat. Clim. Chang.* **3**, 802–806 (2013).
13. Kemp, A. C. & Horton, B. P. Contribution of relative sea-level rise to historical hurricane flooding in New York City. *J. Quat. Sci.* **28**, 537–541 (2013).
14. Reed, A. J. *et al.* Increased threat of tropical cyclones and coastal flooding to New York City during the anthropogenic era. *Proc. Natl Acad. Sci. USA* **112**, 12610–12615 (2015).
15. Brown, S., Nicholls, R. J., Lowe, J. A. & Hinkel, J. Spatial variations of sea-level rise and impacts: an application of DIVA. *Clim. Change* **134**, 403–416 (2013).
16. Hinkel, J. *et al.* Sea-level rise impacts on Africa and the effects of mitigation and adaptation: an application of DIVA. *Reg. Environ. Chang.* **12**, 207–224 (2011).
17. Hinkel, J., Nicholls, R. J., Vafeidis, A. T., Tol, R. S. J. & Avagianou, T. Assessing risk of and adaptation to sea-level rise in the European Union: an application of DIVA. *Mitig. Adapt. Strateg. Glob. Chang.* **15**, 703–719 (2010).
18. Hinkel, J. & Klein, R. J. T. Integrating knowledge to assess coastal vulnerability to sea-level rise: The development of the DIVA tool. *Glob. Environ. Chang.* **19**, 384–395 (2009).
19. Vafeidis, A. T. *et al.* A new global coastal database for impact and vulnerability analysis to sea-level rise. *J. Coast. Res.* **244**, 917–924 (2008).
20. Merrifield, M., Genz, A. S., Kontoes, C. P. & Marra, J. J. Annual maximum water levels from tide gauges: Contributing factors and geographic patterns. *J. Geophys. Res. Ocean* **118**, 2535–2546 (2013).
21. Marcos, M., Tsimplis, M. N. & Shaw, A. G. P. Sea level extremes in southern Europe. *J. Geophys. Res.* **114**, C01007 (2009).
22. Bernier, N. B. & Thompson, K. R. Predicting the frequency of storm surges and extreme sea levels in the northwest Atlantic. *J. Geophys. Res.* **111**, C10009 (2006).
23. Haigh, I. D. *et al.* Estimating present day extreme water level exceedance probabilities around the coastline of Australia: tides, extra-tropical storm surges and mean sea level. *Clim. Dyn.* **42**, 121–138 (2013).
24. Kernkamp, H. W. J., Van Dam, A., Stelling, G. S. & de Goede, E. D. Efficient scheme for the shallow water equations on unstructured grids with application to the Continental Shelf. *Ocean Dyn.* **61**, 1175–1188 (2011).
25. Dee, D. P. *et al.* The ERA-Interim reanalysis: configuration and performance of the data assimilation system. *Q. J. R. Meteorol. Soc.* **137**, 553–597 (2011).
26. Carrère, L. *et al.* in *Proc. 20YPRA Symp.*, 3–8 (ESA/CNES, Venice-Lido, Italy, 2012).
27. Rabus, B., Eineder, M., Roth, A. & Bamler, R. The shuttle radar topography mission - A new class of digital elevation models acquired by spaceborne radar. *ISPRS J. Photogramm. Remote Sens.* **57**, 241–262 (2003).
28. Center for International Earth Science Information Network (CIESIN) — Columbia University, International Food Policy Research Institute (IFPRI) — The World Bank, and Centro Internacional de Agricultura Tropical (CIAT). *Global Rural-Urban Mapping Project, Version 1 (GRUMPv1): Population Count Grid* (NASA Socioeconomic Data and Applications Center (SEDAC), Palisades, NY, 2011).
29. Balk, D. L. *et al.* Determining global population distribution: methods, applications and data. *Adv. Parasitol.* **62**, 119–156 (2011).
30. Muis, S., Verlaan, M., Winsemius, H., Aerts, J. C. J. H. & Ward, P. J. Global Tide and Surge Reanalysis (GTSR). *3TU. Datacentrum* https://dx.doi.org/10.4121/uuid:aa4a6ad5–e92c–468e (2016).
31. Menéndez, M. & Woodworth, P. L. Changes in extreme high water levels based on a quasi-global tide-gauge data set. *J. Geophys. Res.* **115**, C10011 (2010).
32. Cid, A., Castanedo, S., Abascal, A. J., Menéndez, M. & Medina, R. A high resolution hindcast of the meteorological sea level component for Southern Europe: the GOS dataset. *Clim. Dyn.* **43**, 2167–2184 (2014).
33. Haigh, I. D. *et al.* Estimating present day extreme water level exceedance probabilities around the coastline of Australia: tropical cyclone-induced storm surges. *Clim. Dyn.* **42**, 139–157 (2013).
34. Flather, R. A., Smith, J. A., Richards, J. D., Bell, C. & Blackman, D. L. Direct estimates of extreme storm surge elevations from a 40-year numerical model simulation and from observations. *Glob. Atmos. Ocean Syst.* **6**, 165–176 (1998).
35. Needham, H. F. & Keim, B. D. A storm surge database for the US Gulf Coast. *Int. J. Climatol.* **32**, 2108–2123 (2012).
36. Mahar, A. *et al.* Devastating storm surges of Typhoon Haiyan. *Int. J. Disaster Risk Reduct.* **11**, 1–12 (2014).
37. Mori, N. *et al.* Local amplification of storm surge by Super Typhoon Haiyan in Leyte Gulf. *Geophys. Res. Lett.* **41**, 1–8 (2014).
38. Holland, G. J. An analytic model of the wind and pressure profiles in hurricanes. *Mon. Weather Rev.* **108**, 1212–1218 (1980).
39. Lin, N., Emanuel, K. A., Oppenheimer, M. & Vanmarcke, E. Physically based assessment of hurricane surge threat under climate change. *Nat. Clim. Chang.* **2**, 462–467 (2012).
40. Emanuel, K. A., Ravela, S., Vivant, E. & Risi, C. A statistical deterministic approach to hurricane risk assessment. *Bull. Am. Meteorol. Soc.* **87**, 299–314 (2006).
41. Woodruff, J. D., Irish, J. L. & Camargo, S. J. Coastal flooding by tropical cyclones and sea-level rise. *Nature* **504**, 44–52 (2013).
42. Haigh, I. D., Nicholls, R. J. & Wells, N. C. A comparison of the main methods for estimating probabilities of extreme still water levels. *Coast. Eng.* **57**, 838–849 (2010).
43. Tsimplis, M. N. & Woodworth, P. L. The global distribution of the seasonal sea level cycle calculated from coastal tide gauge data. *J. Geophys. Res.* **99**, 16031 (1994).
44. Bernier, N. B. & Thompson, K. R. Tide-surge interaction off the east coast of Canada and northeastern United States. *J. Geophys. Res.* **112**, 1–12 (2007).
45. Horsburgh, K. J. & Wilson, C. Tide-surge interaction and its role in the distribution of surge residuals in the North Sea. *J. Geophys. Res.* **112**, C08003 (2007).
46. Wolf, J. Coastal flooding: impacts of coupled wave–surge–tide models. *Nat. Hazards* **49**, 241–260 (2008).
47. Zijl, F., Verlaan, M. & Gerritsen, H. Improved water-level forecasting for the Northwest European Shelf and North Sea through direct modelling of tide, surge and non-linear interaction. *Ocean Dyn.* **63**, 823–847 (2013).
48. Dietrich, J. C. *et al.* Modeling hurricane waves and storm surge using integrally-coupled, scalable computations. *Coast. Eng.* **58**, 45–65 (2011).
49. Bunya, S. *et al.* A high-resolution coupled riverine flow, tide, wind, wind wave, and storm surge model for southern louisiana and mississippi. Part I: model development and validation. *Mon. Weather Rev.* **138**, 345–377 (2010).

50. Wesselink, A., Warner, J. & Kok, M. You gain some funding, you lose some freedom: The ironies of flood protection in Limburg (The Netherlands). *Environ. Sci. Policy* **30,** 113–125 (2013).
51. Muis, S., Güneralp, B., Jongman, B., Aerts, J. C. J. H. & Ward, P. J. Flood risk and adaptation strategies under climate change and urban expansion: A probabilistic analysis using global data. *Sci. Total Environ.* **538,** 445–457 (2015).
52. IPCC. *Managing the Risks of Extreme Events and Disasters to Advance Climate Change Adaptation* (Cambridge University Press, 2012).
53. Zhang, K., Li, Y., Liu, H., Xu, H. & Shen, J. Comparison of three methods for estimating the sea level rise effect on storm surge flooding. *Clim. Change* **118,** 487–500 (2012).
54. Brown, J. M., Wolf, J. & Souza, A. J. Past to future extreme events in Liverpool Bay: model projections from 1960–2100. *Clim. Change* **111,** 365–391 (2011).
55. Lowe, J. A., Gregory, J. M. & Flather, R. A. Changes in the occurrence of storm surges around the United Kingdom under a future climate scenario using a dynamic storm surge model driven by the Hadley Centre climate models. *Clim. Dyn.* **18,** 179–188 (2001).
56. UNISDR. *Global Assessment Report on Disaster Risk Reduction: Revealing Risk, Redefining Development.* 178 (United Nations International Strategy for Disaster Reduction, Geneva, Switzerland, 2011).
57. Cid, A. *et al.* Long-term changes in the frequency, intensity and duration of extreme storm surge events in southern Europe. *Clim. Dyn.* **46,** 1503–1516 (2015).
58. Losada, I. J. *et al.* Long-term changes in sea-level components in Latin America and the Caribbean. *Glob. Planet. Chang.* **104,** 34–50 (2013).
59. Ward, P. J. *et al.* Strong influence of El Niño Southern Oscillation on flood risk around the world. *Proc. Natl Acad. Sci. USA* **111,** 15659–15664 (2014).
60. Wahl, T., Jain, S., Bender, J., Meyers, S. D. & Luther, M. E. Increasing risk of compound flooding from storm surge and rainfall for major US cities. *Nat. Clim. Chang.* **5,** 1093–1097 (2015).
61. Kew, S. F., Selten, F. M., Lengaigne, M. & Hazeleger, W. The simultaneous occurrence of surge and discharge extremes for the Rhine delta. *Nat. Hazards Earth Syst. Sci.* **13,** 2017–2029 (2013).
62. Zheng, F., Westra, S., Leonard, M. & Sisson, S. A. Modeling dependence between extreme rainfall and storm surge to estimate coastal flooding risk. *Water Resour. Res.* **50,** 2050–2071 (2014).
63. Keef, C., Svensson, C. & Tawn, J. A. Spatial dependence in extreme river flows and precipitation for Great Britain. *J. Hydrol.* **378,** 240–252 (2009).
64. van den Hurk, B. J. J. M., van Meijgaard, E., de Valk, P., van Heeringen, K.-J. & Gooijer, J. Analysis of a compounding surge and precipitation event in the Netherlands. *Environ. Res. Lett.* **10,** 035001 (2015).
65. Verlaan, M., De Kleermaeker, S. & Buckman, L. in *Australasian Coasts & Ports Conference 2015: 22nd Australasian Coastal and Ocean Engineering Conference and the 15th Australasian Port and Harbour Conference*, 229–234 (Auckland, New Zealand & Engineers Australia and IPENZ, 2015).
66. Stammer, D. *et al.* Accuracy assessment of global barotropic ocean tide models. *Rev. Geophys.* **52,** 1–40 (2014).
67. IOC, IHO, & BODC. *Centenary Edition of the GEBCO Digital Atlas, Published on CD-ROM on Behalf of the Intergovernmental Oceanographic Commission and the International Hydrographic Organization as Part of the General Bathymetric Chart of the Oceans.* (British Oceanographic Data Centre, Liverpool, UK, 2003).
68. Charnock, H. Wind stress on a water surface. *Q. J. R. Meteorol. Soc.* **81,** 639 (1955).
69. Rijkswaterstaat. TIDEMAT Tidal Analysis. User's Manual 1.01, 30 p. (The Hague, The Netherlands, 2004).
70. Gumbel, E. J. The return period of flood flows. *Ann. Math. Stat.* **12,** 163–190 (1941).
71. Jenkinson, A. F. The frequency distribution of maximum values. *Q. J. R. Meteorol. Soc.* **81,** 158–171 (1954).

Acknowledgements
The research leading to these results has received funding from the European research project RISES-AM- (grant agreement no. 603396). P.J.W. received additional funding from the Netherlands Organisation for Scientific Research (NWO) in the form of a VENI grant (grant no. 863-11-011). J.C.J.H.A. and S.M. received funding from the NWO in the form of a VICI Grant (grant no. 453-14-006). We thank SURFsara (www.surfsara.nl) for the support in using the Lisa Computer Cluster.

Author contributions
S.M. was responsible for carrying out the model runs and the validation of GTSR, including the tidal analysis of the UHCL data set. M.V. developed the GTSM model, provided a script to apply the FES2012 model, and assisted with carrying out the model runs. H.C.W. was responsible for the post-processing of the data. S.M., M.V., H.C.W. and P.J.W. were involved in the development of the data set and the data analysis. All authors discussed the results and implications and were involved in the writing of the paper.

Additional information

Competing financial interests: The authors declare no competing financial interests.

22

Gains and losses of coral skeletal porosity changes with ocean acidification acclimation

Paola Fantazzini[1,2], Stefano Mengoli[3], Luca Pasquini[1], Villiam Bortolotti[4], Leonardo Brizi[1,2], Manuel Mariani[1,2], Matteo Di Giosia[5], Simona Fermani[5], Bruno Capaccioni[6], Erik Caroselli[7], Fiorella Prada[7], Francesco Zaccanti[7], Oren Levy[8], Zvy Dubinsky[8], Jaap A. Kaandorp[9], Pirom Konglerd[9], Jörg U. Hammel[10], Yannicke Dauphin[11], Jean-Pierre Cuif[11], James C. Weaver[12], Katharina E. Fabricius[13], Wolfgang Wagermaier[14], Peter Fratzl[14], Giuseppe Falini[5] & Stefano Goffredo[7]

Ocean acidification is predicted to impact ecosystems reliant on calcifying organisms, potentially reducing the socioeconomic benefits these habitats provide. Here we investigate the acclimation potential of stony corals living along a pH gradient caused by a Mediterranean CO_2 vent that serves as a natural long-term experimental setting. We show that in response to reduced skeletal mineralization at lower pH, corals increase their skeletal macroporosity (features $>10\,\mu m$) in order to maintain constant linear extension rate, an important criterion for reproductive output. At the nanoscale, the coral skeleton's structural features are not altered. However, higher skeletal porosity, and reduced bulk density and stiffness may contribute to reduce population density and increase damage susceptibility under low pH conditions. Based on these observations, the almost universally employed measure of coral biomineralization, the rate of linear extension, might not be a reliable metric for assessing coral health and resilience in a warming and acidifying ocean.

[1] Department of Physics and Astronomy, University of Bologna, Viale Berti Pichat 6/2, 40127 Bologna, Italy. [2] Centro Enrico Fermi, Piazza del Viminale 1, 00184 Rome, Italy. [3] Department of Management, University of Bologna, Via Capo di Lucca 34, 40126 Bologna, Italy. [4] Department of Civil, Chemical, Environmental, and Materials Engineering, University of Bologna, Via Terracini 28, 40131 Bologna, Italy. [5] Department of Chemistry 'G. Ciamician', University of Bologna, Via F. Selmi 2, 40126 Bologna, Italy. [6] Department of Biological, Geological and Environmental Sciences, Section of Geology, University of Bologna, Piazza di Porta S. Donato 1, 40126 Bologna, Italy. [7] Marine Science Group, Department of Biological, Geological and Environmental Sciences, Section of Biology, University of Bologna, Via F. Selmi 3, 40126 Bologna, Italy. [8] The Mina and Everard Goodman Faculty of Life Sciences, Bar-Ilan University, Ramat-Gan 52900, Israel. [9] Section Computational Science, Faculty of Science, University of Amsterdam, Science Park 904, room C3.147, 1090 GE Amsterdam, The Netherlands. [10] Institute of Materials Research, Helmholtz-Zentrum Geesthacht, Outstation at DESY, Building 25c Notkestr. 85, D-22607 Hamburg, Germany. [11] Micropaléontologie, UFR TEB Université P. & M. Curie, 75252 Paris, France. [12] Wyss Institute for Biologically Inspired Engineering at Harvard University, 60 Oxford Street, Cambridge, Massachusetts 02138, USA. [13] Australian Institute of Marine Science, PMB 3, Townsville, 4810 Queensland, Australia. [14] Department of Biomaterials, Max Planck Institute of Colloids and Interfaces, 14424 Potsdam, Germany. Correspondence and requests for materials should be addressed to P.F. (email: paola.fantazzini@unibo.it) or to G.F. (email: giuseppe.falini@unibo.it) or to S.G. (email: s.goffredo@unibo.it).

Climate change is among the biggest threats to the health of marine ecosystems. Rising atmospheric carbon dioxide partial pressure (pCO_2)[1] increases ocean surface pCO_2 due to CO_2 diffusion across the air-water interface, leading to ocean acidification (OA)[2]. Since global warming and OA are coupled and are predicted to act synergistically[3,4], the future health of marine ecosystems and their corresponding long-term economic impacts on human coastal populations remain uncertain[5–7]. It is therefore of great interest to understand how increasing atmospheric CO_2 concentrations will affect these marine habitats and the species that inhabit them.

Since the early 1800s, ocean pH has decreased from 8.2 by *ca.* 0.1 U[8] and, if CO_2 emissions continue at their current rates, the average sea surface pH is predicted to drop to 7.8 by the year 2100 (ref. 1). The Mediterranean Sea, with its closed circulation patterns and limited water exchange with the adjacent Atlantic Ocean, has already undergone a larger decrease in surface pH compared with the global average[9], making it an ideal site for OA studies[10–13].

Near Panarea Island, off the southwestern coast of Italy, lies a series of active volcanic vents in the seabed releasing CO_2 emissions that acidify the surrounding seawater, making this location an ideal natural laboratory for OA studies. The underwater CO_2 emissions generate a stable pH gradient with levels matching several Intergovernmental Panel on Climate Change (IPCC) sea surface pH predictions associated with different atmospheric CO_2 emission scenarios for the end of the century[1].

The present study investigates the effects of environmental pH on skeletal structures and growth at multiple length scales in the solitary scleractinian coral *Balanophyllia europaea* living along the pH gradient. We studied 74 corals of similar age (mean age of 12 years) that had spent their lives at the CO_2-pH gradient. Using a combination of scanning electron microscopy (SEM), atomic force microscopy (AFM), small-angle X-ray scattering (SAXS), micro computed tomography (μCT), nanoindentation, hydrostatic weight measurement and time-domain nuclear magnetic resonance (TD-NMR), we document the skeletal mass, bulk volume, pore volume, porosity, biomineral density, bulk density, hardness, stiffness (ratio between elastic stress and strain), biometry data, net calcification rate and linear extension rate for each coral. Weight measurements combined with TD-NMR data represent a non-destructive technique for quantifying skeletal porosity over length scales spanning from tens of nanometres to tens of micrometres[14,15], whereas μCT analysis permits a detailed large-scale quantitative 3D analysis of skeletal architecture, including the interseptal regions.

We show that in response to depressed calcification at lower pH, corals increase their skeletal porosity maintaining constant linear extension rate, which is important for reaching critical size at sexual maturity. However, higher skeletal porosity and reduced bulk density and stiffness may contribute to reduced mechanical strength, increasing damage susceptibility, which could result in increased mortality in an acidic environment.

Results

Study site and seawater carbonate chemistry. The four sites around the main vent are reported in Fig. 1. Site 1 (S1) has an

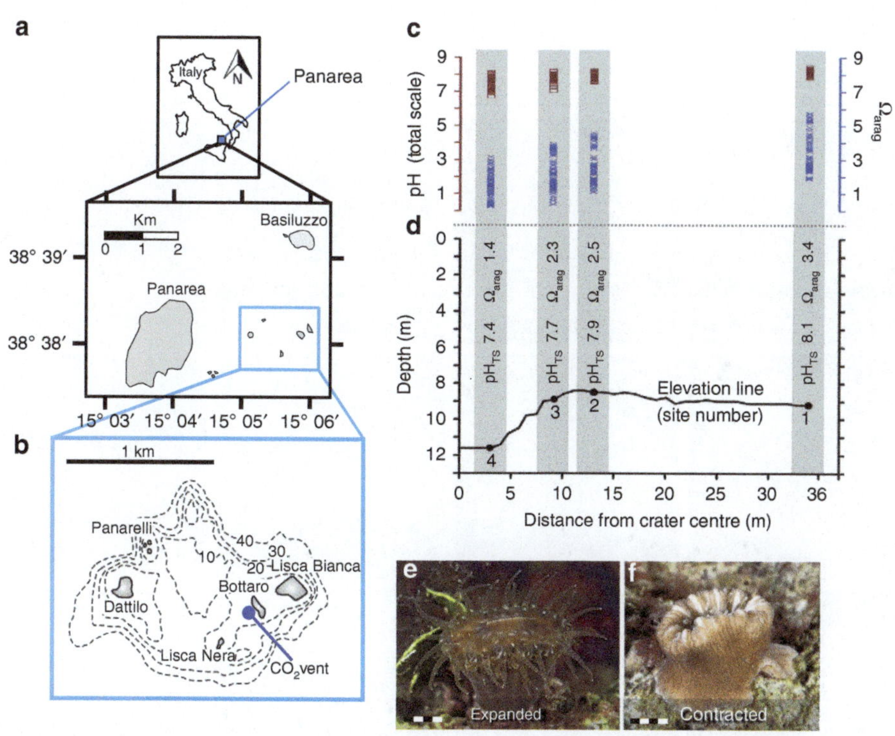

Figure 1 | Study location. Located off the southwestern coast of Italy (**a**), near Panarea Island (**b**), there are underwater volcanic vents releasing persistent gaseous emissions (98–99% CO_2 without instrumentally detectable toxic compounds), resulting in a stable pH gradient. Four sites at various distances from the primary vent were initially selected for study. No temperature difference exists among the four study sites throughout the year. (**c**) Ranges of measured pH_{TS} (number of observations [n] = 103–110 per site) and Ω_{arag} (n = 96–104 per site) values at the four sites, showing consistent increases in both pH_{TS} and Ω_{arag} from the vent crater to its periphery. (**d**) Bathymetric profile of the sites with associated mean pH_{TS} and Ω_{arag} values. No corals were found at Site 4, characterized by the lowest pH (mean pH_{TS} 7.4). Living specimens of *Balanophyllia europaea*, photographed at night with expanded [illegible] (**f**); marker 5 mm.

average total scale pH (pH_{TS}) of 8.1, equivalent to the average surface pH of modern oceans. S2's average pH_{TS} of 7.9 aligns with IPCC's predictions of a conservative CO_2 emissions scenario (Representative Concentration Pathway (RCP6.0)), and the average pH_{TS} of 7.7 for S3 fits the predictions of the 'business-as-usual' CO_2 emissions scenario (RCP8.5). Since no corals were found at S4 (within the vent crater, pH_{TS} 7.4), only S1–3, which had growing coral populations and pH_{TS} values ranging from 8.1 to 7.7, were included in the present study. Of the measured parameters at the three sites along the pH gradient (pH_{TS}, total alkalinity, temperature and salinity), only pH_{TS} differed significantly across sites ($N=103$–110; $P<0.001$, Kruskal–Wallis χ^2-test). The changing pH was accompanied by significant shifts in carbonate–bicarbonate equilibria, with aragonite saturation (Ω_{arag}) decreasing by >30% from S1 to S3 (Fig. 1, Supplementary Table 1, Supplementary Note 1).

Multi-scale analysis of coral skeletal properties. Combined results of SEM, μCT, AFM and TD-NMR skeletal analyses of corals growing at each study site revealed the detailed, multi-scale structural organization of the skeletons of *B. europaea* (Fig. 2, Supplementary Figs 1–3).

At the macroscale (relating to feature sizes >10 μm), linear extension rate (averaging *ca.* 1 mm per year) did not vary among sites (Supplementary Table 2), whereas net calcification rate ($N=44$; $P<0.01$, robust *t*-statistics test) and skeletal bulk density ($N=44$, $P<0.001$, robust *t*-statistics test) significantly declined (Fig. 3, Supplementary Table 3) and skeletal porosity and macroscale porosity increased ($N=44$; $P<0.001$, robust *t*-statistics test, Fig. 3, Supplementary Fig. 1, and Supplementary Table 3). The differences between S1 and S3 were *ca.* −18% for net calcification rate, *ca.* −7% for bulk density, *ca.* +21% for porosity and *ca.* +30% for macroscale porosity (Supplementary Table 2). The corallite interseptal volume fraction from μCT (Supplementary Fig. 4) showed a difference among sites ($N=30$; $P<0.05$, F and Kruskal–Wallis χ^2-tests, Supplementary Table 4) but no significant dependence on pH (Supplementary Fig. 5). The biometric data for the corallites (Supplementary Table 5) did not vary among sites.

At the micro/macroscale, skeletal stiffness significantly declined with decreasing pH ($n=122$; $P<0.001$, robust *t*-statistics test, Fig. 3, Supplementary Fig. 6, Supplementary Tables 4 and 6), whereas skeletal microscale porosity did not vary among sites (Supplementary Table 2 and Supplementary Fig. 1).

At smaller length scales (at the micro and nanoscales), SEM and AFM showed that the organization of the aragonite fibre bundles (Fig. 2c,g,k) and the morphology and packing of the constituent mineral grains (Fig. 2d,h,l) appeared similar among corals from the three different sites, confirming that the basic biomineralization products were not affected by reduced pH[13] (Fig. 2 and Supplementary Fig. 2). Skeletal biomineral hardness did not change among sites (Supplementary Fig. 7, Supplementary Table 4). Also, skeletal biomineral density values were similar across sites and consistent with those measured in previous studies[16] (Supplementary Table 2). Similar results were obtained from both SAXS and TD-NMR analyses, which revealed that nanoscale porosity did not change significantly with changing pH (Supplementary Figs 1e and 8–10 and Supplementary Tables 2 and 4).

The principal component analysis identified three major components: growth, skeletal porosity and biomineral density (Supplementary Tables 7 and 8). Only the first two components depended significantly on pH (Supplementary Table 9).

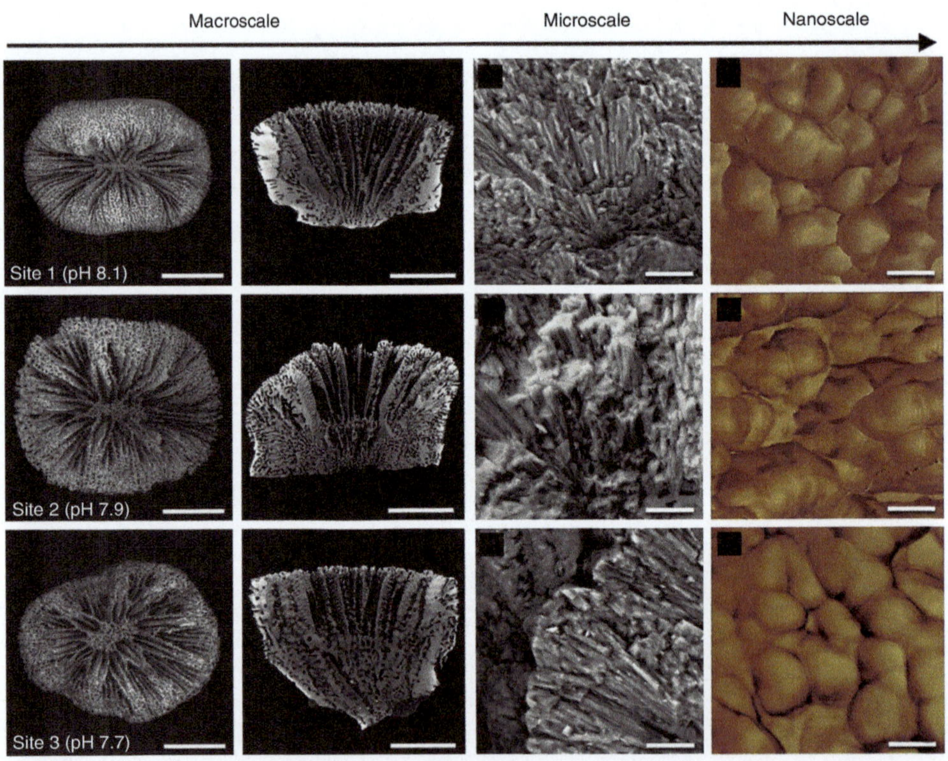

Figure 2 | Skeletal morphology of *Balanophyllia europaea* growing under different pH conditions from the macroscale to the nanoscale. Each row in the figure corresponds to a different study site and sample age is 9–11 years. Images are representative of all observed skeletons. (**a,e,i**) Low magnification SEM images of coral skeletons, marker 5 mm. (**b,f,j**) Internal sections of corallites from μCT images, marker 5 mm. (**c,g,k**) SEM images of entire skeletal fibres from fractured septae, marker 10 μm. (**d,h,l**) AFM images of mineral grains on the skeletal fibre surfaces, marker 50 nm.

In summary, skeletal nano and microstructural features, linear extension rate, interseptal volume fraction and corallite biometry of *B. europaea* did not change significantly with decreasing pH, despite a clear reduction in net calcification rates. This reduction in net calcification rate was accompanied by an increase in skeletal porosity (Fig. 3f, $N = 44$; $P < 0.001$, robust *t*-statistics test) and a consequent decrease in skeletal bulk density and stiffness.

Discussion

Results of the present study complement previous research on *B. europaea* at this same vent site, which revealed no changes in skeletal calcium carbonate polymorph, organic matrix content, aragonite fibre thickness and skeletal hardness in corals growing along the pH gradient[13]. There was, however, a significant reduction in population density along the pH gradient, decreasing by a factor of 3 with increasing proximity to the vent crater centre (that is, from S1 to S3)[13].

Figure 4 summarizes these results at the ocean, population, macro, micro and nanoscales for *B. europaea*. At the macroscale, increasing acidity was associated with a reduction in net calcification rate and a parallel increase in skeletal porosity, coupled with a decrease in skeletal bulk density. Linear extension rate and corallite shape (biometry and interseptal volume fraction) did not depend on pH, probably as a result of the compensation of reduced net calcification rate by increased skeletal porosity. At the micro/macroscale, the declining skeletal stiffness with decreasing pH could be coupled to an increased volume fraction of pores having a size comparable to the indentation area (that is, at the border between the micro and macroscales). At the nanoscale, porosity, biomineral hardness and density were not significantly affected by pH. These results, bolstered by qualitative SEM and AFM analyses, suggest that the 'building blocks' produced by the biomineralization process are substantially unaffected, but the increase in skeletal porosity is both a gain and a loss for the coral. In fact, in an acidic

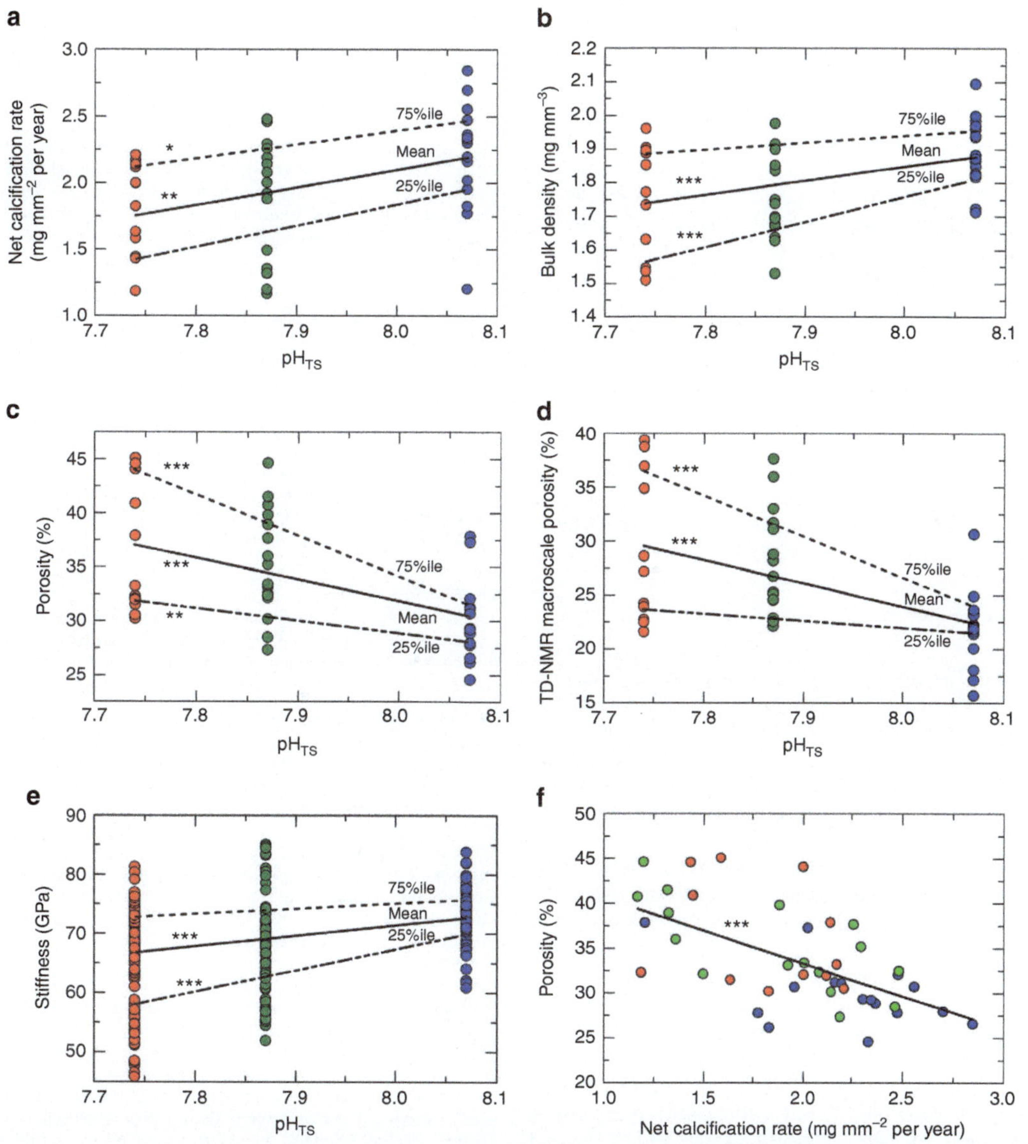

Figure 3 | Scatterplots of skeletal parameters, and correlation analysis between porosity and net calcification rate. Site 1 = blue, Site 2 = green, Site 3 = red. Straight lines represent the best-fit linear regression (mean, solid line), 25% quantile and 75% quantile (dashed lines). (**a-e**) Skeletal parameters (y-axes) plotted against pH$_{TS}$. (**f**) Scatterplot of porosity (P_A) versus net calcification rate in corals from Sites 1 to 3. For **a-d**,**f** $N = 44$; ... 0.05, robust *t*-statistics test.

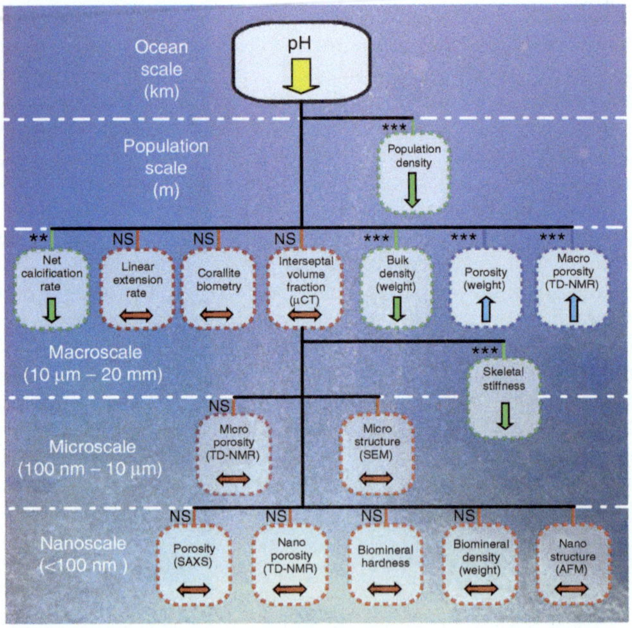

Figure 4 | Summary of responses in *Balanophyllia europaea* skeletal parameters as a function of pH from the ocean to the nanoscale. The significant decline in population density with pH was measured in previously published research[13]. Net calcification rate and bulk density decrease with decreasing pH, whereas porosity (P_A) increases, preserving the linear extension rate and corallite shape (biometry and interseptal volume fraction as measured by μCT do not correlate with pH). The increase in porosity is associated with a decrease in skeletal stiffness. At the nanoscale, the 'building blocks' (the fundamental structural components of the coral skeleton, aragonite fibre bundles and their constituent mineral grains) produced by the biomineralization process are substantially unaffected by increased acidity. Green boxes denote parameters found to have a direct relationship with pH, blue boxes denote parameters that have an inverse relationship and red boxes denote parameters found to have no relationship with pH. The significances of the regression of each parameter (dependent variable) with pH (independent variable) are indicated; ***$P<0.001$; **$P<0.01$; NS indicates no significance. Micro and nanoscale structure observations by SEM and AFM represent qualitative data.

environment, where the net calcification is depressed, enhanced macroporosity keeps linear extension rate constant, potentially meeting functional reproductive needs (for example, the ability to reach critical size at sexual maturity); however, it also reduces the mechanical strength of the skeletons, increasing damage susceptibility, which could result in increased mortality and the observed population density decline[13].

While the results reported here for *B. europaea* may not be representative of the generalized response of all coral species to OA, they are consistent with field observations made on other reef-building scleractinians. For example, while maintaining constant skeletal linear extension, decreased rates of calcification and losses in bulk skeletal density as a function of reduced aragonite saturation have been observed in *Montasraea annularis*[17] and *Porites astreoides*[18]. While low aragonite saturation as a sole driver for the observed reduction in coral calcification has been discussed[19], our conclusions regarding a balance between reduced net calcification rate and increased macroporosity to maintain constant linear extension can explain the outcomes of those studies[17,18]. In fact bulk density depends both on biomineral density and porosity. Our multi-scale analysis shows that all the skeletal features at the nano and microscales, including biomineral density, are unchanged. The decrease of bulk density with decreasing pH depends on the increase of macroporosity, leaving the linear extension rate constant.

Our findings, together with the well-described detrimental effects of heat stress on the scleractinian zooxanthellate coral *B. europaea*[16,20–23], provide several independent and consistent clues regarding the sensitivity of this species to global climate change predicted for the coming decades. Together with results from previous studies[24], we demonstrate that the almost universally employed measure of coral biomineralization, namely the rate of linear extension, might not be a reliable metric for assessing coral health and resilience in a warming and acidifying ocean. Indeed, although the coral's ability to maintain linear extension rate and gross skeletal morphology under conditions of decreasing oceanic pH could allow it to reach sexual maturity, it could reduce skeletal resistance to environmental challenges, affecting the long-term survivability of the species.

Methods

Study site. Off the southwestern coast of Italy, near the island of Panarea, an area delimited by the islets of Dattilo, Bottaro, Lisca Nera and Lisca Bianca (Fig. 1) is characterized by underwater volcanic CO_2 vents. The main vent, a crater measuring $20\,m \times 14\,m$ and $\sim$10-m deep, generates a persistent column of CO_2 bubbles that rise from the seabed to the sea surface. In this hydrothermally stable setting with ambient temperature, CO_2 emissions establish a pH gradient that extends $\sim 34\,m$ from the centre of this crater to its periphery[13]. Three sites along this pH gradient were chosen for study. Distances (d) from the main crater centre and mean pH_{TS} values of the three sites from which corals were collected are: S1 (the control site), $d=34\,m$, $pH_{TS}=8.07$; S2, $d=13\,m$, $pH_{TS}=7.87$; S3, $d=9\,m$, $pH_{TS}=7.74$. Water depth varies from 11.6 m within the crater to 8.8 m at the crater edge (S2) to 9.2 m at the outer margins (S1), where the local pH matches that of the surrounding seawater. The study site has stable hydrothermal–chemical properties[25].

The corals. *B. europaea* (Fig. 1) is a solitary zooxanthellate scleractinian coral endemic to the Mediterranean Sea found at depths ranging from 0 to 50 m[26]. Specimens of *B. europaea* were randomly collected by SCUBA diving at the three study sites along the pH gradient (26 from S1, 26 from S2 and 22 from S3) between November 2010 and May 2013. This sample size was chosen to limit damage on the natural population, which significantly diminishes in the most acidic sites[13], and was considered suitable for properly describing the skeletal properties, as shown in previous studies[13,14]. Biometric data were recorded for the specimens (that is, width-to-length, width-to-height and height-to-length ratios, Supplementary Table 5). Nanoindentation, hydrostatic weight measurement, SEM, AFM, SAXS, μCT and TD-NMR analyses were performed on a subsample of specimens from each site.

Coral sample cleaning. Coral skeletons were submerged in 1% solution of sodium hypochlorite for 3 days to dissolve polyp tissue. After washing with deionized water and drying at 50 °C for 3–4 days, each coral was examined under a binocular microscope to remove fragments of sediment, rock and encrusting organisms[16].

Weight measurements. The buoyant method, usually applied for corals[27–29], was used to measure the total volume occupied by the coral skeleton (called bulk volume, V_B), the bulk density (d_b, ratio of the mass to V_B), the biomineral density (d_r, ratio of the biomineral mass to biomineral volume, excluding pore volume connected to the external surface, also called real density or micro-density) and the apparent porosity (or effective or connected porosity, P_A)[30] (ratio of the pore volume connected to the external surface (V_A) to V_B). This method is based on the Archimedean principle applied to a specimen after full saturation with the same fluid in which it is submerged (water in our case). It is worth to underline that the pore volume (V_A) measured by the buoyant method is only the volume of the pores that can be saturated with water, that is, connected with the external surface. Pores inside the biomineral that are not connected to the external surface (occluded pores) are not measured. An estimate of the occluded porosity gave a negligible maximum value of $\sim$3% (in porosity units) (Supplementary Note 2), which was homogeneous among sites. For coherence with coral literature, for the apparent porosity we use the term porosity.

The skeletons were weighed with a precision balance to determine the dry mass (m) and then placed inside a dryer chamber and evacuated with a rotary mechanical pump down to a vacuum of 10^{-2} mbar. After 6 hours, water was gently introduced to fully saturate the samples. The pump was switched off, the chamber was vented to the ambient atmosphere, and the masses of the fully water-saturated samples (m_s) determined. With a hydrostatic balance, the masses of saturated samples fully immersed in water (m_h) were determined. The skeletal parameters were calculated (Supplementary Table 2) by means of the following operational definitions: $\rho_w=$ water density, $V_A=(m_s-m)/\rho_w$; $V_B=(m_s-m_h)/\rho_w$; $P_A=V_A/V_B=(m_s-m)/(m_s-m_h)$; $d_b=m/V_B$; $d_r=m/(V_B-V_A)$.

After weight measurement, the fully saturated samples were removed from the water container and rapidly placed on wet paper to remove excess water on the external surface. Each specimen was then placed in the bottom of a glass tube, which was then sealed for TD-NMR measurements.

TD-NMR method and parameter definitions. This technique (Supplementary Note 3) was applied to obtain skeletal 'pore-size' distributions by the analysis of quasi-continuous distributions of the transverse relaxation time T_2 (ref. 14) by means of the algorithm UPEN[31], implemented in the UpenWin software[32]. Supplementary Figure 1a shows examples of T_2 distributions for specimens of *B. europaea*. In all cases, the slope of the distribution presented a strong increase at a specific T_2 value, the 'cutoff', revealing a sharp boundary between two distinct pore classes. This point was chosen as the boundary between smaller (shorter T_2) and larger (longer T_2) pores. The ratio of the area under the distribution for T_2 larger than the cutoff to the total area under the distribution, called 'macroscale pore volume fraction' (Supplementary Fig. 1b), indicates the fraction as a percentage of the pore volume in the macroscale. On the basis of the comparison with mercury intrusion porosimetry data, the sizes of pores in the macroscale pore volume fraction, corresponding to the major fraction of pores, are >10–$20\,\mu m$. The tail in the distribution is about 3 orders of magnitude long and should correspond to smaller pore sizes up to tens of nanometres[14]. The distribution with $T_2 < 3$ ms was chosen to identify the signal corresponding to water in the smallest pores or size-scale, which we refer to here as 'nanoscale pore volume fraction'. The remaining fraction of the distribution is called 'microscale pore volume fraction'. The nano, micro and macroscale pore volume fractions multiplied by the skeletal porosity (Supplementary Table 2) produced the TD-NMR nano, micro and macroscale porosities, respectively.

TD-NMR apparatus and data acquisition. A homebuilt relaxometre based on a 0.2 T permanent magnet operating at 8 MHz was used for acquisition of the transverse relaxation curve of the ^1H nuclei of water molecules saturating the cleaned coral skeletons. The relaxometer was equipped with a coil ~ 2 cm in diameter to analyse the entire coral and a Spinmaster console (Stelar, Mede, Pavia, Italy) for automatic pulse sequence transmission and data acquisition. The transverse relaxation data were acquired by using the Carr–Purcell–Meiboom–Gill sequence[33] with a 200 μs echo time. Sixteen corals from S1, 16 corals from S2 and 12 corals from S3 were analysed.

SEM images. To investigate the meso and macroscale organization of the aragonite fibre bundles, SEM analyses were performed. The cleaned coral skeletons were mounted (uncoated) to conductive carbon tape and examined with a Tescan Vega3 GMU (Czech Republic) variable pressure scanning electron microscope. Three corals per site were analysed.

Evaluation of net calcification rate. The age of each coral was determined based on growth band analysis via computerized tomography[22,34]. The linear extension rate was obtained by dividing the length of each sample (maximum axis of the oral disc) by its age. The net calcification rate was calculated as: net calcification rate = bulk density × linear extension rate[21,35] (Supplementary Table 2).

μCT analysis. To compute the corallite interseptal volume fraction (ratio of the total pore space volume inside the corallite to the total volume occupied by the corallite itself), μCT analyses were acquired and the convex hull approach was followed. μCT scans of coral skeletons were performed with a GE phoenix X-ray Nanotom S. The voxel sizes in the scans varied from 2.91 to 13.25 μm depending on the actual size of the investigated coral sample. All image processing steps in the analysis were performed with the open source software FIJI[36]. Supplementary Fig. 3 shows different views of the 3D volume renderings obtained from three corals of the same age from each of the three study sites. Ten corals from each site were analysed.

Nanoindentation. To analyse mechanical properties of the coral skeletons, skeletal stiffness (Young's modulus) and hardness were determined by Oliver–Pharr analysis of load-depth curves[37], using a Nanoindentation Tester, model NHT-TTX by CSM Instruments, Switzerland, equipped with a Berkovich diamond tip (opening angle = 142.3°). On each sample, a minimum of 10 indentation tests were carried out, in the oral region, with a minimum distance of 30 μm between two tests (the total number of tests for each site is reported in Supplementary Table 4). To avoid influence from the embedding resin, a minimum distance of 100 μm from the skeleton/resin boundaries was always maintained.

SAXS analysis. To obtain information on skeletal porosity at the nanoscale, SAXS analyses (Supplementary Note 4) were performed on three corals per site. Sections were mounted onto a laboratory SAXS apparatus with a rotating anode X-ray generator (Bruker, AXS, Karlsruhe, Germany) operating with a copper anode. Light [illegible] of the three specimens are shown in Supplementary Fig. 8

AFM analysis. To study the morphology and packing of skeletal mineral grains at the nanoscale, AFM images were acquired. Samples were lightly polished using diamond paste, cleaned with milli-Q water and observed using a Veeco AFM Dimension 3100 Nanoscope III, Plainview, NY. The probe consisted of a cantilever with an integrated silicon nitride tip. Samples were imaged at room temperature and in air using tapping mode phase contrast imaging.

Statistical analysis. Statistical analyses (analysis of variance, multivariate, principal components and quantile analysis) were performed using the Statistical Package STATA 9.0 (StataCorp LP). To test the significance of the differences among sites, parametric (F) or non-parametric (Kruskal–Wallis χ^2) tests were run. Non-parametric tests were performed for data that did not assume normal distributions. Multivariate regression analyses were performed to investigate the relationships between one dependent and one or several independent variables, using ordinary least squares robust to outliers. The model is described by the equation (1):

$$y_i = a + \sum_{j=1}^{M} b_j x_{i,j} + \varepsilon_i \quad (1)$$

where the index i refers to the i-observation, x_j is an independent variable, y_i is the value of the dependent variable and ε_i is the corresponding error. The constants a, b_j ($j = 1, M$) are the best-fit parameters, to be determined by the best fit. Quantile analysis was performed to study the previous relationships for homogeneous groups of values of the dependent variable. This analysis was used to give a more comprehensive picture of the effect of the independent variable (pH) on the dependent variables of Fig. 3 and Supplementary Fig. 1, as it can show different effects of the independent variable in different ranges of the dependent variable.

References

1. Stocker, T. F. *et al.*in *Climate Change 2013: The Physical Science Basis* (Cambridge Univ. Press, 2013).
2. Sabine, C. L. *et al.* The oceanic sink for anthropogenic CO_2. *Science* **305,** 367–371 (2004).
3. Six, K. D. *et al.* Global warming amplified by reduced sulphur fluxes as a result of ocean acidification. *Nat. Clim. Change* **3,** 975–978 (2013).
4. Rodolfo-Metalpa, R. *et al.* Coral and mollusc resistance to ocean acidification adversely affected by warming. *Nat. Clim. Change* **1,** 308–312 (2011).
5. Hoegh-Guldberg, O. *et al.* Coral reefs under rapid climate change and ocean acidification. *Science* **318,** 1737–1742 (2007).
6. Dove, S. *et al.* Future reef decalcification under a business-as-usual CO_2 emission scenario. *Proc. Natl Acad. Sci. USA* **110,** 15342–15347 (2013).
7. Pandolfi, J. M., Connolly, S. R., Marshall, D. J. & Cohen, A. L. Projecting coral reef futures under global warming and ocean acidification. *Science* **333,** 418–422 (2011).
8. Orr, J. C. *et al.* Anthropogenic ocean acidification over the twenty-first century and its impact on calcifying organisms. *Nature* **437,** 681–686 (2005).
9. Touratier, F. & Goyet, C. Impact of the Eastern Mediterranean transient on the distribution of anthropogenic CO_2 and first estimate of acidification for the Mediterranean Sea. *Deep-Sea Res. I* **58,** 1–15 (2011).
10. Yilmaz, A. *et al.* Impact of acidification on biological, chemical and physical systems in the Mediterranean and Black Sea Mediterranean. *Mediterranean Sci. Committee (CIESM), Monograph Ser.* **36,** 124 (2008).
11. Calvo, E. *et al.* Effects of climate change on Mediterranean marine ecosystems: the case of the Catalan Sea. *Climate Res.* **50,** 1–29 (2011).
12. Ziveri, P. Research turns to acidification and warming in the Mediterranean Sea, IMBER (Integrated Marine Biogeochemistry and Ecosystem Research). http://www.imber.info/index.php/News/Newsletters/Issue-n-20-May-2012 (2012).
13. Goffredo, S. *et al.* Biomineralization control related to population density under ocean acidification. *Nat. Clim. Change* **4,** 593–597 (2014).
14. Fantazzini, P. *et al.* Time-Domain NMR study of Mediterranean scleractinian corals reveals skeletal-porosity sensitivity to environmental changes. *Environ. Sci. Technol.* **47,** 12679–12686 (2013).
15. Laine, J., Labady, M., Albornoz, A. & Yunes, S. Porosities and pore sizes in coralline calcium carbonate. *Mater. Charact.* **59,** 1522–1525 (2008).
16. Caroselli, E. *et al.* Environmental implications of skeletal micro-density and porosity variation in two scleractinian corals. *Zoology* **114,** 255–264 (2011).
17. Carricart-Ganivet, J. P. Sea surface temperature and the growth of the West Atlantic reef-building coral *Montastraea annularis. J. Exp. Marine Biol. Ecol.* **302,** 249–260 (2004).
18. Crook, E. D., Cohen, A. L., Rebolledo-Vieyra, M., Hernandez, L. & Paytan, A. Reduced calcification and lack of acclimatization by coral colonies growing in areas of persistent natural acidification. *Proc. Natl Acad. Sci. USA* **110,** 11044–11049 (2013).
19. Iglesias-Prieto, R., Galindo-Martinez, C. T., Enriquez, S. & Carricart-Ganivet, J. P. Attributing reductions in coral calcification to the saturation state of

aragonite, comments on the effects of persistent naturals acidification. *Proc. Natl Acad. Sci. USA* **111,** E300–E301 (2014).
20. Goffredo, S., Caroselli, E., Mattioli, G., Pignotti, E. & Zaccanti, F. Variation in biometry and demography of solitary corals with environmental factors in the Mediterranean Sea. *Mar. Biol.* **152,** 351–361 (2007).
21. Goffredo, S. *et al.* Inferred level of calcification decreases along an increasing temperature gradient in a Mediterranean endemic coral. *Limnol. Oceanogr.* **54,** 930–937 (2009).
22. Goffredo, S., Caroselli, E., Mattioli, G., Pignotti, E. & Zaccanti, F. Relationships between growth, population structure and sea surface temperature in the temperate solitary coral *Balanophyllia europaea* (Scleractinia, Dendrophylliidae). *Coral Reefs* **27,** 623–632 (2008).
23. Airi, V. *et al.* Reproductive efficiency of a Mediterranean endemic zooxanthellate coral decreases with increasing temperature along a wide latitudinal gradient. *PLoS ONE* **9,** e91792 (2014).
24. Wooldridge, S. A. Assessing coral health and resilience in a warming ocean: Why looks can be deceptive. *Bioessays* **36,** 1041–1049 (2014).
25. Capaccioni, B., Tassi, F., Vaselli, O., Tedesco, D. & Poreda, R. Submarine gas burst at Panarea Island (southern Italy) on 3 November 2002: A magmatic versus hydrothermal episode. *J. Geophys. Res.* **112,** B05201 (2007).
26. Zibrowius, H. Les scleractiniaires de la Méditerranée et del'Atlantique nord-oriental. *Mem. Inst. Oceanogr* **11,** 1–284 (1980).
27. Jokiel, R. L., Maragos, J. E. & Franzisket, L. In: *Coral Reef Research Methods.* (eds Stoddart, D. R. & Johannes, R. E.) **5** 529–542 (UNESCO, 1978).
28. Davies, P. S. Short-term growth measurements of corals using an accurate buoyant weighing technique. *Mar. Biol.* **101,** 389–395 (1989).
29. Walsh, S.-J. in Proceedings of 12th International Coral Reef Symposium, Cairns, Australia, 9–13 July 2012.
30. Manger, G. E. Porosity and bulk density of sedimentary rocks. *Geol. Survey Bull.* **1144-E** (1963).
31. Borgia, G. C., Brown, R. J. S. & Fantazzini, P. Examples of marginal resolution of NMR relaxation peaks using UPEN and diagnostics. *Magn. Reson. Imaging* **19,** 473–475 (2001).
32. Bortolotti, V., Brown, R. J. S. & Fantazzini, P. *UpenWin: a Software to Invert Multi-Exponential Decay Data;* http://wwwunibo.it/PortaleEn/Research/Services + for + companies/UpenWin.htm (2009).
33. Cowan, B. *Nuclear Magnetic Resonance and Relaxation* (Cambridge Univ. Press, 1997).
34. Goffredo, S., Mattioli, G. & Zaccanti, F. Growth and population dynamics model of the Mediterranean solitary coral *Balanophyllia europaea* (Scleractinia Dendrophylliidae). *Coral Reefs* **23,** 433–443 (2004).
35. Lough, J. M. & Barnes, D. J. Environmental controls on growth and of the massive coral *Porites. J. Exp. Mar. Biol. Ecol.* **245,** 225–243 (2000).
36. Schindelin, J. *et al.* Fiji: an open-source platform for biological-image analysis. *Nat. Methods* **9,** 676–682 (2012).
37. Oliver, W. C. & Pharr, G. M. An improved technique for determining hardness and elastic modulus using load and displacement sensing indentation experiments. *J. Mater. Res.* **7,** 1564–1583 (1992).

Acknowledgements

The research leading to these results has received funding from the European Research Council under the European Union's Seventh Framework Programme (FP7/2007–2013)/ERC grant agreement n° 249930—CoralWarm: Corals and global warming: the Mediterranean versus the Red Sea. Bartolo Basile, Francesco Sesso and Eolo Sub diving center assisted in the field. Francesca Gizzi and Giorgia Polimeni helped during preparation and participated in field surveys. The Scientific Diving School collaborated with the underwater activities. We are grateful to Ingrid Zenke (MPI Potsdam) for help with the SAXS measurements, Michela Reggi for sample preparation, Gianni Neto and Francesco Sesso for the images in the Panarea site.

Author contributions

S.G., G.F., P.F., L.P. and Z.D. conceived and designed the research. E.C., F.P., S.G. and B.C. collected the samples and performed the diving field work. L.B., M.M., M.D.G., E.C., F.P., P.K., J.U.H., Y.D., J-P.C., J.C.W., W.W. and G.F. performed the laboratory analyses. P.F., S.M., L.P., V.B., L.B., M.M., E.C., F.P., J.A.K., P.K., J.U.H., Y.D., J-P.C., J.C.W., W.W., P.F., G.F. and S.G. analysed the data. All authors wrote the manuscript and participated in the scientific discussion.

Additional information

Competing financial interests: The authors declare no competing financial interests.

23

Radiocarbon evidence for enhanced respired carbon storage in the Atlantic at the Last Glacial Maximum

E. Freeman[1], L.C. Skinner[1], C. Waelbroeck[2] & D. Hodell[1]

The influence of ocean circulation changes on atmospheric CO_2 hinges primarily on the ability to alter the ocean interior's respired nutrient inventory. Here we investigate the Atlantic overturning circulation at the Last Glacial Maximum and its impact on respired carbon storage using radiocarbon and stable carbon isotope data from the Brazil and Iberian Margins. The data demonstrate the existence of a shallow well-ventilated northern-sourced cell overlying a poorly ventilated, predominantly southern-sourced cell at the Last Glacial Maximum. We also find that organic carbon remineralization rates in the deep Atlantic remained broadly similar to modern, but that ventilation ages in the southern-sourced overturning cell were significantly increased. Respired carbon storage in the deep Atlantic was therefore enhanced during the last glacial period, primarily due to an increase in the residence time of carbon in the deep ocean, rather than an increase in biological carbon export.

[1] Godwin Laboratory for Palaeoclimate Research, Department of Earth Sciences, University of Cambridge, Cambridge CB2 3EQ, UK. [2] LSCE/IPSL, Laboratoire CNRS-CEA-UVSQ, Domaine du CNRS, bât 12 91198 Gif-sur-Yvette, France. Correspondence and requests for materials should be addressed to E.F.

Numerous proxies suggest that ocean circulation at the Last Glacial Maximum (LGM) was different from modern, especially in the Atlantic Ocean[1–4]. Despite this, a consensus has yet to emerge on the LGM ocean circulation[5], primarily because of limitations associated with the various proxies, as well as a paucity of data. In agreement with stable oxygen isotope data[6], $\delta^{13}C$ (stable carbon isotopic ratio) and Cd/Ca data suggest that North Atlantic Deep Water (NADW) shoaled by up to 1,000 m during the LGM relative to today[3,4,7]. However, the combination of conservative and non-conservative impacts on nutrient proxies has permitted alternative interpretations, such as greater nutrient accumulation along the flow path[8] (because of greater export production or longer transit times), or variable contributions from northern- and southern-sourced waters between 2 and 4 km water depth[9] (without a complete elimination of NADW in the deep Atlantic). Although $^{231}Pa/^{230}Th$ has been used to reconstruct past rates of overturning circulation[10,11], its interpretation presents difficulties, as it represents an integrated flow speed across the water column and may be affected by scavenging[12].

Additional proxy data are therefore needed to reconstruct the pattern and rate of ocean circulation at the LGM, and to assess its impact on the marine carbon cycle. Radiocarbon-based ventilation ages could prove particularly useful in this regard as they provide an indication of the mean timescale for carbon exchange between the ocean and the atmosphere, including the effects of air–sea gas (and isotope) exchange in the regions of origin of a given parcel of water, and the mean time elapsed since that water was in the mixed layer. Furthermore, by combining ventilation ages with indicators of respired carbon accumulation, such as stable carbon isotopes, changes in the efficiency of the biological carbon pump can be estimated. This is not possible using reconstructions of flow rates or respired/total nutrient levels alone, as the efficiency of the biological carbon pump is controlled by both the export productivity (that is, the flux of carbon into the deep ocean) and the 'leakage' of this carbon out of the deep ocean (via all the processes that contribute to mass transfer from the deep ocean interior to the surface ocean mixed layer where air–sea exchange occurs).

Here we investigate the Atlantic Ocean overturning circulation at the LGM using both radiocarbon and stable carbon isotopes. We find a shallow well-ventilated northern-sourced cell overlying a poorly ventilated, predominantly southern-sourced cell. Furthermore, we find that the storage of respired carbon in the deep Atlantic Ocean was enhanced because of a long residence time, with a constant flux of respired carbon to the deep Atlantic.

Results

Radiocarbon and stable carbon isotopes. Here we present new radiocarbon-based ventilation ages along with stable carbon isotopes from 1,000 to 3,500 m on the Brazil Margin and from 1,100 to 4,700 m on the Iberian Margin (Supplementary Table 1). Ventilation ages are based on benthic-planktonic (B-P) radiocarbon age offsets, to which the estimated surface reservoir age is added to provide an estimate of the deep ocean versus atmospheric radiocarbon age offset (B-Atm). Given the proximity of cores in each of our study regions, surface reservoir ages are expected to be very similar for all cores in each transect. Therefore, despite some uncertainty in the surface reservoir ages, vertical gradients in ventilation ages within each transect will remain accurate.

Radiocarbon-based ventilation ages provide an indication of the extent of isotopic disequilibrium between the marine- and atmospheric carbon reservoirs. More radiocarbon-depleted waters (indicated by higher radiocarbon ventilation ages) will reflect less efficient radiocarbon (and therefore CO_2) exchange between the ocean interior and the atmosphere on average, and therefore a longer carbon sequestration time in the ocean interior. For continuous export of biologically fixed carbon to the ocean interior, a greater radiocarbon ventilation age and a longer carbon sequestration time (that is, relative to the atmosphere) should in turn reflect enhanced accumulation of respired carbon in the ocean interior and a more efficient (since less 'leaky') soft-tissue carbon pump[13].

Ventilation of the Brazil Margin. On the Brazil Margin, the radiocarbon ventilation age-depth profile for the LGM is significantly different from that of the modern ocean (Fig. 1). Although the shallowest core (1,000 m), has a similar ventilation age to that of the modern ocean, the deeper cores have considerably higher ventilation ages, with the deepest core being offset from the surface ocean by up to 1,700 years, compared with just 400 years today.

Stable carbon isotopes also had a distinctly different profile at the LGM compared with the modern. $\delta^{13}C$ shows a significant decrease with depth, ranging from ~1.1‰ at 1,000 m to ~0.2‰ in the deep ocean. The steepest gradient (~ −0.5‰/1,000 m) is seen between 2,300 and 3,550 m, whereas in the modern ocean the gradient is <0.02‰/1,000 m between these depths.

Ventilation of the Iberian Margin. The Iberian Margin also shows a very different ventilation age depth-profile at the LGM compared with the modern ocean (Fig. 1). Although the intermediate ocean (<2,100 m) was well ventilated, the deep ocean was poorly ventilated, with B-P age offsets more than 1,000 years greater than in the modern ocean. The shallowest core (1,127 m) has a negative B-P age offset suggesting that the surface was more poorly ventilated than the intermediate ocean (likely influenced by Mediterranean Outflow Water[14]). Beneath this, the B-P age offset increases with depth in the water column. A relatively steep gradient (~950 years/1,000 m) is seen for the upper 2,600 m, with a smaller increase in B-P age offset with depth below (~200 years/1,000 m). A similar profile is also seen in $\delta^{13}C$ with a steep gradient (~ −0.4‰/1,000 m) between 2 and 3.7 km but with relatively little change (< −0.1‰/1,000 m) with depth below ~3,700 m. The range of $\delta^{13}C$ (0–0.8‰) at the LGM is significantly greater than the range in the modern North Atlantic (1.0–1.1‰) between 2 and 4.7 km. $\delta^{13}C$ at 4.7 km are more than 0.9‰ lower at the LGM than in the modern ocean. However, extremely low $\delta^{13}C$, previously observed in both the deep South Atlantic (< −0.8‰)[15,16] and in the abyssal western North Atlantic (−0.3 to −0.5‰)[17,18], are not observed in any of the Iberian Margin cores.

Discussion

We combined our new ventilation ages with available published data in order to tentatively reconstruct the distribution of radiocarbon in the Atlantic Ocean during the LGM[18–25] (Fig. 2 and Supplementary Table 2). These radiocarbon-based ventilation ages show that, regardless of the uncertainties in LGM surface reservoir ages, the entire deep Atlantic Ocean (>2 km) was less ventilated during the LGM than the modern ocean (Fig. 3). Today, the deep North Atlantic is well ventilated by newly formed NADW, whereas the deep South Atlantic is primarily influenced by southern-sourced waters (Lower Circumpolar Deep Water, LCDW, and Antarctic Bottom Water, AABW), which have greater ventilation ages because of reduced air–sea gas exchange in the AABW formation regions[26] and because of mixing and entrainment of 'pre-aged' circumpolar deep water into AABW[27]. During the LGM this North–South

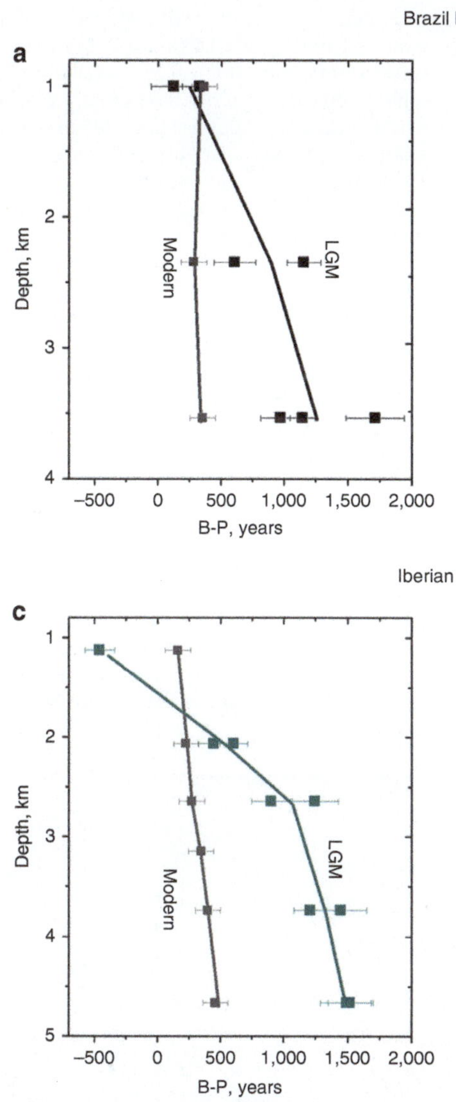

Figure 1 | LGM depth profiles for the Brazil Margin and the Iberian Margin. The benthic-planktonic radiocarbon age offset (B-P; **a**, **c**) and the stable carbon isotopic ratio, $\delta^{13}C$ (**b**, **d**). Modern B-P values (calculated by subtracting the modern surface reservoir age from the bottom water ventilation age) are show in grey (GLODAP[38]). Modern $\delta^{13}C$ of DIC is also shown in grey (GEOSECS[38]). Data are given in Supplementary Table 3.

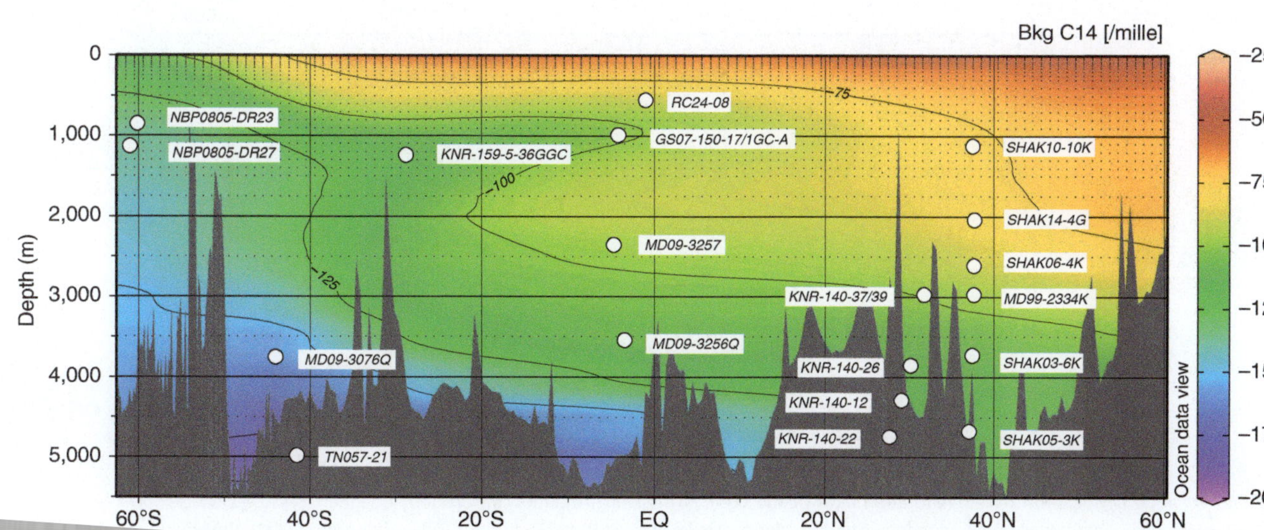

...ations with modern background $\Delta^{14}C$ (GLODAP[38]), using Ocean Data View[71].

ventilation age gradient in the deep Atlantic was significantly reduced or even eradicated (in the abyss), suggesting a distinctly different circulation must have prevailed (Figs 3 and 4). Based on the apparent northward expansion of poorly ventilated deep waters from the glacial South Atlantic to the deep equatorial and North Atlantic, the LGM circulation appears to have been characterized primarily by a greatly reduced influence of NADW in the deepest Atlantic (>2,000 m). It is notable that abyssal waters in the western North Atlantic and on the northern Brazil Margin had a similar ventilation age to that of the abyssal South Atlantic (Fig. 3). The lack of an increase in ventilation age along the flow path between the South Atlantic and the equatorial and North Atlantic suggests that neither the deep eastern Atlantic nor the deep western Atlantic contained pure southern-sourced waters. A persistent, albeit greatly reduced, northern source of radiocarbon was therefore provided to depths of at least 4.7 km, either via mixing with an overlying well-ventilated water mass or via direct episodic input of dense well-ventilated (for example, overflow) waters.

In the modern ocean, radiocarbon-based ventilation ages are well correlated with DIC and anti-correlated with $\delta^{13}C$ owing to the remineralization of organic carbon as waters flow from the North Atlantic to the North Pacific, via the Southern Ocean (Fig. 5). DIC thus increases in the modern ocean by $\sim 115\,\mu mol\,kg^{-1}$ (ref. 13), whereas the $\delta^{13}C$ of DIC decreases by ~0.8‰ (Fig. 5), for every 10% decrease in the deep ocean's radiocarbon content relative to the atmosphere (that is, for every ~850 ^{14}Cyr ventilation age increase). A similar gradient in $\delta^{13}C$ versus ventilation age is seen in the North and equatorial Atlantic

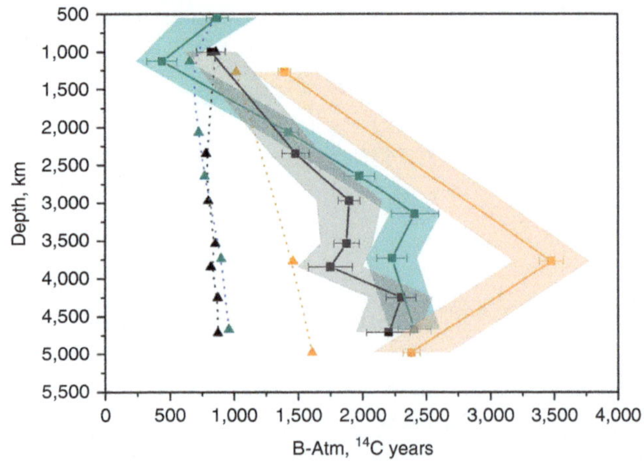

Figure 3 | Atlantic radiocarbon-based ventilation age depth profiles. Triangles: modern ventilation ages (dotted lines). Squares: LGM ventilation ages (solid lines)[18-24,34]. Error bars indicate the analytical error on the radiocarbon measurements. The shaded area represents the uncertainty of the surface reservoir age. Blue, East Atlantic; black, West Atlantic; orange, South Atlantic.

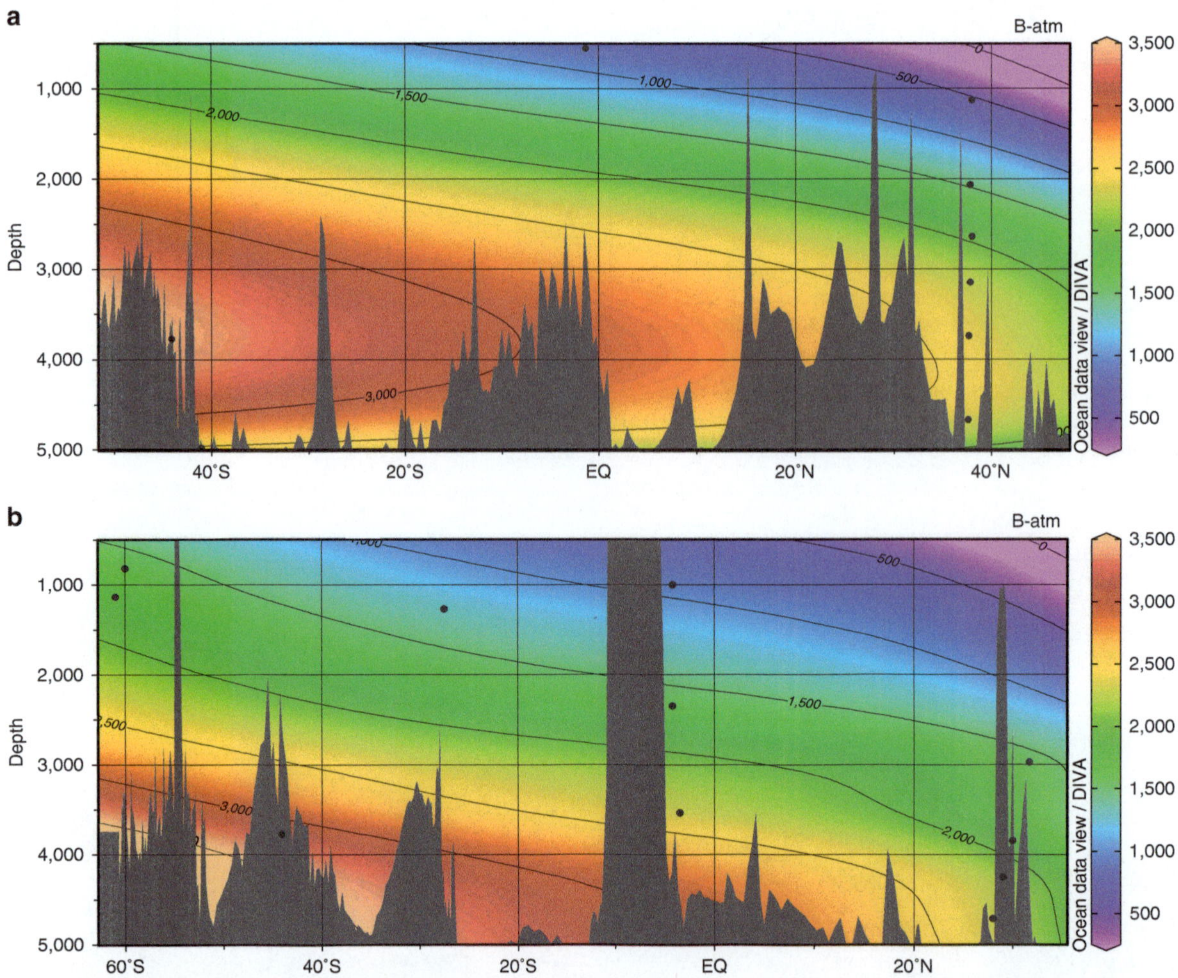

Figure 4 | Atlantic radiocarbon based ventilation ages at the LGM. (**a**) Eastern Atlantic. (**b**) Western Atlantic[18-25,34]. MD07-3076Q (44°S, 14°W, 3770 m) is included in both the eastern and the western profiles as it is located on the Mid-Atlantic Ridge. The dots indicate the position of the cores where ventilation ages have been determined. Gridded in ODV[71] using DIVA gridding with a signal-to-noise ratio of 72. Data are given in Supplementary Table 4.

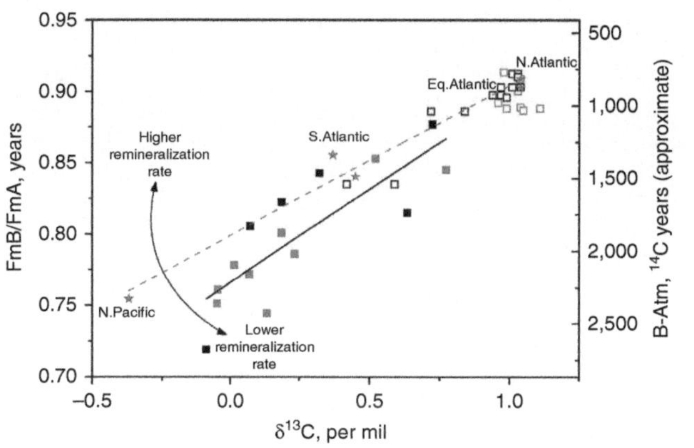

Figure 5 | Crossplot of ventilation age and stable carbon isotopic ratio. Ventilation ages are plotted as FmB/FmA, where FmB/FmA = exp(B-Atm/-8033), and Fm = fraction modern, because B-Atm ages do not mix linearly due to the exponential decay of radiocarbon. Approximate B-Atm values are shown as a guideline only. Modern whole ocean data ($\sim$2.5 km water depth; grey stars) is show along with modern data for the entire water column >2 km (open squares) on the Iberian Margin (blue) and Brazil Margin (black). LGM data are shown (filled squares) for the Iberian Margin (blue) and the Brazil Margin (black; Modern data are from GEOSECS and GLODAP[38]). The modern gradient is 0.11 ± 0.01 and the LGM gradient is 0.13 ± 0.02, equivalent to 1.1 ± 0.1‰/kyr and 0.9 ± 0.2‰/kyr in the modern and in the LGM, respectively. For a given rate of remineralization, an increase in residence time (that is, lower FmB/FmA) results in a greater amount of carbon storage.

at the LGM, suggesting that the remineralization rate of organic carbon in the deep Atlantic did not change significantly. On this basis, the LGM rate of DIC increase per year of radiocarbon decay (because of the soft-tissue pump alone) would also be expected to have remained roughly constant, as hypothesized in ref. 13. All else being equal, the ventilation age increase that we observe across the deep Atlantic would therefore imply a significant increase in the respired carbon inventory of the deep ocean at the LGM. This increase was proposed in ref. 8 but later questioned in ref. 9 because of an updated LGM state estimate. Here we not only propose an increase in the remineralized carbon inventory of the deep ocean (based on stable carbon isotope evidence, similar to that advanced previously), but also attribute it to an increase in residence time rather than an increase in biological carbon export rates.

Our results demonstrate a distinct difference in the distribution of radiocarbon in the Atlantic at the LGM relative to today. The deep ocean was more poorly ventilated at all latitudes, whereas the intermediate ocean remained as well ventilated as in the modern ocean. This strongly suggests a shoaling of northern-sourced waters, without a major reduction in their overturning strength, and an expansion of southern-sourced waters at depth. This is in agreement with other proxy reconstructions, which our radiocarbon data help to link more directly to carbon cycle impacts. The concentration of dissolved cadmium inferred from Cd/Ca measured on benthic foraminifera show a clear decrease in the upper $\sim$2.2 km of the water column at the LGM relative to today, with the opposite sense of change in the deep ocean[4] (Supplementary Fig. 1a). This would suggest that high-nutrient waters expanded at depth while low-nutrient northern-sourced waters dominated in the intermediate ocean, consistent with $\delta^{13}C$ data that also show the opposite sense of change above and below $\sim$2.2 km (ref. 4; Supplementary Fig. 1b). In principle, this change in nutrient distributions could have been driven by enhanced ... residence times.

As indicated above, our combined radiocarbon and stable isotope data indicate that the latter mechanism was dominant. Our interpretation that this change in circulation was primarily associated with an expansion of southern-sourced waters in the deepest Atlantic is further supported by εNd data, which indicate an increase in southern-sourced waters at depth with an increase in northern-sourced waters above[28] (Supplementary Fig. 1c). $^{231}Pa/^{230}Th$ ratios also display a distinct change of the opposite sign above and below $\sim$2.2 km water depth[10] (Supplementary Fig. 1d). These $^{231}Pa/^{230}Th$ ratios suggest that the North Atlantic overturning circulation cell was just as strong during the LGM as it is today, but that it penetrated to shallower depths. Although this contrasts with many numerical modelling studies that tend to find a slower overturning when NADW shoals[29,30], our ventilation ages also suggest that the upper North Atlantic overturning rate was not reduced so much that it significantly decreased the supply radiocarbon to the intermediate ocean. The radiocarbon data are therefore consistent with the proposal that a significant shoaling of NADW may not necessarily imply a major reduction in the NADW formation/transport rate at the LGM.

However, while our results support the existence at the LGM of a shallow NADW cell overlying an expanded, radiocarbon-depleted AABW below, they also suggest that the deep southern-sourced cell was not completely isolated. In both the eastern and the western North Atlantic, a northern source of relatively radiocarbon-enriched and high $\delta^{13}C$ waters was mixed into the deep cell. This could imply a more complex Atlantic circulation than one characterized by two completely separate overturning cells at the LGM[31], where the northern overturning cell is either strong and deep or weak and shallow.

As in the landmark study of ref. 3, our stable isotope data demonstrate the existence of enhanced gradients in $\delta^{13}C$ between $\sim$2,000 and 3,500 m in the equatorial and North Atlantic. This pattern of $\delta^{13}C$ in the Atlantic has been interpreted primarily to demonstrate a reduction in the proportion of northern-sourced waters with depth, and an increase in the total nutrient content of the deep Atlantic. Our new radiocarbon data demonstrate that this did not primarily reflect an increase in 'preformed' nutrients (that is, nutrients that were advected to the ocean interior, and that would be associated with carbon that was well-equilibrated with the atmosphere[32]), but rather an increase in respired nutrients (and therefore respired carbon). Alternatively, and equivalently for ocean versus atmosphere carbon inventory changes, the radiocarbon data might also indicate an increase in the 'disequilibrium' carbon inventory of the deep Atlantic[33]. However, in this case, a very large decrease in air–sea CO_2 exchange efficiency would need to be invoked in parallel with an increase in preformed nutrients, in order to account for the paired stable carbon isotope (and nutrient) reconstructions. The observed expansion of southern-sourced deep waters at the LGM was therefore very likely associated with an increase in the efficiency of the soft-tissue carbon pump, primarily due to an increase in the residence time of carbon in the deep Atlantic, rather than an increase in biological export productivity (that is, at low/mid-latitudes). Although global data coverage and better constraints on surface reservoir ages are required to assess the precise magnitude of the atmospheric CO_2 drawdown that is implied by the observed ocean circulation changes, our findings support a direct contribution from the Atlantic overturning circulation via its impact on the deep ocean's respired carbon inventory.

Methods

Radiocarbon and stable carbon isotope measurements. Radiocarbon ages were measured on benthic and planktonic foraminifera from a series of cores recovered from the Brazil Margin by the R/V *Marion Dufresne* (MD09-3256Q, MD09-3257) and the Iberian Margin by the R/V *James Cook* (SHAK03-6K, SHAK05-3K, SHAK06-4K, SHAK10-10K and SHAK14-4G; Fig. 2). Results from these cores are combined with previously published data from the same transect regions (MD99-2334K and GS07-150-17/1GC-A)[22,34]. Core locations are given in Supplementary Table 1. Benthic stable carbon isotopes were also measured in these cores on *Cibicidoides wuellerstorfi*. No stable carbon isotopes were measured for SHAK10-10K because of the lack of any specimens of *Cibicidoides*.

Foraminifera were picked from the $>212\,\mu m$ size fraction and where necessary from the 150–212 μm fraction. Samples of *Globigerinoides ruber* (Brazil Margin) or *Globigerina bulloides* (Iberian Margin) and samples of mixed benthic foraminifera (excluding agglutinated species) were picked and graphitized in the Godwin Radiocarbon Laboratory at the University of Cambridge. Stable carbon isotope measurements were conducted on *Cibicidoides wuellerstorfi* and on DIC at the Godwin Laboratory, Cambridge.

Radiocarbon measurements. Samples were cleaned on a glass plate with deionized water to remove any loose material. Samples were then dried and acidified. The CO_2 produced was converted to graphite using a standard hydrogen/iron catalyst reduction method[35]. AMS-^{14}C dates were obtained for the graphite samples at the ^{14}Chrono Centre, Queens University Belfast. All dates are reported as conventional radiocarbon ages following[36].

Stable carbon isotope measurements on foraminifera. Stable carbon isotope measurements were conducted on *Cibicidoides wuellerstorfi* in the Godwin Laboratory. Each measurement was run on 1–3 individuals with a combined mass of 50–180 μg. The foraminifera were transferred into sample vials, crushed and then dried in an oven at 50 °C. Samples were reacted with orthophosphoric acid (100%) and the CO_2 produced was cryogenically dried and then admitted to the dual inlet mass spectrometer for isotopic analysis by comparison with a reference gas. Each run of 30 samples was accompanied by 10 reference carbonates and 2 control samples. The results are reported with reference to the international standard VPDB and the precision is better than $\pm 0.06‰$ for $^{12}C/^{13}C$ and $\pm 0.08‰$ for $^{16}O/^{18}O$.

Stable carbon isotope DIC measurements. DIC measurements were made using a Thermo Gas Bench attached to a Delta V Mass Spectrometer. Three or four drops of orthophosphoric acid (100%) were preloaded into a reaction vial, which was capped, sealed and the headspace flushed with Helium gas. Approximately 1.5 ml of sample water was injected into the vial through the butyl rubber septa using a syringe and left to react for 1 h. The sample tubes were transferred to the Gas Bench and CTC CombiPal Autosampler and the resulting CO_2 in the headspace analysed using a Thermo Delta V Mass Spectrometer. A series of standards and reference samples distributed throughout the run were used to calibrate to the international standard VPDB. Results have a reproducibility of better than 0.1 per mille.

Ventilation ages. Here we report ventilation ages as either deep to shallow sub-surface radiocarbon age offsets, based on benthic-planktonic radiocarbon age differences (B-P), or as deep-atmospheric radiocarbon age offsets (B-Atm), which represent the B-P plus the surface reservoir age (that is, the shallow sub-surface versus atmosphere radiocarbon disequilibrium) that applies to the planktonic radiocarbon age. Although B-P offsets provide useful (and very accurate) information regarding vertical radiocarbon gradients at various locations in the ocean, B-Atm estimates are advantageous by virtue of referencing radiocarbon activities throughout the ocean interior relative to a single reference point: the radiocarbon activity of the atmosphere. They also maintain a constant scaling with respect to radiocarbon disequilibria, unlike relative isotopic offset metrics such as $\Delta^{14}C$ (ref. 37). When mapped throughout the ocean, B-Atm offsets may thus provide a coherent means of inferring the patterns and rates of ocean-atmosphere CO_2 exchange and transport in the ocean interior[38]. However, the uncertainty associated with LGM surface reservoir age estimates makes the derivation of accurate B-Atm offsets challenging.

Surface reservoir ages at the LGM. To accurately reconstruct the distribution of radiocarbon in the Atlantic Ocean at the LGM, benthic radiocarbon activities must be referenced to a single (atmospheric) reference point. Therefore, we need to not only consider the deep- to surface age differences but also the surface to atmosphere age offsets that apply in each instance (alternatively each benthic radiocarbon date can be directly compared with the contemporary atmospheric radiocarbon age, given independent calendar ages). The planktonic foraminifera used in this study calcify in the upper surface ocean (<100 m deep), but even at these depths the water can be in significant disequilibrium relative to the atmosphere. Today surface reservoir ages range from around 400 to 600 years on the Brazil and Iberian Margins (GLODAP[38]). However, these would have varied in the past due to changes in pCO_2 (affecting air–sea carbon isotope exchange at a given CO_2 solubility), changes in ocean circulation (influencing the mixing/upwelling of 'aged' waters into the surface ocean mixed layer) and, for example, changes in high latitude temperature, salinity and sea-ice cover (affecting CO_2 solubility and gas exchange in sub- mixed-layer source regions). However, determining surface reservoir ages in the past is challenging, and not possible in contexts where independent calendar age or contemporaneous atmospheric radiocarbon age constraints are lacking. The modern surface reservoir age is therefore often used as a best guess, under the tentative assumption that physical conditions affecting radiocarbon exchange and transport have not changed over time[24,39,40].

To accurately determine surface reservoir ages, a calendar age model that is independent of radiocarbon dating and a history of atmospheric radiocarbon variability are required. Although U/Th dating can be used to provide calendar ages for corals[43], chronostratigraphic (including tephrochronological) approaches must be used for sediment cores. In high latitudes, cores can be linked stratigraphically to independently dated ice cores[21,22,41]. However, for low latitude, cores in which chronostratigraphic signals are typically more subdued and synchrony with high-latitude climate changes may be questionable, it is much more difficult to obtain an independent calendar chronostratigraphy. Yet, it is important to note that even at low latitudes, especially in regions with deep mixed layers or upwelling regimes, shallow sub-surface radiocarbon disequilibria are very likely to have changed across the last deglaciation[42,43].

Although it is possible in principle to use model simulations to infer past surface reservoir ages, this approach requires knowledge of past changes in ocean circulation and ocean–atmosphere gas exchange efficiency (for example, because of changes in sea-ice extent or mixed layer depths), which strictly we do not have (indeed, this is typically what we are seeking). Despite this fundamental limitation, efforts have been made to model surface reservoir ages in the past, for example, assuming a constant ocean circulation[44,45]. These studies have shown that at the LGM the pCO_2 difference alone would have caused surface reservoir ages to be around 250 years higher than under modern conditions—a passive response in the ocean that does not reflect the ocean's impact on atmospheric CO_2 but rather the atmosphere's impact on the ocean. Additional changes in the mean exchange rate of CO_2 between the ocean and the atmosphere, because of, for example, increased sea-ice cover and changes in the large-scale overturning circulation, are likely to have further increased these values, especially at high latitudes. Indeed, where shallow sub-surface reservoir age estimates are available for the LGM, they are typically significantly higher than the modern values[21,22,25,42,46–48].

Despite the challenges involved in estimating past shallow sub-surface reservoir age variability, it is clear that ignoring them entirely is not a viable option, especially in a study that seeks to constrain past ocean circulation changes.

Recent work on core MD09-3257, from the Brazil Margin, showed that an independent calendar age model could be established for that core using U-Th dated speleothem records on the adjacent continent[49], based on the fact that increased precipitation is marked by a decrease in $\delta^{18}O$ in South American speleothems[50,51] and increased sedimentary Ti/Ca ratios in marine sediment cores off the northeastern Brazilian coast[52]. However, because of a lack of available calendar age tie points for the LGM, surface reservoir ages based on this calendar age-scale remain sparse. We therefore adopt an estimate for LGM surface reservoir ages on the Brazil Margin based on the modern reservoir age corrected for pCO_2-dependent air–sea gas exchange effects[44,45]. A reservoir age of 750 years (250 years greater than modern), which would have arisen due to globally reduced air–sea gas exchange rates, is used. Because it is unclear how ocean circulation changes may have affected this region we apply large uncertainties to this estimate of ± 250 years. By combining high-resolution radiocarbon dating with calendar age models of centennial precision, it may be possible to reduce this uncertainty in the future.

Because the Iberian Margin cores contain a very clear event stratigraphy[53,54], these cores were stratigraphically aligned to the uranium-series dated speleothem records from Hulu Cave[55] and the layer counted (GICC05) NGRIP dust record[56,57] using the Zr/Sr ratio determined using X-ray fluorescence (XRF; see Supplementary Figs 2 and 3). The cores were aligned to both the Hulu speleothem and NGRIP simultaneously using a series of tie points including two tie-points within the LGM (18–23 1000 years before present (kyrs BP)). Shallow sub-surface reservoir ages were determined directly by subtracting the planktonic age from the contemporaneous atmospheric age, using the calendar ages obtained at tie-points, and the atmospheric radiocarbon calibration curve IntCal13 (ref. 58; Supplementary Table 3). LGM surface reservoir ages on the Iberian Margin are thus estimated to be around 900 years with upper and lower limits of 1,100 years and 700 years, respectively. These ages are consistent with the lowest estimates determined for the Iberian Margin by ref. 22 (that is, excluding their tie-points in HS1 and the LGM where the event stratigraphy is arguably more equivocal).

Age models. The planktonic radiocarbon ages were used to construct age models for each of the cores. In order to do so, an estimated surface reservoir age was subtracted from each planktonic radiocarbon date before conversion to calendar ages using BChron version 3.1.5 (ref. 59) and the IntCal13 calibration curve[58]. These calendar ages were then used to construct sediment depth-age models using the Markov Chain Monte-Carlo method, also using Bchron. As LGM surface reservoir age estimates have not been directly estimated on the Brazil Margin, a 'best guess' value of 750 years, based on isotope exchange constraints backed up by model simulations[44,45], was used. For the Iberian Margin, a surface reservoir age of 900 years was used for the LGM based on the stratigraphic alignments described above.

For compiled data, for cores with age models that do not depend on radiocarbon data, surface and deep reservoir ages were determined by subtracting the contemporary atmospheric radiocarbon age (based on Intcal13) from the planktonic and benthic radiocarbon ages, respectively. For sites with no independent age model, the pCO_2 corrected modern reservoir ages (that is, modern plus 250 years at the LGM) were used and added to the benthic-planktonic age offset to determine the deep ocean-atmospheric age offset (Supplementary Table 2).

Stratigraphic alignments. The cores were stratigraphically aligned to the Hulu speleothem $\delta^{18}O$ record[55] and the NGRIP dust record[57] (on the GICC05 age model[56,60]). The NGRIP dust content changes rapidly and synchronously with changes in ice core $\delta^{18}O$ (ref. 61). NGRIP dust, which is predominantly sourced from East Asian deserts[62–65], increases during cold periods[57]. These increases are caused by atmospheric changes[66,67] and/or intensified sources[68], but the exact mechanism is unconstrained. The Hulu speleothem $\delta^{18}O$ record resembles East Asian monsoon changes and is highly correlated with millennial scale Greenland temperature variations[55,69].

The sediment cores were aligned using elemental ratios determined using high-resolution XRF. Zr/Sr ratios were used to align the Iberian Margin cores. Zr/Sr anti-correlates strongly with Ca/Ti in these cores. The Zr/Sr and Ca/Ti ratios reflect the relative input of biogenic (Ca, Sr) and detrital (Zr, Ti) material, and have been shown to correlate strongly with planktonic $\delta^{18}O$ and alkenone sea-surface temperatures on the Iberian Margin, as well as with millennial scale variations in Greenland ice core $\delta^{18}O$ (ref. 70). Zr/Sr values are generally higher during stadials and lower during interstadials, whereas Ca/Ti ratios show the opposite trend.

A master core, SHAK03-6K was selected for the Iberian Margin through which the other cores were aligned. SHAK03-6K was chosen as the master core based on its high quality (no discontinuities) and age range (SHAK14-4G was also of very high quality but only reached back to ~20 kyrs). SHAK03-6K was aligned to both the Hulu speleothem and the NGRIP dust record using a series of tie points (Supplementary Fig. 2). No tie-points were used in the interval 0–11.4 kyrs BP because of the lack of clear signals in the records. The calendar ages at tie points in SHAK03-6K were then transferred to all the other Iberian Margin cores (Supplementary Fig. 3).

XRF data. The Iberian Margin cores were scanned at the University of Cambridge using an Avaatech XRF core scanner (2nd generation). The surface of the cores was scraped clean then covered with a 4-μm SPEXCertiPrep Ultralene foil to avoid contamination and to prevent the cores drying out and cracking. Each section was measured at three different voltages and currents: 10 kV and 750 μA, 30 kV and 500 μA, and at 50 kV and 1,000 μA. The entire length of each core was analysed at 5-mm resolution with an irradiated surface length and width of 5 mm (downcore) and 12 mm (cross core), respectively. The count time was 60 s for each measurement. Element intensities were obtained by post-processing of the XRF spectra using the Canberra WinAxil software with standard software settings and spectrum-fit models.

References

1. Boyle, E. A. & Keigwin, L. D. Deep Circulation of the North Atlantic over the Last 200,000 Years: geochemical Evidence. *Science* **218,** 784–787 (1982).
2. Duplessy, J. C. *et al.* Deep water source variations during the last climatic cycle and their impact on the global deep water circulation. *Paleoceanography* **3,** 343–360 (1988).
3. Curry, W. & Oppo, D. Glacial water mass geometry and the distribution of $\delta^{13}C$ of Sigma CO_2 in the western Atlantic Ocean. *Paleoceanography* **20,** PA1017 (2005).
4. Marchitto, T. M. & Broecker, W. S. Deep water mass geometry in the glacial Atlantic Ocean: a review of constraints from the paleonutrient proxy Cd/Ca. *Geochem. Geophys. Geosystems* **7,** Q12003 (2006).
5. Lynch-Stieglitz, J. *et al.* Atlantic meridional overturning circulation during the Last Glacial Maximum. *Science* **316,** 66–69 (2007).
6. Lund, D. C., Mix, A. C. & Southon, J. Increased ventilation age of the deep northeast Pacific Ocean during the last deglaciation. *Nat. Geosci.* **4,** 771–774 (2011).
7. Oppo, D. W. & Lehman, S. J. Mid-depth circulation of the subpolar North Atlantic during the Last Glacial Maximum. *Science* **259,** 1148–1152 (1993).
8. Gebbie, G. How much did Glacial North Atlantic Water shoal? *Paleoceanography* **29,** 190–209 (2014).
9. Gebbie, G., Peterson, C. D., Lisiecki, L. E. & Spero, H. J. Global-mean marine ^{13}C and its uncertainty in a glacial state estimate. *Quat. Sci. Rev.* **125,** 144–159 (2015).
10. ... geometry of the glacial Atlantic Meridional
11. McManus, J. F., Francois, R., Gherardi, J.-M., Keigwin, L. D. & Brown-Leger, S. Collapse and rapid resumption of Atlantic meridional circulation linked to deglacial climate changes. *Nature* **428,** 834–837 (2004).
12. Keigwin, L. D. & Boyle, E. A. Did North Atlantic overturning halt 17,000 years ago? *Paleoceanography* **23,** PA1101 (2008).
13. Sarnthein, M., Schneider, B. & Grootes, P. M. Peak glacial ^{14}C ventilation ages suggest major draw-down of carbon into the abyssal ocean. *Clim. Past* **9,** 2595–2614 (2013).
14. Zahn, R., Sarnthein, M. & Erlenkeuser, H. Benthic isotope evidence for changes of the Mediterranean outflow during the Late Quaternary. *Paleoceanography* **2,** 543–559 (1987).
15. Charles, C. D. & Fairbanks, R. G. Evidence from Southern Ocean sediments for the effect of North Atlantic deep-water flux on climate. *Nature* **355,** 416–416 (1992).
16. Ninnemann, U. S. & Charles, C. D. Changes in the mode of Southern Ocean circulation over the last glacial cycle revealed by foraminiferal stable isotopic variability. *Earth Planet. Sci. Lett.* **201,** 383–396 (2002).
17. Boyle, E. & Keigwin, L. North Atlantic thermohaline circulation during the past 20,000 years linked to high-latitude surface temperature. *Nature* **330,** 35–40 (1987).
18. Keigwin, L. D. Radiocarbon and stable isotope constraints on Last Glacial Maximum and Younger Dryas ventilation in the western North Atlantic. *Paleoceanography* **19,** PA4012 (2004).
19. Cléroux, C., Demenocal, P. & Guilderson, T. Deglacial radiocarbon history of tropical Atlantic thermocline waters: absence of CO_2 reservoir purging signal. *Quat. Sci. Rev.* **30,** 1875–1882 (2011).
20. Barker, S., Knorr, G., Vautravers, M. J., Diz, P. & Skinner, L. C. Extreme deepening of the Atlantic overturning circulation during deglaciation. *Nat. Geosci.* **3,** 567–571 (2010).
21. Skinner, L. C., Fallon, S., Waelbroeck, C., Michel, E. & Barker, S. Ventilation of the deep southern ocean and deglacial CO_2 rise. *Science* **328,** 1147–1151 (2010).
22. Skinner, L. C., Waelbroeck, C., Scrivner, A. E. & Fallon, S. J. Radiocarbon evidence for alternating northern and southern sources of ventilation of the deep Atlantic carbon pool during the last deglaciation. *Proc. Natl Acad. Sci. USA* **111,** 5480 (2014).
23. Sortor, R. N. & Lund, D. C. No evidence for a deglacial intermediate water [DELTA]^{14}C anomaly in the SW Atlantic. *Earth Planet. Sci. Lett.* **310,** 65 (2011).
24. Keigwin, L. D. & Schlegel, M. A. Ocean ventilation and sedimentation since the glacial maximum at 3 km in the western North Atlantic. *Geochem. Geophys. Geosystems* **3,** 1034 (2002).
25. Burke, A. & Robinson, L. F. The Southern Ocean's role in carbon exchange during the last deglaciation. *Science* **335,** 557–561 (2012).
26. Matsumoto, K. Radiocarbon-based circulation age of the world oceans. *J. Geophys. Res. Oceans* **112,** C09004 (2007).
27. Orsi, A. H., Johnson, G. C. & Bullister, J. L. Circulation, mixing, and production of Antarctic Bottom Water. *Prog. Oceanogr.* **43,** 55–109 (1999).
28. Gutjahr, M., Frank, M., Stirling, C. H., Keigwin, L. D. & Halliday, A. N. Tracing the Nd isotope evolution of North Atlantic Deep and intermediate waters in the western North Atlantic since the Last Glacial Maximum from Blake Ridge sediments. *Earth Planet. Sci. Lett.* **266,** 61–77 (2008).
29. Otto-Bliesner, B. L. *et al.* Last Glacial Maximum ocean thermohaline circulation: PMIP2 model intercomparisons and data constraints. *Geophys. Res. Lett.* **34,** L12706 (2007).
30. Chikamoto, M. O. *et al.* Variability in North Pacific intermediate and deep water ventilation during Heinrich events in two coupled climate models. *Deep Sea Res. Part II* **61-64,** 114–126 (2012).
31. Ferrari, R. *et al.* Antarctic sea ice control on ocean circulation in present and glacial climates. *Proc. Natl. Acad. Sci. USA* **111,** 8753–8758 (2014).
32. Ito, T. & Follows, M. J. Preformed phosphate, soft tissue pump and atmospheric CO_2. *J. Mar. Res.* **63,** 813–839 (2005).
33. Ito, T. & Follows, M. J. Air-sea disequilibrium of carbon dioxide enhances the biological carbon sequestration in the Southern Ocean. *Glob. Biogeochem. Cycles* **27,** 1129–1138 (2013).
34. Freeman, E. *et al.* An Atlantic-Pacific ventilation seesaw across the last deglaciation. *Earth Planet. Sci. Lett.* **424,** 237–244 (2015).
35. Freeman, E., Skinner, L. C., Reimer, R., Scrivner, A. E. & Fallon, S. Graphitization of small carbonate samples for palaeoceanographic research at the Godwin Radiocarbon Laboratory, University of Cambridge. *Radiocarbon* **58,** 89–97 (2015).
36. Stuiver, M. & Polach, H. Reporting of ^{14}C data. *Radiocarbon* **19,** 355–363 (1977).
37. Cook, M. S. & Keigwin, L. D. Radiocarbon profiles of the NW Pacific from the LGM and deglaciation: evaluating ventilation metrics and the effect of uncertain surface reservoir ages. *Paleoceanography* **30,** 174–195 (2015).
38. Key, R. M. *et al.* A global ocean carbon climatology: results from Global Data Analysis Project (GLODAP). *Glob. Biogeochem. Cycles* **18,** GB4031 (2004).

39. De Pol-Holz, R., Keigwin, L., Southon, J., Hebbeln, D. & Mohtadi, M. No signature of abyssal carbon in intermediate waters off Chile during deglaciation. *Nat. Geosci.* **3**, 192–195 (2010).
40. Max, L. *et al.* Pulses of enhanced North Pacific Intermediate Water ventilation from the Okhotsk Sea and Bering Sea during the last deglaciation. *Clim. Past* **10**, 591–605 (2014).
41. Govin, A. *et al.* Evidence for northward expansion of Antarctic Bottom Water mass in the Southern Ocean during the last glacial inception. *Paleoceanography* **24**, PA1202 (2009).
42. de la Fuente, M., Skinner, L., Calvo, E., Pelejero, C. & Cacho, I. Increased reservoir ages and poorly ventilated deep waters inferred in the glacial Eastern Equatorial Pacific. *Nat. Commun* **6**, 7420 (2015).
43. Lindsay, C. M., Lehman, S. J., Marchitto, T. M. & Ortiz, J. D. The surface expression of radiocarbon anomalies near Baja California during deglaciation. *Earth Planet. Sci. Lett.* **422**, 67–74 (2015).
44. Butzin, M., Prange, M. & Lohmann, G. Readjustment of glacial radiocarbon chronologies by self-consistent three-dimensional ocean circulation modeling. *Earth Planet. Sci. Lett.* **317-318**, 177–184 (2012).
45. Galbraith, E. D., Kwon, E. Y., Bianchi, D., Hain, M. P. & Sarmiento, J. L. The impact of atmospheric pCO_2 on carbon isotope ratios of the atmosphere and ocean. *Glob. Biogeochem. Cycles* **29**, 307–324 (2015).
46. Sikes, E. L., Samson, C. R., Guilderson, T. P. & Howard, W. R. Old radiocarbon ages in the southwest Pacific Ocean during the last glacial period and deglaciation. *Nature* **405**, 555–559 (2000).
47. Waelbroeck, C. *et al.* The timing of the last deglaciation in North Atlantic climate records. *Nature* **412**, 724–727 (2001).
48. Skinner, L. *et al.* Reduced ventilation and enhanced magnitude of the deep Pacific carbon pool during the last glacial period. *Earth Planet. Sci. Lett.* **411**, 45–52 (2015).
49. Burckel, P. *et al.* Atlantic Ocean circulation changes preceded millennial tropical South America rainfall events during the last glacial. *Geophys. Res. Lett.* **42**, 411–418 (2015).
50. Cheng, H. *et al.* Climate change patterns in Amazonia and biodiversity. *Nat. Commun.* **4**, 1411–1411 (2013).
51. Cruz, F. W. *et al.* Orbitally driven east-west antiphasing of South American precipitation. *Nat. Geosci.* **2**, 210–214 (2009).
52. Jaeschke, A., Rühlemann, C., Arz, H., Heil, G. & Lohmann, G. Coupling of millennial-scale changes in sea surface temperature and precipitation off northeastern Brazil with high-latitude climate shifts during the last glacial period. *Paleoceanography* **22**, PA4206 (2007).
53. Shackleton, N. J. The 100,000-year ice-age cycle identified and found to lag temperature, carbon dioxide, and orbital eccentricity. *Science* **289**, 1897–1902 (2000).
54. Skinner, L. C., Elderfield, H. & Hall, M. in *Ocean Circulation: Mechanisms and Impacts'Past and Future Changes of Meridional Overturning* 197–208 (American Geophysical Union, 2007).
55. Southon, J., Noronha, A. L., Cheng, H., Edwards, R. L. & Wang, Y. A high-resolution record of atmospheric 14C based on Hulu Cave speleothem H82. *Quat. Sci. Rev.* **33**, 32 (2012).
56. Andersen, K. K. *et al.* The Greenland Ice Core Chronology 2005, 15ka. Part 1: constructing the time scale. *Quat. Sci. Rev.* **25**, 3246–3257 (2006).
57. Ruth, U., Wagenbach, D., Steffensen, J. P. & Bigler, M. Continuous record of microparticle concentration and size distribution in the central Greenland NGRIP ice core during the last glacial period. *J. Geophys. Res. Atmospheres* **108**, 4098 (2003).
58. Reimer, P. *et al.* IntCal13 and Marine13 Radiocarbon Age Calibration Curves 0–50,000 Years cal BP. *Radiocarbon* **55**, 1869–1887 (2013).
59. Parnell, A. C., Haslett, J., Allen, J. R. M., Buck, C. E. & Huntley, B. A flexible approach to assessing synchroneity of past events using Bayesian reconstructions of sedimentation history. *Quat. Sci. Rev.* **27**, 1872–1885 (2008).
60. Rasmussen, S. O. *et al.* A new Greenland ice core chronology for the last glacial termination. *J. Geophys. Res. Atmospheres* **111**, D06102 (2006).
61. Ruth, U. *et al.* Ice core evidence for a very tight link between North Atlantic and east Asian glacial climate. *Geophys. Res. Lett.* **34**, L03706 (2007).
62. Biscaye, P. *et al.* Asian provenance of glacial dust (stage 2) in the Greenland Ice Sheet Project 2 Ice Core, Summit, Greenland. *J. Geophys. Res. Oceans* **102**, 26765–26781 (1997).
63. Kahl, J. D. W. *et al.* Air mass trajectories to Summit, Greenland: a 44-year climatology and some episodic events. *J. Geophys. Res.* **102**, 26861–26875 (1997).
64. Svensson, A., Biscaye, P. & Grousset, F. Characterization of late glacial continental dust in the Greenland Ice Core Project ice core. *J. Geophys. Res.-Atmospheres* **105**, 4637–4656 (2000).
65. Bory, A. J., Biscaye, P. E., Svensson, A. & Grousset, F. E. Seasonal variability in the origin of recent atmospheric mineral dust at NorthGRIP, Greenland. *Earth Planet. Sci. Lett.* **196**, 123–134 (2002).
66. Petit, J. R., M, B. & A, R. Ice age aerosol content from East Antarctic ice core samples and past wind strength. *Nature* **293**, 391–394 (1981).
67. Petit, J. R. *et al.* Palaeoclimatological and chronological implications of the Vostok core dust record. *Nature* **343**, 56–58 (1990).
68. Fuhrer, K., Wolff, E. & Johnsen, S. Timescales for dust variability in the Greenland Ice Core Project (GRIP) ice core in the last 100,000 years. *J. Geophys. Res. Atmospheres* **104**, 31043–31052 (1999).
69. Wang, Y. J. *et al.* a high-resolution absolute-dated late pleistocene monsoon record from Hulu Cave, China. *Science* **294**, 2345–2348 (2001).
70. Hodell, D. *et al.* Response of Iberian Margin sediments to orbital and suborbital forcing over the past 420 ka. *Paleoceanography* **28**, 185–199 (2013).
71. Schlitzer, R. Ocean Data View http://odv.awi.de (2016).

Acknowledgements

We thank James Rolfe and Ian Mather for stable isotope measurements, Simon Crowhurst for XRF core scanning, and Ron Reimer for doing the AMS measurements. This work was made possible by NERC support for the cruise that collected the JC89-SHAK cores (NE/J00653X/1) and by NERC grant NE/L006421/1, as well as support from the Royal Society and the Isaac Newton Trust. MD09 cores were collected on board R/V Marion Dufresne during RETRO cruise III, supported by ESF EUROMARC project RETRO, IPEV and ANR project ANR-09-BLAN-0347. C.W. acknowledges support from the European Research Council grant ACCLIMATE/no 339108. This is LSCE contribution 5579.

Author contributions

E.F. and L.C.S. designed the study. L.C.S., D.H. and C.W. collected the core material. E.F. compiled the published data and obtained all the new data. E.F. and L.C.S. analysed the data. E.F. and L.C.S. wrote the manuscript with contributions from all authors.

Additional information

Competing financial interests: The authors declare no competing financial interests.

24

Mesoscale atmosphere ocean coupling enhances the transfer of wind energy into the ocean

D. Byrne[1,2], M. Münnich[1], I. Frenger[3,4] & N. Gruber[1,2]

Although it is well established that the large-scale wind drives much of the world's ocean circulation, the contribution of the wind energy input at mesoscales (10–200 km) remains poorly known. Here we use regional simulations with a coupled high-resolution atmosphere–ocean model of the South Atlantic, to show that mesoscale ocean features and, in particular, eddies can be energized by their thermodynamic interactions with the atmosphere. Owing to their sea-surface temperature anomalies affecting the wind field above them, the oceanic eddies in the presence of a large-scale wind gradient provide a mesoscale conduit for the transfer of energy into the ocean. Our simulations show that this pathway is responsible for up to 10% of the kinetic energy of the oceanic mesoscale eddy field in the South Atlantic. The conditions for this pathway to inject energy directly into the mesoscale prevail over much of the Southern Ocean north of the Polar Front.

[1] Environmental Physics, Institute of Biogeochemistry and Pollutant Dynamics, ETH Zurich, CHN E 23.2, Universitatstrasse 16, Zürich 8092, Switzerland. [2] Center for Climate Systems Modeling (C2SM), ETH Zurich, Zürich 8092, Switzerland. [3] Princeton University, Princeton, New Jersey 08544, USA. [illegible] Centre for Ocean Research, Kiel 24105, Germany. Correspondence and requests for materials should be addressed to D.B. (email:

It has become increasingly apparent that oceanic mesoscale processes are a crucial element of the large-scale oceanic circulation with major implications for climate and ocean biogeochemistry[1–5]. This is especially the case for the Southern Ocean, where eddies and other mesoscale processes such as meanders and filaments partially compensate the wind-driven circulation via an eddy-induced overturning circulation, whose balance determine the net Meridional Overturning Circulation[6–8]. This compensation is dependent on the kinetic energy (KE) of the mesoscale eddy field[7]. Thus, changes to this eddy KE potentially have substantial implications for the uptake of carbon[9–12] and heat[13,14] in the Southern Ocean. One strategy to include mesoscale processes in weather and climate models is to resolve them directly by increasing the resolution of the ocean component. This captures the main energy pathway into the ocean mesoscale eddy field, that is, the large-scale potential energy generation by wind and buoyancy, followed by baroclinic instability. However, in such an approach, any energy flux into the ocean emerging from the mesoscale interaction between the ocean and atmosphere remains unresolved.

A well-studied mesoscale energy pathway involves the mechanical coupling between oceanic mesoscale structures and the overlying atmosphere[15–18]. This pathway emerges, as the wind stress exerted by the atmosphere on the ocean depends not only on the motion of the air but also on the relative motion of the surface ocean[15]. Relative to a situation where this mechanical coupling is not included, the momentum flux into the ocean is reduced whenever the ocean current and the wind are aligned. Correspondingly, the flux is enhanced when the wind and the ocean currents oppose each other. Owing to the quadratic dependence of the wind stress on the wind speed, the net effect is a reduction of the energy flux into the ocean, an effect that has been termed 'mechanical damping'[15–17,19]. Regional model simulations suggest that this reduction may be as large as 17–27%, depending on the region[16,18].

Here we demonstrate that this mechanical damping can be outweighed by a second mesoscale energy pathway that involves the thermodynamic coupling between oceanic mesoscale features and the overlying atmosphere[20–26]. This thermodynamic pathway, which has received comparatively little attention so far[2,27], is driven by the sea-surface temperature (SST) anomalies associated with the oceanic mesoscale field. Through their anomalous heat fluxes, positive SST anomalies destabilize the atmosphere above them, causing more momentum from higher up in the atmospheric boundary layer to be mixed downward, leading to stronger surface winds[22] (Fig. 1). Cold SST anomalies have the opposite effect, stabilizing the overlying atmosphere and reducing the near-surface winds. Our results indicate that this thermodynamic coupling provides an energy conduit that modifies the oceanic mesoscale energy field substantially. Provided a wind gradient of the right sign, this thermodynamic pathway can fully compensate for the mechanical damping effect and may actually contribute up to 10% of the KE in the mesoscale field.

Results

Mesoscale air–sea coupling hypothesis. One potential reason for the lack of consideration of this thermodynamic pathway in the past is the common assumption that at the scale of an ocean mesoscale feature, the background wind is constant. Under this prevailing view, the ocean mesoscale-induced thermodynamic response of the atmosphere results in only little net change in the energy transfer. In the case of a perfectly symmetric eddy, that is, a coherent mesoscale vortex, the net flux is actually zero, as the increased work on one side of the eddy is exactly balanced by the reduced work on the other side. However, ocean eddies rarely live in an area of constant background wind, in particular not in the Southern Ocean[28] (Fig. 2a).

The potential impact of the consideration of a lateral wind gradient over the scale of an ocean eddy (compared with the no

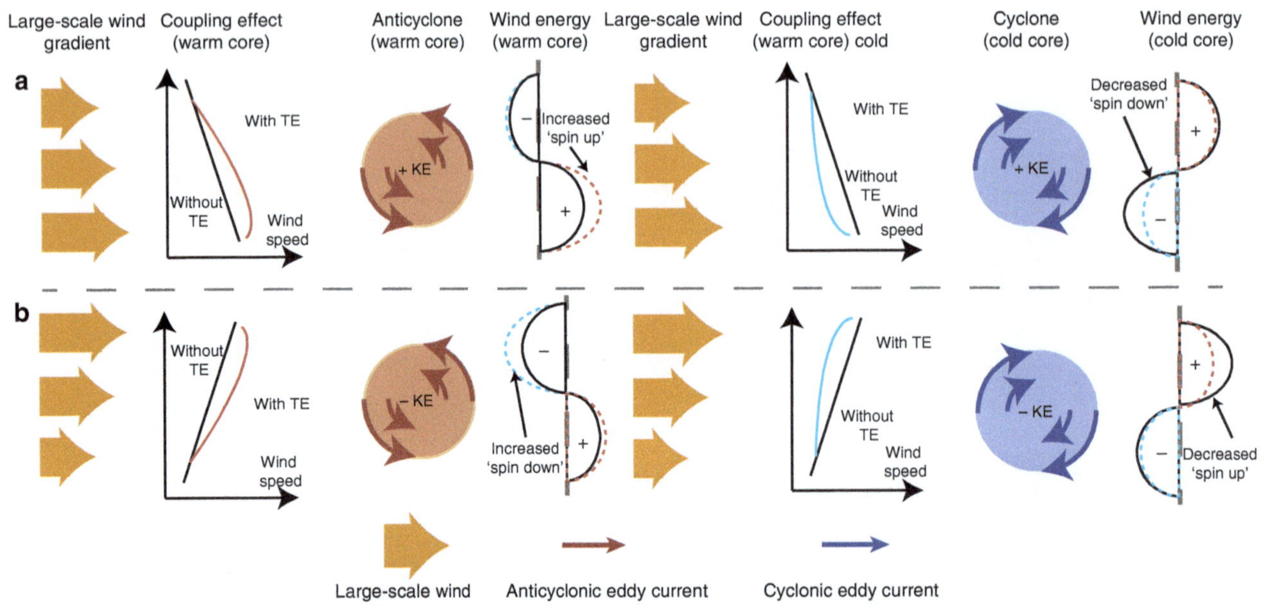

Figure 1 | Proposed modification of the exchange of wind energy between the atmosphere and ocean. (**a**) Cases for a negative wind gradient (increasing from north to south). (**b**) Cases for a positive wind gradient. TE, thermodynamic effect. (**a**) Response for a warm-core anticyclonic eddy and for a cold-core cyclonic eddy separately (both in the southern hemisphere). The eddy-associated anomalies in SST modify the overlying winds and hence the wind stress exerted onto the ocean (coupling effect). In the case of a warm-core eddy lying within a negative wind gradient (**a**), this enhances the energy transfer induced by the mechanical coupling between the eddy current and the wind on the south side of the eddy, whereas it affects the mechanical spin down on the northern side little (wind energy). The net effect is thus an increase in the net energy transfer into the ocean. The other cases operate analogously, resulting in regions with a negative wind gradient tending to energize the oceanic eddies (**a**), whereas regions with a positive gradient having the opposite effect (**b**).

gradient case; see, for example, Fig. 1 in ref. 17) is conceptually illustrated in Fig. 1. A negative wind gradient (defined as an increase in wind speed from north to south) results in either a 'spin-up' (anticyclonic) or 'spin down' (cyclonic) forcing that is active even without taking into account the feedback of the SST anomalies on the atmosphere. If we now add the change in the winds caused by the anomalous SSTs associated with the underlying eddies, this effect gets altered further. The net result is an increase in KE for both warm and cold core eddies that reside in a negative wind gradient and a decrease in KE when they are located in a positive wind gradient.

The wind gradient over the scale of an ocean eddy also has an effect on the mechanical damping, as this effect will now be stronger on the side of the eddy with the stronger wind. However, we propose that as long as the background wind across the eddy changes more than the ocean surface current $\sim O\ (0.5\ \mathrm{m\,s^{-1}})$, the thermodynamic effect will dominate the energy transfer at these scales.

Modelling the mesoscale air–sea coupling. To test our hypothesized thermodynamic pathway associated with a background wind speed gradient, we employed a high-resolution (10 km) coupled atmosphere–ocean regional model (COSMO[29]–Regional Oceanic Modeling System (ROMS)[30])[22] in the South Atlantic, a region with high eddy activity. We undertook three numerical model experiments, to quantify the thermodynamic effect and to contrast it with the mechanical dampening one. In the first simulation (mesoscale mechanically coupled, MMC), we only consider the mechanical damping effect, that is, the ocean surface velocity is sent to the atmospheric model every time step for the calculation of the wind stress, but the atmosphere is not aware of any eddy-scale SST anomalies. In the standard simulation (mesoscale fully coupled, MFC), we additionally allow for the thermodynamic effect, that is, the wind changes in response to the underlying SST anomalies. Here, the ocean model provides at each time step the SST and ocean surface velocity to the atmosphere, and the atmosphere returns the net heat, freshwater and momentum fluxes. To isolate the role of the mechanical energy pathway, we conducted a third simulation (mesoscale thermodynamically coupled, MTC), where only the thermodynamic response of the atmosphere is allowed, that is, the ocean surface current is omitted from the wind stress calculation. All simulations produce a realistic distribution of eddies, with several hundred warm- and cold-core eddies of varying sizes (20–300 km) populating the South Atlantic domain (Fig. 1b) at any given time, in excellent agreement with observations[22].

The simulations reveal that the thermodynamic effect, which we isolate by contrasting the fully coupled with the mechanically-only coupled simulation (MFC–MMC), results in a substantial change of the oceanic mesoscale energy field. This is illustrated in Fig. 2c by the difference of the ocean KE power spectra from these two simulations, calculated over the entire western South Atlantic (the region defined by the red box (Fig. 2a)). This difference reveals that the fully coupled run contains ~10% more KE in the mesoscale range (20–400 km) compared with the mechanically coupled run. In addition, the change in wind forcing between the two simulations, as shown from the time-averaged, power spectrum of the wind stress (Fig. 2b), reveals that changes to the wind forcing are mostly restricted to spatial scales of ~20–600 km. This indicates that the ocean KE change is driven by changes in the mesoscale wind forcing and not by changes in the transfer of energy from larger scales.

In contrast, our simulations reveal only a modest dampening effect from the mechanical energy pathway, isolated by contrasting the fully coupled with the thermodynamically-only coupled simulation (MFC–MTC). In this case, we find a 3% reduction in the energy contained in the oceanic mesoscale field. Thus, in our

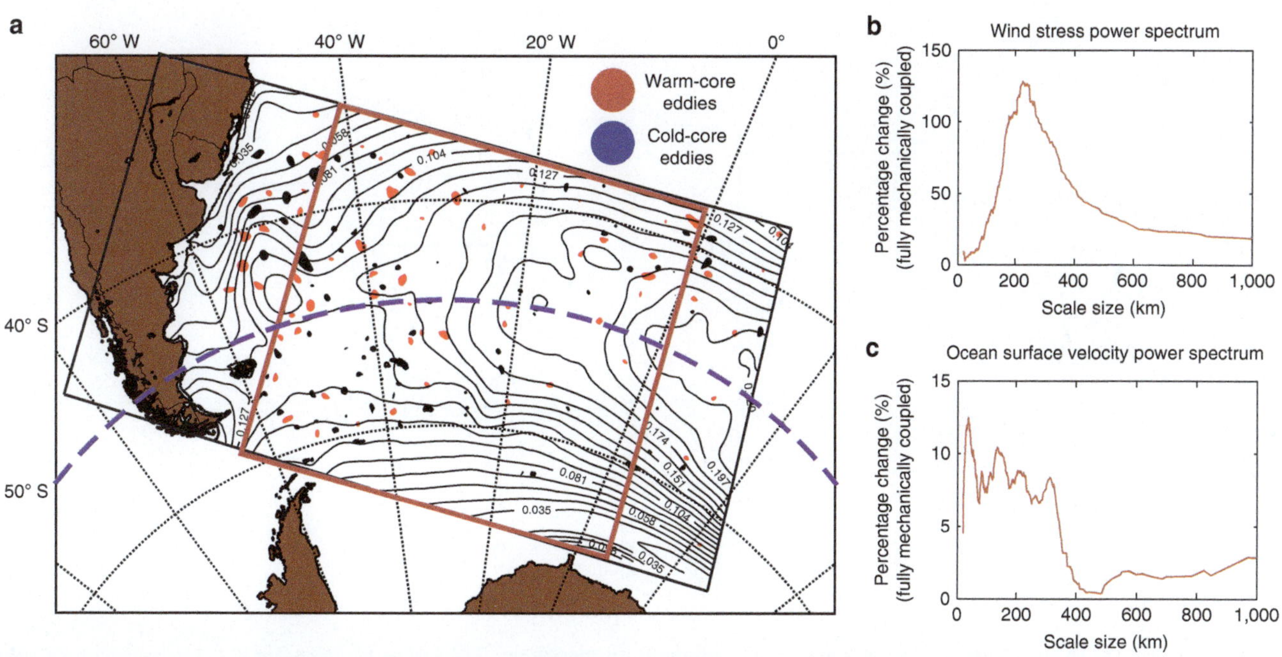

Figure 2 | Model domain wind stress and ocean KE power spectrum. (**a**) Map of the model domain, showing a snapshot of the detected warm (red) and cold (blue) core eddies. The contours depict the austral winter climatological wind stress magnitude taken from ERA interim. The red square indicates the region used for the spectral analysis. The dashed line at 50°S separates the northern region characterized by a negative wind gradient from the southern [illegible] wind gradient. (**b**) Relative change of the wind stress power spectrum between the fully coupled and the mechanically coupled [illegible] for the relative change of the ocean KE power spectrum.

South Atlantic domain, the thermodynamic pathway clearly outweighs the mechanical one.

Thermodynamic coupling at the ocean eddy scale. To understand the mechanisms behind the thermodynamic coupling effect energizing the oceanic mesoscale, we first focus on the contribution of eddies. This choice is justified by eddies being highly abundant in our analysis region[28] and also by our hypothesis having been developed for such coherent vortices. We will demonstrate that eddies largely explain the effect, suggesting that the contribution of the other mesoscale features, such as filaments, are of lesser importance.

Our conceptual model for how the thermodynamic coupling at the mesoscale has an impact on the net energy transfer from the atmosphere to the ocean (Fig. 1) suggests that the overall gain of eddy KE in our South Atlantic domain must come from a prevalence of spin-up conditions over spin-down conditions. The climatology of the wind stress magnitude in Fig. 2a (black contours) reveals that both conditions occur in our domain, that is, a region of negative meridional wind gradient north of ~50°S, representing a 'spin-up' region, and a region of positive wind gradient south of ~50°S, forming a 'spin down' region. Thus, the domain-wide increase indicates that the energy gain in the 'spin-up' region exceeds the energy loss in the 'spin-down' region, probably due to more eddies populating the North.

To test this explanation and our hypothesis, we split the domain into the two distinct regions: north and south of 50°S, that is, the 'spin-up' and 'spin-down' regions (purple dashed line Fig. 2a). In each of these two domains, we disentangle the ocean eddy response by using a semi-Lagrangian frame of reference. Mean composites of the wind energy input, and KE are computed from a rectangle centred on each eddy we identified in the model using two methods[21,22]. Each eddy is scaled to a common size and rotated so that the background wind falls along the *x* direction[21].

The wind stress composites associated with warm- and cold-core eddies (Fig. 3a) clearly show the positive and negative wind stress gradients across eddies for both the 'spin-up' and 'spin-down' regions. The dipole structure of the mean energy input reflects the respective rotation directions of the eddies and their interaction with the background westerlies (Fig. 3b).

In light of our proposed mechanism, we expect the difference in energy input to follow the strength of the background wind[22] (Fig. 1). Figure 3c shows this difference along the chord (south to north) indicated by the dashed line in Fig. 3b. For the negative

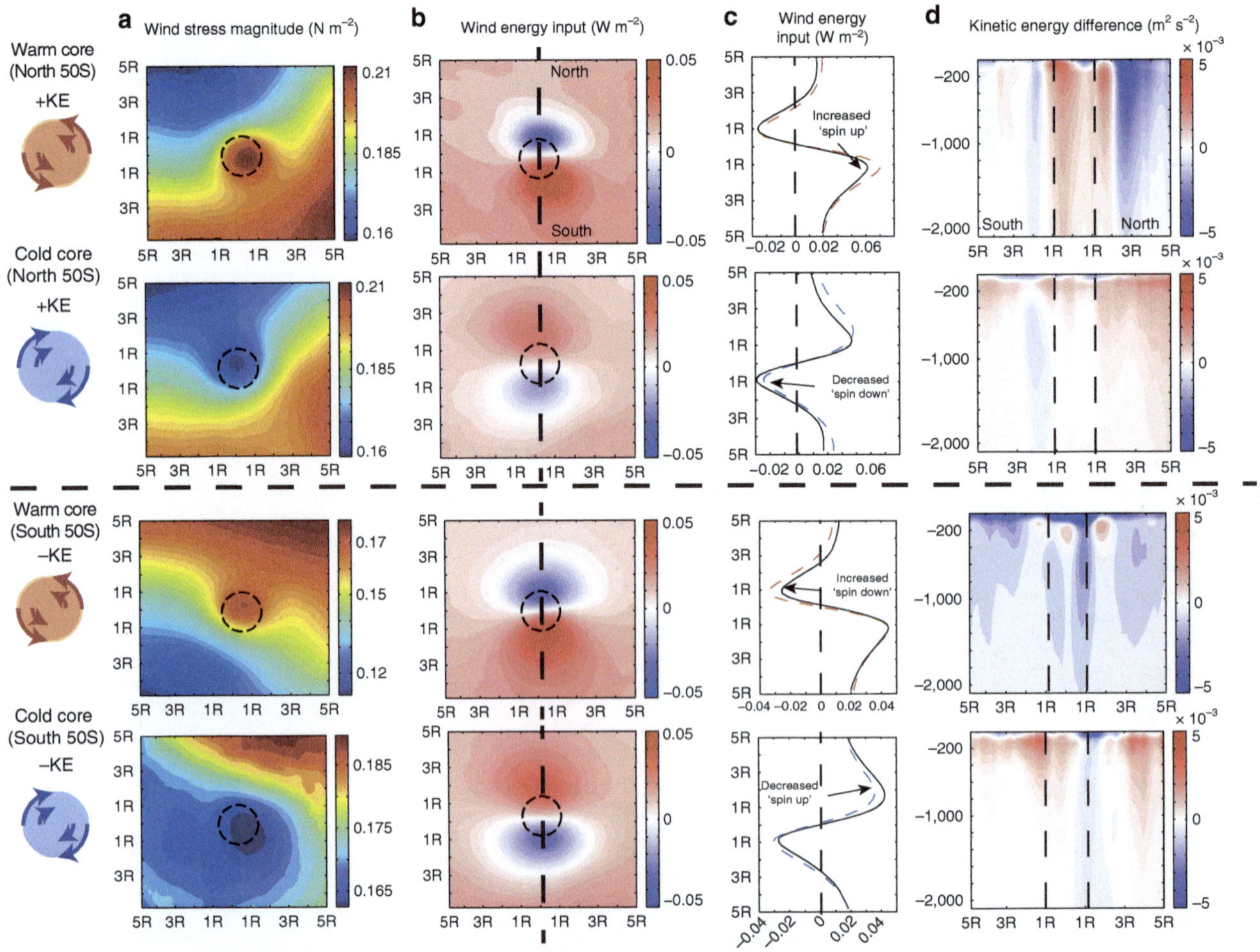

Figure 3 | Model results of the modification of the wind energy transfer at the mesoscale. Composite eddy maps, separated by wind stress gradient (north and south of 50°S) and cold- and warm-core eddies. (**a**) Eddy-induced wind stress change above the eddies from the fully coupled simulation (MFC). (**b**) Wind energy input for the fully coupled simulation showing the dipole structure resulting from the coupling between the ocean and atmosphere; (**c**) wind energy input along the chord (vertical line in **b**) separately for the fully coupled (MFC dashed red) and the mechanically coupled simulation (MMC solid black). (**d**) Vertical section of the difference in ocean KE between the fully and mechanically coupled simulation (MFC-MMC) (along the same vertical line in **b**), showing an increase for both warm- and cold-core eddies in the northern and a decrease in the southern wind gradient region.

wind gradient north of 50°S, there is an increase of the energy input for warm-core (spin-up) and a decrease of energy loss (spin down) for the cold-core eddies (dashed relative to solid line in Fig. 3c). This leads to an intensification for both eddy types north of 50°S, that is, an increase in their KE, as predicted (Fig. 3d). The vertical sections in Fig. 3d additionally reveal that this enhancement is surface intensified, supporting the hypothesis of a direct scale-to-scale forcing from the overlying wind. South of 50°S, the positive wind gradient produces the opposite effect, increasing the 'spin down' of warm-core and decreasing the 'spin-up' of cold-core eddies, resulting in a reduction in the KE (Fig. 3c,d). We note the weaker signal for the cold core eddy in the second row of Fig. 3d. This is mainly due to a peak in the cold-core eddy abundance residing in no wind gradient (Fig. 4a), resulting in a weaker signal compared with the background (also seen in the wind energy input difference). A stronger signal is recovered if one only chooses eddies that reside in the positive wind gradient (Supplementary Fig. 1).

We tested the robustness of these composites by subsampling from all identified eddies a set of every odd or even number, resulting in only marginal changes in the curves (Supplementary Fig. 2a,b). The results are also independent of the eddy detection method[22].

Mesoscale energy budget. To demonstrate quantitatively that the diagnosed increase/decrease in the KE at the eddy scale stems from the thermodynamic energy pathway, we consider the local energy budget difference between the fully coupled and mechanically coupled simulations. Given our previous finding that the change in eddy KE stems directly from the energy input at the mesoscale (see Fig. 2b,c), we interpret that the difference of KE we detect in the ocean eddy composite (Fig. 3d) is a direct response to the difference in the wind energy input (Fig. 3c). We furthermore assume that changes to potential energy, bottom friction and energy transport would not play a major role. This assumption is justified given that all simulations started from the same initial conditions and ran over a period of 3 months only, that is, long enough to fully establish the mesoscale atmosphere–ocean interactions of interest here, but short enough for the other contributions to remain largely unchanged. Thus, we would expect that the difference in the local KE at the eddy scale between the two simulations should be largely explained by change in the wind forcing only, that is,

$$<\Delta \mathrm{KE}> = \Delta\overline{\boldsymbol{\tau}_{\mathrm{w}} \cdot \mathbf{v}_{\mathrm{os}}}\Delta T \quad (1)$$

where ΔKE is the KE difference between the fully coupled and mechanically coupled simulation (MFC–MMC). The expression $\boldsymbol{\tau}_{\mathrm{w}} \cdot \mathbf{v}_{\mathrm{os}}$ is the wind power input computed from the dot product of the wind stress $\boldsymbol{\tau}_{\mathrm{w}}$ and the ocean surface velocity $\mathbf{v}_{\mathrm{os}}$ and ΔT represents the average lifetime of an eddy. Angled brackets and over bar denote volume and surface integrals, respectively.

We estimate $\Delta\overline{\boldsymbol{\tau}_{\mathrm{w}} \cdot \mathbf{v}_{\mathrm{os}}}$ from the difference in the wind energy input composite and integrate over an area πR^2 in eddy-normalized coordinates. R is the average eddy size chosen here as $R = 35$ km for the region north of 50°S and $R = 25$ km south of 50°S. For the eddy lifetime, we take a range of 30-60 days. To compute $<\Delta \mathrm{KE}>$, we take the energy difference from the eddy composites and integrate it over the volume of one eddy, that is, $\pi R^2 h$, where h is a depth of 2,000 m.

With these inputs and assumptions, we find a good closure of the mesoscale energy flux budget over the average eddy in both the spin-up and spin-down regions. In the case of warm-core eddies, the energy input difference, $\Delta\overline{\boldsymbol{\tau}_{\mathrm{w}} \cdot \mathbf{v}_{\mathrm{os}}}$, amounts to +4.2 MW in the spin-up region and −2.1 MW in the spin-down region. These values compare very well with the [illegible] in the KE of the average eddy over this time period, that is, $\frac{<\Delta \mathrm{KE}>}{\Delta T}$ of +2.9 MW (ΔT =60 days) to +5.9 MW (ΔT =30 days), respectively, for the spin-up region and a value $\frac{<\Delta \mathrm{KE}>}{\Delta T}$ of −1.4 to −3.0 MW for the spin-down region. The absolute numbers depend highly on our estimates of the mean eddy parameters, but the large number of analysed eddies (~20,000) and the low sensitivity to subsampling gives us some confidence in our results. Bearing in mind the uncertainties, the fact that we find a good closure confirms in both regions our hypothesis that the eddy-induced anomalous thermodynamic energy flux can account for the majority of the change in the KE in the oceanic mesoscale.

Dominance of the thermodynamic pathway. Given the dominant contribution of eddies to the thermodynamic energy pathway, the South Atlantic domain-wide ~5–10% increase in KE must stem then largely from differences in the eddy abundances in the two regions. Indeed, there are substantially more warm- and cold-core eddies in the northern spin-up region, compared with the southern spin-down one (Fig. 4a). As a result, the thermodynamic pathway has a clear signal in the 'spin-up' region and increases the KE by ~+12%. However, in the 'spin-down' region, the signal is less clear and although the spin-down effect is detectable at the eddy scale in this region, there is no significant signal for the thermodynamic pathway for the integrated KE in the 'spin-down' region (solid green line, Supplementary Fig. 3).

There are several reasons for the reduced effect in the high-latitude, South Atlantic spin-down region. First, the eddies are smaller in size compared with eddies in the northern region. Second, during this period ice cover starts to encroach on this region and hinders the extraction of energy directly from the mesoscales via this mechanism. Third, owing to the simple partitioning of north and south of 50°S, there is still a portion of the domain that remains 'spin up' south of 50°S compensating for the 'spin down' (see Fig. 4b).

In contrast to the thermodynamic pathway, the mechanical energy pathway reduces the mesoscale energy content by ~3% in both regions, that is, it is strongly outweighed by the thermodynamic pathway in the spin-up region (dashed purple line, Supplementary Fig. 3). Additional support for the dominance of the thermodynamic pathway over the mechanical pathway in the South Atlantic comes from the analysis of the wind stress curl pattern above eddies and their comparison with satellite observations. The mechanical pathway leads to a monopole curl pattern, as this response is determined largely by the ocean surface current only. This monopole signal is dwarfed by the dipole structure that results from the additional consideration of the thermodynamic effect in the fully coupled simulation. In addition, it is exactly these dipole structures that are found by a corresponding analysis of the wind stress curl pattern over eddies from satellite observations in the South Atlantic and the entire Southern Ocean[22,31] (Supplementary Fig. 4a,b).

Finally, we compare the magnitude of this new energy pathway with the standard baroclinic instability route by analysing the coherence between wind and surface currents ($C_{\tau v_o}$, see Methods). Assuming a linear response between the wind and ocean velocity, the coherence spectrum estimates the scale-to-scale power transfer. Supplementary Fig. 5 shows that with consideration of the thermodynamic energy pathway (red line), the energy contained at scales 50–150 km that may be explained via wind doubles compared with a case with the mechanical damping only (black line). The figure also reveals a strong scale dependence of the thermodynamic pathway. Although this pathway explains only ~2% of the energy at the 50–150 km scale, it is responsible for up to 8% at scales larger than ~100 km (the remaining

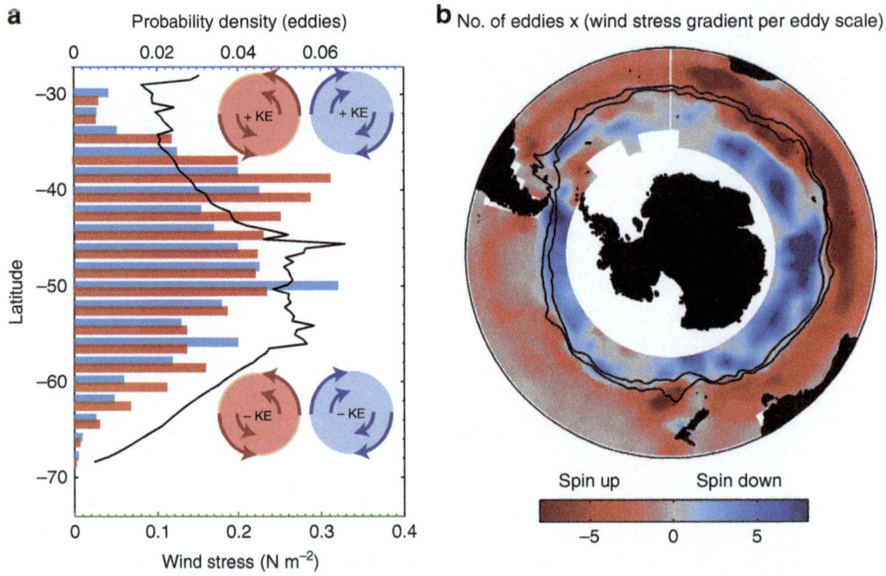

Figure 4 | Satellite observations and regions for the mesoscale thermodynamic pathway. (**a**) Normalized probability distribution of warm (red)- and cold (blue)-core eddies as a function of latitude compared with the mean, time-integrated wind stress (black solid line). (**b**) Southern Ocean map of the product of the average wind stress gradient across an eddy within each 4° (longitude) times 2° (latitude) grid box and the number of eddies within this box. The wind stress was computed from SeaWinds on QuikSCAT and the eddies were identified from satellite sea surface height anomaly data (AVISO) over the period June 2002 to November 2009 (see Methods); shown are only values that are significantly different from 0 at the 1% confidence level (as judged by a *t*-test). The mean positions of the Subantarctic and Polar Fronts are shown as solid black lines; reddish colours show anticyclonic (negative gradient) wind shear situations, bluish colours show cyclonic (positive gradient) wind shear situations; the former represent 'spin-up' conditions with respect to the coupling effect discussed here and the latter represent 'spin-down' conditions.

percentages are assumed to have been supplied via the standard energy pathway, that is, downward cascade from the larger scales).

The fact that our new thermodynamic mechanism is scale dependent is consistent with our proposed hypothesis. Indeed, subsampling only eddies >75 km in diameter when computing the composites in Fig. 3 results in very little change to the result, indicating that the majority of the signal is coming from larger eddies. Further, by binning the magnitude of the asymmetry in wind energy input across eddies as a function of the size of the eddy, we see a clear scale dependence as well. In this case, larger eddies have an increased asymmetry in their wind forcing (not shown).

So far, by focusing exclusively on eddies, we have neglected the possible contribution of other ocean mesoscale structures in the 20–400 km scale range such as meanders, filaments and jet-like structures. Our analyses suggest that their contribution is small, as our proposed mechanism acting on eddies offers a plausible explanation for the majority of the differences in KE. The eddies account for the KE difference between the spin-up/down regions of the domain and they also explain the integrated total change in energy across the domain. Thus, we have good reasons to believe that our eddy-focused mechanism is the dominant effect. We leave it to future studies to determine the specific role played by the other mesoscale phenomena.

Regions for the mesoscale thermodynamic pathway. The thermodynamic energy pathway is important in any region where the abundance of eddies is high and where these eddies are embedded in a strong lateral gradient in the wind stress. Both conditions are met over much of the Southern Ocean. To quantify this effect more precisely, we analysed >600,000 eddies (with lifespans of 2 weeks or more) identified from satellite sea-surface height anomaly data (AVISO) over the period June 2002 to November 2009 (see Methods). Approximately 30% of the eddies (~200,000) exist in a strong gradient with an absolute value exceeding 0.01 $N\,m^{-2}$ per eddy diameter, similar to the average model eddy in Fig. 3a, that is, they experience across their area a change of wind stress of roughly 10% and thus provide an excellent conduit for the mesoscale exchange of energy between the atmosphere and the ocean (see Supplementary Fig. 6a–c). The spatial extent is illustrated in Fig. 4b, which shows the product of the number of eddies instances per 4° (longitude) times 2° (latitude) bin with the average lateral wind gradient over these eddies. This reveals that spin-up conditions (red areas) prevail over large regions of the Southern Ocean north of the Polar Front, and that the energy transfer is likely to be the highest in the Indian Ocean sector given the high number of eddies there[28]. In contrast, spin-down conditions are mostly located south of the Polar Front and, if we extrapolate our results from the South Atlantic to the other sectors, we similarly expect that the integrated reduced energy flux in these regions is much smaller than the enhanced energy flux in the spin-up regions to the north. Thus, it is clear that this mechanism will be important for large regions of the Southern Ocean and most probably anywhere where a strong gradient in the background wind field exists.

Discussion

So far, direct atmosphere–ocean interactions at the mesoscale were generally believed to play little role and, if anything, lead to a reduction of the energy transfer from the atmosphere to the ocean through mechanical damping[15–17,19]. Here we propose a revision to this view, as we have demonstrated that in the Southern Ocean, owing to the presence of large wind gradients, our newly uncovered thermodynamic pathway is outweighing the negative effect of the mechanical damping and may actually result in a net powering of the mesoscale field.

The thermodynamic pathway we described is likely to get stronger in the future with the projected increases and shifts in Southern Ocean wind[32–34]. 'Spin-up' and 'spin down' zones will intensify with increased winds, thus further energizing the oceanic mesoscale in the regions north of the Antarctic Circumpolar Current, whereas dampening the mesoscale activity in the Antarctic Zone south of the Polar Front. Furthermore, more eddies may populate the 'spin-up' side of the wind stress gradient with the projected poleward shift of the wind maxima by the end of the century[34].

To quantify this new pathway we have investigated the transient response of the ocean. However, the question remains as to how this mesoscale energy input at the ocean surface affects the large-scale ocean circulation, which is difficult to predict given the many nonlinear interactions that transfer and dissipate energy among different scales. However, these changes will probably modify vertical mixing and potentially the degree to which the eddy-induced counter flow compensates the increased northward Ekman flow[7,11], which greatly affects the oceanic uptake of CO_2 and heat from the atmosphere[10–14].

Methods

Coupled regional atmosphere-ocean model. We employed a coupled regional atmosphere–ocean model[22], COSMO[29], for the atmospheric component and the ROMS[30] for simulating the ocean. Both the atmosphere and the ocean models used the same horizontal grid with rotated coordinates in reference to a north pole geographic location of 37N, 10W. The model domain spans ~7,000 km by 3,500 km and has a resolution of $(0.09° \times 0.09°) \approx 10\,km^2$. The atmosphere was configured with 40 model layers in the vertical, whereas the ocean model has 42 vertical levels.

The atmospheric model component was initialized with ERA-Interim[35] reanalysis data for 1 June 2004. The ocean part was initialized with climatological January temperature and salinity based on SODA[36] and spun up from rest for 2 years with 6-hourly surface flux forcing (wind stress, net heat and fresh water) from ERA-Interim 2004 reanalysis[35] (2004 forcing repeats during spin-up). Ocean lateral boundary conditions were provided from 30 year mean climatological fields from SODA. After spin-up, the atmosphere and ocean models were coupled. Each of the three 3-month simulations started on 1 June.

Synchronized exchange of data fields between the component models was handled via the OASIS3 coupler[37]. Exchanged fields were sent at every time step (60 s) with no grid or time interpolation between model components. The exchanged fields for each simulation are described below.

MMC, ocean surface velocity was sent to the atmospheric model every time step for calculation of the wind stress. Monthly mean, SST from ERA-Interim was used by the atmospheric model for the calculation of net heat, freshwater and momentum fluxes at each time step. This results in the thermodynamic energy pathway being excluded.

MFC, standard fully coupled setup, with the ocean model sending SST and ocean surface velocity, and the atmosphere returning net heat, freshwater and momentum fluxes at each time step.

MTC, as in MFC, but the ocean surface velocity is not included in the calculation of the wind stress returned by the atmospheric component model, that is, the mechanical energy pathway is excluded.

We employed an updated version of the COARE (Coupled Ocean–Atmosphere Response Experiment)[38] algorithm for the calculation of the turbulent fluxes of heat, moisture and momentum between the atmosphere and ocean, where ocean SST and surface velocity fields are used for the wind stress calculations.

Vertical mixing of momentum and scalars in the ocean was parameterized via a KPP scheme[39].

Eddy identification. We employed two automated eddy detection methods: the first is based on the Okubo–Weiss parameter[40,41], where eddies are identified from sea-level height anomalies, and the second is a vector geometry based eddy detection method where eddies are identified based on their velocity sign reversal[42].

To create the eddy composites for the observed atmospheric and oceanic variable, we used the results from the Okubo–Weiss detection scheme and computed the fields by averaging across all detected eddies after scaling and rotating them[21]. The observed wind speed data stem from collocated satellite data provided by SeaWinds on QuikSCAT. Ocean velocities were estimated from geostrophy for wind energy input composites.

Model-based atmospheric and oceanic variable composites were computed [illegible] fields using both Okubo–Weiss and vector-geometry

One-dimensional wavenumber spectra calculation. The power spectrum displayed in Fig. 1b,c were calculated as follows (here described for the wind stress; however, the same calculation is performed for the ocean with wind stress,τ, replaced by the ocean velocity,**v**.

The power spectrum is calculated for the region denoted by the red box in Fig. 2. From the autocorrelation $|F_i|^2 = F_i(k_m, k_n)F_i^*(k_m, k_n)$ where *denotes the complex conjugate and n,m are gridpoints such that $n = 1,2,3\ldots,N$ $m = 1,2,3\ldots,M$ with $N = M$ in this case. F_i is the Fourier transform of each component of the wind stress field denoted $i = x,y$.

The two-dimensional power spectrum is then calculated as $\tau(k_x, k_y) = |F_x|^2 + |F_y|^2$ from which the one-dimensional power spectrum is computed by binning and averaging all components N_k with the same wavenumber $K = (k_x^2 + k_y^2)^{1/2}$

$$\tau_K = \frac{1}{N_k}\sum \tau(k_x, k_y) \qquad (2)$$

The one-dimensional power spectrum is calculated for each time step and then averaged over the total simulation time.

Coherency magnitude squared. The magnitude-squared coherence was computed as $C_{\tau v_o} = \frac{|G_{\tau v_o}|^2}{G_{\tau\tau}G_{v_o v_o}}$, where $G_{v_o}\tau$ is the cross spectral density, and $G_{\tau\tau}$and $G_{v_o v_o}$ are the autospectrum of the wind stress and ocean surface velocity, respectively. Assuming a linear response between the wind and ocean velocities, $C_{\tau v_o}$estimates the power transfer, where $0 \leq C_{\tau v_o} \leq 1$, and a value of 1 indicates that the ocean velocity can be predicted entirely from the wind stress via a linear function.

Code availability. The atmospheric component of the coupled model COSMO is available via http://www.cosmo-model.org with licenses depending on the intended application of the model. The ocean model (ROMS) is available, depending on the intended application of the model from the lead author D.B. on request. The coupler (OASIS) is available from https://verc.enes.org/oasis

References

1. Holland, W. R. The role of mesoscale eddies in the general circulation of the ocean—numerical experiments using a wind-driven quasi-geostrophic model. *J. Phys. Oceanogr.* **8,** 363–392 (1978).
2. Hogg, A. M. C., Dewar, W. K., Berloff, P., Kravtsov, S. & Hutchinson, D. K. The effects of mesoscale ocean-atmosphere coupling on the large-scale ocean circulation. *J. Clim.* **22,** 4066–4082 (2009).
3. Gruber, N. *et al.* Eddy-induced reduction of biological production in eastern boundary upwelling systems. *Nat. Geosci.* **4,** 787–792 (2011).
4. Rintoul, S. R. & Naveira Garabato, A. C. in *Ocean Circulation and Climate: A 21st Century Perspective* 2nd edn (eds Siedler, G. *et al.*), 471–492 (Academic, 2013).
5. McGillicuddy, Jr. D. J. Mechanisms of physical-biological-biogeochemical interaction at the oceanic mesoscale. *Ann. Rev. Mar. Sci.* **8,** 125–159 (2016).
6. Hallberg, R. & Gnanadesikan, A. The role of eddies in determining the structure and response of the wind-driven southern hemisphere overturning: results from the modeling eddies in the Southern Ocean (MESO) Project. *J. Phys. Oceanogr.* **36,** 2232–2252 (2006).
7. Meredith, M. P., Naveira Garabato, A. C., Hogg, A. M. & Farneti, R. Sensitivity of the overturning circulation in the Southern Ocean to decadal changes in wind forcing. *J. Clim.* **25,** 99–110 (2012).
8. Viebahn, J. & Eden, C. Towards the impact of eddies on the response of the Southern Ocean to climate change. *Ocean Model.* **34,** 150–165 (2010).
9. Ito, T., Woloszyn, M. & Mazloff, M. Anthropogenic carbon dioxide transport in the Southern Ocean driven by Ekman flow. *Nature* **463,** 80–83 (2010).
10. Dufour, C. O. *et al.* Eddy compensation and controls of the enhanced sea-to-air CO2 flux during positive phases of the Southern Annular Mode. *Glob. Biogeochem. Cycles* **27,** 950–961 (2013).
11. Sallée, J.-B., Matear, R. J., Rintoul, S. R. & Lenton, A. Localized subduction of anthropogenic carbon dioxide in the Southern Hemisphere oceans. *Nat. Geosci.* **5,** 579–584 (2012).
12. Frölicher, T. L. *et al.* Dominance of the Southern Ocean in anthropogenic carbon and heat uptake in CMIP5 models. *J. Clim.* **28,** 862–886 (2015).
13. Wolfe, C. L., Cessi, P., McClean, J. L. & Maltrud, M. E. Vertical heat transport in eddying ocean models. *Geophys. Res. Lett.* **35,** 1–5 (2008).
14. Morrison, A. K., Saenko, O. A., Hogg, A. M. & Spence, P. The role of vertical eddy flux in Southern Ocean heat uptake. *Geophys. Res. Lett.* **40,** 5445–5450 (2013).
15. Duhaut, T. H. A. & Straub, D. N. Wind stress dependence on ocean surface velocity: implications for mechanical energy input to ocean circulation. *J. Phys. Oceanogr.* **36,** 202–211 (2006).

16. Dawe, J. T. & Thompson, L. A. Effect of ocean surface currents on wind stress, heat flux, and wind power input to the ocean. *Geophys. Res. Lett.* **33,** 1–5 (2006).
17. Zhai, X., Johnson, H. L., Marshall, D. P. & Wunsch, C. On the wind power input to the ocean general circulation. *J. Phys. Oceanogr.* **42,** 1357–1365 (2012).
18. Zhai, X. & Greatbatch, R. J. Wind work in a model of the northwest Atlantic Ocean. *Geophys. Res. Lett.* **34,** L04606 (2007).
19. Hughes, C. W. & Wilson, C. Wind work on the geostrophic ocean circulation: an observational study of the effect of small scales in the wind stress. *J. Geophys. Res.* **113,** 1–10 (2008).
20. Chelton, D. B., Schlax, M. G., Freilich, M. H. & Milliff, R. F. Satellite measurements reveal persistent small-scale features in ocean winds. *Science* **303,** 978–983 (2004).
21. Frenger, I., Gruber, N., Knutti, R. & Münnich, M. Imprint of Southern Ocean eddies on winds, clouds and rainfall. *Nat. Geosci.* **6,** 608–612 (2013).
22. Byrne, D., Papritz, L., Frenger, I., Münnich, M. & Gruber, N. Atmospheric response to mesoscale sea surface temperature anomalies: assessment of mechanisms and coupling strength in a high-resolution coupled model over the South Atlantic. *J. Atmos. Sci.* **72,** 1872–1890 (2014).
23. Small, R. J. *et al.* Air-sea interaction over ocean fronts and eddies. *Dyn. Atmos. Ocean* **45,** 274–319 (2008).
24. Spall, M. A. Midlatitude wind stress-sea surface temperature coupling in the vicinity of oceanic fronts. *J. Clim.* **20,** 3785–3801 (2007).
25. O'Neill, L. W., Esbensen, S. K., Thum, N., Samelson, R. M. & Chelton, D. B. Dynamical analysis of the boundary layer and surface wind responses to mesoscale SST perturbations. *J. Clim.* **23,** 559–581 (2010).
26. O'Neill, L., Chelton, D. & Esbensen, S. Observations of SST-induced perturbations of the wind stress field over the Southern Ocean on seasonal timescales. *J. Clim.* **16,** 2340–2354 (2003).
27. Gaube, P., Chelton, D. B., Samelson, R. M., Schlax, M. G. & O'Neill, L. W. Satellite observations of mesoscale eddy-induced Ekman pumping. *J. Phys. Oceanogr.* **45,** 104–132 (2015).
28. Frenger, I., Münnich, M., Gruber, N. & Knutti, R. Southern Ocean eddy phenomenology. *J. Geophys. Res.* **120,** 7413–7449 (2015).
29. Steppeler, J. *et al.* Meso-gamma scale forecasts using the nonhydrostatic model LM. *Meteorol. Atmos. Phys.* **82,** 75–96 (2003).
30. Shchepetkin, A. F. & McWilliams, J. C. The regional oceanic modeling system (ROMS): a split-explicit, free-surface, topography-following-coordinate oceanic model. *Ocean Model.* **9,** 347–404 (2005).
31. Frenger, I., Gruber, N., Knutti, R. & Münnich, M. Imprint of Southern Ocean eddies on winds, clouds and rainfall. *Nat. Geosci.* **6,** 608–612 (2013).
32. Thompson, D. W. J. & Solomon, S. Interpretation of recent Southern Hemisphere climate change. *Science* **296,** 895–899 (2002).
33. Marshall, G. J. Trends in the Southern Annular Mode from observations and reanalyses. *J. Clim.* **16,** 4134–4143 (2003).
34. Swart, N. C. & Fyfe, J. C. Observed and simulated changes in the Southern Hemisphere surface westerly wind-stress. *Geophys. Res. Lett.* **39,** 6–11 (2012).
35. Dee, D. P. *et al.* The ERA-Interim reanalysis: configuration and performance of the data assimilation system. *Q. J. R. Meteorol. Soc.* **137,** 553–597 (2011).
36. Carton, J. & Chepurin, G. A simple ocean data assimilation analysis of the global upper ocean 1950-95. Part I: Methodology. *J. Phys.* **30,** 294–309 (2000).
37. Valcke, S. The OASIS3 coupler: a European climate modelling community software. *Geosci. Model Dev.* **6,** 373–388 (2013).
38. Fairall, C. W., Bradley, E. F., Rogers, D. P., Edson, J. B. & Young, G. S. Bulk parameterization of air-sea fluxes for Tropical Ocean-Global Atmosphere Coupled-Ocean Atmosphere Response Experiment difference relative analysis. *J. Geophys. Res.* **101,** 3747–3764 (1996).
39. Large, W. G., Mcwilliams, J. C. & Doney, S. C. Oceanic vertical mixing—a review and a model with a nonlocal boundary-layer parameterization. *Rev. Geophys.* **32,** 363–403 (1994).
40. Okubo, A. Horizontal dispersion of floatable particles in the vicinity of velocity singularities such as convergences. *Deep Sea Res. Oceanogr. Abstr.* **17,** 273–294 (1970).
41. Weiss, J. The dynamics of enstrophy transfer in two dimensional hydrodynamics. *Phys. D Nonlinear Phenom.* **48,** 273–294 (1991).
42. Nencioli, F., Dong, C., Dickey, T., Washburn, L. & McWilliams, J. C. A vector geometry-based eddy detection algorithm and its application to a high-resolution numerical model product and high-frequency radar surface velocities in the Southern California Bight. *J. Atmos. Ocean. Technol.* **27,** 564–579 (2010).

Acknowledgements

We acknowledge the technical support by Stephan Pfahl and Lukas Papritz, and by the Center for Climate Systems Modeling (C2SM). D.B. received funding through ETH Research Grant CH2-01 11-1. I.F. was supported by the 848 SNSF Early Postdoc Mobility Fellowship P2EZP2-152133.

Author contributions

D.B., M.M. and N.G. conceived and designed the simulation experiments. D.B. developed the coupled model. D.B., I.F. and M.M. performed the analysis. All co-authors contributed to the writing of the paper and the interpretation of the results.

Additional information

Competing financial interests: The authors declare no competing financial interests.

25

Mechanisms of increased *Trichodesmium* fitness under iron and phosphorus co-limitation in the present and future ocean

Nathan G. Walworth[1], Fei-Xue Fu[1], Eric A. Webb[1], Mak A. Saito[2], Dawn Moran[2], Matthew R. McIlvin[2], Michael D. Lee[1] & David A. Hutchins[1]

Nitrogen fixation by cyanobacteria supplies critical bioavailable nitrogen to marine ecosystems worldwide; however, field and lab data have demonstrated it to be limited by iron, phosphorus and/or CO_2. To address unknown future interactions among these factors, we grew the nitrogen-fixing cyanobacterium *Trichodesmium* for 1 year under Fe/P co-limitation following 7 years of both low and high CO_2 selection. Fe/P co-limited cell lines demonstrated a complex cellular response including increased growth rates, broad proteome restructuring and cell size reductions relative to steady-state growth limited by either Fe or P alone. Fe/P co-limitation increased abundance of a protein containing a conserved domain previously implicated in cell size regulation, suggesting a similar role in *Trichodesmium*. Increased CO_2 further induced nutrient-limited proteome shifts in widespread core metabolisms. Our results thus suggest that N_2-fixing microbes may be significantly impacted by interactions between elevated CO_2 and nutrient limitation, with broad implications for global biogeochemical cycles in the future ocean.

[1] Marine and Environmental Biology, Department of Biological Sciences, University of Southern California, 3616 Trousdale Parkway, Los Angeles, California
... Chemistry and Geochemistry Department, Woods Hole Oceanographic Institution, Woods Hole, Massachusetts 02543, USA.
... D.H. (email: dahutch@usc.edu).

Biological atmospheric nitrogen (N_2) fixation by cyanobacteria including filamentous *Trichodesmium* spp. is a globally significant biogeochemical process, as it contributes a major fraction of the new N supporting food webs in ocean basin-scale ecosystems[1–4]. Although prevailing N limitation of the central gyre ecosystems provides an important ecological niche for diazotrophic cyanobacteria, field data have demonstrated iron (Fe) or phosphorus (P) limitation of N_2 fixation in both the Atlantic and Pacific Oceans[1,5–9].

Traditionally, biomass limitation by the single nutrient in shortest supply (that is, Liebig limitation[10]) has been invoked as the controlling mechanism for marine primary production and carbon sequestration[11,12] (see Supplementary Note 1 for more discussion). Depletion of this primary limiting nutrient can then lead to a secondary limitation by the next most limiting nutrient.

More recently, several studies have demonstrated nutrient co-limitation, whereby two nutrients can limit growth simultaneously rather than sequentially[1,13–16]. These observations suggest that marine microbes persistently experience periods of selective pressure under widespread nutrient co-limitation[17], which may have favoured the evolution of specific metabolic responses to co-limiting conditions. For instance, diazotrophic cyanobacteria simultaneously limited by Fe and P (Fe/P co-limitation) grow and fix N_2 faster than when limited by either nutrient alone, suggesting that they may possess adaptations specific to co-limited oligotrophic environments[18].

Despite the apparent importance of co-limitation in marine systems, we know little about the molecular mechanisms employed by co-limited microbes and how they may respond to a changing ocean environment[9,13,18]. For instance, increasing anthropogenic carbon dioxide (CO_2) is decreasing seawater concentrations of both hydroxide and carbonate ions (OH^- and CO_3^{2-}), thereby reducing ocean pH (ref. 19). Hence, long-term ocean acidification is likely to have major consequences for key nutrient biogeochemical processes, including N_2 fixation[20]. Past work has observed divergent responses of *Trichodesmium* spp. isolates to CO_2 enrichment, suggesting that temporal CO_2 fluctuations throughout Earth's history, perhaps combined with regional physicochemical forcings, could have resulted in differential ecotypic selection. This niche specialization relative to CO_2 may have in turn influenced current relative abundances and biogeographic distributions of diazotrophic cyanobacteria[21].

In a preceding long-term CO_2 study, one cell line of IMS101 was split into one low (380 μatm) and one high (750 μatm) CO_2 treatment and experimentally selected at each CO_2 concentration for ~4.5 years[22]. Intriguingly, constitutive increases in growth and N_2 fixation rates in all high CO_2-selected cell lines (750 selected) were observed following 4.5 years of selection, even after they were switched back to 380 μatm CO_2 for 2 years. These findings help to reveal the potential responses of a key nitrogen biogeochemical cycle process to the evolutionary consequences of natural selection by future ocean acidification. However, virtually nothing is known about how adaptation of *Trichodesmium* to changing CO_2 will in turn interact with the pervasive, long-term Fe and P co-limitation implied by *in situ* observations[1].

To begin to address these issues, we examine the cellular responses of *Trichodesmium erythraeum* strain IMS101 (hereafter IMS101) to Fe and/or P (co)-limitation using a global proteomics approach in the context of long-term adaptation to both current CO_2 concentrations and projected future ocean acidification conditions[23]. We subjected both the aforementioned 380- and 750-selected cell lines[22] to long-term (~1 year) Fe/P co-limitation selection, followed by either Fe or P additions to Fe/P co-limitation-selected subcultures, to generate steady-state Fe and P single-limitation treatments. Our results demonstrate a complex response of cellular metabolism specific to Fe/P co-limitation, which includes increased growth rates, broad proteome restructuring and cell size reductions relative to growth limited by a single nutrient. This global cellular response may have resulted from long-term selection by widespread Fe/P co-limitation, whereby cell size reductions help to relieve both diffusion and ligand-exchange kinetic limitation, thereby facilitating increased growth[24]. Reduced elemental quotas of smaller cells may also allow cells to maintain more rapid division rates when multiple resources are limiting[22]. Furthermore, elevated CO_2 interacting with Fe/P co-limitation induces additional proteome shifts relative to the present day CO_2, characterized by increased abundances of proteins involved in broad cellular metabolic functions. Together with increased growth rates, this restructuring reveals a unique co-limited phenotype under 'balancing' co-limitation, in which simultaneous limitation by two nutrients may be more advantageous than 'imbalanced' nutrient supply consisting of severe limitation by one resource and an excess of the other. This response fundamentally alters traditional interpretations of interactive nutrient limitations[10] and their consequences for key global biogeochemical processes in both the present and future ocean.

Results

Growth and cell size. We generated Fe/P co-limited treatments from the 380- and 750-selected IMS101 cell lines, using semi-continuous culturing methods in biological triplicate for >1 year at each CO_2 level (Methods). Following this extended co-limitation growth period, either Fe or P was added to subsamples of the Fe/P co-limited cell lines, which were then allowed to acclimate for ~2 months before sampling, to create steady-state, triplicate Fe and P single-limitation treatments at both CO_2 levels (Fig. 1a). This experimental design enables us to examine the effects of both short- and long-term nutrient limitation by Fe and/or P on protein biochemistry of *Trichodesmium* following ~1,000 (380 selected) to ~1,500 (750 selected) generations of selection by CO_2. It has been observed that the mid-latitude oligotrophic oceans may be persistently Fe/P co-limited, a situation which is episodically relieved by pulses of either Fe or P from sources suh as atmospheric dust or advection[1,9]. Accordingly, our experimental design was intended to mimic these processes using diazotrophs adapted to both the present and future CO_2 concentrations. Hence, we used these two CO_2 concentrations to generate a hypothetical chronological progression examining interactions between Fe and/or P (co)-limitation scenarios and changing CO_2 levels in *Trichodesmium*, from the present day up to the year 2100.

As previously observed, growth rates were significantly increased in high CO_2 in both nutrient replete and P-limitation conditions[22] (Fig. 1b). Conversely, no differences in growth rates were observed between the two CO_2 regimes under either Fe limitation or Fe/P co-limitation (see Supplementary Note 2 for more discussion). This lack of elevated growth rates in high CO_2/low Fe relative to low CO_2/low Fe supports the results of prior studies with both a unicellular N_2 fixer and *Trichodesmium*[25,26] (see below and Supplementary Note 3 for more discussion).

Fe/P co-limited growth rates of IMS101 adapted to 380 and 750 μatm CO_2 were lower than those of Fe- and P-replete cells, but were significantly increased relative to both Fe and P single-limitation treatments (by 100–110% and 22–43%, respectively, $P<0.05$) (Fig. 1b top). This growth response is consistent with previous results at current CO_2 levels[18]. In addition, in the Fe/P co-limited cell lines, particulate organic

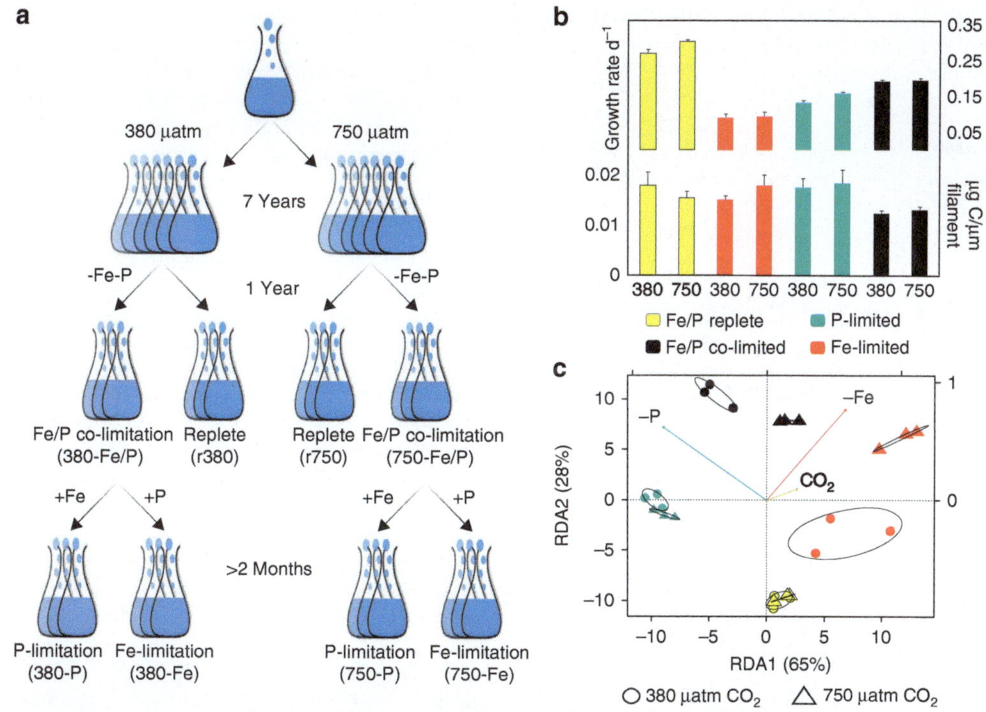

Figure 1 | Experimental design with cell physiology and global proteome redundancy analysis. (**a**) Experimental design is displayed. (**b**) Cell-specific growth rates (top panel) and cell sizes (assessed using the proxy carbon content per filament length (µg C µm^{-1})) (bottom panel) are shown with error bars being s.e. (**c**) Redundancy (RDA) analysis of the global proteome. Colour key applies to both **b** and **c**. Ellipses are 95% confidence limits.

carbon per unit of trichome length was 19–31% (380 selected) and 15–29% (750 selected) lower than in Fe-limited, P-limited and replete treatments, also similar to Garcia *et al.*[18] (Fig. 1b bottom). When compared with either single-limitation treatment, this distinctive co-limitation phenotype suggests a large but unexplained advantage under a two nutrient 'balancing limitation' regime, which is associated with reductions in cell size and volume. This demonstration of a substantial reproductive fitness advantage under Fe/P co-limiting conditions contrasts to the lower growth rate and biomass production typically seen under single nutrient limitations, and have been the focus of most prior work on diazotroph physiology.

Interaction of CO_2 and nutrient limitation on the proteome. Proteome analysis detected 1,908 proteins using a <1% false discovery rate (0.3% actual; minimum of 2 peptides per protein; see Methods), resulting in 307,509 spectra containing identified peptides from 24 discrete samples. Mapping to the ~5,076 proteins predicted in the IMS101 genome[27] yielded 37% coverage of the potential proteome (1,908/5,076). Relative protein abundances were measured by normalized spectral counts relative to total spectra collected across all 24 samples, with the normalization reflecting a very minor change.

To test for nutrient-specific proteome variation, nonmetric multidimensional scaling (Supplementary Fig. 1) and redundancy analysis (Fig. 1c) were applied to the normalized protein abundances, revealing consistent, nutrient-limited abundance patterns across replicate cultures for Fe/P replete, Fe-limited, P-limited, Fe/P co-limited, and/or high and low CO_2 interactive profiles. Permutational multivariate analysis of variance was used to test for statistically significant correlations between limiting nutrient concentrations and proteome variation[28], revealing [illegible] individually significant significant effect on the proteome in nutrient-replete treatments ($P > 0.05$), despite inducing significant growth rate increases as previously observed[22] (Fig. 1b and see Supplementary Note 2 for more discussion). However, the interactive effects of CO_2 with each nutrient scenario (Fe, P or Fe/P) were all significant ($P < 0.05$), suggesting that increased CO_2 concentrations will significantly interact with nutrient-limited proteomes in the future ocean. Hence, the proteomes examined in this study strongly grouped by treatment and were significantly correlated to limiting nutrient concentrations, thus reflecting specific nutrient-limited or co-limited metabolisms. In particular, the distinctive segregation of Fe/P co-limited proteomes away from other single-limitation treatments (Fig. 1c and Supplementary Fig. 1) suggests a broad biochemical response underlying the concurrent increase in growth and decrease in cell size.

To identify proteins responding to specific nutrient/CO_2 treatments, we tested for pairwise changes in protein abundances in each nutrient limitation scenario (P-limited, Fe limited and Fe/P co-limited) relative to either the replete 380 µatm CO_2-selected (r380) or the replete 750 µatm CO_2-selected (r750) controls using the Power Law Global Error Model[29] with an estimated false positive rate of 10^{-3} (Fig. 2). In this hypothetical CO_2 timeline, comparison of the r380 versus each nutrient-limited treatment in the 380 µatm CO_2-selected cultures (n380s) reflects proteome changes due to nutrient (co)-limitation in the absence of elevated CO_2 interactions, as in the present day scenario. Similarly, comparison of the r380 versus each nutrient treatment in the 750 µatm CO_2-selected cell lines (n and r750s) illuminates how nutrient (co)-limited proteomes may interact with the shift to future higher CO_2 conditions (*ca.* year 2100)[23]. Finally, analyses of the nutrient-replete and nutrient-limited, high CO_2-selected cell lines (r750 versus the n750s) reflect proteome nutrient (co)-limitation responses between cell lines already adapted to high CO_2 as in a future ocean. In addition, we [illegible] Fe- and P-limited proteomes with either the 380 or

750 μatm Fe/P co-limitation (380-Fe/P or 750-Fe/P, respectively) treatments, to capture steady-state co-limited proteome changes in response to single nutrient inputs (Fe or P; see below).

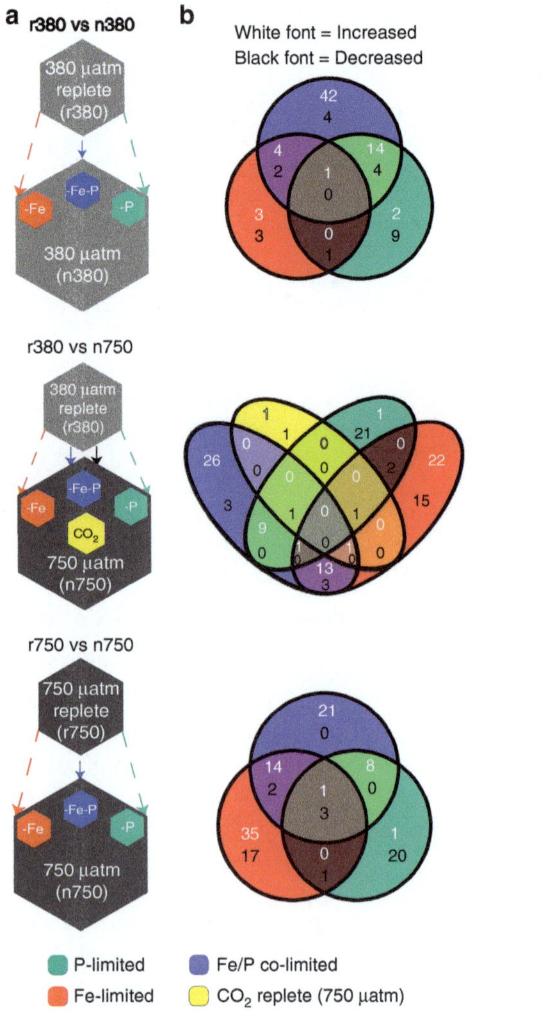

Figure 2 | Experimental treatment comparative matrices and Venn diagrams of differential protein abundances. (**a**) Depiction of the method of differential abundance analysis (see Methods) with smaller hexagons (top) representing the replete reference treatments (r380 and r750) being compared with the nutrient limitation treatments (n380s and n750s, bottom). (**b**) Venn diagrams denote the relationships between the differentially abundant proteins from the analysis in **a** with proteins represented by the numbers therein. White font corresponds to proteins with significantly increased abundances relative to replete conditions and black font denotes significant decreases. The colour key applies to both **a** and **b**.

By assessing global proteome changes in this manner, we are able to track the effect of CO_2 on the proteomes in each of the three nutrient limitation scenarios (Fe, P and Fe/P), while also elucidating the molecular mechanisms underlying cell size decreases and increased growth of the co-limited (Fe/P) phenotype relative to each of the two single-limitation treatments (P-limited or Fe-limited).

In the 'r380 versus n380' scenario (Fig. 2a,b top), Fe/P co-limitation (380-Fe/P) retained the largest fraction of significantly increased protein abundances (~93% of total increased protein abundances), whereas P limitation saw the largest amount of decreased abundances (~61%). However, the interaction of elevated CO_2 and nutrient limitation (for example, r380 versus n750s; Fig. 2a,b middle) drastically increased differentially abundant proteins in Fe limitation (750-Fe), suggesting increased CO_2 to intensify Fe limitation as previously noted[22] (also see below). In addition, numerous proteins (see below) that only increased in Fe/P co-limitation under 380 μatm CO_2 (380-Fe/P) also increased in 750-Fe. Accordingly, in the 'r750 versus n750' high-CO_2 scenario (Fig. 2a,b bottom), Fe limitation dominated the increased protein fraction (~63% of total increased protein abundances), whereas Fe and P limitation made up similar fractions of the decreased protein abundance pool (~53% and 56%, respectively). Taken together, increasing CO_2 in Fe limitation induces widespread changes in protein abundances, whereas in P limitation (750-P), decreases in protein abundances were primarily observed for this subset of the proteome.

Proteins involved in nutrient transport and cell size. The ability to sustain Fe/P co-limited, increased growth relative to single limitation appears to be facilitated through general reductions in cell size, thereby alleviating high cell elemental quota requirements and physical nutrient acquisition limitations on large cells imposed by ligand-exchange kinetics and diffusion. As uptake rate per unit volume will vary inversely with cell diameter[24], an increased surface area to volume quotient should help to relieve these limitations via increased transporter density per unit area, which can be assessed when transporter protein abundances are normalized to a proxy for cell size ($\mu g\,C\,\mu m^{-1}$ filament length). Table 1 shows the average per cent changes in abundance for detected Fe and P stress proteins between the r380 and 380-Fe/P, the r380 and 750-Fe/P, and the r750 and 750-Fe/P before (circle) and after (asterisk) normalizing to $\mu g\,C\,\mu m^{-1}$ filament length. Protein names in italics indicate protein abundances significantly affected by CO_2, that is, from 380- to 750-Fe/P co-limitation conditions (Supplementary Fig. 2). Owing to cell size reductions seen under Fe/P co-limitation, transporter protein abundances increase per unit surface area relative to replete and single-limited cell sizes, which could explain the significant increases in co-limited but not

Table 1 | Per cent changes in Fe and P stress proteins normalized to cell size.

	r380 to 380-Fe/P		r380 to 750-Fe/P		r750 to 750-Fe/P	
	●	★	●	★	●	★
IdiA	2	36	16	53	0.68	15
PstB	340	515	302	430	853	1,036
PhnD	388	603	137	223	121	158
SphX	193	306	84	142	104	138
PstS	21	70	25	68	29	50

Shown are the absolute average percent changes in Fe and P stress protein abundance going from the replete 380 to 380-Fe/P co-limited, the replete 380 to 750-Fe/P co-limited and replete 750 to 750-Fe/P co-limited before (circle) and after (asterisk) normalizing to cell size. Italicized protein names indicate significant changes in protein abundance under increased CO_2 in Fe/P co-limitation.

single-limited growth rates. Hence, decreasing cell size under co-limitation enables cells to either increase or maintain uptake rates by either conserving or only marginally increasing particular transporter abundances (for example, IdiA and PstS) per unit area, thus facilitating energy and material reallocation away from protein synthesis and towards other processes.

Biochemical evidence for cell size reduction specific to Fe/P co-limitation derives from a hypothetical protein (Tery_1090), which was found to include an EzrA (pfam06160) domain (web-based BLASTx)[30]. An EzrA-containing protein is required for regulating cell size in *Staphylococcus aureus*, as average cell diameter significantly increases after deletion of the EzrA protein[31]. This result is consistent with the lower cellular mass (Fig. 1b) and concurrent significant increase in Tery_1090 abundance exclusively under Fe/P co-limitation (Fig. 3a and Methods). Homologues ($\geq 95\%$ of the protein length with e-value $< 1e-10$; see Methods) for Tery_1090 were detected in a handful of colony-forming bacteria based on a search of the current NCBI database (NCBI search using BLASTx; see Methods) and, of these, maximum likelihood phylogenetic analysis places Tery_1090 homologues among the small fraction of the colony-forming cyanobacterial diazotrophs (Fig. 3b). However, the EzrA domain itself is found in proteins distributed across a broad phylogenetic range of both unicellular and colony-forming bacteria whose global sequence homology to Tery_1090 (for example, *Staphylococcus* EzrA protein) fell well below our homologue threshold, suggesting adjacent sequences have considerably diverged over time, whereas the domain itself remained functionally conserved. The exact mechanism involved in the proliferation of the conserved EzrA domain remains to be determined but, nonetheless, the use of different and/or divergent cell size protein machinery with conserved domains suggests that

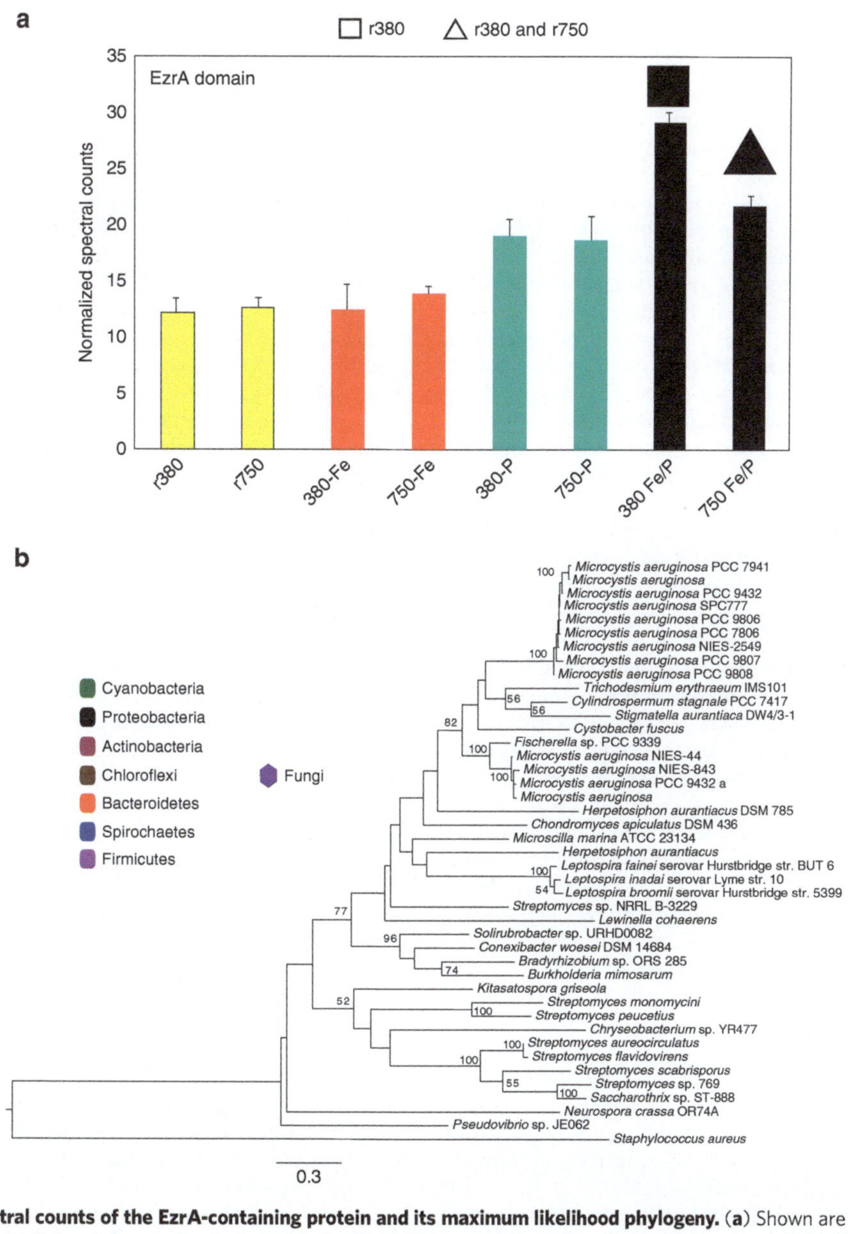

Figure 3 | Normalized spectral counts of the EzrA-containing protein and its maximum likelihood phylogeny. (**a**) Shown are the EzrA-containing protein ... across treatments. Colour key as in Fig. 1; squares indicate significance relative to the replete 380 and triangles relative to both the replete 750 ... phylogeny of detected protein homologues in NCBI (Methods) to the IMS101 EzrA-containing protein, there is strong selective pressure to maintain cell-size reduction capabilities in different habitats among a variety of distantly related bacteria. These data are consistent with prior observations of cell-size reductions with various nutrient-limiting treatments in other microbial systems[32].

In addition, abundance profiles of other IMS101 orthologues to cell size/division proteins are generally consistent with prior observations in other bacterial systems[32]. For instance, cell size increases in *Escherichia coli* when the rod shape-determining gene, *mreB*, is inhibited[33]. Accordingly, an MreB protein homologue (Tery_1150) significantly increased in abundance in our IMS101 Fe/P co-limited cell lines, consistent with decreased cell size. Although the MreB protein and another cell division regulator, MinD, showed increased expression under Fe/P co-limitation, their expression also significantly increased under Fe single limitation, indicating strong control by Fe limitation despite there being no observed change in cell size under this single-limitation scenario (Supplementary Fig. 3 and Supplementary Data 1). Thus, although IMS101 does not significantly reduce cell size under Fe limitation alone, limiting Fe concentrations still impart some control over certain cell size/division machinery.

Sunda *et al.*[24] observed decreases in both growth rates and cell size with decreasing Fe concentrations across a range of eukaryotic phytoplankton. In contrast, we saw reductions in growth, but not cell size, under Fe single limitation in *Trichodesmium*. Cell size only decreased once IMS101 was co-limited by both Fe and P, as in the unicellular diazotrophic cyanobacterium *Crocosphaera*[18], whereas growth rate simultaneously increased relative to Fe-limited growth. Both this unicellular group and colony-forming N_2-fixing cyanobacteria (*Trichodesmium*) share some cell size/division homologues, including MinD, which we observed was more abundant in IMS101 under both Fe/P co-limitation and Fe single limitation (above), despite the lack of cell-size changes under Fe limitation alone. Hence, either unknown mechanisms in diazotrophic cyanobacteria maintain cell size in the face of decreased growth under Fe limitation and/or nutrient-controlled mechanisms governing cell size reductions are only triggered under co-limiting conditions, leading to increased growth. It remains to be seen whether this coordination is specific to Fe/P co-limitation or whether other forms of co-limitation induce a similar response.

Nutrient (co)-limited proteome profiles under increasing CO_2. Both the single- (Fe or P) and co-limited (Fe/P) cell lines shared analogous abundance profiles of several well-characterized Fe or P nutrient stress proteins relative to the replete treatments, respectively (see below, and Supplementary Fig. 2 and Supplementary Data1). Intriguingly, once either Fe or P was added to co-limited cell cultures to achieve new single-limitation steady states, 71–86% and 97–100% of the differentially abundant Fe- and P-limited proteins exhibited significantly reduced abundances relative to their corresponding Fe/P co-limited abundances, respectively (Supplementary Data 2). These substantial fractions of reduced protein abundances following additions of either Fe or P to co-limited cells are accompanied by concurrent decreases in growth and increases in cell size in both Fe- and P-limited cells (Fig. 1b). Hence, the drastic decrease in protein abundances in cells that transitioned from Fe/P co-limited to single-limited steady states (Fe or P) may either be a product of reduced growth and/or a reallocation of energy towards cell-size increases at the cost of reduced growth. To tease apart the respective influences of P and Fe limitation on both the present day and future Fe/P co-limited protein biochemistry, we examined differences in protein composition between single and co-limited scenarios following selection by increasing CO_2.

P limitation versus Fe/P co-limitation under increasing CO_2. P stress proteins (see below and Supplementary Fig. 2) were significantly more abundant in both the P-limited and Fe/P co-limited treatments relative to the replete cell lines, signalling P limitation under both conditions (see Methods). However, Fe/P co-limited cells significantly increased a large protein complement specific to co-limitation (Fig. 2b and see below). This difference between P limitation and Fe/P co-limitation steady states further highlights a broad, coordinated transition between single and co-limited states, with P limitation still persisting in both conditions.

There were also large differences between the 380-Fe/P and 750-Fe/P co-limited proteomes. For instance, adaptation to the interaction of elevated CO_2 with Fe/P co-limitation (750-Fe/P) induced significant reductions in the abundance of particular P stress proteins, relative to 380-Fe/P (see Methods and Supplementary Fig. 2). Log_2 fold changes ranged from -1.5 to -3.3 going from 380- to 750-Fe/P co-limitation for proteins involved in phosphonate acquisition (PhnD, PhnL, PhnK and PhnM), inorganic phosphate (P_i) binding (SphX) and one protein of unknown function (Tery_3845) containing the P response regulator SphR motif, which is an orthologue to PhoB in *E. coli*[34]. In contrast, the high-affinity P_i uptake subunits of the Pst transporter complex, PstB (ATP binding) and PstS (P_i binding), remained unchanged in 750-Fe/P, as did the exopolyphosphatase enzyme, SurE (Supplementary Fig. 2). The *sphX* gene encoding the SphX subunit is an additional P_i binding subunit of the Pst transporter complex only found in a handful of cyanobacteria[34] and is located upstream of the Pst transporter complex operon in IMS101, suggesting its regulation to be independent of the other Pst subunits. Specific reasons for both the reduction of the additional SphX subunit but not other Pst subunits, as well as other subunits of the phosphonate transporter complex in 750-Fe/P need further investigation. Nonetheless, this divergence in P-stress protein abundance suggests that increased CO_2 may have varying effects on P stress complexes under Fe/P co-limitation, thus potentially affecting uptake efficacy of different forms of P.

Fe limitation versus Fe/P co-limitation under increasing CO_2. Similar to published studies[35,36], several Fe stress proteins (Fe starvation-induced protein A (IsiA and IsiB)) were enriched in both Fe-limited and Fe/P co-limited cell lines relative to the replete treatments (see Methods), thereby signalling general Fe limitation (Supplementary Fig. 2 and Supplementary Data1). The smaller amount of differentially abundant proteins observed in Fe single limitation under the present day CO_2 (380-Fe) relative to a previous Fe-limitation study conducted at the present day CO_2 (ref. 36) is likely to be a product of the different methods employed to generate Fe limitation, with Fe stress proteins still signalling Fe limitation in both studies. In particular, Fe single limitation in the present study was generated via P additions to cultures already acclimated to Fe and P co-limiting conditions for 1 year, which may impart fundamentally different physiological pressure relative to abruptly removing Fe from replete cultures as was done in Snow *et al.*[36] and most other previous lab-based experiments[35] (see Supplementary Note 3 for more discussion). Once increased CO_2 interacted with Fe limitation (750-Fe), similar proteomic trends to Snow *et al.*[36] involving major energy, carbon and nitrogen pathways were observed including a decreased photosystem (PSI:PSII) ratio (Fig. 4a and see below) and reduced fructose-1,6-bisphosphate aldolase abundance. In terms of nitrogen metabolism, AbrB (nitrogen uptake regulator), NifE (nitrogenase MoFe cofactor), NifH (nitrogenase) and glutamine synthetase all exhibited significant decreases (Supplementary Fig. 4 and see

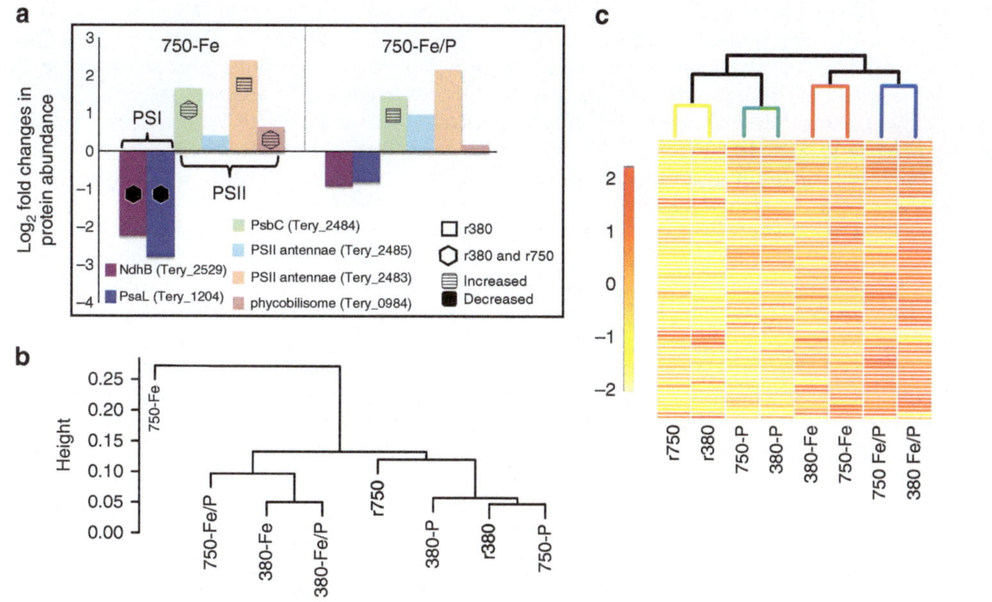

Figure 4 | Log2-fold changes of photosystem proteins and hierarchical clustering of treatments using photosystem proteins or Fe/P complement proteins. (**a**) Shown is a bar graph of log2-fold changes of photosystem proteins in the 750-Fe and 750-Fe/P treatments, respectively. The symbols denote statistical significance relative to either the replete 380 (square) or both the replete 380 and 750 (hexagon). The striped fill denotes statistically significant increases in abundance relative to the replete condition(s) and significant decreases for the black fills. (**b**) Hierarchical clustering of treatments coloured by nutrient using Bray-Curtis dissimilarities calculated from photosystem protein abundances in **a**. (**c**) Hierarchical clustering of treatments with branches coloured by nutrient of Bray-Curtis dissimilarities calculated from protein abundances exclusively responding to Fe/P co-limitation (Fe/P protein complement). Heatmap scale bars represent individual protein abundances standardized across treatments.

Supplementary Note 4 for more discussion). In addition, IsiA, IsiB and IdiA significantly increased in abundance in both 750-Fe and 750-Fe/P proteomes relative to the corresponding low CO_2 treatments (see Methods and Supplementary Fig. 2). Taken together, these proteome shifts suggest that elevated CO_2 intensified Fe limitation[22] in *Trichodesmium* N_2-fixing metabolism, irrespective of P concentration.

Fe limitation in high CO_2-selected cell lines (750-Fe) caused significant decreases in abundance of all detected (PSI) proteins, as seen in previous present day CO_2 Fe-limitation studies[36,37], whereas all detected (PSII) proteins either increased or maintained abundance (Fig. 4a). However, under 750-Fe/P co-limitation, all PSI proteins exhibited abundances statistically indistinguishable from both the replete 380 and 750 treatments, thereby indicating PSI recovery under long-term co-limitation. In addition, 75% of detected, significantly increased PSII proteins in 750-Fe reduced their average abundances. Hence, the interaction of Fe limitation with high CO_2-selected cell lines at steady state significantly reduces the PSI:PSII ratio, but this phenomenon is largely remedied under a Fe/P co-limiting regime. Hierarchical clustering of all detected photosystem components segregated Fe-limited treatments (Fe and Fe/P) away from both replete and P-limited conditions, implicating Fe limitation as a primary driver for the decreased PSI:PSII protein ratio even as P availability varies (Fig. 4b).

Hierarchical clustering of protein abundances exhibiting significant changes solely under co-limitation (the Fe/P protein complement; Supplementary Data 3) groups Fe single and Fe/P co-limitation treatments together, suggesting the Fe/P protein complement to be strongly influenced by Fe as well (Fig. 4c). In addition, numerous proteins that changed abundances exclusively under 380-Fe/P co-limitation also ... in abundance under 750-Fe (Fig. 2 and ... in conjunction

Fe/P co-limitation (Fig. 2 and Supplementary Data 1 and 3) suggest the interaction of high CO_2 adaptation and Fe limitation to be a primary driver of cell-size reduction and increased growth characterizing the Fe/P phenotype.

Exclusive Fe/P co-limitation response. Of the differentially abundant proteins in 380-Fe/P co-limitation relative to the r380, 65% ($n = 46$) were unresponsive in either Fe or P limitation alone (Fig. 2b top), which is evidence that this subset of proteins responds exclusively to co-limitation (Supplementary Data 3). Proteins with increased abundances constituted the majority of the Fe/P protein complement under all three comparative CO_2 scenarios (86, 88 and 90%, respectively) with most showing no changes in abundance in single-limitation treatments (Fig. 2b). Although inherent growth rate-dependent differences between the nutrient limitations may contribute to these proteome shifts, the notably large amount of differentially abundant proteins ($n = 46$) unique to 380-Fe/P co-limitation relative to the corresponding 380 single-limitation treatments (Fig. 2b, top) suggests the possibility of an evolutionarily conserved, coordinated biochemical response controlled by unknown regulatory systems underlying balanced limitation. This complex, distinct co-limitation response may have evolved due to intense selection by global Fe/P co-limitation regimes[1,14].

Interestingly, although no growth rate differences were observed between the low and high CO_2 treatments in Fe-limited (Fe and Fe/P) scenarios (Fig. 1b), large proteomic differences were observed between the 380-Fe and 750-Fe, as well as between the 380-Fe/P and 750-Fe/P (Figs 1c and 2b, Supplementary Fig. 1 and Supplementary Data 1 and 3). These differences suggest that these Fe-limited, CO_2-induced proteome shifts are not growth rate driven, but instead are likely to be CO_2 specific. In fact, they may represent the cellular compensatory mechanism(s) that allow ... maintain similar growth rates under Fe-limited conditions, regardless of changing CO_2 levels. Taken together, these proteome shifts provide mechanistic insights into the departure of the co-limiting from the single limiting response as outlined in the Liebig model[10]. They also reveal changes to proteome architecture mediated by the interactions of CO_2 and Fe limitation that are independent of growth rate.

COG (Cluster of Orthologous Genes) categories were assigned to the distinctive Fe/P co-limited protein complement (~72% of the proteins; Supplementary Data 3 and see Methods). Although full pathway characterization was not possible, possibly due to potential limitations in protein detection, mapping these COG-assigned proteins to Kyoto Encyclopedia of Genes and Genomes pathways[38] revealed them to reside in widespread cellular metabolisms (Supplementary Data 3). These included membrane stability and biogenesis, carbon catabolism and storage, cofactor biosynthesis, carbon fixation, photosynthesis and various precursor metabolisms, all of which together indicate a broad coordinated shift in numerous general cellular processes consistent with cell-size decreases and growth-rate increases.

More specifically, the enrichment of proteins mapping to metabolisms involved in cofactor and precursor biosynthesis in concert with the increased abundances of both P- and Fe-limitation stress proteins suggests cellular reallocation under co-limitation to the biosynthesis of versatile precursor biomolecule residing at various metabolic junctures, possibly allowing for greater metabolic flexibility (Fig. 5). Cells experiencing (co)-limitation typically respond rapidly to nutrient additions, which involves global changes to cellular metabolism reflected in cell-size and growth-rate changes[39]. Hence, biosynthesis of general precursor molecules that can potentially be used by multiple pathways when nutrient fluxes are persistently variable may enable greater cellular plasticity and energy usage efficiency.

For example, increased protein abundance of isopentenyl pyrophosphate isomerase (IDI; Tery_1589) and squalene synthase (SQS; Tery_2043) suggests increased isoprenoid biosynthesis, which serve as critical components in various biochemical functions including quinones in electron transport chains, membrane components, photosynthetic pigments and others[40]. These isoprenoid enzyme increases are consistent with the concurrent increase in the protochlorophyllide reductase subunit ChlL (Tery_1532) and the protoporphyrinogen oxidase HemY (Tery_2218), where ChlL is involved in precursor production for chlorophyll (Chla) biosynthesis[41] and HemY is involved in precursor production for both Chla and haems important for electron transport[42]. Furthermore, increased abundance of enzymes such as ornithine carbamoyltransferase (Tery_1323) involved in arginine/cyanophycin biosynthesis and hence nitrogen storage[43] (Supplementary Fig. 3 and Supplementary Note 4), Glga (glycogen synthase; Tery_2147) involved in carbon storage, ManC and RfaE (lipopolysaccharide biosynthesis enzymes; Tery_1856, Tery_3495) involved in membrane stability and NadE (NAD synthetase; Tery_1984) involved in cofactors for photosynthesis and respiration, all corroborate the re-apportionment of cellular energy towards synthesizing flexible precursors and intermediates involved in a variety of pathways tied to core carbon metabolism (Supplementary Data 3).

Increased abundance of many precursor pathway proteins in Fe/P co-limited cells relative to Fe-limited and P-limited steady-state treatments could be related to the higher growth rates observed in the former condition (Fig. 1b). However, the replete treatments had significantly higher growth rates than Fe/P co-limited cultures (Fig. 1b), which were not reflected by higher levels of precursor pathway proteins, suggesting that this cannot be explained as a simple growth rate-driven phenomenon. Further investigations (for example, metabolite analyses) are necessary to validate the increased production of these intermediates under Fe/P co-limitation, but the increased abundance of these precursor biosynthesis proteins involved in various pathways looks to be a direct product of a metabolic shift under balancing Fe/P co-limitation not seen under single limitations.

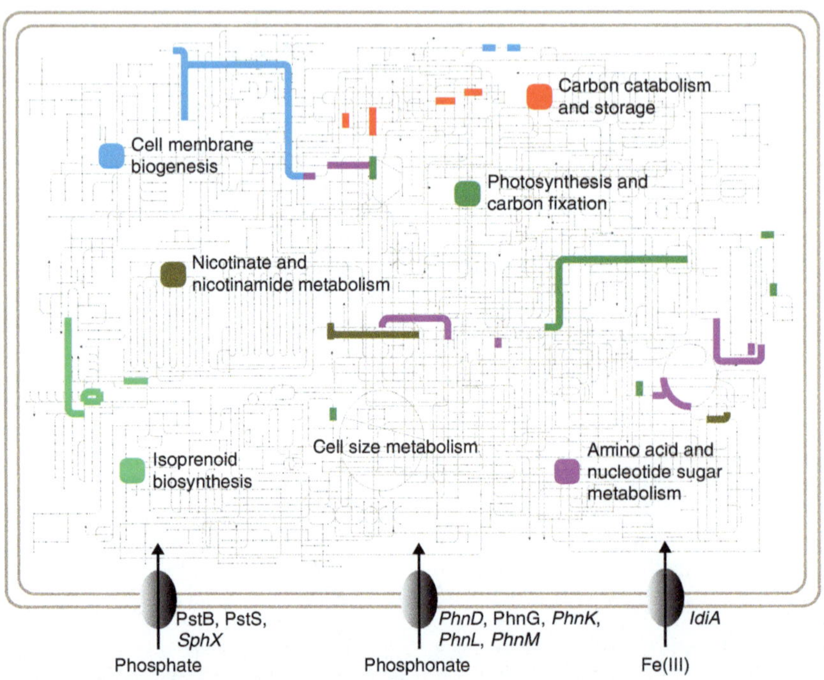

Figure 5 | COG assigned proteins increased exclusively in Fe/P co-limitation mapped to Kyoto Encyclopedia of Genes and Genomes (KEGG) pathways. 'Cell size metabolism' does not exist in KEGG pathways, but was added to acknowledge the cell size protein (Fig. 3a) exhibiting increased abundance in conjunction with these broad metabolic pathways under Fe/P co-limitation. Detected P and Fe transporter proteins are shown, indicating both Fe- and P-limitation, and italicized names are components affected by increased CO_2 (see main text).

The cellular regulation controlling these broad, coordinated proteomic shifts in widespread metabolic pathways under co-limitation is potentially controlled by upstream regulatory mechanisms at a whole systems level. For example, the switching of different RNA polymerase sigma factors (for example, 'sigma switching') has been shown to aid in both stress and adaptive responses via transcriptional initiation of gene sets that are specific to particular environmental or internal cellular changes[44]. Predicted IMS101 sigma factors in the genome were either undetectable, not expressed or below our analysis threshold, thus preventing confident analysis of their differential abundances. However, future efforts can include more targeted studies looking at these proteins under nutrient limitation. In addition, other mechanisms such as DNA modifications[45] (for example, epigenetics) and transposition[46] have also been shown to aid in stress and adaptation, which in turn affect downstream transcription and translation. Hence, widespread changes to the proteome may be a product of coordinated changes from a smaller number of upstream regulatory systems, each controlling numerous biochemical pathways contingent on environmental stimuli. The results described here offer insight into downstream biochemical pathways affected by both independent and interactive nutrients and CO_2, thus providing a foundation for future investigations of the regulatory mechanisms governing these biochemical changes.

The biochemical/physiological coordination described here offers the first molecular and mechanistic insight into the underlying cellular mechanisms governing the 'balancing limitation' phenotype selected for by simultaneous Fe and P co-limitation. Looking ahead, the interaction of future increasing CO_2 with single and multiple nutrient limitation scenarios induces fundamental metabolic shifts away from those seen under the present day CO_2, including many responses that are clearly not simply growth rate driven. Surprisingly, numerous proteins (for example, photosynthetic proteins) that are differentially abundant under either Fe or P limitation are maintained at similar abundances to replete conditions under Fe/P co-limitation at high CO_2. Concurrently, Fe/P co-limitation also induces a significant increase of a wholly different protein complement involved in core cellular functions including a variety of precursor metabolisms. The exact nature of this coordinated molecular and physiological response under Fe/P co-limitation needs further study, but the increased abundance of the Ezra-containing protein along with concurrent cell size reductions provides mechanistic insight into achievement of increased fitness (growth) relative to single-limitation scenarios.

Widespread co-limited oceanic conditions may have selected for a master regulatory pathway specifically evolved to sense co-limiting regimes, which may then affect downstream pathways to produce the co-limited phenotype. Alternatively, sovereign Fe and P regulatory mechanisms may initially respond independently to their respective limiting nutrients, followed by subsequent cellular coordination to produce the observed Fe/P co-limitation proteome profiles and phenotype. Regardless, it remains to be determined how widespread this response is among broad ranges of both prokaryotic and eukaryotic photoautotrophs. If other taxonomic groups exhibit similar co-limited responses, many classic nutrient limitation studies may need to be revisited, to determine what the ultimate consequences will be for controls on key global biogeochemical processes in the present day and future ocean.

Methods

Culturing methods. Stock cultures of *Trichodesmium* strain IMS101 were ... the University of Southern California, Los Angeles, California, USA, in ... They were grown at 26 °C ... 120 μmol photons per $m^2 s^{-1}$ incident irradiance. Experimental cultures were maintained in 0.2 μm-filtered, microwave-sterilized artificial seawater. Artificial seawater and Aquil nutrient stocks (except for the trace metal stock) were passed through a Chelex-100 column, to remove contaminating Fe before medium preparation.

Semi-continuous culturing methods using optically thin cultures were conducted to avoid nutrients becoming depleted before the next dilution[25,47,48]. All experimental conditions used three biological replicate bottles and each replicate was diluted individually according to growth rates calculated daily for that bottle using *in vivo* chlorophyll fluorescence measurements with a Turner 10 AU fluorometer[21,48]. For all experiments, final sampling occurred once steady-state growth (no significant difference in growth rates) was reached for at least ten generations and reported growth rates were calculated based on microscopic cell counts (see below).

To examine interactive effects of Fe and P limitation on growth of the cyanobacteria, *T. erythraeum* were grown in four treatments as follows: (1) Nutrient replete, 10 μM PO_4^{3-} and 250 nM Fe; (2) P-limited, 0.25 μM PO_4^{3-} and 250 nM Fe; (3) Fe-limited, 10 nM Fe and 10 μM PO_4^{3-}; and Fe/P co-limited, 0.25 μM PO_4^{3-} and 10 nM Fe. EDTA concentrations were 25 μM irrespective of Fe conditions. Each of these four experimental nutrient treatments were generated using adapted cell lines that had been previously selected under 380 or 750 p.p.m. CO_2 for ~7 years[22], for a total of 8 treatments. The replete and Fe/P co-limited cultures were grown in steady-state semi-continuous cultures for ~12 months at each CO_2 level. Following this long-term incubation, either Fe or P concentration was increased in subcultures of the Fe/P co-limited cell lines, thus creating cultures limited by either P or Fe alone, respectively. These two sets of single-nutrient-limited cultures were then grown at each pCO_2 level for ~2 months before being sampled together with the replete and Fe/P co-limited cultures.

Seawater medium was bubbled with 0.2-μm-filtered prepared air/CO_2 mixtures (Praxair) to maintain stable targeted CO_2 concentration treatments of 380 and 750 p.p.m. In-line high-efficiency particulate air filters were employed to avoid Fe contamination from particles in the gas tanks or lines and pH was monitored daily with dissolved inorganic carbon (DIC) being measured at the final sampling. Once steady-state growth was achieved, *Trichodesmum* filament abundance and lengths were measured in a 1-ml phytoplankton counting chamber using epifluorescence microscopy and significant differences between treatments were calculated using two-way analysis of variance along with Tukey's test. For proteome analysis, cultures were swiftly and gently filtered onto 5-μm polycarbonate filters (Whatman) during the middle of the photoperiod, immediately flash frozen and stored in liquid nitrogen until protein extraction.

Carbonate buffer system analysis. DIC and pH were measured according to standard protocols[49]. DIC samples were collected from the experimental cultures and immediately poisoned with 200 μl of a saturated $HgCl_2$ solution in 25 ml combusted borosilicate glass bottles. Samples were stored at room temperature until analysis on a UIC CO_2 coulometer conducted in triplicate, as in Fu *et al.*[25]. Certified reference materials were used to calibrate total DIC measurements obtained from A. Dickson (UCSD). To confirm carbonate system equilibration, pH measurements were used for real-time monitoring and made using an Orion model 8102 combination electrode.

Proteome analysis. Protein extraction, label-free mass spectrometry and spectral count normalization for global proteome analysis were performed as previously described[22]. The proteins and their normalized spectral counts can be found in Supplementary Data 4.

Multivariate and pairwise analyses. Filtering of the proteome was performed before multivariate analysis, to eliminate consistently low normalized spectral counts across all treatments. In many cases where a protein contained zero counts in at least half of the treatments, the normalized spectral counts in the remaining treatments also proved to be uninformatively low (<3), which could skew ordination methods. However, certain proteins were only detected when a specific nutrient was limiting (for example, Tery_2498 and Tery_0463) in which case these proteins had zero spectral counts in half of the treatments as above but now consistently higher (>5) counts in treatments where that nutrient was limiting (for example, Fe). Thus, to remove proteins with little to no expression across all treatments, while simultaneously retaining proteins that showed substantial expression only in treatments where a particular nutrient was limiting, for each protein we summed the spectral counts across all replicates of all treatments (24 libraries) and divided by half (12) of the number of libraries. If this quotient was ≥ 5 normalized spectral counts, the protein was retained in the ordination analysis.

Nonmetric multidimensional scaling on normalized spectral counts of the filtered proteome was conducted in 'R' (R Core Team 2014) using the 'metaMDS' function from the vegan package[50] with default settings (Bray–Curtis dissimilarities computed using Wisconsin double standardization) except for 'autotransform = FALSE'. For the permutational multivariate analysis of variance, Bray–Curtis dissimilarities were calculated from the normalized spectral counts using the 'vegdist' function and subsequently input into the 'adonis' function in vegan with 2,000 permutations.

To analyse well-characterized nutrient stress proteins as indicators of cellular ...rient status, Welch's one-tailed *t*-tests assuming heteroscedasticity (unequal variances)[51] were conducted in Microsoft Excel. Well-characterized nutrient stress proteins (Supplementary Fig. 2) are those supported by independent lines of empirical evidence[34–36] representing a priori knowledge of their responses to either Fe or P limitation. Given this prior knowledge, statistically speaking these proteins represent 'planned comparisons' defined as a selected subset of all possible comparisons that are not suggested by the current results but are predefined by prior theory in which an appropriate pairwise testing is a Welch's *t*-test[52]. In contrast, most other proteins lack a priori characterization; thus, their responses to environmental factors are unknown and require control of false-positive rates due to many independent exploratory tests, as in the PLGEM method (see below). *T*-tests were also conducted to test for significant differences between mean values of nutrient stress proteins that changed going from 380-Fe/P to 750-Fe/P.

Exploratory pairwise differential abundance tests were administered using a power law global error model provided in the PLGEM package with default settings[29] and a delta (estimation of false-positive rate) of 0.001. Tests were conducted on normalized spectral counts between the nutrient-replete treatments and nutrient-limited treatments at each respective CO_2. Following each pairwise test, each statistically significant protein within the differentially expressed pool was manually checked.

Metabolic map. All proteins displaying differential expression exclusive to Fe/P co-limitation in each of the three CO_2 scenarios (see main text) were matched to their corresponding COG IDs as per the IMG (Integrated Microbial Genomes) annotation (https://img.jgi.doe.gov). These COG IDs were used as input into the iPath2.0 tool[38], to generate the metabolic map and beautified in Adobe Illustrator.

Hierarchical clustering of protein expression. For all heatmaps, Bray–Curtis dissimilarities were calculated on normalized spectral counts using the 'vegdist' function in the vegan package and hierarchical clustering was performed with the Heatplus package[53] using the average (UPGMA) clustering method.

Maximum likelihood phylogeny. The Tery_1090 protein containing the Ezra domain (see main text) was searched against the NCBI non-redundant database using the web-based BLASTx tool[30] and hits with e-value $< 1e-10$ and high scoring pairs covering $>95\%$ of the original query length were kept, to ensure robust homologue identification and multiple sequence alignments. The *S. aureus*[31] Ezra protein was used as the outgroup. MAFFT was used for amino acid multiple sequence alignment[54] with default settings and MEGA v6.06 (ref. 55) was used to construct maximum likelihood phylogeny with 100 bootstrap replicates and the following eight parameters: (1) substitution type = amino acid; (2) substitution model = wag with freqs. (+F) model; (3) rates among sites = gamma distributed (G); (4) no of discrete gamma categories = 5; (5) gaps/missing data treatment = complete deletion; (6) ML heuristic method = nearest-neighbour interchange; (7) initial tree for ML = NJ/BioNJ; and (8) branch swap filter = very strong.

References

1. Sohm, J. A., Webb, E. A. & Capone, D. G. Emerging patterns of marine nitrogen fixation. *Nat. Rev. Microbiol.* **9,** 499–508 (2011).
2. Moisander, P. H. *et al.* Unicellular cyanobacterial distributions broaden the oceanic N2 fixation domain. *Science* **327,** 1512–1514 (2010).
3. Capone, D. G., Zehr, J. P., Paerl, H. W., Bergman, B. & Carpenter, E. J. *Trichodesmium*, a globally significant marine cyanobacterium. *Science* **276,** 1221–1229 (1997).
4. Mulholland, M. R. The fate of nitrogen fixed by diazotrophs in the ocean. *Biogeosciences* **4,** 37–51 (2007).
5. Rueter, J. G., Hutchins, D. A., Smith, R. W. & Unsworth, N. L. in *Marine Pelagic Cyanobacteria: Trichodesmium and other Diazotrophs* **362,** 289–306 (Springer Netherlands, 1992).
6. Sañudo-Wilhelmy, S. A. *et al.* Phosphorus limitation of nitrogen fixation by *Trichodesmium* in the central Atlantic Ocean. *Nature* **411,** 66–69 (2001).
7. Hynes, A. M., Chappell, P. D., Dyhrman, S. T., Doney, S. C. & Webb, E. A. Cross-basin comparison of phosphorus stress and nitrogen fixation in *Trichodesmium*. *Limnol. Oceanogr.* **54,** 1438–1448 (2009).
8. Chappell, P. D., Moffett, J. W., Hynes, A. M. & Webb, E. A. Molecular evidence of iron limitation and availability in the global diazotroph *Trichodesmium*. *ISME J.* **6,** 1728–1739 (2012).
9. Moore, C. M. *et al.* Processes and patterns of oceanic nutrient limitation. *Nat. Geosci.* **6,** 701–710 (2013).
10. de Baar, H. J. W. von Liebig's law of the minimum and plankton ecology (1899–1991). *Prog. Oceanogr.* **33,** 347–386 (1994).
11. Falkowski, P. G. Evolution of the nitrogen cycle and its influence on the biological sequestration of CO2 in the ocean. *Nature* **387,** 272–275 (1997).
12. Tyrrell, T. The relative influences of nitrogen and phosphorus on oceanic primary production. *Nature* **400,** 525–531 (1999).
13. Saito, M. A., Goepfert, T. J. & Ritt, J. T. Some thoughts on the concept of colimitation: three definitions and the importance of bioavailability. *Limnol. Oceanogr.* **53,** 276–290 (2008).
14. Mills, M. M., Ridame, C., Davey, M., LaRoche, J. & Gelder, R. J. Iron and phosphorus co-limit nitrogen fixation in the eastern tropical North Atlantic. *Nature* **429,** 292–294 (2004).
15. Saito, M. A. & Goepfert, T. J. Zinc–cobalt colimitation of *Phaeocystis antarctica*. *Limnol. Oceanogr.* **53,** 1–11 (2008).
16. Saito, M. A. *et al.* Multiple nutrient stresses at intersecting Pacific Ocean biomes detected by protein biomarkers. *Science* **345,** 1173–1177 (2014).
17. Harpole, W. S. *et al.* Nutrient co-limitation of primary producer communities. *Ecol. Lett.* **14,** 852–862 (2011).
18. Garcia, N. S., Fu, F., Sedwick, P. N. & Hutchins, D. A. Iron deficiency increases growth and nitrogen-fixation rates of phosphorus-deficient marine cyanobacteria. *ISME J.* **9,** 238–245 (2014).
19. Caldeira, K. & Wickett, M. E. Oceanography: anthropogenic carbon and ocean pH. *Nature* **425,** 365–365 (2003).
20. Hutchins, D. A., Mulholland, M. R. & Fu, F.-X. Nutrient cycles and marine microbes in a CO2-enriched ocean. *Oceanography* **22,** 128–145 (2009).
21. Hutchins, D. A., Fu, F.-X., Webb, E. A., Walworth, N. & Tagliabue, A. Taxon-specific response of marine nitrogen fixers to elevated carbon dioxide concentrations. *Nat. Geosci.* **6,** 1–6 (2013).
22. Hutchins, D. A. *et al.* Irreversibly increased nitrogen fixation in *Trichodesmium* experimentally adapted to elevated carbon dioxide. *Nat. Commun.* **6** (2015).
23. Aamaas et al. IPCC 2013. *The Physical Science Basis. Contribution of working group I to the fifth assessment report of the intergovernmental panel on climate change.* (Cambridge University Press, 2013) http://www.ipcc.ch/report/ar5/wg1/.
24. Sunda, W. G. & Huntsman, S. A. Iron uptake and growth limitation in oceanic and coastal phytoplankton. *Marine Chem.* **50,** 189–206 (1995).
25. Fu, F. -X. *et al.* Interactions between changing pCO2, N2 fixation, and Fe limitation in the marine unicellular cyanobacterium *Crocosphaera*. *Limnol. Oceanogr.* **53,** 2472–2484 (2008).
26. Shi, D., Kranz, S. A., Kim, J.-M. & Morel, F. M. M. Ocean acidification slows nitrogen fixation and growth in the dominant diazotroph *Trichodesmium* under low-iron conditions. *Proc. Natl Acad. Sci. USA* **109,** E3094–E3100 (2012).
27. Walworth, N. *et al.* *Trichodesmium* genome maintains abundant, widespread noncoding DNA in situ, despite oligotrophic lifestyle. *Proc. Natl Acad. Sci. USA* **112,** 4251–4256 (2015).
28. Sohm, J. A. *et al.* Co-occurring *Synechococcus* ecotypes occupy four major oceanic regimes defined by temperature, macronutrients and iron. *ISME J.* **10,** 333–345 (2016).
29. Pavelka, N. *et al.* A power law global error model for the identification of differentially expressed genes in microarray data. *BMC Bioinformatics* **5,** 203 (2004).
30. Altschul, S., Gish, W., Miller, W., Myers, E. W. & Lipman, D. J. Basic local alignment search tool. *J. Mol. Biol.* **215,** 403–410 (1990).
31. Jorge, A. M., Hoiczyk, E., Gomes, J. P. & Pinho, M. G. EzrA contributes to the regulation of cell size in *Staphylococcus aureus*. *PLoS ONE* **6,** e27542 (2011).
32. Chien, A.-C., Hill, N. S. & Levin, P. A. Cell size control in bacteria. *Curr. Biol.* **22,** R340–R349 (2012).
33. Tropini, C. *et al.* Principles of bacterial cell-size determination revealed by cell-wall synthesis perturbations. *Cell Rep.* **9,** 1520–1527 (2014).
34. Su, Z., Olman, V. & Xu, Y. Computational prediction of Pho regulons in cyanobacteria. *BMC Genomics* **8,** 156 (2007).
35. Chappell, P. D. & Webb, E. A. A molecular assessment of the iron stress response in the two phylogenetic clades of *Trichodesmium*. *Environ. Microbiol.* **12,** 13–27 (2010).
36. Snow, J. T. *et al.* Quantifying integrated proteomic responses to iron stress in the globally important marine diazotroph *Trichodesmium*. *PLoS ONE* **10,** e0142626 (2015).
37. Raven, J. A. Iron acquisition and allocation in stramenopile algae. *J. Exp. Bot.* **64,** 2119–2127 (2013).
38. Yamada, T., Letunic, I., Okuda, S., Kanehisa, M. & Bork, P. iPath2.0: interactive pathway explorer. *Nucleic Acids Res.* **39,** W412–W415 (2011).
39. Arrigo, K. R. Marine microorganisms and global nutrient cycles. *Nature* **437,** 349–355 (2005).
40. Lange, M. B., Rujan, T., Martin, W. & Croteau, R. Isoprenoid biosynthesis: the evolution of two ancient and distinct pathways across genomes. *Proc. Natl Acad. Sci. USA* **97,** 13172–13177 (2000).
41. Fujita, Y., Tsujimoto, R. & Aoki, R. Evolutionary aspects and regulation of tetrapyrrole biosynthesis in cyanobacteria under aerobic and anaerobic environments. *Life* **5,** 1172–1203 (2015).
42. Kobayashi, K., Masuda, T., Tajima, N., Wada, H. & Sato, N. Molecular phylogeny and intricate evolutionary history of the three isofunctional enzymes involved in the oxidation of protoporphyrinogen IX. *Genome Biol. Evol.* **6,** 2141–2155 (2014).

43. Stephan, D. P., Ruppel, H. G. & Pistorius, E. K. Interrelation between cyanophycin synthesis, L-Arginine catabolism and photosynthesis in the cyanobacterium *Synechocystis sp.* strain PCC 6803. *Z. Naturforsch. C* **55,** 927–942 (2000).
44. Camsund, D. & Lindblad, P. Engineered transcriptional systems for cyanobacterial biotechnology. *Front. Bioeng. Biotechnol.* **2,** 40 (2014).
45. Kronholm, I. & Collins, S. Epigenetic mutations can both help and hinder adaptive evolution. *Mol. Ecol.* **25,** 1856–1868 (2015).
46. Casacuberta, E. & González, J. The impact of transposable elements in environmental adaptation. *Mol. Ecol.* **22,** 1503–1517 (2013).
47. Fu, F. -X., Warner, M. E., Zhang, Y., Feng, Y. & Hutchins, D. A. Effects of increased temperature and CO2 on photosynthesis, growth, and elemental ratios in marina *Synechococcus* and *Prochlorococcus* (Cyanobacteria). *J. Phycol.* **43,** 485–496 (2007).
48. Hutchins, D. A. *et al.* CO2 control of *Trichodesmium* N2 fixation, photosynthesis, growth rates, and elemental ratios. *Limnol. Oceanogr.* **52,** 1293–1304 (2007).
49. Cooley, S. R. & Yager, P. L. Physical and biological contributions to the western tropical North Atlantic Ocean carbon sink formed by the Amazon River plume. *J. Geophys. Res.* **111 (2006).**
50. Okasanen, J., Blanchet, F. G., Kindet, R. & Legendre, P. vegan: Community Ecology Package, R package version 2.0-4 (2012).
51. Zhang, B. *et al.* Detecting differential and correlated protein expression in label-free shotgun proteomics. *J. Proteome Res.* **5,** 2909–2918 (2006).
52. Day, R. W. & Quinn, G. P. Comparisons of treatments after an analysis of variance in ecology. *Ecol. Monographs* **59,** 433–463 (1989).
53. Ploner, A. *Heatplus: heatmaps with row and/or column covariates and colored clusters. (R package version,* 2012).
54. Katoh, K. & Standley, D. M. MAFFT multiple sequence alignment software version 7: improvements in performance and usability. *Mol. Biol. Evol.* **30,** 772–780 (2013).
55. Tamura, K., Dudley, J., Nei, M. & Kumar, S. MEGA4: Molecular Evolutionary Genetics Analysis (MEGA) software version 4.0. *Mol. Biol. Evol.* **24,** 1596–1599 (2007).

Acknowledgements

Grant support was provided by U.S. National Science Foundation OCE 1260490 to D.A.H., E.A.W. and F.-X.F., and OCE OA 1220484 and G.B. Moore Foundation 3782 and 3934 to M.A.S.

Author contributions

Experiments were conceived and carried out by D.A.H., F.-X.F., N.G.W., E.A.W. and M.A.S., with experimental and analytical contributions from D.M., M.R.M. and M.D.L. N.G.W. wrote the manuscript, with assistance from D.A.H., E.A.W., M.A.S., M.D.L. and F.-X.F.

Additional information

Competing financial interests: The authors declare no competing financial interests.

Spatial and temporal changes in cumulative human impacts on the world's ocean

Benjamin S. Halpern[1,2,3], Melanie Frazier[3], John Potapenko[4], Kenneth S. Casey[5], Kellee Koenig[6], Catherine Longo[3], Julia Stewart Lowndes[3], R. Cotton Rockwood[7], Elizabeth R. Selig[6], Kimberly A. Selkoe[3,8] & Shaun Walbridge[9]

Human pressures on the ocean are thought to be increasing globally, yet we know little about their patterns of cumulative change, which pressures are most responsible for change, and which places are experiencing the greatest increases. Managers and policymakers require such information to make strategic decisions and monitor progress towards management objectives. Here we calculate and map recent change over 5 years in cumulative impacts to marine ecosystems globally from fishing, climate change, and ocean- and land-based stressors. Nearly 66% of the ocean and 77% of national jurisdictions show increased human impact, driven mostly by climate change pressures. Five percent of the ocean is heavily impacted with increasing pressures, requiring management attention. Ten percent has very low impact with decreasing pressures. Our results provide large-scale guidance about where to prioritize management efforts and affirm the importance of addressing climate change to maintain and improve the condition of marine ecosystems.

[1] Bren School of Environmental Science and Management, University of California, Santa Barbara, California 93106, USA. [2] Imperial College London, Silwood Park Campus, Buckhurst Road, Ascot SL57PY, UK. [3] National Center for Ecological Analysis and Synthesis, 735 State St Suite 300, Santa Barbara, California 93101, USA. [4] Department of Geography, University of California, Santa Barbara, California 93106, USA. [5] NOAA National Centers for Environmental Information, Silver Spring, Maryland 20910, USA. [6] Betty and Gordon Moore Center for Science and Oceans, Conservation International, Arlington, Virginia 22202, USA. [7] Scripps Institution of Oceanography, University of California, San Diego, California 92093, USA. [8] Hawaii Institute of Marine Biology, University of Hawaii, Kaneohe, Hawaii HI 97644, USA. [9] ESRI, Boston Office, Middleton, Massachusetts 01949, USA. Correspondence and requests for materials should be addressed to B.S.H. (email: halpern@bren.ucsb.edu).

The ocean is crowded with human uses[1]. As human populations continue to grow and migrate to the coasts, demand for ocean space and resources is expanding, increasing the individual and cumulative pressures from a range of human activities. Marine species and habitats have long experienced detrimental impacts from human stressors[2,3], and these stressors are generally increasing globally[4]. However, the spatial patterns of these stressors are varied and the amount of recent change is largely unknown. In many places, we know little about which stressors are having the biggest impact on ecosystem condition, their cumulative effect or how the composition of pressures is changing over time.

Quantifying and mapping local- and global-scale stressors in a standardized, comparable manner offers a powerful means to assess both the spatial pattern and temporal change of individual human pressures, as well as their total impact on natural systems across highly variable geographies[2]. Quantitative methods to map cumulative human impacts were recently developed and initially applied to marine ecosystems globally[2]. To date, these methods have been applied to marine and freshwater regions around the world to assess spatial patterns of cumulative impact[2,5–9], and to explore how cumulative impacts affect or relate to other ecological processes or conservation needs (for example, refs 10,11). These efforts have helped identify which areas and ecosystem types are relatively pristine or heavily impacted, where hotspots of biodiversity and impacts overlap, and which stressors dominate human impact[12–14], in turn informing biodiversity conservation, threat mitigation and spatial planning decision processes (for example, ref. 15). Missing from these studies is an assessment of the location and intensity of change in cumulative impacts over time. Such temporal assessments would illuminate where and to what degree stressors are increasing or decreasing in intensity and impact, thus providing a means to assess management efficacy and adaptively respond to change. They can also support proactive management by informing our expectation of future states by tracking current trajectories.

Here we calculate and map the cumulative impact of 19 different types of anthropogenic stress on 20 global marine ecosystem types using best available global-scale data as of 2013 (Supplementary Tables 1–2). For 12 of these anthropogenic stressors, we used equivalent methods and data sources in the current and previous (5 years before) time periods, allowing assessment of the 5-year change in their individual and cumulative impacts (see Supplementary Methods). To help identify regions with different management and conservation needs, we identify areas experiencing the greatest and least cumulative impact and highest or lowest amount of change.

Results

Change in cumulative impact. Nearly 66% of the ocean experienced increases in cumulative impact over the 5-year study span

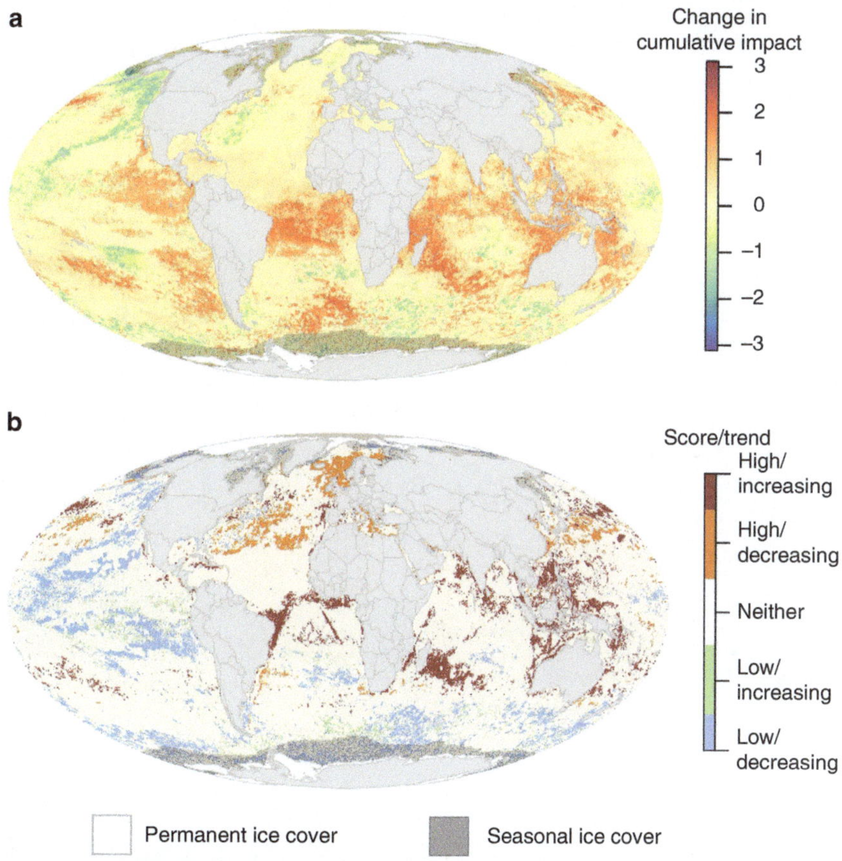

Figure 1 | Change in cumulative human impact to marine ecosystems. (**a**) Absolute difference between current (as of 2013) and earlier (as of 2008) per-pixel cumulative impact scores based on 12 anthropogenic stressors that could be compared across time (max cumulative impact score for both periods = 11.1). Positive scores represent an increase in cumulative impact. (**b**) Extreme combinations of cumulative impact and impact trend include areas with combinations of the highest (top quartile) and lowest (bottom quartile) impact and increasing (top quartile) and decreasing (bottom quartile) impact. [illegible] shaded white and the area within maximum sea ice extent is shaded to indicate where scores are less certain [illegible]

(Fig. 1a, Supplementary Fig. 1). Increases tended to be located in tropical, subtropical and coastal regions, with average increases in 77% of all exclusive economic zones (EEZs; Supplementary Data 1–2). In contrast, only 13% of the ocean experienced decreases over the study span (Supplementary Data 3), with these areas concentrated in the Northeast and Central Pacific and Eastern Atlantic (Fig. 1a). The Southern Ocean showed a patchy mix of increases and decreases. The largest average increases were in French territorial holdings in the Indian Ocean, Tanzania and the Seychelles. The greatest average decreases were within the EEZs of several remote South Pacific islands, the Alaskan coast and several European countries (Slovenia, Albania, Denmark and the Netherlands; Fig. 1a and 2, Supplementary Fig. 2; Supplementary Data 2 and 4). Change in cumulative impact was uncorrelated with current cumulative impact (Supplementary Table 3). Overall, countries with greater increases in coastal population had larger 5-year changes in cumulative impacts. Absolute coastal population size was unrelated to change in cumulative impact. Nevertheless, many places that are largely uninhabited or have relatively low population densities still experienced large increases in impacts (Fig. 2), suggesting that population size may not always drive decreases in ecological condition.

Globally, increases in climate change stressors (sea surface temperature anomalies, ocean acidification and ultraviolet radiation) drove most of the increase in cumulative impact (Fig. 3a, Supplementary Fig. 3; Supplementary Table 4), confirming the need to address climate change to maintain and sustain marine ecosystems globally. Commercial fishing impacts increased in <10% of the ocean for any type of commercial fishing, and on average in only 40 (17%) of 239 EEZs (Fig. 3c,d, Supplementary Fig. 3). In fact, impacts from four of the five types of commercial fishing decreased in 70–80% of the ocean, consistent with results suggesting global catch has stabilized or is declining in most parts of the ocean[16] and that well-managed fisheries are achieving sustainable yields[17]. However, we used fisheries catch as a measure of impact on ecosystems, which does not account for potential longer-lasting impacts of overfishing. We also had to assume that the proportion of catch per gear type remained constant within each EEZ, and so we may underestimate the impact of fishing. In addition, legacy effects of overfishing would not be captured by this analysis, and are likely greatest with habitat-modifying gear and long-lived species that primarily occur along continental shelves and pelagic waters, respectively. Such legacy effects may also be pronounced for invasive species, where current shipping intensity (and associated ballast water release and hull fouling) does not reflect past exposure and establishment of invasive species. Land-based stressors all increased globally (Fig. 3b, Supplementary Fig. 3; Supplementary Table 4), but these increases were concentrated in coastal areas of only 27–52% of all EEZs (depending on type of stressor; Supplementary Data 1).

Current cumulative impact. This updated assessment of cumulative impact confirms previous findings[2] that no part of the global ocean is without human influence. Nearly the entire ocean (97.7%) is affected by multiple stressors. Several 'hotspots' of cumulative impact exist where nearly all stressors overlap, most notably in the North Sea and South and East China Seas (Fig. 4, Supplementary Fig. 4). The many stressors associated with climate change

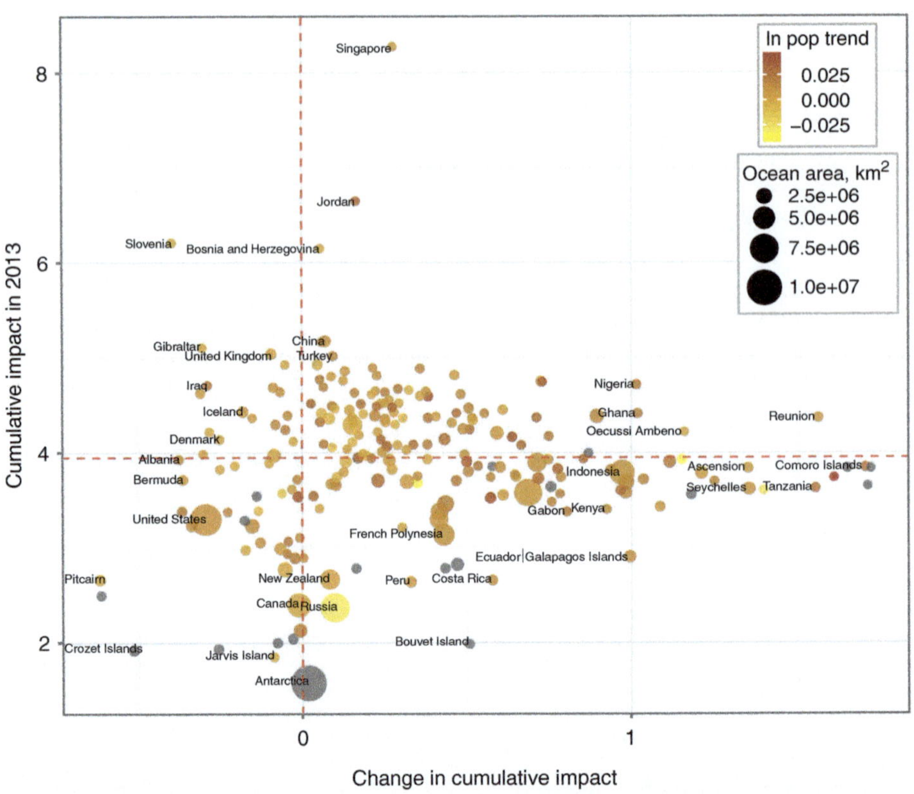

Figure 2 | Current cumulative impact versus change in impact. The relationship between current cumulative impact (as of 2013) and 5-year change in impact from 5 years before for each country's EEZ (200 nm) is shown based on the 12 common stressors. Bubbles are scaled to the area (ln) of each country's EEZ and colour-coded by the change in the log of coastal population (25 miles inland) per year from 2008 to 2013; a subset of countries is labelled. Grey bubbles are nearly uninhabited. Horizontal dashed red line is the global median cumulative impact score in 2013; vertical line is no change over time. See Supplementary Tables 1 and 2 for data for all countries.

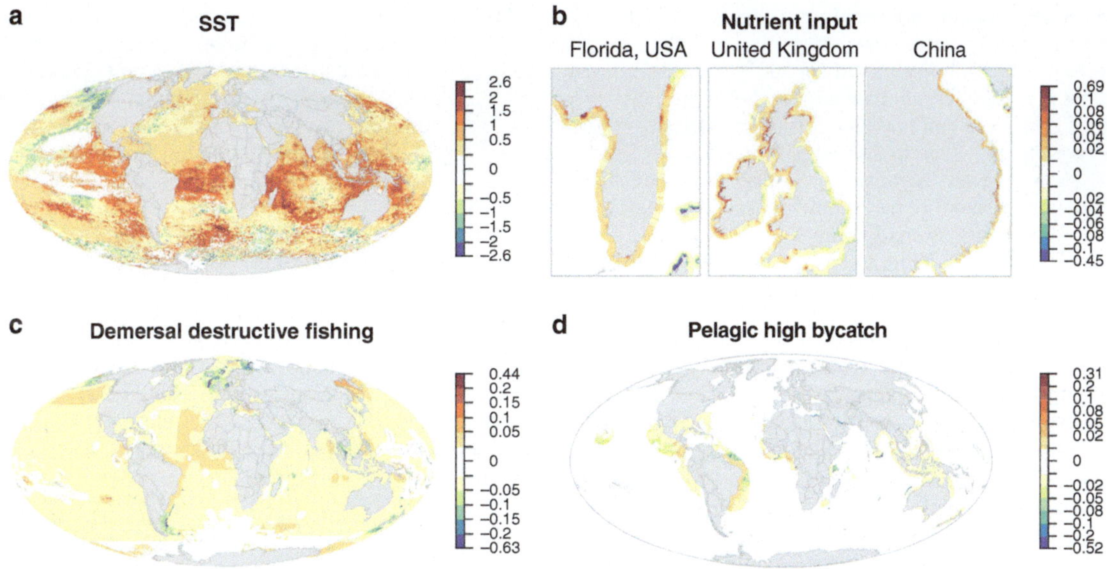

Figure 3 | Absolute difference in 2013 versus 2008 per-pixel stressor intensities for four representative stressors. (**a**) Sea surface temperature anomalies, (**b**) nutrient input, (**c**) demersal destructive fishing and (**d**) pelagic high bycatch fishing. Positive scores represent an increase in stressor intensity. Note that colour scales differ among panels and are nonlinear.

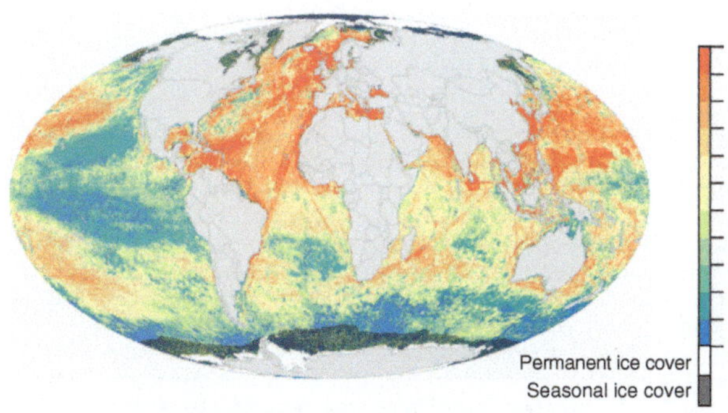

Figure 4 | Cumulative human impact to marine ecosystems as of 2013. Impact scores are based on all 19 anthropogenic stressors. Colours are assigned to 10-quantiles in the data, except the highest scores which are the top 5% of scores. Areas of permanent sea ice are shaded white and the area within maximum sea ice extent is masked to indicate where scores are less certain because change in sea ice extent could not be included (see Supplementary Methods).

(anomalously high sea surface temperatures, ocean acidification and increasing ultraviolet radiation) dominate humanity's footprint on the open ocean, but commercial fishing and shipping also cover large areas of the oceans and contribute significantly to overall impact (Supplementary Figs 5–7; Supplementary Data 5 and 6). In nearshore coastal waters, stressor combinations are more complex and varied (Supplementary Figs 4 and 8; Supplementary Data 7 and 8). National waters currently experiencing highest estimated impacts include those off Singapore, Jordan, Slovenia and Bosnia (Supplementary Fig. 5), while the most impacted coastal ecoregions[18] include the Faroe Islands, Eastern Caribbean, Cape Verde and Azore islands (Supplementary Data 1 and 3). Least impacted geographic areas are primarily in the poles, but also include relatively large areas in the central Pacific like the waters surrounding Jarvis Island and Palmyra Atoll (USA) and the Line Group of Kiribati, as well as temperate ecoregions around ... (Fig. 4; Supplementary Data 1

Discussion

These patterns of change in pressures over time offer guidance on where mitigation efforts are most needed (that is, where cumulative impacts are strongly increasing) and where past management actions may be paying dividends (impacts are high but strongly decreasing; Fig. 1a). Furthermore, overlaying the current (best available data as of 2013) cumulative impact map with 5-year changes in cumulative impact (Fig. 1b) reveals two scenarios of particular importance to management: areas of high and increasing impact, and areas of low and decreasing impact. The former scenario (5% of the ocean) merits immediate management action, focusing on pressure mitigation; the latter scenario (10% of the ocean) may be a lower priority, although controlling or decreasing pressures on already low impact areas could be strategic (Fig. 1b; Supplementary Table 5). Several areas of very high impact (North Sea, Mediterranean Sea and East China Sea; Fig. 4) experienced decreases in cumulative impact, while many offshore regions in the subtropical Atlantic and Indian Oceans that previously had relatively low impact scores saw large increases (Fig. 1b).

Decreases in individual stressors were generally relatively small on average and limited in area, but occurred for each stressor type and included areas of notable decrease (Supplementary Fig. 3). For example, demersal destructive (for example, trawl) fishing decreased significantly in many European countries, pelagic high bycatch (for example, longline) fishing decreased in several parts of the Middle East, sea surface temperature anomalies decreased in the Line Islands in the Pacific and around Alaska, USA and nutrient input decreased in the Adriatic Sea. Because of legacy effects of overfishing, decreases in catch may not translate into improved ecosystem condition or sustainable yields. Few stressors expanded their global footprint (Supplementary Table 4), primarily because their extent was already nearly global[2]. In short, even where some stressors show signs of decreasing, cumulative impact across all stressors is generally increasing, especially in coastal areas where human uses of the ocean are the greatest.

Our results do not account for potential losses in habitat which ... with high intensity of multiple overlapping stressors, especially within intertidal and nearshore habitats. Habitat extent is poorly known for most marine habitats, and change in extent essentially unknown for all; our analyses used the same habitat extent for all habitats for both time periods. While change in sea ice extent is well mapped via satellites, we did not include sea ice as a habitat because it is naturally highly variable, or as a stressor because its impact on ecosystems remains poorly understood[19] (see Supplementary Methods); we represent this uncertainty as shaded areas on mapped results of cumulative human impact. Trends in sea ice extent over the 5 years of our analyses varied spatially but many locations had already lost significant amounts of ice; where significant loss occurred during the time period of our analyses, our estimates of cumulative impact (and climate stressor impact in particular) would be significantly underestimated.

Cumulative impact assessment currently relies on assumptions of linear and additive responses of natural systems to stressors[20]. However, marine ecosystems may exhibit threshold responses to intense and cumulative stress[21], creating nonlinear relationships of cumulative impact to ecological condition. Recent syntheses show that nonlinear responses of ecosystems to increases in single stressors are common but difficult to predict[22]. Emerging evidence also suggests that stressor interactions are more commonly synergistic and mitigative than additive[23]. Currently, insufficient information exists to allow incorporation of these relationships into the cumulative impact assessment, but once available they can be accommodated. Furthermore, several known stressors to marine systems could not be included because of insufficient global data, including offshore energy infrastructure (for example, wind farms, submerged pipes and cables, deep sea mining, marine debris). Nevertheless, cumulative impact assessments remain one of the few comprehensive quantitative tools to measure how humans are affecting natural systems, and how actions targeting specific stressors may be expected to improve or exacerbate overall impacts. Because the approach allows direct comparisons, it is possible to measure change through time, allowing for a detailed view of where individual and cumulative human impacts are increasing or decreasing and which stressors are most important for driving those changes. This analysis of change over 5 years cannot fully account for natural longer term climate variations (such as decadal oscillations), but it provides a strong indication of direction (and location) of human-caused change. This assessment is thus constructive both to set management priorities and assess effectiveness of past actions, and is particularly useful for marine spatial planning and ecosystem-based management that must address the cumulative impact of multiple human uses of the ocean[24,25].

Empirically, measuring overall condition of natural systems remains difficult and resource intensive. Few, if any, approaches exist that allow direct comparison of condition globally and across scales. Our assessment of cumulative impact, although a prediction rather than measurement of condition, is highly valuable because of its global, scalable and quantitative comparability. Furthermore, previous global assessments[2] and a regional comparison of cumulative impact scores and ecosystem condition in the Baltic Sea[26] suggest modelled impact scores describe actual condition reasonably well. Our approach also moves beyond assessing change in intensity of stressors (that is, 'ecological footprints') by accounting for vulnerability or resilience of ecosystems to different stressor types[27]. In other words, our approach accounts for the reality that increases in stressor intensity may not lead to changes in ecosystem condition, while in other cases (that is, more sensitive ecosystems) small increases in intensity could cause large changes in condition.

Our results offer guidance for most management and conservation strategies, both proactive and reactive. For example, results can support prioritization of regions or stressors of concern globally (for example, as is done by the Global Environment Facility branch of the World Bank) and nationally (for example, through US's National Ocean Policy), track progress towards meeting management and policy objectives (for example, as mandated by the European Union's Marine Strategy Framework Directive) and potentially even set targets for total acceptable cumulative human impact on ecosystems in support of broader ecosystem-based management goals. If the ocean is going to continue to support and sustain human values and needs, addressing and mitigating cumulative impacts must become standard. Our finding that the majority of global waters are currently experiencing large and increasing cumulative impact of human activities brings urgency to addressing this need.

Methods

General model. Calculation of cumulative impacts followed and built on the approach developed previously[2,6]. Cumulative impact (I_C) is the per-pixel average of the habitat vulnerability-weighted stressor intensities (see Supplementary Table 1 for list of stressors and habitats), where weights ($\mu_{i,j}$) are determined by the vulnerability of each *i*... *m* habitat (*E*) to each *j*... *n* stressor (*D*), such that:

$$I_C = \sum_{j=1}^{n} \frac{1}{m} \sum_{i=1}^{m} D_i \times E_j \times \mu_{i,j} \quad (1)$$

In the previous global analyses[2], the sum of weighted intensities was used to account for the three-dimensionality of the ocean; here we calculate the average (following ref. 6) to produce a single two-dimensional map. The previous approach (and results) was not used for temporal comparisons (see 'temporal comparisons' below). We used ecosystem vulnerability weights ($\mu_{i,j}$) developed previously[27] for all stressor–habitat combinations, including new ones added here, as all stressor–habitat combinations were assessed in that study.

Habitat and stressor data. For nearshore areas, we assumed benthic habitats are well-mixed with the water column above and so treat them as a single depth layer, as done elsewhere[6]. At depths >60 m, we treat the surface waters as a separate pelagic habitat, and at depths >200 m we assume three distinct depth layers (benthic, deep pelagic and surface pelagic). For offshore waters (>60-m depth), fully overlapping habitats from benthic and pelagic systems lead to imperfect representations of three-dimensional impact in a two-dimensional representation; in nearshore coastal areas there is only a single depth layer, removing this issue.

We updated most stressor data layers used previously[2], and used newly developed or significantly improved data sources for four layers (nutrient and organic land-based pollution, commercial shipping and port volume, which is used for invasive species, and ocean-based pollution), as well as data for two stressors new to this analysis (light pollution and sea level rise; see Supplementary Table 1 for full list of data). The only data layers that could not be updated were inorganic pollution from land-based sources, artisanal fishing and ocean acidification, and thus in those cases we used the exact same data as used in the previous analyses. Habitat data are infrequently updated and improved, and so all of the habitat data remain the same as those used previously[2]. As such, changes in cumulative impact scores are entirely due to changes in stressor intensities.

Normalization of stressor data. We first $\log[X+1]$ transformed each stressor data layer, except benthic structures. Benthic structures were treated as binary data since an oil rig either exists or does not. The transformation of data appropriately reduces the effect of extreme outliers when rescaling the data to assign the relative impact of different levels of the anthropogenic stressors considered here[28]. All data layers were then rescaled between 0 and 1, with the highest per-pixel transformed value for each stressor across either time period set = 1. We rescaled data in this way to ensure comparability across time periods (that is, using the same reference point across time). If stressor intensities increase in the future beyond this reference value, then analyses across all years of analyses would need to be redone. Rescaling allows for direct comparison among drivers with dramatically different units of measurement.

Temporal comparisons. We recalculated previous (2008) scores using updated methods to allow direct comparison with current (best available data as of 2013) results. Because some data sources were new or were developed using new methods that could not be applied to past data, we restricted temporal analyses to only those data layers that could be directly compared across time. This left 12 stressor layers and all habitat data (see Supplementary Table 1).

To help address potential management priorities, we classified each pixel as high, medium or low current (2013) cumulative impact and as increasing, no change and decreasing impact across the 5-year time frame of the study. High and low impact categories were classified as the top and bottom 25% of values, respectively, with all other values categorized as medium. This led to cutoff values of >4.02 (high impact) and <2.739 (low impact). Increasing and decreasing impact were similarly classified as the top and bottom 25% of values, respectively, with cutoff values of >0.602 (increasing) and < −0.045 (decreasing).

Data projection and representation. We used the same land-sea mask (and derived coastline) as we developed previously[2]. As was also done for that study, all data were represented at ~1 km^2 resolution, even though several layers had native resolutions at coarser scales. In doing so, we assumed the coarse-scale value was evenly distributed across all 1 km^2 cells within that region. For climate change drivers (sea surface temperature (SST) and ultraviolet anomalies and changes in ocean acidification), this assumption is reasonable given the scale at which those drivers act. The land-based drivers, human population data and benthic structures data were all available or produced at ~1 km^2 resolution, and spread of the impact of these stressors into the ocean at the same resolution is reasonable. Regardless, when coarse-scale data are distributed equally to all 1 km^2 cells within the larger area, the coarser scale pattern is essentially recreated while the finer resolution information is preserved where and when it is appropriate. Finally, before all analyses, we converted all data to the Mollweide projection with a WGS84 datum as it is an accurate single global projection that preserves area and allows data transfer and analysis among operating systems and software.

Summarizing results. To help aid decision making at regional, national and sub-national scales, we summarized individual and cumulative impact of stressors, and recent change in impact, by EEZs (using international standards for boundary delineation; ref. 29), marine ecoregions[18], large marine ecosystems and Food and Agriculture Organization (FAO) high seas regions. In each case, we averaged per-pixel values (current impact and change in impact), allowing direct comparison among regions despite large differences in size.

Input data. Methods for preparing stressor data that were unchanged from the previous analyses (Supplementary Table 1) are described in detail elsewhere[2]. Stressors with updated data were prepared using more recent years from the same data source. In these cases, we describe the new data but do not elaborate methods. We primarily focus on describing those layers where updating required new methods. Data for all habitats were unchanged from previous analyses[2]. The 20 different habitats included are listed in Supplementary Table 1.

Vulnerability weights. We used nearly identical vulnerability weights as developed and used before[27]. Because global data layers used here do not perfectly match the categories used in these vulnerability studies, we made the following adjustments. Commercial activity was equated with our shipping layer, non-point source weights were used for our pollution layers, nutrient input was the average of oligotrophic and eutrophic weights, and demersal nondestructive low bycatch commercial fishing weights were determined by multiplying high bycatch values by 0.75. The non-zero weights for sea level rise in deeper habitats and pelagic waters did not make sense and so were set equal to zero ($N = 5$). Light pollution weights had to be derived new for this study; peer-reviewed literature and our own expert judgment were used to set these values. Supplementary Table 2 provides the full set of vulnerability weights.

Methodological comparisons. To compare results from current methods and updated data sources to those from past methods[2], we correlated per-pixel output for 2008 from past and current approaches (results shown in Supplementary Fig. 9). Differences are expected for at least two reasons. First, the past approach summed rather than averaged impact scores across habitats within a pixel. This method was changed, following previous methods[6], to account for imperfect habitat extent data. For pixels with only a single habitat, the two methods produce identical results. In coastal areas, and in particular intertidal areas, multiple habitats typically exist within a single pixel, and so differences in scores would be expected to be concentrated in these areas. This is in fact what we found (see Supplementary Fig. 9).

Second, differences in how stressor data were normalized should lead to very minor differences when maximum stressor intensity has increased over time, which it has for many stressors. Previously[2], stressors were normalized to the maximum value for that time period, whereas in the current analysis comparing current to previous time periods, stressors were normalized to the maximum across both time periods. We controlled for this when doing the temporal analyses in this study, but did not (and could not) control for this when comparing previously published results to our current results.

References

1. Crowder, L. B. *et al.* Sustainability—resolving mismatches in US ocean governance. *Science* **313,** 617–618 (2006).
2. Halpern, B. S. *et al.* A global map of human impact on marine ecosystems. *Science* **319,** 948–952 (2008).
3. Kappel, C. V. Losing pieces of the puzzle: threats to marine, estuarine, and ... *Front. Ecol. Environ* **3,** 275–282 (2005).
...
5. Ban, N. & Alder, J. How wild is the ocean? Assessing the intensity of anthropogenic marine activities in British Columbia, Canada. *Aquat. Conserv.* **18,** 55–85 (2008).
6. Halpern, B. S. *et al.* Mapping cumulative human impacts to California Current marine ecosystems. *Conserv. Lett.* **2,** 138–148 (2009).
7. Selkoe, K. A. *et al.* A map of human impacts to a 'pristine' coral reef ecosystem, the Papahanaumokuakea Marine National Monument. *Coral Reefs* **28,** 635–650 (2009).
8. Allan, J. D. *et al.* Joint analysis of stressors and ecosystem services to enhance restoration effectiveness. *Proc. Natl Acad. Sci. USA* **110,** 372–377 (2013).
9. Micheli, F. *et al.* Cumulative human impacts on mediterranean and black sea marine ecosystems: assessing current pressures and opportunities. *PLoS ONE* **8,** e79889 (2013).
10. Tittensor, D. P. *et al.* Global patterns and predictors of marine biodiversity across taxa. *Nature* **466,** 1098–U1107 (2010).
11. Maxwell, S. M. *et al.* Cumulative human impacts on marine predators. *Nat. Commun.* **4,** 2688 (2013).
12. Selig, E. R. *et al.* Global priorities for marine biodiversity conservation. *PLoS ONE* **9,** e82898 (2014).
13. Halpern, B. S., Lester, S. E. & McLeod, K. L. Placing marine protected areas onto the ecosystem-based management seascape. *Proc. Natl Acad. Sci. USA* **107,** 18312–18317 (2010).
14. Halpern, B. S. *et al.* Global hotspots of land-based impacts to coastal marine ecosystems. *Conserv. Lett.* **2,** 189–196 (2009).
15. E.U. European Commission Legislation. Oceans, Seas and Coasts. Legislation: The marine directive http://ec.europa.eu/environment/marine/eu-coast-and-marine-policy/marine-strategy-framework-directive/index_en.htm, last updated: 22 April 2015 (2012).
16. Pauly, D. *et al.* Towards sustainability in world fisheries. *Nature* **418,** 689–695 (2002).
17. Worm, B. *et al.* Rebuilding global fisheries. *Science* **325,** 578–585 (2009).
18. Spalding, M. D. *et al.* Marine ecoregions of the world: a bioregionalization of coast and shelf areas. *Bioscience* **57,** 573–583 (2007).
19. Post, E. *et al.* Ecological consequences of sea-ice decline. *Science* **341,** 519–524 (2013).
20. Halpern, B. S. & Fujita, R. Assumptions, challenges, and future directions in cumulative impact analysis. *Ecosphere* **4,** 131 (2013).
21. Hughes, T. P., Bellwood, D. R., Folke, C., Steneck, R. S. & Wilson, J. New paradigms for supporting the resilience of marine ecosystems. *Trends Ecol. Evol.* **20,** 380–386 (2005).
22. Hunsicker, M. E. *et al.* Characterizing driver-response relationships in marine pelagic ecosystems for improved ocean management. *Ecol. Appl* (in press).
23. Crain, C. M., Kroeker, K. & Halpern, B. S. Interactive and cumulative effects of multiple human stressors in marine systems. *Ecol. Lett.* **11,** 1304–1315 (2008).
24. McLeod, K. L. & Leslie, H. M. *Ecosystem Based Management for the Oceans* (Island Press, 2009).
25. Halpern, B. S., McLeod, K. L., Rosenberg, A. A. & Crowder, L. B. Managing for cumulative impacts in ecosystem-based management through ocean zoning. *Ocean Coast. Manage.* **51,** 203–211 (2008).
26. Andersen, J. H., Halpern, B. S., Korpinen, S., Murray, C. & Reker, J. Baltic Sea biodiversity status relates to cumulative human pressures: implications for ecosystem-based management. *Estuar. Coast Shelf Sci* **161,** 88–92 (2015).
27. Halpern, B. S., Selkoe, K. A., Micheli, F. & Kappel, C. V. Evaluating and ranking the vulnerability of global marine ecosystems to anthropogenic threats. *Conserv. Biol.* **21,** 1301–1315 (2007).
28. Malczewski, J. On the use of weighted linear combination method in GIS: common and best practice approaches. *Transact. GIS* **4,** 5–22 (2000).
29. VLIZ. Maritime Boundaries Geodatabase, version 8.0. Available at http://www.vliz.be/vmdcdata/marbound (2014).

Acknowledgements

All data are described in the Supplementary Information and are freely available at doi:10.5063/F19Z92TW. This paper is a contribution to Imperial College's Grand Challenges in Ecosystems and the Environment initiative. This manuscript has been prepared with support from *Arctic Options: Holistic Integration for Arctic Coastal Marine Sustainability*, with funding from the National Science Foundation, Division of Polar Programs (Grant No. PLR-1263819).

Author contributions

B.S.H. and E.R.S. conceived the project. M.F., J.P., K.S.C., R.C.R. and S.W. analysed the data. All authors contributed to writing the manuscript.

Additional information

... The authors declare no competing financial interests.

PERMISSIONS

All chapters in this book were first published in NC, by Nature Publishing Group; hereby published with permission under the Creative Commons Attribution License or equivalent. Every chapter published in this book has been scrutinized by our experts. Their significance has been extensively debated. The topics covered herein carry significant findings which will fuel the growth of the discipline. They may even be implemented as practical applications or may be referred to as a beginning point for another development.

The contributors of this book come from diverse backgrounds, making this book a truly international effort. This book will bring forth new frontiers with its revolutionizing research information and detailed analysis of the nascent developments around the world.

We would like to thank all the contributing authors for lending their expertise to make the book truly unique. They have played a crucial role in the development of this book. Without their invaluable contributions this book wouldn't have been possible. They have made vital efforts to compile up to date information on the varied aspects of this subject to make this book a valuable addition to the collection of many professionals and students.

This book was conceptualized with the vision of imparting up-to-date information and advanced data in this field. To ensure the same, a matchless editorial board was set up. Every individual on the board went through rigorous rounds of assessment to prove their worth. After which they invested a large part of their time researching and compiling the most relevant data for our readers.

The editorial board has been involved in producing this book since its inception. They have spent rigorous hours researching and exploring the diverse topics which have resulted in the successful publishing of this book. They have passed on their knowledge of decades through this book. To expedite this challenging task, the publisher supported the team at every step. A small team of assistant editors was also appointed to further simplify the editing procedure and attain best results for the readers.

Apart from the editorial board, the designing team has also invested a significant amount of their time in understanding the subject and creating the most relevant covers. They scrutinized every image to scout for the most suitable representation of the subject and create an appropriate cover for the book.

The publishing team has been an ardent support to the editorial, designing and production team. Their endless efforts to recruit the best for this project, has resulted in the accomplishment of this book. They are a veteran in the field of academics and their pool of knowledge is as vast as their experience in printing. Their expertise and guidance has proved useful at every step. Their uncompromising quality standards have made this book an exceptional effort. Their encouragement from time to time has been an inspiration for everyone.

The publisher and the editorial board hope that this book will prove to be a valuable piece of knowledge for researchers, students, practitioners and scholars across the globe.

LIST OF CONTRIBUTORS

Jörn Callies
MIT/WHOI Joint Program in Oceanography, Massachusetts Institute of Technology, Building 54-1615, 77 Massachusetts Avenue, Cambridge/Woods Hole, Massachusetts 02139, USA

Raffaele Ferrari
Massachusetts Institute of Technology, Cambridge, Massachusetts 02139, USA

Jody M. Klymak
University of Victoria, Victoria, British Columbia V8W 3P6, Canada

Jonathan Gula
University of California, Los Angeles, California 90095, USA

Sönke Dangendorf and Jürgen Jensen
Department of Civil Engineering, Research Institute for Water and Environment, University of Siegen, Paul-Bonatz-Strasse 9-11, 57076 Siegen, Germany

Marta Marcos
IMEDEA (UIB-CSIC), Miquel Marquès, 21, E-07190 Esporles, Spain

Alfred Müller and Kevin Berk
Department of Mathematics, University of Siegen, Walter-Flex-Strasse 3, 57072
Siegen, Germany

Eduardo Zorita
Helmholtz-Centre Geesthacht, Max-Planck-Strasse 1, 21502 Geesthacht, Germany

Riccardo Riva
Departement of Geoscience and Remote Sensing, Delft University of Technology, Stevinweg 1, 2628 Delft, Netherlands

Luis J. Gilarranz and Jordi Bascompte
Department of Integrative Ecology, Estación Biológica de Doñana, (EBD-CSIC), Calle Américo Vespucio s/n, E-41092 Sevilla, Spain

Camilo Mora
Department of Geography, University of Hawaii, Honolulu, Hawaii 96822, USA

Hongmei Li, Tatiana Ilyina, Wolfgang A. Müller and Frank Sienz
Max Planck Institute for Meteorology, Bundesstraβe 53, 20146 Hamburg, Germany

Nicholas P. Foukal and M. Susan Lozier
Nicholas School of the Environment, Duke University, Durham, North Carolina 27708, USA

Mathew Koll Roxy
Centre for Climate Change Research, Indian Institute of Tropical Meteorology, Pashan, Pune 411008, India

Kapoor Ritika
Centre for Climate Change Research, Indian Institute of Tropical Meteorology, Pashan, Pune 411008, India
Department of Environmental Sciences, Fergusson College, University of Pune, Pune 411004, India

Pascal Terray
Sorbonne Universites (UPMC, Univ Paris 06)-CNRS-IRD-MNHN, LOCEAN Laboratory, 4 place

Jussieu, F-75005 Paris, France
Indo-French Cell for Water Sciences, IISc-IITM-NIO-IRD Joint International Laboratory, IITM, Pune 411008, India

Raghu Murtugudde
ESSIC, University of Maryland, College Park, Maryland 20740, USA

Karumuri Ashok
Centre for Climate Change Research, Indian Institute of Tropical Meteorology, Pashan, Pune 411008, India
Centre for Earth and Space Sciences, University of Hyderabad, Hyderabad 500046, India

B.N. Goswami
Centre for Climate Change Research, Indian Institute of Tropical Meteorology, Pashan, Pune 411008, India
Earth and Climate Science Department, Indian Institute of Science Education and Research, Pune 411008, India

H.E. Elderfield
Department of Earth Sciences, University of Cambridge, Cambridge CB2 3EQ, UK

E.W. Wolff
Department of Earth Sciences, University of Cambridge, Cambridge CB2 3EQ, UK
British Antarctic Survey, High Cross, Madingley Road, Cambridge CB3 0ET, UK

T.M. Conway
Department of Earth Sciences, University of Cambridge, Cambridge CB2 3EQ, UK
British Antarctic Survey, High Cross, Madingley Road, Cambridge CB3 0ET, UK
Department of Earth Sciences, Institute for Geochemistry and Petrology, ETH Zürich, 8092 Zürich, Switzerland (T.M.C.)

R. Röthlisberger
British Antarctic Survey, High Cross, Madingley Road, Cambridge CB3 0ET, UK
Federal Office for the Environment, Bern, Switzerland (R.R.)

R. Mulvaney
British Antarctic Survey, High Cross, Madingley Road, Cambridge CB3 0ET, UK

Ruediger Stein and Wilfried Jokat
Alfred Wegener Institute (AWI) Helmholtz Centre for Polar and Marine Research, Am Alten Hafen 26, Bremerhaven 27568, Germany
Department of Geosciences (FB5), University of Bremen, Klagenfurter Strasse 4, Bremen 28359, Germany

Kirsten Fahl, Gregor Knorr, Frank Niessen, Catalina Gebhardt, Laura Jensen and Jens Matthiessen
Alfred Wegener Institute (AWI) Helmholtz Centre for Polar and Marine Research, Am Alten Hafen 26, Bremerhaven 27568, Germany

Michael Schreck
Arctic Research Centre, Korea Polar Research Institute, 26 Songdomirae-ro, Yeonsu-gu, Incheon 406-840, Korea

Matthias Forwick
Institute of Geology, University of Tromsø—The Arctic University of Norway, P O Box 6050
Langnes, Tromsø 9037, Norway

Michael Kaminski
Geosciences Department, College of Petroleum Engineering & Geosciences, King Fahd University of Petroleum & Minerals, Dhahran 31261, Saudi Arabia

Achim Kopf
Department of Geosciences (FB5), University of Bremen, Klagenfurter Strasse 4, Bremen 28359, Germany
MARUM-Center for Marine Environmental Sciences, University of Bremen, Leobener Strasse, Bremen 28359, Germany

Gerrit Lohmann
Alfred Wegener Institute (AWI) Helmholtz Centre for Polar and Marine Research, Am Alten Hafen 26, Bremerhaven 27568, Germany
MARUM-Center for Marine Environmental Sciences, University of Bremen, Leobener Strasse, Bremen 28359, Germany

Volker Roeber and Jeremy D. Bricker
International Research Institute of Disaster Science, Tohoku University 468-1-E304 AzaAoba, Aramaki, Aoba-ku, Sendai 980-0845, Japan

Jan de Leeuw and Joris T. Eggenhuisen
Faculty of Geosciences, Utrecht University, PO Box 80021, 3508TA Utrecht, The Netherlands

Matthieu J.B. Cartigny
National Oceanography Centre, European Way, Southampton, Hampshire SO14 3ZH, UK

Maria de la Fuente and Eva Calvo
Departament de Biologia Marina i Oceanografia, Institut de Ciències del Mar, CSIC, Passeig Marítim de la Barceloneta 37–49, Barcelona 08003, Spain

Luke Skinner
Godwin Laboratory for Palaeoclimate Research, Department of Earth Sciences, University of Cambridge, Downing Street, Cambridge CB2 3EQ, UK.

Carles Pelejero
Departament de Biologia Marina i Oceanografia, Institut de Ciències del Mar, CSIC, Passeig Marítim de la Barceloneta 37–49, Barcelona 08003, Spain
Institució Catalana de Recerca i Estudis Avançats, Barcelona, Spain

Isabel Cacho
Grup de Recerca de Geociències Marines, Departament d'Estratigrafia, Paleontologia I
Geociències Marines, Universitat de Barcelona, C/ Martí i Franquès, Barcelona 08028, Spain

Hanno Sandvik
Centre for Biodiversity Dynamics, Department of Biology, Norwegian University of Science and Technology, 7491 Trondheim, Norway

Robert T. Barrett
Department of Natural Sciences, Tromsø University Museum, PO Box 6050 Langnes, 9037 Tromsø, Norway

Geir Helge Systad and Tone K. Reiertsen
Norwegian Institute for Nature Research, FRAM—High North Research Centre for Climate and the Environment, 9296 Tromsø, Norway

Kjell Einar Erikstad
Centre for Biodiversity Dynamics, Department of Biology, Norwegian University of Science and Technology, 7491 Trondheim, Norway
Norwegian Institute for Nature Research, FRAM—High North Research Centre for Climate and the Environment, 9296 Tromsø, Norway

Jofrid Skarjhamar, Mette Skern-Mauritzen, Mari S. Myksvoll and Frode Vikebø
Institute of Marine Research and Hjort Centre for Marine Ecosystem Dynamics, PO Box 1870 Nordnes, 5817 Bergen, Norway

Nigel G. Yoccoz
Norwegian Institute for Nature Research, FRAM—High North Research Centre for Climate and the Environment, 9296 Tromsø, Norway
Department of Arctic and Marine Biology, University of Tromsø, PO Box 6050 Langnes, 9037 Tromsø, Norway

Tycho Anker-Nilssen and Svein-Håkon Lorentsen
Norwegian Institute for Nature Research, PO Box 5685 Sluppen, 7485 Trondheim, Norway

S.M. Glenn, T.N. Miles, G.N. Seroka, R.K. Forney, H. Roarty, O. Schofield and J. Kohut
Center for Ocean Observing Leadership, Department of Marine and Coastal Sciences, School of Environmental and Biological Sciences, Rutgers University, 71 Dudley Road, New Brunswick, New Jersey 08901, USA

Y. Xu
State Key Laboratory of Estuarine and Coastal Research, East China Normal University, 3663 Zhongshan Road North, Shanghai 200062, China

F. Yu
Institute of Oceanology, Chinese Academy of Sciences, 7 Nanhai Road, Qingdao 266071, China

Ellen E. Martin
Department of Geological Sciences, University of Florida, Gainesville, Florida 32611, USA

Chandranath Basak
Department of Geological Sciences, University of Florida, Gainesville, Florida 32611, USA
Lamont-Doherty Earth Observatory, 61 Route 9W, PO Box 1000, Palisades, New York 10964, USA (C.B.)

Keiji Horikawa
Department of Geological Sciences, University of Florida, Gainesville, Florida 32611, USA
Graduate School of Environmental Studies, Nagoya University,Furo-cho, Chikusa-ku, Nagoya 464-8601,

Saburo Sakai and Jonaotaro Onodera
Japan Agency for Marine-Earth Science and Technology, 2-15 Natsushima-cho, Yokosuka 237-0061, Japan

Tatsuhiko Sakamoto
Japan Agency for Marine-Earth Science and Technology, 2-15 Natsushima-cho, Yokosuka 237-0061, Japan
Graduate School and Faculty of Bioresources, Mie University, 1577 Kurimamachiya-cho, Tsu, Mie 514-8507, Japan (T.S.)

Kimitaka Kawamura and Osamu Seki
Institute of Low Temperature Science, Hokkaido University, N19W8, Kita-ku, Sapporo, Hokkaido 060-0819, Japan

Minoru Ikehara
Center for Advanced Marine Core Research, Kochi University, B200 Monobe, Nankoku 783-8502, Japan

T.A. Ronge, R. Tiedemann, F. Lamy and P. Köhler
Alfred-Wegener-Institut Helmholtz-Zentrum für Polar- und Meeresforschung, Department for Marine Geology, PO Box 120161, Bremerhaven 27515, Germany

B.V. Alloway
School of Geography, Environment and Earth Sciences, Victoria University of Wellington, PO Box 600, 6012 Wellington, New Zealand

R. De Pol-Holz
GAIAAntárctica Universidad de Magellanes, Department of Paleclimatology, Oceanography, Punta Arenas 01855, Chile

K. Pahnke
Max Planck Research Group—Marine Isotope Geochemistry, Institute for Chemistry and Biology of the Marine Environment, Department of Marine Isotope Geochemistry, Carl von Ossietzky University, PO Box 2503, Oldenburg 26111, Germany

J. Southon
School of Physical Science, Department of Earth Science, University of California, Irvine, California 92697-4675, USA

L. Wacker
Laboratory of Ion Beam Physics (HPK), Eidgenössische Technische Hochschule, Schafmattstrasse 20, Zürich 8093, Switzerland

Lixiao Xu, Peiliang Li, Qinyu Liu, Cong Liu and Wendian Gao
Physical Oceanography Laboratory/Qingdao Collaborative Innovation Center of Marine Science and Technology, Ocean University of China, Qingdao 266100, China

Shang-Ping Xie
Physical Oceanography Laboratory/Qingdao Collaborative Innovation Center of Marine Science and Technology, Ocean University of China, Qingdao 266100, China
Scripps Institution of Oceanography, University of California San Diego, La Jolla, California 92093, USA

Wei Liu and Shang-Ping Xie
Climate, Atmospheric Sciences, and Physical Oceanography (CASPO), Scripps Institution of Oceanography, University of California, San Diego, 9500 Gilman Drive, La Jolla, California 92093, USA

Jian Lu
Pacific Northwest National Laboratory, 902 Battelle Blvd, Richland, Washington 99354, USA

Sev Kender and Melanie J. Leng
Centre for Environmental Geochemistry, School of Geography, University of Nottingham, Nottingham NG7 2RD, UK
British Geological Survey, Keyworth, Nottingham NG12 5GG, UK

Erin L. McClymont
Department of Geography, Durham University, Durham DH1 3LE, UK

Aurora C. Elmore
Department of Geography, Durham University, Durham DH1 3LE, UK
Department of Earth Sciences, University of Cambridge, Cambridge CB2 3EQ, UK

Henry Elderfield
Department of Earth Sciences, University of Cambridge, Cambridge CB2 3EQ, UK

Dario Emanuele
DST—Università del Sannio, Via dei Mulini 59a, Benevento 82100, Italy

Francisco M. Cornejo-Castillo, Ana M. Cabello, Josep M. Gasol, Ramon Massana, Silvia G. Acinas and Guillem Salazar
Department of Marine Biology and Oceanography, Institut de Ciències del Mar (ICM), CSIC, Pg. Marítim de la Barceloneta 37-49, 08003 Barcelona, Spain

Patricia Sánchez-Baracaldo
School of Geographical Sciences, University of Bristol, Bristol BS8 1SS, UK

Gipsi Lima-Mendez and Jeroen Raes
Department of Microbiology and Immunology, Rega Institute KU Leuven, Herestraat 49, 3000 Leuven, Belgium
VIB Center for the Biology of Disease, VIB, Herestraat 49, 3000 Leuven, Belgium
Department of Applied Biological Sciences (DBIT), Vrije Universiteit Brussel, Pleinlaan 2, 1050 Brussels, Belgium

Pascal Hingamp
Aix Marseille Université CNRS IGS UMR 7256, 13288 Marseille, France

Adriana Alberti
CEA-Institut de Génomique, Genoscope, Centre National de Séquençage, 2 rue Gaston Crémieux, CP5706, F-91057 Evry, France

Shinichi Sunagawa
European Molecular Biology Laboratory, Structural and Computational Biology Unit, Meyerhofstrasse 1, 69117 Heidelberg, Germany

Peer Bork
European Molecular Biology Laboratory, Structural and Computational Biology Unit, Meyerhofstrasse 1, 69117 Heidelberg, Germany
Max-Delbrück-Centre for Molecular Medicine, 13092 Berlin, Germany

Colomban de Vargas
CNRS, UMR 7144, Station Biologique de Roscoff, Place Georges Teissier, 29680 Roscoff, France
Sorbonne Universités, UPMC Université Paris 06, UMR 7144, Station Biologique de Roscoff, Place Georges Teissier, 29680 Roscoff, France

Chris Bowler
Ecole Normale Supérieure, PSL Research University, Institut de Biologie de l'Ecole Normale Supérieure (IBENS), CNRS UMR 8197, INSERM U1024, 46 rue d'Ulm, F-75005 Paris, France

Patrick Wincker
CEA-Institut de Génomique, Genoscope, Centre National de Séquençage, 2 rue Gaston Crémieux, CP5706, F-91057 Evry, France
Université d'Evry, UMR 8030, CP5706 Evry, France
Centre National de la Recherche Scientifique (CNRS), UMR 8030, CP5706 Evry, France

Jonathan P. Zehr
Department of Ocean Sciences, University of California, Santa Cruz, California 95064, USA

Jacob N.W. Howe and Alexander M. Piotrowski
Department of Earth Sciences, University of Cambridge, Cambridge CB2 3EQ, UK

Taryn L. Noble
Department of Earth Sciences, University of Cambridge, Cambridge CB2 3EQ, UK
Institute for Marine and Antarctic Studies (IMAS), University of Tasmania, Hobart, Tasmania 7001, Australia

Stefan Mulitza
MARUM-Center for Marine Environmental Sciences, University of Bremen, Leobener Strasse, D-28359 Bremen, Germany

Cristiano M. Chiessi
School of Arts, Sciences and Humanities, University of São Paulo, Av. Arlindo Bettio 1000, CEP03828-000 São Paulo SP, Brazil

Germain Bayon
Institut Français de Recherche pour l'Exploitation de la Mer (IFREMER), Unité de Recherche Géosciences Marines, F-29280 Plouzané, France

Sanne Muis, Jeroen C.J.H. Aerts and Philip J. Ward
Institute for Environmental Studies (IVM), Vrije Universiteit Amsterdam, 1081 HV Amsterdam, The Netherlands

Hessel C. Winsemius
Deltares, 2600 MH Delft, The Netherlands

Martin Verlaan
Deltares, 2600 MH Delft, The Netherlands
Mathematical Physics, Electrical Engineering, Mathematics and Computer Science (EEMCS), TU Delft, 2628 CD Delft, The Netherlands

Andrew M.W. Newton, Mads Huuse and Simon H. Brocklehurst
School of Earth, Atmospheric and Environmental Sciences, University of Manchester, Manchester M13 9PL, UK

E. Freeman, L.C. Skinner and D. Hodell
Godwin Laboratory for Palaeoclimate Research, Department of Earth Sciences, University of Cambridge, Cambridge CB2 3EQ, UK

C. Waelbroeck
LSCE/IPSL, Laboratoire CNRS-CEA-UVSQ, Domaine du CNRS, bât 12 91198 Gif-sur-Yvette, France

M. Münnich
Environmental Physics, Institute of Biogeochemistry and Pollutant Dynamics, ETH Zurich, CHN E 23.2, Universitatstrasse 16, Zürich 8092, Switzerland

D. Byrne and N. Gruber
Environmental Physics, Institute of Biogeochemistry and Pollutant Dynamics, ETH Zurich, CHN E 23.2, Universitatstrasse 16, Zürich 8092, Switzerland
Center for Climate Systems Modeling (C2SM), ETH Zurich, Zürich 8092, Switzerland

I. Frenger
Princeton University, Princeton, New Jersey 08544, USA.
GEOMAR, Helmholtz Centre for Ocean Research, Kiel 24105, Germany

Nathan G. Walworth, Fei-Xue Fu, Eric A. Webb, Michael D. Lee and David A. Hutchins
Marine and Environmental Biology, Department of Biological Sciences, University of Southern California, 3616 Trousdale Parkway, Los Angeles, California 90089, USA

Mak A. Saito, Dawn Moran and Matthew R. McIlvin
Marine Chemistry and Geochemistry Department, Woods Hole Oceanographic Institution, Woods Hole, Massachusetts 02543, USA

Benjamin S. Halpern
Bren School of Environmental Science and Management, University of California, Santa Barbara, California 93106, USA
Imperial College London, Silwood Park Campus, Buckhurst Road, Ascot SL57PY, UK
National Center for Ecological Analysis and Synthesis, 735 State St Suite 300, Santa Barbara, California 93101, USA

Catherine Longo, Julia Stewart Lowndes and Melanie Frazier
National Center for Ecological Analysis and Synthesis, 735 State St Suite 300, Santa Barbara, California 93101, USA

John Potapenko
Department of Geography, University of California, Santa Barbara, California 93106, USA

Kenneth S. Casey
NOAA National Centers for Environmental Information, Silver Spring, Maryland 20910, USA

Kellee Koenig and Elizabeth R. Selig
Betty and Gordon Moore Center for Science and Oceans, Conservation International, Arlington, Virginia 22202, USA

R. Cotton Rockwood
Scripps Institution of Oceanography, University of California, San Diego, California 92093, USA

Kimberly A. Selkoe
Hawaii Institute of Marine Biology, University of Hawaii, Kaneohe, Hawaii HI 97644, USA

Shaun Walbridge
ESRI, Boston Office, Middleton, Massachusetts 01949, USA

Index

A

Air-sea Fluxes, 1, 111, 208

Anomalies, 15, 23-25, 28-34, 38-43, 85-86, 101, 130, 142, 148, 200-203, 207-208, 222-223, 225

Anthropogenic, 9-10, 14, 17-20, 24, 28, 36-37, 42-44, 51, 57, 94, 111, 133, 140-142, 144-145, 147-149, 184, 191, 207, 210, 218, 221, 223-225

Anthropogenic Co2, 28, 51, 94, 140, 191

Anthropogenic Footprints, 9, 20

Anthropogenic Influences, 18-19

Anthropogenic Warming, 36, 141, 144-145

Atmospheric Component, 9, 11, 28, 43, 207

Atmospheric Dust, 46-48, 51, 210

Atmospheric Fe Fluxes, 46-47, 49-50

Atmospheric Gases, 1

B

Baroclinic Instability, 3, 5-8, 140, 205

Bioenergetic Model, 21

Breaking Waves, 68

Buoyancy Gradients, 2, 7

C

Carbon Storage, 46-47, 122, 168, 173, 193, 197, 216

Climate Models, 37, 44, 57, 132-133, 138, 150, 185, 199, 202

Climate Variations, 10, 224

Cmip5 Models, 37, 42-44, 149, 207

Co2 Sink, 23, 94

Colonial Breeding, 95-96, 101

D

Decadal Prediction System, 23, 25

Dipolar Rainfall Pattern, 38

E

Earth System Models, 23-24

Eastern Equatorial Pacific, 84-86

Environmental Degradation, 18

F

Fish Larvae, 95, 97-99

Fishing Effort, 18-22

Food-web Persistence, 18-20, 22

G

Glacial Carbon Pool, 121-122, 127

Glacial Intervals, 46-47, 51

Glacial-interglacial, 46-47, 51, 53, 88, 92, 113, 116, 130, 154, 156, 167-168, 171, 174

Global Environmental Change, 18-19, 156

Global Thermohaline Circulation, 124

Gulf Stream, 2-3, 7-8, 30-31, 34, 140

H

Heat Penetration, 141-143

Heat Transport, 24-25, 30-31, 66, 118, 141-142, 147-148, 168, 207

High-latitude Warming, 55

I

Ice Ages, 46, 130

Inter-gyre, 30-31, 34

Ipcc Ar5, 37

K

Kinetic Energy, 2-8, 201, 204

L

Lagrangian Analyses, 30-31

Land-sea Thermal, 36-40, 42-43

Land-sea Thermal Gradient, 36, 38-39

Larval Drift Model, 95-96, 99-100

Last Deglaciation, 53, 84, 86-89, 91, 93-94, 115-116, 118, 121-122, 129, 168, 173-174, 199-200

Last Glacial Maximum, 46-47, 53, 85-86, 93-94, 115-116, 130, 167-169, 172-174, 193, 199

Late Miocene, 55, 57, 59-66, 116, 120

Local Food Web, 19

M

M2 Glaciation, 112-116

Macroecological Pattern, 95, 99

Marine Food Webs, 18, 20, 22

Mesoscale, 1-8, 53, 104, 110-111, 132-133, 137-140, 201-208

Mesoscale Eddy, 2-3, 132-133, 140, 201-202, 208

Meteotsunami, 69

Miocene Sediments, 55, 60

Moderate Storms, 68, 72

Modern Climate, 55

Modern Seafloor, 77, 80

Monotonic Warming, 38-39

Morphodynamics, 77-78, 82

Multi-channel Seismic, 57-58, 67

Multi-decadal Oscillation, 37

N

Numerical Modelling, 2, 140, 197

O

Ocean Mixing, 23, 25
Ocean Sst, 38, 41, 43-44, 56
Ocean Surface, 1-2, 27-28, 34, 57, 60, 66, 92, 105, 111, 140, 187, 203, 205, 207-208
Ocean Volume, 85, 92
Ocean Warming, 28, 36-38, 40-44, 147-149
Oceanic Carbon Sink, 28
Oscillation, 17, 23-24, 28-29, 31, 34, 37, 43, 74, 93, 119, 130, 142, 148, 185

P

Pliocene, 57-60, 63-64, 66, 112-113, 115-120, 156-157, 174
Polarstern, 55, 57-59, 64, 66-67, 131

Q

Quasi-geostrophic, 2, 5-8, 207

R

Radiocarbon, 54, 84-94, 121-122, 124-130, 173-174, 193-200
Rain-fed, 36-38

S

Sea Level Pressure, 11, 106-107, 146
Sea Surface Temperature, 3, 8, 21, 23-24, 29, 34, 38, 43-44, 86, 93, 108, 111, 120, 142, 191-192, 200, 208, 222-223, 225
Sea-ice Cover, 55, 57, 59-63, 65-66, 198
Sea-surface Temperature, 30-31, 63, 66, 151-152, 156, 201-202
Seasonality, 2-4, 7, 151, 154-157
Socioeconomic Impacts, 21
Species Biomass, 18, 20
Species Richness, 18-22, 151
Storm Surge, 69-70, 73, 75-76, 108, 183-185
Submarine Channels, 77-78, 82-83
Submesoscale Flows, 1-3, 5-6
Submesoscale Turbulence, 1-3, 5-8
Subpolar Gyre, 23-24, 28-31, 34
Surface Mixed Layer, 1, 30, 93, 146

T

Thermodynamic Forcing, 39
Tide Gauge, 10-13, 15, 176, 184
Topography, 8, 69, 71, 73, 75, 81, 83, 95-96, 99-100, 114, 133, 138-140, 184, 208
Tracking Ocean, 141
Trophic Interactions, 22
Tropical Oceans, 38, 43
Tropical Rainfall, 36
Tropospheric Temperature, 39-40, 43
Tsunami-like Wave, 68-69, 72
Turbidity Currents, 77-78, 80-83
Typhoon Haiyan, 68-70, 72-76, 181, 184

W

Warm Sst, 40, 43, 106-107
Warming Hiatus, 141-144, 148-149
Warming Troposphere, 40
Wave Spectrum, 71, 75